Spinoza

The International Library of Critical Essays in the History of Philosophy
Series Editor: Tom Campbell

Titles in the Series:

Hume: Moral and Political Philosophy
Rachel Cohon

Adam Smith
Knud Haakonssen

Grotius, Pufendorf and Modern Natural Law
Knud Haakonssen

Immanuel Kant, Vols I & II
Heiner F. Klemme and Manfred Kuehn

Hegel, Vols I & II
David Lamb

Marx
Scott Meikle

Locke's Moral, Political and Legal Philosophy
J.R. Milton

Hume: General Philosophy
David W.D. Owen

Bentham: Moral, Political and Legal Philosophy, Vols I & II
Gerald J. Postema

The General Philosophy of John Stuart Mill
Victor Sanchez-Valencia

Spinoza
Gideon Segal and Yirmiyahu Yovel

Hobbes
Robert Shaver

Descartes
Tom Sorell

Mill's Moral, Political and Legal Philosophy
C.L. Ten

Locke
Udo Thiel

Nietzsche
Richard White

Leibniz
Catherine Wilson

Rousseau
Robert Wokler

Spinoza

Edited by

Gideon Segal

Holon Academic Institute of Technology, Israel
The Hebrew University of Jerusalem, Israel

and

Yirmiyahu Yovel

The Hebrew University of Jerusalem, Israel
New School University, New York

Routledge
Taylor & Francis Group
LONDON AND NEW YORK

First published 2002 by Ashgate Publishing

Reissued 2018 by Routledge
4 Park Square, Milton Park, Abingdon, Oxon OX14 4RN
605 Third Avenue, New York, NY 10017

Routledge is an imprint of the Taylor & Francis Group, an informa business

A Library of Congress record exists under LC control number: 00051060

ISBN 13: 978-1-138-73122-6 (hbk)
ISBN 13: 978-1-138-73119-6 (pbk)
ISBN 13: 978-1-315-18798-3 (ebk)

Contents

Acknowledgements

The editors and publishers wish to thank the following for permission to use copyright material.

Brill Academic Publishers for the essays: Emilia Giancotti (1991), 'On the Problem of Infinite Modes', in Yirmiyahu Yovel (ed.), *God and Nature: Spinoza's Metaphysics* (Spinoza by 2000, Vol. 1) Leiden: E.J. Brill, pp. 97–118. Copyright © 1991 E.J. Brill; Alexandre Matheron (1993), 'Ideas of Ideas and Certainty in the *Tractatus de Intellectus Emendatione* and in the *Ethics*', in Yirmiyahu Yovel (ed.), *Spinoza on Knowledge and the Human Mind* (Spinoza by 2000, Vol. 2), Leiden: E.J. Brill, pp. 83–91. Copyright © 1993 E.J. Brill; Yirmiyahu Yovel (1993), 'The Second Kind of Knowledge and the Removal of Error', in Yirmiyahu Yovel (ed.), *Spinoza on Knowledge and the Human Mind* (Spinoza by 2000, Vol. 2), Leiden: E.J. Brill, pp. 93–110. Copyright © 1993 E.J. Brill.

Bar-Ilan University Press for the essay: William K. Frankena (1977), 'Spinoza on the Knowledge of Good and Evil', *Philosophia*, **7**, pp. 15–44. Copyright © 1977 Bar-Ilan University, Ramat-Gan, Israel.

History of Philosophy Quarterly for the essays: Michael Della Rocca (1991), 'Causation and Spinoza's Claim of Identity', *History of Philosophy Quarterly*, **8**, pp. 265–76; Mark A. Kulstad (1996), 'Spinoza's Demonstration of Monism: A New Line of Defense', *History of Philosophy Quarterly*, **13**, pp. 299–316.

Imprint Academic for the essay: Michael A. Rosenthal (1997), 'Why Spinoza Chose the Hebrews: The Exemplary Function of Prophecy in the *Theological-Political Treatise*', *History of Political Thought*, **XVIII**, pp. 207–41. Copyright © 1997 Imprint Academic, Exeter, UK.

Iyyun: The Jerusalem Philosophical Quarterly for the essay: Heidi M. Ravven (1998), 'Spinoza's Individualism Reconsidered: Some Lessons from the *Short Treatise on God, Man, and His Well-Being*', *Iyyun, The Jerusalem Philosophical Quarterly*, **47**, pp. 265–92. Copyright © 1998 Iyyun, The Jerusalem Philosophical Quarterly.

Journal of Social Philosophy/Midwest Studies in Philosophy for the essay: Margaret D. Wilson (1983), 'Infinite Understanding, *Scientia Intuitiva*, and *Ethics* I.16', *Midwest Studies in Philosophy*, **8**, pp. 181–92.

Journal of the History of Philosophy for the essays: Richard E. Aquila (1978), 'The Identity of Thought and Object in Spinoza', *Journal of the History of Philosophy*, **16**, pp. 271–88; Gideon Segal (2000), 'Beyond Subjectivity: Spinoza's Cognitivism of the Emotions', *British Journal for the History of Philosophy*, **8**, pp. 1–19. Copyright © 2000 Taylor & Francis.

Modern Schoolman for the essay: Lee C. Rice (1991), 'Tanquam Naturae Humanae Exemplar: Spinoza on Human Nature', *The Modern Schoolman*, **LXVIII**, pp. 291–303.

Monist for the essay: Errol E. Harris (1971), 'Spinoza's Theory of Human Immortality', *Monist*, **55**, pp. 668–85. Copyright © 1971, THE MONIST, La Salle, Illinois 61301. Reprinted by permission.

Oxford University Press for the essay: Edwin Curley (1994), 'Spinoza on Truth', *Australasian Journal of Philosophy*, **72**, pp. 1–16.

Philosophical Papers for the essay: Jonathan Bennett (1986), 'Spinoza on Error', *Philosophical Papers*, **XV**, pp. 59–73.

Philosophical Review for the essays: Spencer Carr (1978), 'Spinoza's Distinction between Rational and Intuitive Knowledge', *Philosophical Review*, **LXXXVII**, pp. 241–52. Copyright © 1978 Cornell University; Daisie Radner (1971), 'Spinoza's Theory of Ideas', *Philosophical Review*, **LXXX**, pp. 338–59. Copyright © 1971 Cornell University. Reprinted by permission of the publisher.

Philosophy Research Archives for the essays: Harold Zellner (1986), 'Spinoza's Causal Likeness Principle', *Philosophy Research Archives*, **XI**, pp. 453–62; Harold Zellner (1988–89), 'Spinoza's Temporal Argument for Actualism', *Philosophy Research Archives*, **XIV**, pp. 303–09. Used with permission of the Philosophy Documentation Center, publisher of the Philosophy Research Archives.

Southern Journal of Philosophy for the essay: Charles Jarrett (1991), 'Spinoza's Denial of Mind-Body Interaction and the Explanation of Human Action', *Southern Journal of Philosophy*, **XXIX**, pp. 465–85.

Studia Spinozana for the essays: Andrew Collier (1991), 'The Materiality of Morals: Mind, Body and Interests in Spinoza's "*Ethics*"', *Studia Spinozana*, **7**, pp. 69–92; J. Thomas Cook (1995), 'Did Spinoza Lie to his Landlady?', *Studia Spinozana*, **11**, pp. 15–37; Herman De Dijn (1993), 'Knowledge, Anthropocentrism and Salvation', *Studia Spinozana*, **9**, pp. 247–61; Yirmiyahu Yovel (1985), 'Spinoza: The Psychology of the Multitude and the Uses of Language', *Studia Spinozana*, **1**, pp. 305–33.

University of Calgary for the essay: Diane Steinberg (1981), 'Spinoza's Theory of the Eternity of the Mind', *Canadian Journal of Philosophy*, **XI**, pp. 35–68.

Every effort has been made to trace all the copyright holders, but if any have been inadvertently overlooked the publishers will be pleased to make the necessary arrangement at the first opportunity.

Series Preface

The International Library of Critical Essays in the History of Philosophy is designed to bring together, in an accessible form, the best journal essays in English on the history of philosophy. The series makes readily available, in a systematic manner, the most important essays in the history of philosophy, selected and presented by volume editors who are highly respected in their fields. The original essays are reproduced in full which helps to make the series an invaluable reference tool for all scholars interested in the history of philosophy. It is of particular assistance to advanced undergraduate and graduate students in the history of philosophy as well as those who research in this area. The series focuses primarily on the philosophical traditions of Germany, France, Britain and the United States.

The dominant approach of the series is philosophical in that the essays are primarily accounts and critiques from the perspective on the philosophical issues of the time. Some essays trace philosophical influences and elucidate the context of the major philosophical writings. Other essays offer commentary on the texts in the light of current theories of interpretation and historical understanding.

The overriding criteria for the selection of essays are their quality and importance. The essays overall are chosen to ensure a systematic coverage of all important scholarly debates but they also reflect the interests and perspectives of the individual editors which gives each volume a distinctive flavour. I am very grateful to all the editors for the enthusiasm and experience they have brought to the difficult task of selecting essays which bring out the central controversies over the interpretation and understanding of the work of the enduring figures and schools in the history of philosophy.

The International Library of Critical Essays in the History of Philosophy complements the highly successful International Research Library of Philosophy, edited by Professor John Skorupski, which now runs to over 20 volumes and is widely used throughout the world.

The series owes much to the vision and persistence of John Irwin and the great work done by Valerie Saunders, Sonia Hubbard and the Ashgate editorial team.

TOM CAMPBELL
Series Editor
Centre for Applied Philosophy and Public Ethics
Charles Sturt University

Introduction

This collection of essays aims to present a wide range of interpretations of central themes in Spinoza's philosophy. Philosophical interpretations of Spinoza divide into three general categories. The first sets Spinoza within what is taken to be his historical context. Special emphasis is laid here on aspects of his teaching that seem to bear the influence of Spinoza's own education (and self-education), either through concepts assimilated into his own thinking, or those he undertook to refute and replace. A second interpretative approach uses analytic tools in an attempt to critically reconstruct Spinozistic issues and theories. Finally there are philosophers who explore Spinoza's texts in their own terms, attempting to present a coherent picture of one or more aspects of Spinoza's teaching. Given the broad span of issues with which Spinoza deals, and the intricacies of his language, the latter often proves to be the most difficult track to follow – so much so that some Spinoza scholars, such as the late Margaret Wilson, wonder at times whether the inconsistencies and idiosyncrasies in the text deserve the painstaking analysis they require, while at other times they are impressed with the powerful and startlingly original nature of Spinoza's underlying conceptions (see Wilson, Chapter 13, this volume). The 25 essays in this collection exemplify these three attitudes to Spinoza interpretation, although most avail themselves of more than one. In making the selection we preferred studies that treat their subject as a viable, endurable philosophical issue, whether the writer accepts Spinoza's presentation or highlights his difficulties. On each issue the essays critically analyse the texts, rather than simply portraying the Spinozistic ideas they express.

The following is a selective survey of some points and outlooks offered in this volume. Not being a full summary, it must necessarily leave out equally important points and can only delineate its subject, not flesh it out. The wealth of intricate connections, agreements and contrasts in the 25 essays offered here must be left to the reader to discover and assess.

Basic Metaphysical Issues

Following the structure of Spinoza's *Ethics*, Part I focuses on central metaphysical themes: monism, the ontology of substance, attributes and modes, causality and the meaning of determinism. Substantial monism, its ontological premises and its ramifications are treated in the first three contributions. Richard Aquila and Michael Della Rocca approach the subject through the issue of the identity of the mental and the physical, whereas Mark Kulstad concentrates exclusively on arguments regarding the relation of attributes to each other and to the substance.

According to Richard Aquila (Chapter 1) Spinoza's argument for the identity of thought and its object rests on the identity of the attribute of thought with the attribute of extension. But, he claims, the thought–object identity neither rigorously follows from, nor is compatible with, the identity of the attributes. Aquila advances an alternative explanation: Spinoza reached the identity thesis by adopting a Cartesian conceptual framework with regard to ideas, while offering an original, non-Cartesian interpretation of the ontological bearing of that framework. He preserved the distinction between a thing's (including idea's) formal and objective reality, while his original ontology denied reality to an idea other than what is constituted by its being an idea of some object. Thus, although Aquila does not expressly say so, it seems that, in Spinoza's concept of idea, nothing is left as the denotation of an idea's *formal* reality. This, according to Aquila, is the basis of Spinoza's identification of thought and object: 'Ideas are their objects regarded with respect to the attribute of thought because an idea is nothing but the state of affairs of some object *being thought* . . . [and so] they are states of affairs involving their objects – *and nothing else besides*' (p.14). This relation between thought and object, Aquila concludes, is clearly distinguished from the one holding between modes of different attributes, although, he claims, Spinoza 'confused the two cases'.

The relation between a mode of extension and its idea is again treated by Michael Della Rocca (Chapter 2), who unequivocally espouses an interpretation of this relation as numerical identity. He then goes on to grapple with claims that, although such an interpretation draws from an outright statement in the *Ethics* (E2p7s), it leads to conclusions that clash head-on with basic Spinozistic doctrines. First, if a mode of extension is numerically identical with a mode of thought, then whatever that mode of extension causes is equally caused by the mode of thought identical with it. Such cross-attribute causality – that is, a mode under one attribute being a cause of a modification in another – is clearly inconsistent with what is often called Spinoza's doctrine of parallelism. However, Della Rocca contends that for a mode to be the cause of another does not mean it is its cause under any description. Causal contexts are not *referentially transparent* – that is, the truth of a claim that A causes B is not independent of how we describe A and B. A related objection was offered by Jonathan Bennett who claims that a numerical identity theory would entail that, if we accept numerical identity between modes of extension and their ideas, then (the concept of) modes of extension would involve the concept of the attribute of thought and vice versa. In reply, Della Rocca shows that, in Spinoza, the relation of conceptual involvement is coextensive with causal relation. Therefore, inasmuch as a cross-attribute causality is not entailed by the numerical identity theory, neither does cross-attributive conceptual involvement.

The relation between modes of extension and modes of thought, which is central to the above essays, is an aspect of Spinoza's monism, stating that all of God's infinite attributes – of which we know only two, thought and extension – are constituents of one and the same substance; and, further, each of these qualitative ontological kinds exhaust all reality, so that every singular thing is a mode of the one substance-God under each and every attribute. Most readers of Spinoza take his monism to mean that God is the only existing substance – that is, '*except for God*, no substance can be or be conceived' (*Ethics* I, Prop. 14, Curley's translation). Mark Kulstad, in his essay 'Spinoza's Demonstration of Monism' (Chapter 3) follows a different line, according to which 'no substance can be or be conceived *external to God*' (Shirley's translation). Kulstad claims that the logic of Spinoza's demonstration of the existence of God (E1p11d) implies the necessary existence of other substances, in apparent violation of E1p5: 'in nature there cannot be two or more substances of the same nature or attribute.' In an attempt to solve this, Kulstad takes E1p5 as referring only to substances constituted of one attribute and, more importantly, interprets Spinoza's monism to mean that each of God's attributes is a substance, but not one *external* to God. Thus, the logic of E1p11d entails the existence of other substances apart from (but not external to) God, without thereby violating E1p5. In his unconventional (and somewhat Cartesian) interpretation, Kulstad reads E1p5 as denying the possibility of there being two or more substances of one attribute, yet accepting the single-attribute extended substance as one of God's constituting *substances* (and similarly with other attributes–substances that constitute God's essence).

Since Hegel's (actually, Solomon Maimon's) characterization of Spinoza's system as 'acosmism', it has been often contended that the *Ethics* leads to an ontological dead-end, as it fails to adequately derive (or show the deductibility of) the finite from the infinite, and the many from the one. Emilia Giancotti's essay (Chapter 5) claims that, although Spinoza himself seems to have hinted at such failure, this is a rash judgement. She shows this through an analysis of the concept of infinite modes. The one mediated infinite mode Spinoza mentions by name is the Face of the Whole Universe which Giancotti construes as the infinite plurality of finite modes taken together, regulated and moved by Motion-and-Rest in the attribute of extension and by God's Intellect in the attribute of thought. Thus Motion-and-Rest and God's Intellect, which are the immediate infinite modes in Extension and Thought respectively, provide principles of individuation of particular finite modes under the two attributes. However, what enables this process is the immanent nature of the divine causality, and the idea that 'Multiplicity is at the heart of substance'. Giancotti's interpretation posits Spinoza's metaphysics in opposition to that of Descartes – for, unlike Spinoza, Descartes holds Extension-matter to be a mass in a state of rest – as well as to the Neo-Platonic tradition of emanation. Her reading also stands in contrast with any non-secular interpretation of Spinoza's conception of the relation between God and individual things. Giancotti stresses that, when Spinoza calls the infinite intellect 'son of God', rather than aiming at an assimilation of elements of positive Christian religion to his thought, he attempts to bring the sacred back to a natural sphere.

Harold Zellner's contribution on 'causal likeness' – that is, the principle that causality implies that at least one property is shared between cause and effect, finds it to stem from a doctrine said be implied in *Ethics* I, Axiom 4, viz., the Transmission Theory of Causality (see Chapter 4). This is the claim that causality is a 'passing along' of some property from the cause to the effect. In clear contrast to Humean causality, in Spinoza, causal relations (and generally anything in nature) cannot be a matter of 'brute facts'. Unless you accept that the effect somehow springs *ex nihilo*, or is inexplicably correlated with some property in the cause, you are led to the position that the affected property is 'passed along' from the cause. One result of this interpretation is Zellner's proposed solution to a fault he finds in Spinoza's demonstration of the Affect Imitation doctrine, which is a basic principle in Spinoza's social theory. This is the claim that if we perceive someone like us affected in a certain way, we become similarly affected. Interestingly, Zellner suggests that Spinoza's proof of this principle can be rescued through the transmission theory.

In his second contribution, 'Spinoza's Temporal Argument for Actualism' (Chapter 6), Zellner tackles Spinoza's movement from determinism to actualism – that is, the doctrine that this is the only possible world. (The issue is often discussed under the name *necessitarianism* rather than actualism.) Zellner offers a reconstruction of Spinoza's not quite explicit argument for actualism and, using the language of counterfactuals, analyses possible improvements to an argument that, in his judgement, fails to yield the required result.

Knowledge, Truth and Error

Moving from metaphysics to the theory of knowledge, the second section treats Spinoza's theory of ideas, the kinds of knowledge and the concepts of truth and error. In Chapter 7 Daisie Radner responds to critics of Spinoza's theory of ideas by presenting some aspects of it as attempts to solve difficulties in Descartes. First, Descartes' concept of knowledge as (picture-like) representation is incongruent with his sharp thought-extension dualism. Spinoza's main progress lies in his breaking down of Descartes' conflation of two relations – between thought and what it represents, and between thought and its object. Knowledge, according to Spinoza, is a cognitive process in which one's body is affected by an external body; by having this affection as its *object*, one's mind *represents* the external body. Thus, as the object of the mind is *one's body's affection*, the mind represents the *external body* producing that affection. Radner claims, however, that although the affection of the body by external bodies is free from incongruity of the kind that seemed to fail Cartesian theory, the problem is only pushed one step higher to the relation between mind and body. For one wonders how, according to this Spinozistic theory, can the body's affection, being an occurrence in a mode of extension, be the object of the mind, which is a mode of thought.

To be sure, most of the time representation is inadequate, and here another Cartesian weakness is said to be tackled – the one regarding the criterion of adequacy of ideas. Radner's interpretation leads to the controvertible conclusion that the only adequate objects of knowledge according to Spinoza are: (1) the affections of the knower's body (since only in their case the object of the idea is or includes the thing the idea represents); and (2) the essence of God. (A further illumination can be gathered from Heidi Ravven's interpretation, in Part III.)

Jonathan Bennett's 1986 essay on Spinoza's concept of error (Chapter 10) presents what seems to be a formidable difficulty in Spinozism. According to Bennett, there is no room left in Spinoza for false beliefs. Due to the mind–body identity thesis, each idea, being part of the mind, is strictly *identical* with its object, which is nothing but an affection of its body. Thus an idea always accords with its object and therefore cannot be claimed false. Errors, according to Spinoza, are cases in which our ideas are lacking – in other words, cases of ignorance – and Bennett judges this theory to be totally mistaken. Rather than a solution to this impasse, Bennett provides an explanation as to how it befell and what, within Spinoza's system, obstructs a possible solution to it. It should be pointed out that it is the truth of ideas understood in terms of a correspondence theory that finds no satisfactory account in Spinoza's system under Bennett's interpretation.

That Spinoza himself affirms in *Ethics* I, Axiom 6, that '[a] true idea must agree with its object' has presumably been a major prompt for the ascription to him of a correspondence theory of truth by scholars. Edwin Curley's essay (Chapter 9) treats the evidence for such interpretation and for another one, construing Spinoza's concept of truth to be a coherence theory. Curley's conclusion does not amount to a decisive verdict on the issue. Rather, he finds in Spinoza's texts an unsolved conflict between two tendencies regarding the meaning and terms of truth, although, he says, Spinoza does not appear to have been aware of this discordance.

In Chapter 8 Alexandre Matheron draws attention to a seeming contradiction between the *Treatise on the Emendation of the Intellect* (TIE) and the *Ethics* concerning reflexive knowledge – that is, in Spinoza's terms, 'ideas of ideas'. In the TIE Spinoza apparently holds that knowledge is prior to our reflection on it, and, moreover, that certainty in one's own true ideas is reached independently of reflexive epistemological considerations. Matheron takes this to mean that epistemology is not a part of science. However, it also seems that, in the TIE, Spinoza holds certainty of one's knowledge to be *prior* to reflection. Matheron's analysis of Spinoza's treatment of this subject in the *Ethics*, where, in apparent contrast with the TIE, the sheer concomitance of the idea and reflection upon it is affirmed, leads to the following conclusion: although certainty inheres in one's having a true idea, and although having such idea is necessarily and simultaneously accompanied by an idea *of it* (that is, by reflection), it is still possible that the reflective level is implicit rather than fully conscious. Matheron suggests that this may ensue from one's wavering between a true conception of the nature of true ideas, which takes them as intellectual acts of understanding – hence self-sufficient, perfect knowledge, logically independent of reflection, and a false conception, regarding them as 'pictures on a tablet'. Perfect knowledge is prior to, and independent of, reflective awareness to its nature. Nevertheless, reflection inheres in knowledge in a germinal form, and, Matheron adds, 'that is exactly why methodology always comes *after* science, even if it then in turn helps science to make further progress' (p. 123).

The essays by Spencer Carr, Yirmiyahu Yovel and Margaret Wilson represent the extensive literature on the theory of the three kinds of knowledge. Two of the issues touched upon – namely, the distinction between the second and the third kinds of knowledge, and the meaning of the third kind, *scientia intuitiva* – belong to the most intensely debated topics in the literature.

In Chapter 12 Spencer Carr argues against an interpretation that had long prevailed regarding the second and the third kinds of knowledge. Central to this interpretation is the conviction that, in the *Ethics*, the distinction between *ratio* and *scientia intuitiva* is one between rational knowledge of general truths and intuitive knowledge of individual things. Carr claims that the crucial difference between the two kinds of knowledge rests rather on the ordering of one's thought process in each. Whereas in *ratio* we move from individual things to their inferred causes, in intuitive knowledge we begin with the cause (ultimately, with God-Substance), and arrive at adequate cognition of the essences of individual things. Contrary to an entrenched conviction among readers, Carr also contends that *scientia intuitiva* is an inferential cognition. The only exception is our knowledge of the attributes of God which are the foundation of our (intuitive) knowledge of all other things, and therefore our intuitive knowledge of them is necessarily direct rather than inferential. Carr's claims seem to be logically independent, so one may differ from him regarding the nature of the distinction between intuitive and rational knowledge, and still share his claim that intuitive knowledge applies (also) to general principles, or his other claim that *ratio* applies to individual things. These seem to be Wilson's and Yovel's views correspondingly (and Wilson also seems to accept that intuitive knowledge can be inferential), although their arguments have different focuses than Carr's.

Yirmiyahu Yovel (Chapter 11) works out a naturalistic genesis for the pivotal 'common notions', which are said to be the foundations of rationality. Although they function *a priori*, as in Descartes, they have a different origin altogether. Spinoza considers Descartes' 'natural light' supernatural, a miraculous power of the disembodied mind. To be truly natural, the 'light' of reason must be compatible with the body's being immersed in the rest of nature and affected by its causality – though not in the way of the empiricists. Spinoza's solution: all external causes affecting the body share certain natural uniformities which they imprint on the body. The accumulated marks of these uniformities form a substrate of which the mind, by special attention, becomes reflexively aware – thus forming the 'common notions' which provide the first principles of reason.

Yovel also provides an analysis of error and how it is overcome on the level of *ratio* by general 'explanatory schemes' which suffice to remove error even without providing concrete particular explanations. Although, ideally, Hegel's dictum 'the true is the whole', applies to Spinoza, we should be excessively pedantic to aspire to such a practical impossibility. Rational knowledge can remedy our confusions and superstitions by yielding topic-specific explanatory schemes, working as skeletal cognitive patterns to be further fleshed out and particularized to the degree required by the knower's interests. (Yovel thus holds, with Carr and against the dominant view that *ratio* involved knowledge of particular things.)

Margaret Wilson's essay (Chapter 13) focuses on the crucial Proposition 16 in Part I, which states that infinitely many things, 'i.e., everything that can fall under an infinite intellect', follow in infinite ways from the necessity of the divine nature. The gist of her argument relates to the meaning of 'follow' which, Wilson claims, is a case in which intuitive knowledge operates on the level of infinity. What follows by necessity from the divine nature are the essences of things, which the infinite intellect infers from an adequate idea of God's attributes. Given this result, Wilson claims that E1p16 should not be taken as saying that everything that falls under the divine intellect is necessarily actualized *at a certain time and place*, and this position narrows the gap between Spinoza's divine necessitarianism and Leibniz's divine voluntarism. Moreover, Wilson suggests a heightened sense of the thematic unity of Spinoza's system by concluding that *scientia intuitiva*, a concept whose importance emerges in discussing human liberation and salvation in Part V, is already present in Part I, Proposition 16, where basic metaphysical issues regarding the relation between the Substance and modes are dealt with.

Anthropology and Ethics

Charles Jarrett's essay (Chapter 14) takes us from metaphysics to anthropology and ethics, and thus to Parts III and VI of the *Ethics*. Beginning with issues dealt with in the ontological section of our volume – such as Della Rocca (Part I) – he claims that, to avoid contradiction between mind–body identity and non-interactionism, causality must be taken as referentially opaque and then goes on to inquire into the implication of this parallelism on the explanation of human action. The upshot seems to be that the *conceptual* distinction between bodily changes and mental events, which Spinoza maintains alongside their *ontological* identity, is essential to the possibility of explaining human behaviour, for teleological and functional explanations only become possible once we move to the body's mental correlate. Thus *Ethics* III, with the concept of *conatus* giving rise to goals, desires and ethical ideas, is only possible on the basis of mind–body conceptual distinction. Considering this within a broader ontological context, Jarrett concludes that the thought-extension (and mind–body) conceptual division must be taken as an irreducible one. However, as he notes, having reached this conclusion, 'most of the real work remains' (some of which is taken up by other contributions in this and the following Part).

In treating Spinoza's concept of human nature, Lee Rice (Chapter 15) alludes to a query raised by Matheron (1986) – one that puts into question the title of our present section. Is there a Spinozistic anthropology? Matheron asks, and answers that, in strict theoretical terms, Spinoza does not know what humans are, and he willingly forgoes such knowledge, for he does not need it in order to construct his system. ('Spinoza, en toute rigueur et théoriquement, ne sait pas ce que c'est que l'homme, et il s'en passe très bien: il n'a pas besoin de le savoir pour édifier son système' (Matheron, 1986, p. 21). See Rice's essay, Chapter 15, this volume, fn 41, p. 234.) The issue is pivotal, cached in Spinoza's treatment, in *Ethics* IV, of the ethical basis of sociability. The latter is expectedly anchored in ontology. A crucial concept here is that human beings have some things *in common* – that is, agreeing with each other's nature (E4p30–31). Rice interprets commonality in terms of similarity, thereby denying, in resemblance to Matheron's position, the possibility of a clear-cut delineation of a human essence.

One consequent issue pertains to humans' conduct toward animals. As Rice blatantly puts it, 'at least some members of the species *homo sapiens* may certainly display fewer human features (e.g., those with genetic defects, etc.) than some higher animals' (p. 234). However, Spinoza seems to ignore this, since he allows that we make use of animals for our advantage as we please (E4p37s1). Rice views this as a difficulty to his nominalist interpretation of Spinoza's concept of human nature, as being but a name for certain similarities between strict individual modes. That these similarities exist is necessary but is not sufficient for there to be a metaphysically rooted human nature. Consequently, the socializing of humans, rather than an expression of their being a species, is due to these similarities, leading them to form their groupings as an optimal realization of their *conatus*. However, Rice wonders, are there not enough similarities between humans and animals, to make the latter share in human-developed well-being too?

Andrew Collier (Chapter 18) and Heidi Ravven (Chapter 16) each present an interpretation that is antithetical to the Rice's individualism. Ravven discusses the issue within the context of the metaphysics involved in Spinoza's ethical doctrine. She claims that '[s]oul and body have and manifest boundaries that are constantly shifting in relation to what at first appears to be beyond them' (p. 257). However, an expanded individual can be inadequately localized, its awareness spanning and reflecting a limited interactive expanse – this is the case treated in the TTP and TP – or it can overcome its separateness from all things beyond it, and its limited awareness to the body and its relations, ever expanding towards the reality of a perfected individual. It seems, although Ravven does not expressly say so, that this ever-expanding individual has the infinite being and knowledge of God as its regulative existential horizon, and that, further, what she describes as the soul's 'seamless unity and identity with the infinite thought and extension' (p. 263) is what Spinoza conceives as the mind's eternity. (Here is one out of many examples in this volume of the unavoidable arbitrariness of the division to sections of any collection on Spinoza's thought.) It is worth noting that Ravven's essay is almost unique in the present volume, in drawing extensively on the *Short Treatise on God, Man and His Well-Being*, joining a tendency in recent years – also attested in Curley's contribution (Part II) – to reappraise the place of this earlier work in Spinoza's mature thinking.

Spinoza's theory of the affects, which forms a major part in his psychological teaching in *Ethics* III, analyses them on the level of body as well as on that of the mind. In Chapter 17 Gideon Segal examines the meaning and ontological status of the affects in their mental aspects, to which he assigns the term 'emotions'. Taking as his starting point a renowned text from the preface to Part III of the *Ethics*, Segal claims that, in the study of the affects, Spinoza endorsed a major methodological principle that Descartes had employed in his metaphysics of matter – namely, the exclusion of non-rationalizable content from the ultimate report of the studied object. With Descartes this line led to the relegation of sensible qualities of bodies to the realm of (what Locke later termed) secondary qualities. Spinoza's application of a similar principle to the affects resulted in the exclusion of any non-cognitive content from the ultimate report of the emotions. Consequently, Segal claims that Spinoza's cognitivism of the affects banishes the inner *feeling* involved in emotions to the no-man's land of psychological epiphenomena, saving as subjects worthy of his science only their cognitive content subsumable under common notions, and so amenable to discursive treatment.

Andrew Collier's essay on the materiality of morals (Chapter 18) links the anthropological issues treated by Rice and Ravven to moral issues, treated in the present volume by William Frankena (Chapter 19). Collier distinguishes between a human body 'in the narrow sense' and an extended human body that includes whatever interacts with that narrow-sense body. That human bodies are extendible beyond the mere organ regularly so called, stems, according to Collier, from Spinoza's claims that: (a) the mind's power of understanding extends to ideas of things that interact with its (narrow-sense) body; and (b) one's body is the object of the complex idea which is the mind. Given these two claims, one ought to admit that the human body is extendible beyond the mere organism enclosed in its skin, to include external bodies it interacts with. Although Spinoza has not expressly drawn that conclusion (at least in the *Ethics* – an issue amply discussed in Heidi Ravven's essay), given the inner logic of his mind–body doctrine, Collier says, 'he ought to have done so'.

An interesting result is that what are normally taken to be different human beings in fact partly share each other's expanse. Having shown this, Collier proceeds to implications of this concept of extendible body to the meaning of *conatus* (the striving of each thing to persevere in its own being), and thereby to implications for moral theory. Given one's body's expanded span, his *conatus* extends to whatever is included in it, and so he becomes interested in the well-being of other people and inanimate parts of the world as well.

Moral theory is at the focus of William Frankena's contribution (Chapter 19). He examines a series of metaethical issues raised in connection with Spinoza's major claim in *Ethics* IV, Proposition 8, that 'the knowledge of good and evil in nothing but the affect of Joy or Sadness, insofar as we are conscious of it'. Taken at its face value, E4p8 might seem as proclaiming ethical emotivism or hedonism. However, Frankena interprets it as a non-analytic, non-definitional equation of one's being conscious of having an affect of joy or sadness with one's knowledge that some things are conducive and others harmful to one's perfection or power. On this interpretation, Spinoza's ethical naturalism is adjoined by what Frankena identifies (in line with Segal's essay in this section) as Spinoza's cognitivistic position with regard to the affects. Having joy is, or includes, perceiving some good, since it is, or includes, being cognizant (if occasionally confusedly so) of an affection of one's body in which its perfection is increased or helped. However, analysing Spinoza's other statements in the *Ethics* that deal with, or have implications on, the issue of knowledge of good and evil, Frankena shows (among other things) that the fact that an affect, qua cognition, is, or includes, knowledge of good or evil does not mean that it is the only kind of such knowledge, and although the latter seems to be Spinoza's claim in E4p8, other passages in the *Ethics* provide significant exceptions.

Salvation and the Mind's Eternity

While a moral viewpoint is an essential characteristic of human life, salvation is – or at least it has been considered within certain religious traditions to be – the ultimate and thorough answer to human plight. Indeed, Spinoza's *Ethics* is structured so as to give the reader the idea that the whole intellectual course he passes while reading aims at the last, fifth Part of the book, where the author's theory of the eternal love of God and the state of Beatitude attained therewith is portrayed. Part IV of this volume presents the reader with issues involved in the three major dimensions of Spinoza's doctrine of salvation – namely, the epistemological, ethical and ontological dimensions.

The central issue in Herman De Dijn's essay (Chapter 20) is how Spinoza's scientific, non-anthropocentric metaphysics, can lead to salvation. De Dijn rejects different versions of a rationalistic interpretation to Spinoza's ethics, which assign to reason the task of either relieving a person of passions or rationalizing her or his self-knowledge and activity. Among other faults, such interpretations are mistaken since they are only seemingly non-anthropocentric, for they retain the illusion of freedom in a world where man is doomed to be subject to Nature's insurmountable forces. Moreover, the dictates of reason within the new non-anthropocentric world-view cannot but fail to pervade human active life. De Dijn goes as far as the following, seemingly paradoxical key claim: it is precisely insofar as the rational person cannot help developing the science of good and bad and the ideal of freedom presented in *Ethics* IV, that he or she is not really free. In aspiring to a human ideal, a rational person is dominated by longing, hope and fear. In contrast, intuitive knowledge involves a metaphysical psychology which 'encapsulates everything that affects us in the crystal-clear, objectifying atmosphere of thinking', forming a 'detached look upon ourselves' similar to the one produced by the 'new science of physics vis-à-vis our contact with nature' (p. 350). Thus our rationalistic illusory view of ourselves is replaced by serene meditative self-knowledge, combined with our awareness of our being but modes of an impersonal power, involving a joyful, celebrating acceptance of the truth about ourselves and a love of God that asks nothing in return. What De Dijn's interpretation seems to clearly imply, among other things, is that the whole discussion of rational life in *Ethics* IV has no practical objective ('practical' in the utilitarian meaning), but is rather intended as a preparatory for a serene detachment (a happy resignation) from what is taken to be the illusion that one can rationalize one's life through reason.

Most scholars affirm that the *Theologico-Political Treatise* advances, alongside eternal salvation attainable for those capable of true philosophy, an alternative salvation for those unable to practise adequate knowledge and who are therefore also banished from the blessing of the intellectual love of God. This duality of routes to salvation opens for Thomas Cook (Chapter 21) several interpretative options, when he examines an incident in Spinoza's life mentioned by biographers. When Spinoza's landlady asked him whether she could be saved by practising her religion, he answered positively, 'provided, whilst you apply yourself to Piety, you live at the same time a peaceable and quiet life' (p. 357). Questioning the truthfulness of this alleged response, Cook examines several meanings of salvation that Spinoza might have had in mind, and the question turns on whether or not he could honestly believe his landlady was capable of any of them. Since he could in no way hold that the kind of salvation involved in intuitive knowledge could be attained by practising the Christian religion, Spinoza apparently meant, in accordance with the TTP, that by leading a pious and peaceable life his landlady might have reached salvation through an alternative route, aimed at in the teachings of the prophets. However, Cook sides with Leo Strauss in interpreting the TTP's approval of prophetic ethics as 'a sop to palliate the pious faithful and a smokescreen to confuse the persecutorially powerful' (p. 366). The only option left to save the sincerity of Spinoza's answer to his landlady is that he referred her to a second-rate salvation, which amounts to peace of mind attained by maximizing the place of love in one's affective life and minimizing the conflict with other people. However, this is not the kind of salvation Spinoza's *Ethics* aims at.

Diane Steinberg (Chapter 23) and Errol Harris (Chapter 22) attempt new solutions to a perennial problem that perplexes readers of the *Ethics*. It stems from a perplexing announcement, at the end of the Scholium to Proposition 20 in Part V, that '. . . it is time now to pass to those things which pertain to the Mind's duration without relation to the body'. For Spinoza's parallelism and, more specifically, the concept of the human mind as the idea of the human body, which entails (in Errol Harris's words) 'that body and mind are inseparable and co-terminous' (p. 383), do not seem to allow any such separate existence of the mind. A traditional reading starting with Joachim (1901) tried to resolve the difficulty by taking eternity to apply only to those aspects of the mind that were not individual. Whereas the individual portion of the mind is a mental correlate of a person's individual body, and hence cannot exist unless the body does, the eternal part of the mind, so this reading goes, encompasses those ideas that, in having them, a person shares in God's Idea (*Idea Dei*) – and the extended correlate of these ideas is the face of the whole universe, rather than individual bodies. A question still remains open whether the eternity of the mind consists in indefinite existence in time, or whether it means an atemporal existence of some kind. Harris and Steinberg treat this issue in ways that differ considerably from the so-called traditional line.

According to Steinberg (Chapter 23), in *Ethics* V Spinoza submits that the human mind consists of two separate parts. What Spinoza refers to as 'the eternal part of the mind' is not an idea of the body as an actually existing thing – which idea exists in time, as does its object – but rather the body's idea 'under a form of eternity'. Spinoza offers this theory, Steinberg claims, in order to allow for the mind's being active – that is, for a person's having adequate knowledge – for, she argues, given Spinoza's theory of the physical world, his concept of activity, based on adequate causality, is inapplicable to the actually existing body and hence to whatever part of the mind that is correlative with parts or affections of the body. Now, Steinberg admits that, in holding the eternal part of the mind to have no parallel in the actually existing body – and necessarily so, since this part is timeless, whereas the body and all its parts are in time – she is denying what most (although not all) commentators see as basic, namely that all thinking modes have their correlates in extension. However, she argues, Spinoza himself seems to have abandoned hermetic parallelism in his doctrine of the eternity of the mind.

In Chapter 22 Errol Harris attempts to avoid the apparent inconsistency between Spinoza's account of the mind–body relation and his doctrine of the immortality of the soul. According to Harris, the eternity of the human mind consists in its capacity for transcendence involved in adequate knowledge. Harris assumes that the mind is eternal specifically in and through expressing to the highest degree its cognitive faculty and *without thereby breaking off its correlation with its body*. In adequately knowing its objects in their spatio-temporal and causal relations, the mind necessarily transcends these relations, which are partial features of the absolute totality that is God, into which the adequately known object systematically integrates. However, pace Joachim, and others who, like him, ascribe the mind's eternity to its non-individual eternal idea in God, in so existing in God, Harris argues, the mind does not differ from any other finite mode – it is the individual mind as such that is capable of eternal life. For to adequately register its objects in their spatio-temporal causal relations, the mind must transcend space and time (in other words, be eternal) 'while yet, paradoxically, being identical in substance with the body, as its idea' (p. 396). Like Steinberg, Harris seems to clearly affirm this eternal existence to be atemporal in nature, rather than an indefinitely continued existence after the dispersion of the body.

Religion and Politics

The essays in Part V, the final section of this volume, extend to Spinoza's social and political ideas and his critique of religion – and thus link the *Ethics* to the *Theologico-Political Treatise*. The TTP, with its textual intricacies and wordings that often, at least at face value, seem to contradict established Spinozistic doctrines, is read by Yirmiyahu Yovel (Chapter 24) within a broad span of language strategies, which use several forms of equivocation addressed simultaneously to the true philosopher and to the larger public – the so-called 'multitude'. Yovel understands both parts of the TTP as united by a single underlying question – 'What is to be done about the multitude?' – which, to Spinoza, is a philosophical question *par excellence*. Multitude is understood to be the large group incapable of attaining a life of reason and intuitive knowledge, and motivated by passive affects and the *imaginatio*. Spinoza's programme is to *reshape* the *imaginatio* as an imitation of *ratio*: to create habits and institutions, backed by respect, obedience and habit (rather than inner understanding) that will redirect people's passions and behaviour from aggression and mutual harm to socially beneficial channels. Ideally, the members of the multitude will be motivated by the passions and the imaginations to act or behave socially in a *semi*-rational way – that is, in the same external way as the persons motivated by reason. This reform of the passions is, according to Yovel, the only remedy to the multitude's inherent tendency to intolerance, unstable behaviour, fanaticism and violence. But a social reform of the passions requires institutional means – primarily a popular religion of reason (distinct from the philosophers') and a reformed semi-rational state, separated from religion and politically superior to it. No less, it needs a new approach to Europe's sacred texts (biblical hermeneutics) and strategies of dual language used for: (1) prudence and defence; (2) undermining established authorities; and (3) enabling the shift from inadequate to adequate meanings of old concepts and goals.

Similarly, pointing at Spinoza's emphasis on the incapacity of rational discourse in mass politics, in Chapter 25 Michael Rosenthal analyses in turn the role he attributed to models and examples (*exemplaria*) appealing to the imagination of people, and shaping it so as to route them to a required normative conduct. Narrative *exemplaria* were essential in prophetic discourse, the intended audience of which, the ancient Hebrews, were in a situation resembling Hobbesian State of Nature. The aim of this imaginative tool was to authorize a set of institutions and so to erect the ancient Hebrew state on the basis of shared norms of conduct. In calling the Hebrews 'chosen', the prophets used imaginational language (the only effective language when addressing common people) to transcend individual interests – which by themselves (for example in the State of Nature) tend to breed conflict and strife – and to create a common standard of judgement and behaviour. However, Rosenthal shows that Spinoza's aims in the TTP ranged wider than the historical case of the Hebrews. For, in writing it, Spinoza consciously tried to perform just the same move vis-à-vis the politically ripped Dutch society of his day. Spinoza used the example of the ancient Hebrew state to illustrate, in terms of the biblical text that the

Dutch were familiar and identified with, a lesson relevant to them as it was for the ancient Hebrews – the fate of a society in which priesthood, misusing prophecy and spreading superstition, oversteps state authority, rather than be kept under its control. In so doing, Spinoza appealed to the imaginative notions that lay at the foundations of his own society and, Rosenthal goes on to claim, so should we do regarding our own society, if we want to make the TTP an exemplar in establishing a free, tolerant political order.

Concluding Remarks

The wealth of topics treated in works on Spinoza is far beyond the scope of what an anthology can represent. The present volume, while attesting to the viability and relevance of Spinoza studies, is another invitation to students and scholars to probe into the various issues invoked by Spinoza's writings. We hope that the 25 essays collected here will provide insights and drives for further inquiry into a philosophy which, we believe, when seriously engaged, will prove a source of a rewarding philosophical experience.

References

Joachim, Harold H. (1901), *A Study of the Ethics of Spinoza*, Oxford: Clarendon Press.

Matheron, Alexandre (1986), 'L'anthropologie spinoziste?', in A. Matheron, *Anthropologie et politique au xviie siècle*, Paris, Vrin, pp. 17–27.

Part I
Basic Metaphysical Issues

[1]

The Identity of Thought and Object in Spinoza

RICHARD E. AQUILA

TWO LINES OF THOUGHT converge in Spinoza's doctrine of the identity of thought and its object. Only one of them is contained in the "official" presentation of Spinoza's argument. It lies in the general conception he offers of the relationship among the various attributes of reality. Despite Spinoza's claims, however, the latter conception does not require acceptance of the doctrine in question. Furthermore, Spinoza's own response to objections concerning that doctrine's *compatibility* with his general conception of the attributes shows that the relation between thought and its object cannot be regarded as an instance of the "identity" obtaining among the attributes generally. Nevertheless, it is possible to see why Spinoza's doctrines led him to see an "identity" in both cases. Thus we may also see how two distinct lines of reasoning became blurred into one in his thought, to the inevitable advantage of the official presentation. The second—"unofficial"—motivation stems from a certain Cartesian assumption concerning the nature of thought and its object, combined with Spinoza's crucial departure from a doctrine essential to Descartes's own implementation of that assumption. The resulting conception, it will be interesting to see, anticipates a view elaborated in Sartre's *Being and Nothingness.*

I. The *Ethics*[1] defends a remarkably original view concerning the relationship between bodily and mental events. Bodily events are (qua bodily) "modes" of the attribute extension; mental events are (qua mental) modes of thought. But

> substance thinking and substance extended are one and the same substance, which is now comprehended under this attribute and now under that. Thus, also, a mode of extension and the idea of that mode are one and the same thing expressed in two different ways. . . . therefore, whether we think of Nature under the attribute of extension, or under the attribute of thought or under any other attribute whatever, we shall discover one and the same order or one and the same connection of causes, that is to say, in every case the same sequence of things. (2, 7, note)

Thus Spinoza defends a form of the "identity theory" with respect to mind and body. Since the human mind is a system of thoughts constituting an "idea" of the body (2, 10–13), a person's mental events are, regarded under the attribute of extension, the

[1] *Spinoza Opera*, ed. Carl Gebhardt, 4 vols. (Heidelberg: Carl Winter, 1925), vol. 2; trans. W. H. White (revised by A. H. Stirling), ed. James Gutmann (New York: Hafner Publishing Co., 1949). References to *Ethica* appear parenthetically by part number followed by proposition number in the text.

very *same* events as those that constitute that person's body. However, Spinoza's view differs in two ways from some other forms of the identity theory. First, most recent defenses of that theory maintain that the concepts involved in a consideration of material events as mental are subject to a "topic-neutral" *analysis.* That is, what those concepts express concerning the events in question can be analyzed in terms that involve a conception neither of specifically mental nor of specifically material events.[2] On the other hand, the concepts involved in regarding those events in *material* terms cannot be analyzed solely in terms involving no specifically material concepts. To this extent there is, on such a view, a certain priority accorded to the material conception of the events in question. Spinoza's view rejects this priority: neither way of regarding events can be explained in more basic terms (1, 10). The second respect in which Spinoza's version differs from other versions of the "identity theory" concerns Spinoza's claim that those events that, regarded materially, constitute a person's body are not simply the *same* events as those that, regarded mentalistically, constitute a person's mind. In addition, the events regarded in the former of these ways are the *objects* of the very ideas with which they are "identical": "a mode of extension and the idea of that mode are one and the same thing expressed in two different ways."

Philosophers other than Spinoza have also espoused theories of mind-body identity embodying the first of these features.[3] However, its combination with the second appears to assign to Spinoza a conception of mind-body relations that is unique in the history of philosophy. It may also appear to reveal, unfortunately, the thoroughly incredible nature of Spinoza's conception. For how are we to make sense of the claim that certain events in one's body are not only the same events that, regarded mentalistically, constitute that person's thoughts but, in addition, have for their object nothing other than themselves? A particular brain state might very well be, regarded mentalistically, a particular thought whose object is also some particular brain state, but can we make sense of the claim that it is its *own* object? And even if we can make sense of it, what reason could there be for thinking it true?[4]

Spinoza presents his doctrine of the identity of thought and its object as a straightforward consequence of his demonstration, in part 1 of the *Ethics*, that there can be (and be conceivable) no more than the single substance (God) that necessarily exists. He first introduces that doctrine (quoted above) with this remark:

> Before we go any further, we must here recall to our memory what we have already demonstrated—that everything which can be perceived by the infinite intellect as constituting the essence of substance pertains entirely to the one sole substance only, and consequently that

---

[2] See J. J. C. Smart, "Sensations and Brain Processes," in V. C. Chappell, ed., *The Philosophy of Mind* (Englewood Cliffs, N.J.: Prentice-Hall, 1962), p. 167. This is a revised version of a paper appearing in *The Philosophical Review* 68 (1959): 141-156.

[3] Spinoza is often classified with P. F. Strawson as a proponent of the "double aspect" theory. See Jerome A. Shaffer, *Philosophy of Mind* (Englewood Cliffs, N.J.: Prentice-Hall, 1968), pp. 52ff; also Douglas Odegard, "The Body Identical with the Human Mind," *The Monist* 55 (1971): 585-586.

[4] A number of commentators deny that Spinoza ever did hold that the "ideal" aspects of the body, which constitute the human mind, have only that body for their object. Thus Henry E. Allison maintains that Spinoza's claim that the human mind is the "idea of the body" means only that it is "related functionally rather than causally" to that body: *Benedict De Spinoza* (Boston: Twayne Publishers, 1975), pp. 89, 94. See G. H. R. Parkinson, *Spinoza's Theory of Knowledge* (Oxford: The Clarendon Press, 1954), pp. 110-111; Wallace Matson, "Spinoza's Theory of Mind," *The Monist* 55 (1971): 577. The soundest view, it seems to me, is that of Daisie Radner, who argues that although Spinoza did indeed maintain that the human mind contains only ideas whose *object* is the body, some of those ideas might also "represent" something other than that body ("Spinoza's Theory of Ideas," *The Philosophical Review* 80 [1971]: 346ff.).

substance thinking and substance extended are one and the same substance, which is now comprehended under this attribute and now under that. (2, 7, note)

Spinoza then proceeds directly from the conclusion that thinking and extended substance are one and the same substance (God) to the further conclusion that any particular *mode* of the one thinking and extended substance can be viewed both as a mode of extension and as a mode of thought.[5] This inference of course requires a particular conception of the relation between substance and its attributes for which Spinoza does not explicitly argue. The conception involves the supposition of some sort of *identification* of substance with each of its attributes. Given this identification, any particular mode of thinking substance or of extended substance would necessarily be a mode of the attributes thought and extension themselves. It is clear that Spinoza does assume some sort of identity between a substance and its attributes. Thus part 1, proposition 4 makes it clear that Spinoza regards the very definition of an "attribute" as implying that substances are "the same thing" as their attributes; part 1, proposition 15 straightforwardly identifies "extended substance" as one of the infinite *attributes* of God; and letters 2 and 9 explicitly confirm the identification of substance and its attributes.[6]

However, even if Spinoza is entitled to proceed from the identity of thinking and extended substance to the conclusion that any particular mode of thought is also a mode of extension (and conversely), it is clear that he is not *ipso facto* entitled to proceed to the doctrine of the identity of thought and its object. That every mode of extension is identical with a mode of thought does not imply that it is identical with a mode of thought whose *object* is that very mode of extension. In one sense, of course, the term "thought" may be employed precisely in order to designate objects of thought, qua objects. In that case Spinoza might simply be taken as claiming that each mode of extension is at the same time the object of some mode of intellectual activity.[7]

---

[5] This conclusion of course requires an additional assumption, beyond the proof that God is the only substance; namely, that thought and extension are attributes of God. Whereas this assumption is presumably argued in 2; 1, 2, the note to the first of these propositions shows that Spinoza considers it independently evident. It rests, according to him, on the fact that we can conceive an infinite thinking and an infinite extended substance, together with the definition of an "attribute" (as what expresses the "essence" of substance: see def. 4 and, for Spinoza's gloss on that definition, 1, 19) and the definition of "God." In fact, however, the assumption requires more than this, since it also presupposes the impossibility of any reductive *analysis* of our conceptions of either thought or extension. Spinoza nowhere argues independently for this.

[6] Gebhardt, 4:7, 46; trans. A. Wolf, *The Correspondence of Spinoza* (London: George Allen & Unwin, 1928). A selection of these letters is also included in John Wild, ed., *Spinoza Selection* (New York: Charles Scribner's Sons, 1930). It is not difficult to see why Spinoza would have required some sort of identification of substance and attribute. Since attributes constitute the various ways in which intellect can comprehend the nature of substance (see def. of "Attribute," pt. 1, and letter 9), if substance is anything over and above its attributes (e.g., some sort of "substratum" in which they inhere), then there will be something in it that eludes all possible powers of reason. Unfortunately, this at most implies that substance is to be identified with the *totality* of its attributes, not with each of them separately. However, Spinoza can appeal here to his arguments against the possibility of "dividing" substance into independent parts (1; 12, 13). If God is identical only with the totality of the attributes, and not with each one separately—each, in the latter case, constituting not merely an essential *part* of the divine nature, but a distinctive intellectual grasp of the *whole* of it—then God would be divisible into several distinct substances, each of which is characterized by less than the totality in question, unless there is some necessary *connection* among its parts corresponding to the distinctions among the attributes. But this would require a necessary connection between distinct systems of modifications of the various attributes, which is impossible on Spinoza's view.

[7] See E. M. Curley, *Spinoza's Metaphysics: An Essay in Interpretation* (Cambridge, Mass.: Harvard University Press, 1969), pp. 121ff. Curley suggests identifying Spinoza's "ideas" with *propositions*; and the relation between idea and object, with that between a proposition and the fact that makes it true. Oddly,

But even granting, in the first place, that the term "thought" may designate objects, and not merely activities, of intellection, it would still appear to be consistent with the identity of thinking and extended substance to maintain that particular modes of extension, considered under the attribute of thought, are simply identical with particular modes of intellectual activity, not with the objects of that activity (qua objects). In any case, secondly, Spinoza himself insists that "ideas," which are—as is "self-evident" (2, 5, dem.)—just particular modes of "thought," are to be regarded precisely *as* modes of mental activity. "By idea I understand a conception of the mind which the mind forms because it is a thinking thing. . . . I use the word "conception" rather than "perception" because . . . the word conception seems to express the action of the mind" (2, def. 3; 2, 49, note). (Insofar as any mode of extension is actually considered under the attribute of thought, of course, it will in fact be an object of intellectual activity, namely, of the activity of thus considering it. But this does not imply that the mode in question must then be regarded as object of the very mode of intellectual activity with which it is regarded as 'identical." Hence an appeal to the definition of an "attribute" as a way in which substance must be *conceived* by the intellect will not help at this point.) The question thus remains, What leads Spinoza to conclude that there is some sort of identity between ideas regarded as modes of mental activity and the objects of those very activities?[8] Taken by itself, the identity of thinking and extended substance at most requires the conclusion that each mode of extension is identical with some mode of intellectual activity; it does not require any particular conclusion concerning the *object* of the intellectual activity with which each mode of extension is identified. Thus it would be compatible with the identity of thinking and extended substance to maintain, à la Leibniz, that while each part of reality must be *apprehending* some part of reality, what it must be apprehending is not simply itself but the whole of reality outside it.

II. In addition to failing to *follow* from his general conception of the "identity" of the attributes, Spinoza's view of the relation between thought and its object also

Curley supports his suggestion with a passage, quoted below, in which Spinoza insists on regarding ideas as modes of mental *activity*, whereas "propositions" appear to be certain *objects* of intellectual activity, namely objects of affirmation and negation.

[8] It is frequently suggested that Spinoza is the victim of confusion generated by his ambiguous use of terms. Thus A. E. Taylor suggests that "the conception and the *conceptum* are confused together by using the same word *idea* indifferently for either" ("Some Incoherencies in Spinozism (1)," *Mind* 46 [1937]: 154). See Celestine Sullivan, "Critical and Historical Reflections on Spinoza's 'Ethics'," *University of California Publications in Philosophy* 32 (1958): 3. Though this particular ambiguity is prominent in Descartes (see Section 3 below), the suggestion seems unlikely in view of Spinoza's insistence in the passage quoted above. Taylor also suggests that Spinoza's confusion may be due to an "equivocation by which 'the idea of Peter' may mean either 'the mental complex which corresponds to Peter's brain and nervous system, the mind of Peter,' or 'the mental complex which exists when Paul thinks of Peter'. . ." ("Some Incoherencies," p. 153). See Frederick Pollock, *Spinoza: His Life and Philosophy*, 2nd. ed. (London: Kegan Paul, 1899), pp. 123ff., and H. Barker, "Notes on the Second Part of Spinoza's *Ethics*," part 2, *Mind* 47 (1938): 295ff. This suggestion seems *ad hoc*, inasmuch as the first of these uses in Spinoza appears to have originally emerged *from his arguments* rather than being already current in the way that the "Cartesian" ambiguity was. For an interesting discussion of this criticism see H. F. Hallett, "On a Reputed Equivoque in the Philosophy of Spinoza," *Review of Metaphysics* 3 (1949): 189-212. (The Taylor, Barker, and Hallett articles are reprinted in S. Paul Kashap, ed., *Studies in Spinoza* [Berkeley: University of California Press, 1972].) Radner also argues, pp. 339-340, that the demonstration of 2, 13 rests on some such ambiguity as these, though her argument appears to be question-begging.

seems *incompatible* with that conception. It seems incompatible, namely, with his doctrine of the conceptual independence of the attributes, the doctrine that "each attribute of a substance must be conceived through itself" (1, 10). Spinoza clearly regards this proposition as implying that the conception of any particular *mode* of a given attribute (qua mode of that attribute) involves the conception of no other attribute: "The formal being of ideas is a mode of thought (as is self-evident). . . . It is a mode, therefore (Prop. 10, Pt. 1), that involves the conception of no other attribute of God. . ." (2, 5, dem.). This seems to imply that each of the attributes involves an independently conceivable way of comprehending the divine essence as it is manifest in particular modifications. Accordingly, if the relationship between a mode of thought and the mode of extension that is its object is merely a particular instance of the relationship between a particular mode regarded with respect to one attribute and that same mode regarded with respect to another, then our conception of any particular idea of a given mode of extension (qua mode of thought) could involve no conception of the attribute of extension itself. Not only has this appeared a completely incredible doctrine to some commentators,[9] but it also appears incompatible with some of Spinoza's own claims.

Spinoza does claim that it is possible to regard any given mode of thought without attention to the *object* of that thought. This, according to Spinoza, is just what is involved in an idea of an *idea* of any particular object of thought, as contrasted with an idea of that object *simpliciter*:

> For, indeed, the idea of the mind, that is to say, the idea of the idea, is nothing but the form of the idea in so far as this is considered as a mode of thought and without relation to the object, just as a person who knows anything by that very fact knows that he knows, and knows that he knows that he knows, and so on *ad infinitum*. (2, 21, note)

But to grant the possibility of considering an idea without regard to the particular object of that idea is not to deny the possibility of also regarding that same idea as an idea of the object in question. In the latter case, however, one is surely not abandoning altogether a consideration of the idea precisely *as* an idea. One is not, for example, regarding the idea merely as a mode of extension; one is simply regarding it as an idea *of* some mode of extension. Granting this would of course require Spinoza's admission that a conception of an idea of some mode of extension, as an idea, may involve in itself the conception of the attribute extension. Despite its incompatibility with his own "official" position, this concession seems in fact to be implicit in some of Spinoza's claims. Thus, for example, after maintaining that the "idea of the mind is united to the mind in the same way as the mind itself is united to the body" (2, 21), Spinoza appears to elaborate on this claim by maintaining that the distinction between an idea and an idea of an idea merely involves a certain act of *abstraction*: the idea of an idea of some object is just that very idea itself, considered without *attention* to its

---

[9] See Barker, p. 290. Matson suggests that, in view of the difficulty in question, "it may be permissible to doubt whether the absolute equality of the attributes is indispensable to Spinoza's philosophy" (p. 577). As I suggest below (see n. 24), it would rather be best to preserve the equality of the attributes while denying that thought is, strictly speaking, one of the attributes in the first place.

particular object (see quotation above). This suggests that distinguishing an idea of a given mode of extension from that mode of extension itself similarly involves an act of abstraction: our idea of the mode of extension, considered "in itself," involves abstractively considering that mode of extension apart from a certain concrete situation that includes it, namely, apart from any attention to the concrete idea *of* that mode of extension. If this is so, of course, then the relation between an idea of a mode of extension and that mode of extension itself will not be strictly assimilable to Spinoza's general conception of the relationship between a given entity considered with regard to one attribute and that same entity considered with regard to another.

Confirmation of this suggestion is found in Spinoza's response to a difficulty raised by von Tschirnhaus. The difficulty concerns Spinoza's claim (1; 9, 11) that there are infinitely many attributes of substance. From this it follows that corresponding to any particular mode of extension there are modes of infinitely many other attributes identical with that given mode, except for being considered with respect to those other attributes. Now my mind, according to Spinoza, is constituted of the ideas of my own body, and it is, according to the official position, *identical* with that body. My own body, however, is not only identical with certain modes of extension; it is also identical with infinitely many modes of attributes other than extension. If, accordingly, an idea's "identity" with the mode of extension that is its object is just the *same* kind of identity as is involved in Spinoza's general conception of the relationship among the infinitely many attributes of substance, then it would seem to follow that the idea that constitutes my mind is also identical with ideas of those infinitely many modes of the various attributes with which my particular body is identical. In that case, however, it would seem to follow that those infinitely many modifications are all equally (though not necessarily equally *conscious*) *objects* of the ideas constituting my mind, since the relationship between an idea and its object is, according to the official position, simply that between a given modification and that same modification regarded under the attribute of thought (letter 65).

Spinoza's response to von Tschirnhaus is instructive:

> In answer to your objection I say that although each thing is expressed in infinite modes in the infinite understanding of God, yet the infinite ideas by which it is expressed cannot constitute one and the same mind of an individual thing, but an infinity of minds: seeing that each of these infinite ideas has no connection with the others, as I explained in the same Scholium to Proposition VII, Part II of the *Ethics*, and as is evident from Proposition X, Part I. If you will pay a little attention to these, you will see that no difficulty remains. . . . (Letter 66)

In this response Spinoza acknowledges that an idea of a given mode of extension is also "expressed," in the "infinite understanding," as an idea of modes of infinitely many attributes. Nevertheless, Spinoza insists, although any idea of a mode of extension is thus identical with infinitely many ideas, it cannot constitute a single *mind* together with those ideas: the change of "regard" involved in the "identity" of a given idea with infinitely many ideas is incompatible with the way in which ideas must be regarded in order to be regarded as constituting one mind.

Spinoza claims that this follows from his arguments in part 2, proposition 7 (note) and in part 1, proposition 10, since, according to those arguments, the infinitely many ideas with which a given idea is identical can have "no connection" with any of those other ideas. Now part 1, proposition 10 simply affirms that "each attribute of a substance must be conceived through itself"; part 2, proposition 7 argues that while a mode of any given attribute is identical with modes of infinitely many attributes (being the same thing as these, but regarded in different terms), nevertheless when things are considered *as* modes of any given attribute they can be "explained" only by relating them to other modes considered as modes of that same attribute:

When things are considered as modes of thought we must explain the order of the whole of Nature or the connection of causes by the attribute of thought alone, and when things are considered as modes of extension, the order of the whole of Nature must be explained through the attribute of extension alone, and so with other attributes.

It is not clear how these claims lead to Spinoza's conclusion.[10] Apparently he is claiming that a given idea cannot constitute an individual mind together with the infinitely many ideas with which it is identical, because the ideas in question cannot be regarded as part of a causally unified whole. (See Spinoza's discussion of the identity of individuals, lemmata following 2, 13.) But why cannot the ideas be regarded as part of a causally unified whole? The only consideration Spinoza offers is that some item *A* cannot be regarded as causally connected with *B* so long as *A* and *B* are regarded as modes of distinct attributes. What we are presumably considering, however, is a whole constituted of various modes of the attribute of *thought.* Why cannot an *idea* of a mode of a given attribute be causally connected with an *idea* of a mode of a distinct attribute, even though those modes themselves cannot be regarded as causally connected with one another?[11]

What Spinoza's response shows is that the relationship between an idea and its object is not, despite the "official" presentation, simply an instance of the kind of "identity" obtaining among the attributes of substance generally. The latter involves the claim that a mode of any given attribute can be (and *is*, in the infinite understanding of God) regarded as a mode of infinitely many attributes, each constituting a conceptually independent way of regarding the entity in question. Spinoza's official position is that an idea of a mode of extension and that mode of extension itself have precisely this kind of identity: the former just *is* the latter, considered not with regard to the attribute of extension but with regard to the attribute of thought. But Spinoza's

[10] Taylor maintains that Spinoza's response rests on a fallacy resulting from the identification of a mind with its "ideas": it moves from the observation that an idea of a mode of an attribute that "corresponds" to a mode of extension is not the same *idea* as an idea of that mode of extension itself to the conclusion that it cannot be thought by the same thinking *mind* (pp. 152–153). Spinoza's reply, however, seems to have quite a different motivation, resting on the assumption of some sort of radical dependence of an idea upon its object. H. H. Joachim simply concludes that, in the light of Spinoza's response to von Tschirnhaus, "no full and coherent account can be given of the stages in the eternal 'descent' from the Infinite Intellect to the single Ideas or Finite Minds. . ." (*Spinoza's Tractatus De Intellectus Emendatione* [Oxford: Clarendon Press, 1940], p. 76).

[11] Of course, a given idea cannot be the cause or effect of the ideas with which it is identical, since cause and effect must be distinct. But so long as it is possible for an idea of a mode of one attribute to be causally connected with ideas of modes of distinct attributes, then an idea might still form part of a "causally unified whole" with the ideas with which it is identical. Thus idea *A* might be regarded as part of the same whole as *B* if they both have some third ideas as a consequence and are both consequences of a fourth. In any case, Spinoza's own argument does not appeal to the impossibility of an idea's being causally related to itself but rests directly on the observation that regarding things as causally connected presupposes regarding them with respect to a single attribute.

response to von Tschirnhaus shows that the matter is not this straightforward. That response seems to imply that to consider a mode of some attribute as the object of an *idea* is to consider that mode in a way that does depend upon a conception of that attribute. The dependence in fact appears to be an especially intimate one: the concept of a mode of any given attribute is included in the concept of an idea *of* that mode, but it is not included simply because the latter is the concept of an idea qua standing in some external *relation* with that particular mode. If that were the case, then there would be no reason why, contrary to Spinoza's reply to von Tschirnhaus, the ideas themselves that stand *in* such external relations with modes of distinct attributes (namely the relation of being "of" them) might not also be regarded as causally related to one another. Thus a mode of any attribute must enter so intimately into the very *being* of an idea of that mode that the identity of the latter as a particular mode is radically dependent upon the former. This distinguishes the relationship between an idea and object from the kind of relationship that obtains between any given mode of one attribute and the infinitely many modes of other attributes with which it is identical.

III. It has been argued that Spinoza's doctrine of the identity of thought and its object neither follows from nor is compatible with his general conception of the identity obtaining among the attributes of God. Nevertheless, it may be possible to see why Spinoza was led to regard the relationship between thought and its object according to the model involved in our conception of the latter sort of identity. The explanation may simply lie in the fact that Spinoza had independent grounds for regarding thought as in some way "identical" with its object. This independent motivation, together with his general argument concerning the identity of the attributes, may have led him to blur an important distinction between the two cases.

Spinoza's account of the relationship between thought and its object presupposes certain elements of the Cartesian approach to that problem. Descartes distinguishes two modes in which objects of thought may exist. Objects of thought, at least when regarded as the "forms" or "essences" of possible entities of which the intellect forms a conception, may exist as the forms or essences of particular entities *actualizing* the possibilities in question, or, in the absence of such actualization, they may exist merely as the "content" of thought itself. The latter existence constitutes what Descartes calls the "objective reality of an idea": "By the *objective reality of an idea* I mean that in respect of which the thing represented in the idea is an entity, in so far as that exists in the idea. . . . For whatever we perceive as being as it were in the objects of our ideas, exists in the ideas themselves objectively."[12] Objects of thought existing in the former way exist not only objectively but "formally":

If the question be, what the idea of the sun is, and the reply is given, that it is the object thought of in so far as that exists objectively in the understanding, he will not understand that it is the sun itself, in so far as that extrinsic attribute is in it . . . but that it is in the mind in the way in which objects are wont to exist there. Hence the idea of the sun will be the sun itself existing in the

[12] Descartes, "Reply to Objections II," *Oeuvres de Descartes*, ed. Charles Adam and Paul Tannery, 12 vols. (Paris: J. Vrin, 1897-1913), 9:124; trans. Elizabeth S. Haldane and G. R. T. Ross in *Philosophical Works*, 2 vols. (Cambridge: Cambridge University Press, 1970), 2:52. These two editions are cited hereafter as AT and HR.

mind, not indeed formally, as it exists in the sky, but objectively, i.e. in the way in which objects are wont to exist in the mind. . . .[13]

Unfortunately, Descartes's own example may be misleading. It may suggest that the distinction between formal and objective reality is a distinction between two different ways in which certain *particulars* (e.g., the sun) might exist, rather than between two modes in which the *natures* of possible particulars might exist. But the following passage makes Descartes's intent clear:

Neither do we have any other idea of substance itself, precisely taken, than that it is a thing in which this something that we perceive or which is present objectively in some of our ideas, exists formally or eminently. For by means of our natural light we know that a real attribute cannot be an attribute of nothing.[14]

As the last two passages indicate, further, Descartes employed the term "idea" in two distinct ways. In one sense the term is used to designate *objects* of thought qua "objective" realities; in another sense Descartes uses the term in order to designate the *acts* of thought in which such objects are present:

In this term *idea* there is here something equivocal, for it may either be taken materially, as an act of my understanding, and in this sense it cannot be said that it is more perfect than I; or it may be taken objectively, as the thing which is represented by this act, which, although we do not suppose it to exist outside of my understanding, may, none the less, be more perfect than I, because of its essence.[15]

But what is the precise *relationship* between an idea regarded as act of thought and an idea as object of thought? Descartes's view appears to be that an act of thought, or an idea regarded "materially," is nothing but the very *process* by which the "form" constituting an idea's objective reality—an idea considered "objectively"—comes to be present in the mind. It can be argued, at least, that this is the view dictated by overall consistency.[16] It also appears to be suggested by the following claim:

Since ideas themselves are forms, and are never composed of any matter, when we take them as representing something, we regard them not *in a material guise* but *formally*; but if we were to consider them not in so far as they represent this or that other thing, but in the respect in which they are operations of the intellect, it might be said that they were taken materially, but then they would have no reference to the truth or falsity of objects.[17]

Any "operation of the intellect," it would seem, may be regarded in two ways. We may regard it while abstracting from the presence in it of the particular "form" by which the mind is aware of one particular sort of object rather than another; or we may regard it, more concretely, *as* that very form insofar as the latter is present in the

[13] "Reply to Objections I," AT, 9:82; HR, 2:10.
[14] "Reply to Objections II," AT, 9:125; HR, 2:53.
[15] *Meditations*, "Preface to the Reader," AT, 7:8; HR, 1:138.
[16] See chap. 1 of my *Intentionality: A Study of Mental Acts* (University Park, Pa.: Pennsylvania State University Press, 1977), or "Brentano, Descartes and Hume on Awareness," *Philosophy and Phenomenological Research* 35 (1974):236ff.
[17] "Reply to Objections IV," AT, 9:180; HR, 2:105.

operation in question. The first of these procedures recalls Spinoza's description of an "idea of an idea": "For, indeed, the idea of the mind, that is to say, the idea of the idea, is nothing but the form of the idea in so far as this is considered as a mode of thought and without relation to the object. . ." (2, 21, note). This procedure, of course, abstracts precisely from what allows us to distinguish the "content" of one idea from that of another. In this respect Spinoza merely follows Descartes's lead: "If ideas are only taken as certain modes of thought, I recognise amongst them no difference or inequality, and all appear to proceed from me in the same manner; but when we consider them as images, one representing one thing and the other another, it is clear that they are very different one from the other."[18] That Descartes, however, is led to this conclusion appears to confirm our suggestion that ideas as acts of thought are nothing but the coming-to-be-present in the mind of certain forms or natures constituting the *objects* of thought; if mental acts were anything more than this, then there would presumably be something remaining in them even after we abstracted from a consideration of those very forms or natures.

Spinoza shared the Cartesian assumption that ideas as modes of mental activity are ideas of particular objects in virtue of themselves *containing* those objects—or at least the "essences" of them—in some unique way:

> The true idea of Peter is the objective essence of Peter [*essentia Petri objectiva*]. . . . Hence it is clear that certainty is nothing else than the objective essence of a thing: in other words, the mode in which we perceive an actual reality is certainty.[19]

> I state that there exists in Nature an infinite power of thought, which in so far as it is infinite, contains in itself objectively the whole of Nature, and its thoughts proceed in the same way as Nature, which, to be sure, is its ideatum. (Letter 32)[20]

It is precisely this assumption, combined with certain crucial departures from Descartes's own implementation of it, that leads to a kind of "identification" of thought and its object in Spinoza. In particular, the ground for Descartes's distinction between formal and objective reality lay, as we have already seen, in his notion of "substance." Objectively real essences, that is, are also formally real just in case they actually comprise part of the nature of some substance.[21] But even when certain essences are merely *objectively* real for Descartes, their ontological status depends upon their presence in a substance. It depends, namely, upon their being "objectively contained" in some particular *mind*. Thus the distinction between objective and formal reality, in Descartes's account, rests on a distinction between two different ways in which objects of thought may be present in a substance. They may be present objectively in mental substances or present formally in either mental or nonmental substances. Ideas as modes of mental activity are, as we have seen, nothing other than the coming-to-be of the former kind of presence.

Spinoza, however, rejected the view that individual minds are substances in which

---

[18] *Meditation* 3, AT, 9:31-32; HR, 1:161-162.

[19] Spinoza, *Tractatus de Intellectus Emendatione*, Gebhardt, 2:14-15; I have modified the translation of R. H. M. Elwes, as contained in Gutmann, p. 12.

[20] Gebhardt, 4:173-174; I have modified Wolf's translation. See also 2, 7, corol.

[21] See n. 14, above.

mental activity occurs. Minds are rather nothing, on Spinoza's view, over and above the ideas that are attributed to them:

> The essence of man is formed (Corol. Prop. 10, pt. 2) by certain modes of the attributes of God, that is to say (Ax. 2, pt. 2), modes of thought. . . . Therefore an idea is the first thing which forms the being of the human mind. . . . Hence it follows that the human mind is a part of the infinite intellect of God, and therefore, when we say that the human mind perceives this or that thing, we say nothing else than that God has this or that idea; not indeed in so far as He is infinite, but in so far as He is manifested through the nature of the human mind, or in so far as He forms the essence of the human mind. . . . (2, 11, dem., corol.)

It follows accordingly, that the distinction between formal and objective reality cannot rest on the sort of ground that Descartes attempted to provide for it.[22] An idea, qua mode of mental activity, is a certain state of affairs involving the *object* of that activity. It is the particular state of affairs that we may call that object's "coming to be objectively present in mind." But the crucial difference between Spinoza and Descartes is that whereas the latter does the former cannot construe that state of affairs as a *relational* one. The grammatical forms we use to describe ideas are simply misleading: "objective presence in mind" cannot, as it can for Descartes, be explicated in terms of some special relation between an object and an entity that *is* the mind. For ideas just *are* the "objective presences" of objects, and the existence of the mind, on Spinoza's view, is ontologically secondary to that of ideas themselves.[23]

It follows that Spinoza is required to construe "objective presence" in a unique way. An idea, that is, qua mode of mental activity, must be a unique sort of state of affairs involving the *object* of that very idea, distinct from any other sort of state of affairs involving the same object. Thus *A* being objectively present in thought and *A* being greater than *B* are two distinct states of affairs involving the same object. But the former differs in an important way from the latter: whereas *A* being greater than *B* is a state of affairs involving *A* in a relation to some other entity, *A* being objectively present in *thought* does not involve any entity other than *A* itself. Or at least—and this is the crucial point—it does not involve any entity other than *A* itself as an actual constituent of itself. In a certain sense, of course, the state of affairs of *A* being present in thought obviously does involve, for Spinoza, something other than *A* itself. Insofar as it is the state of affairs of something being *thought*, namely, it necessarily involves the "attribute"[24] of thought itself. But it does not thereby involve thought as an actual constituent of itself. Rather, it seems to involve it merely as the essential "form" of the state of affairs in question—in much the same way, for example, as the negative state of affairs consisting of *A* being *not* greater than *B* involves "negation" in an essential way, but not as an actual constituent in addition to *A* and *B* (and, if one also countenances relational "universals," the relation *greater-than*) themselves.

---

[22] See Allison, p. 89: In regarding ideas as acts, Spinoza "differs markedly from Descartes, for whom an idea is essentially an object of consciousness, which belongs in the mind more or less as a property belongs to a thing. . . . it is only natural that in rejecting this doctrine, Spinoza likewise found it necessary to provide a different conception of ideas." Allison somewhat overstates the case, however, inasmuch as it turns out, as I shall argue, that in an important sense the object of an idea is *all* there is to an idea for Spinoza.

[23] Nor can it be argued that "objective presence" involves some peculiar relation to God's mind, since the latter is just the attribute of thought modified in particular ideas.

[24] It begins to emerge at this point, I think, that thought is best not regarded as an attribute at all.

Such a view of ideas seems to be implicit in a passage quoted earlier: "The idea of the mind, that is to say, the idea of the idea, is nothing but the form of the idea in so far as this is considered as a mode of thought and without relation to the object" (2, 21, note). Spinoza appears to be saying here that what makes any ideational state of affairs an *ideational* state of affairs is just a matter of the unique *form* exhibited by that state of affairs; it is not merely a matter of some constituent present in it. This, however, does not imply that thought *cannot* enter as a constituent into ideational states of affairs. Presumably this would occur whenever one idea becomes the object of another idea. Unfortunately, as we have already seen, Spinoza himself is particularly confused on the subject of higher-order ideas. Apparently he is so committed to regarding ideas as "nothing" over and above their objects that the idea of an idea simply collapses into the original idea itself—the only remaining ground for a distinction then lying in a distinction between the idea as a particular ideational state of affairs and the "form" of that same state of affairs.[25]

This view, it now becomes clear, provides motivation for asserting that ideas are "identical" with their objects, and are those very objects regarded with respect to the "attribute" of thought, independent of any general considerations concerning the identity of the attributes. Ideas are their objects regarded with respect to the attribute of thought because an idea is nothing but the state of affairs of some object *being thought*. But ideas are also, in a certain sense, *identical* with their objects, because they are states of affairs involving their objects—*and nothing else besides.* Those objects are their sole constituents (which is not, one should observe, to deny that those same objects are at the same time constituents of other states of affairs as well). An object's objective reality (an idea of that object) then consists in that object's involvement in this state of affairs. (On Spinoza's view, of course, all objects are, in the infinite understanding of God, involved in such states of affairs.) That same object's "formal" reality, on the other hand, consists in the multitude of *other* states of affairs in which it is involved (e.g., being greater than *B*).[26]

This interpretation would explain the radical dependence of an idea upon its object, which Spinoza conceded in responding to von Tschirnhaus and which is incompatible with regarding the relationship between thought and its object as a particular instance of the general "identity" connecting the modes of the infinitely many distinct attributes of substance.[27] It also explains the suggestion (contained in the note to 2, 21) that distinguishing an idea from its object involves some sort of intellectual *abstraction*, which is equally incompatible with assimilating the relation between thought and

[25] Barker observes (pt. 1, p. 165, n. 2) that while Spinoza draws a sharp distinction in the *Tractatus* (see Gebhardt, 2:14; Gutmann, pp. 11–12) between an idea and an idea of an idea, the distinction is by no means so sharply drawn in the *Ethics*. Joachim, similarly, notes that the element of self-awareness appears to serve as a kind of *formal* element in an idea, rather than as a distinct intellectual fact: "This awareness of this thinking—the *form* of self-consciousness—is inseparable from all 'ideas' or minds" (*A Study of the Ethics of Spinoza*) [1901; reprint ed., New York: Russell & Russell, 1964]).

[26] It should be noted that this point is compatible with maintaining that the modes of *any* attributes should be regarded as states of affairs rather than as mere particulars. See the next paragraph and n. 28, below.

[27] Barker suggests (pt. 1, p. 165) that Spinoza attempts to avoid this contradiction by regarding the *essentia objectiva* of an extended thing as included in an idea, not the *esse formale* that corresponds to it. He then objects that such "mental duplication" is useless and leads to contradiction anyway (pt. 1, p. 167; pt. 2, pp. 290–293). As I have argued, however, Spinoza's view appears to be that the distinction in question involves two states of affairs involving the *same* reality. See H. F. Hallett, *Benedict de Spinoza* (London: The Athlone Press of the University of London, 1957), pp. 59–60: "When Spinoza describes an idea as the 'objective essence' of its *ideatum* or object, though to the unwary post-Lockeian this may suggest that it is an

object to the kind of identity obtaining among the attributes of substance generally. If my suggestion is correct, then the sense in which an idea is "identical" with its object is *not* the same as what is involved in the general identity of the attributes. The latter involves the claim that a mode of any given attribute may be regarded (and is so regarded, in the infinite understanding of God) as a mode of infinitely many distinct attributes, each constituting a conceptually independent way of regarding an object. Each mode then *is* those other modes, regarded with respect to the attribute in question. This sort of relation, however, has nothing in particular to do with relations between a state of affairs and the entities involved *in* that state of affairs. We may, to be sure, maintain that each of the infinitely many modes that are identical in this way really is, in a certain sense, a state of affairs *itself.*[28] Thus, we may say, some particular mode of the attribute *F* is the state of affairs of *F* being modified in one particular form rather than another; a particular mode of the attribute *G* is the state of affairs of *G* being modified in one particular form rather than another; and so on, for all the attributes. But even so, the crucial difference must remain: regarding any one of these states of affairs as identical with the others is not regarding it as in any way a *constituent* of those others; it is simply regarding it as the same state of affairs, described from a different point of view. The relationship between an idea and its object must be quite different from this.

Nevertheless, it is possible to see why Spinoza might have confused the two cases. Both might, after all, be described as cases in which one and the same thing is "regarded" in two distinct ways. The distinction between the formal and objective reality of an object is a distinction between two states of affairs in which any object is necessarily involved; hence it requires regarding any given object "from two different points of view." Furthermore, as we have seen, the "point of view" specifically involved in attending to the *objective* reality of an object, unlike that involved in considering an object's formal reality, requires, in an important sense, attending to nothing more than that object itself.

Now it may be objected that this account rests on an unintelligible notion, namely, that of a state of affairs with only a single constituent. This notion may appear to be an unintelligible one since a state of affairs, presumably, is characterized by propositional structure, and this necessarily involves a complexity of content. It is not always necessary, of course, in order to render the thought or argument of some particular

---

objective mental thing reproducing in another medium the content of the *ideatum*, the suggestion is false because it ignores the distinct ways in which the abstracted content is related to ideas and to *ideata*, viz. 'objectively' or terminally to the former, and attributively to the latter." Barker's suggestion may, of course, apply to Descartes himself, since for Descartes the *esse formale* of extended reality involves the actualization of the *essentia objectiva* in some external substance that is decidedly *not* a constituent of one's idea of it.

[28] See Curley, pp. 56, 123. In addition to the reason Curley offers for regarding modes as states of affairs (or "facts"), it also follows from the interpretation I have been defending. On that view an idea of a particular mode of extension does not simply contain an entity that happens to be extended in a particular way. Rather it contains an entity precisely *qua* extended in that way. Unless, therefore, we adopt a purely *subjectivist* view of the distinction among the attributes (see Hallett, *Benedict de Spinoza*, p. 41), the object of the idea in question would have to be the very state of affairs of some entity *being* extended in a particular way. Otherwise, since the "something" modalized with respect to extension in that way is also modalized with respect to infinitely many other attributes, there would be no ground for distinguishing the ideas corresponding to those distinct modalizations. (For a discussion of the subjectivist-objectivist controversy, see M. S. Gram, "Spinoza, Substance and Predication," *Theoria* 34 [1968]:222–244; Alan Donagan, "Essence and the Distinction of Attributes in Spinoza's Metaphysics," in Marjorie Grene, ed., *Spinoza: A Collection of Critical Essays* [Garden City, N.Y.: Anchor Books, 1973].)

philosopher "intelligible," that one deal entirely in notions that are *themselves* intelligible. In many cases it may be sufficient to take note of what may have led the philosopher to *think* that the notion in question is an intelligible one. The reflections I have introduced concerning Spinoza's transformation of Descartes are at least relevant in this regard. Strictly speaking, however, the view I have attributed to Spinoza does *not* require the supposition of states of affairs with only a single constituent. What it requires is only the supposition that ideational states of affairs are states of affairs with only their *objects* as constituents. Let us suppose, as has already been suggested, that *all* modes are, on Spinoza's view, states of affairs of some sort. Then any particular mode of extension E_n will be a particular state of affairs. Let us assume that its constituents are $E_1 \ldots E_m$. What Spinoza might simply argue, then, is that an *idea* of E_n is a state of affairs that itself consists of nothing but the constituents $E_1 \ldots E_m$.[29]

It is crucial to notice, in making this suggestion, that a state of affairs is always at least in *some* way "more" than the complete collection of its constituents. Consider, for example, A being greater than B. Its constituents are A and B and, if we adopt a form of realism with regard to relational universals, the relation designated by the words "greater than." But the state of affairs cannot literally be identified with the *collection* of these constituents, since all of the members of that collection might exist whether or not that state of affairs obtained. (If B were greater than A, then the collection consisting of A, B, and the relation *greater-than* would still exist.) Nor could we attempt to avoid this conclusion by introducing, as some philosophers have, a peculiar "nexus" that accounts for the *unity* among the constituents of any state of affairs.[30] For inasmuch as all states of affairs with the same propositional structure will involve precisely the same kind of unity, a "nexus" will amount merely to a peculiar sort of relational universal. In that case it is subject to the same difficulty as the one that involved the original relation *greater-than*. If, on the other hand, a "nexus" is supposed to be *individuated* in the case of each particular state of affairs, then we seem unable to distinguish a "nexus" from a state of affairs in the first place, and its introduction will not have dealt with our original problem. Thus it seems that a state of affairs is always something "more" than its constituents. We might attempt to give a name to this additional something by calling it the "propositional form" of any given state of affairs. We might then also remind ourselves of Spinoza's claim, quoted earlier, that the properly *ideational* element in any ideational state of affairs is to be regarded as a sort of "form" exhibitied by that state of affairs. But we would also need to bear in mind, as not all philosophers have, that the propositional "form" of any state of affairs is not itself an *entity* among the entities actually contained in that state of affairs.

---

[29] Or alternatively (and also preferably, if Spinoza does not permit the inclusion of *properties* as constituents of states of affairs), Spinoza could claim that the idea in question contains nothing but $E_1 \ldots E_m$ together with the original E_n itself. Thus, $E_1 \ldots E_m$ being contained in E_n and $E_1 \ldots E_m$ being *thought* to be contained in E_n would be two states of affairs with just the same constituents. Of course, if Spinoza was unclear concerning the distinction between states of affairs and complex particulars in the first place, then one would expect him to tend to regard the latter of these states of affairs either as containing nothing but E_n itself, or else as containing nothing but $E_1 \ldots E_m$.

[30] See Gustav Bergmann, *Realism: A Critique of Brentano and Meinong* (Madison: University of Wisconsin Press, 1967), p. 9. For an elaboration of the point I am making here, and a specific application of the criticism to Bergmann's own theory of intentionality, see my *Intentionality*, chap. 3, or "Intentionality and Possible Facts," *Nous* V (1971):411-417.

If a state of affairs is never literally identical with the complete collection of its constituents, then there appears to be nothing in principle preventing the existence of two *distinct* states of affairs with precisely the same constituents; what would presumably distinguish them is their propositional "form." In fact, some examples come to mind independently of our present concern with ideational states of affairs. Consider, for example, these states of affairs: "green and yellow not being instantiated by anything" and "green and yellow not being identical with one another." Since being-instantiated and (not-) being-identical are not themselves *constituents* of these states of affairs, the latter appear to involve precisely the same constituents. (It might seem that we could also use an example cited earlier, namely, the distinction between "*A* being greater than *B*" and "*A* not being greater than *B*," since presumably nonbeing is not a constituent of these states of affairs either. But perhaps it might be objected that inasmuch as these states of affairs could differ only to the extent that *A* and *B* have themselves undergone some internal change, the two states of affairs in question must thereby differ in their own internal constitution. This raises some difficult questions, however, which it is best to avoid in the present context. For example, if *C* is a constituent of *B*, is *C* thereby a constituent of any states of affairs into which *B* enters?) There is, therefore, nothing in principle preventing the supposition that the very same $E_1 \ldots E_m$ that constitute the mode of extension E_n also constitute the mode of thought that is an *idea* of E_n. They could not, of course, constitute that mode of thought in virtue of being *literally* identical with it; in that case the distinction between the "attributes" of thought and extension would be blurred in a way that Spinoza could not tolerate. Yet those very same entities might "constitute" an ideational state of affairs in virtue of constituting the *sole constituents* of a state of affairs with the peculiar "form" characteristic of the attribute Thought. The crucial point would simply be: an idea of the mode of extension E_n is, in an important sense, nothing *more* than that mode of extension; it involves no further *entities* than that mode of extension involves. In a *sense*, at least, an idea is "identical" with its object; it is only "formally" distinct from it.[31]

Although Spinoza's doctrine of the identity of thought and its object is not, contrary to his own claims, a corollary of his general theory of the "identity of the attributes," it is thus possible to see that it nevertheless follows in a certain way from principles central to his philosophy. Furthermore, Spinoza's own suggestion concerning the derivation of that doctrine in fact conceals the real *meaning* of the "identity" in question.

IV. Once we have clarified the real nature of the "identity" involved in Spinoza's

[31] It might be objected that, on the account I am suggesting, Spinoza is precisely *not* regarding an idea as a state of affairs involving its object and nothing else besides. On the suggested account, an idea of E_n does not involve E_n itself as a constituent, but merely the constituents *of* E_n. Spinoza, however, would not be the only philosopher to have assumed what this objection must deny, namely, that an idea of a given state of affairs is "really" an idea whose objects are the *constituents* of that state of affairs. This is precisely the point of Bertrand Russell's analysis of judgment and belief (see *The Problems of Philosophy* [London: Oxford University Press, 1943], pp. 196ff.). (One crucial *difference* is that, unlike Spinoza, Russell considers the judging subject *itself* to be in every case one of the constituents of an ideational state of affairs.) The assumption, of course, that an idea of a given state of affairs is really an idea whose objects are the constituents of that state of affairs is not immune to criticism. It may in fact rest on a confusion of "opaque" and "transparent" judgment-ascriptions: "*S* judges that *A* is greater than *B*" *vs.* "*S* judged *of A* and *of B* that the first is greater than the second." It is not unlikely, I think, that some such confusion underlies Spinoza's reasoning.

theory of the identity of thought and its object, it is worth taking note, in conclusion, of a remarkable similarity that thereby reveals itself between Spinoza's view and some of Sartre's claims in *Being and Nothingness.*[32] Sartre, unlike Spinoza, does not generally emphasize the "identity" of consciousness and its object. Rather, he insists upon establishing that consciousness is in some sense "nothing":

Of course the very apprehension of the world as a totality causes the appearance *alongside the world* of a nothingness which sustains and encompasses this totality. . . . But this nothingness *is not* anything except human reality apprehending itself as excluded from being and perpetually beyond being, in commerce with nothing. It amounts to the same thing whether we say, human reality is that by which being is revealed as totality—or human reality is that which causes *there to be* nothing outside of being. (P. 251/230)

At least occasionally, however, Sartre allows himself a more Spinozistic formulation. Consciousness, namely, in some sense *is* its object:

Thus knowledge is the *world.* To use Heidegger's expression, the world and outside of that—nothing. . . . This *nothing* is human reality itself as the radical negation by means of which the world is revealed. (P. 251/230).

Thus nothingness is this hole of being, this fall of the in-itself toward the self, the fall by which the for-itself is constituted. But this nothingness can only "be made-to-be" if its borrowed existence is correlative with a nihilating act on the part of being. This perpetual act by which the in-itself degenerates into presence to itself we shall call an ontological act. Nothingness is the putting into question of being by being—that is, precisely consciousness or for-itself. It is an absolute event which comes to being by means of being. . . . (P. 126/121)

Consciousness for Sartre, being "nothing" but its object, can only be that object *itself* "regarded from a certain point of view"; consciousness, namely, is its object regarded as achieving "presence to itself."

I have argued that Spinoza was misled by the fact that his general theory of the "attributes" introduces an ambiguity into the claim that an idea is just the very *object* of that idea, namely, that object regarded as an act of "ideating" itself. The claim might be taken to deal with a special case of the relationship between something described in one set of terms and that same thing described in conceptually independent terms; or it might be taken to mean that the relation is like that obtaining between an object and a state of affairs *containing* that object and nothing else besides. Though space does not permit a full presentation of the argument,[33] a few words will indicate the basis for concluding that Sartre's claims are to be explicated along the latter lines.

It might appear that the comparison with Sartre is implausible in view of Sartre's insistence that objects of consciousness are never actually "contained" in the consciousness of them: "Consciousness has no 'content'. . . . The first procedure of a philosophy ought to be to expel things from consciousness and to reestablish its true

[32] Paris: Gallimard, 1943; the first part of each parenthetical reference will be to Hazel Barnes's translation (New York: Washington Square Press, 1968), the second to the original edition. The comparison with Sartre is not meant to imply, of course, that all modes of thought are *conscious* for Spinoza.

[33] For a fuller treatment, see my "Two Problems of Being and Non-Being in Sartre's *Being and Nothingness,*" *Philosophy and Phenomenological Research* 38 (1977): 167–186.

connection with the world, to know that consciousness is a positional consciousness *of* the world'' (p. 11/17–18). It seems clear, however, that Sartre intends only to deny that objects of consciousness are contained in consciousness in the way that one concrete *thing* might be contained in another. This is compatible with maintaining that consciousness is a *state of affairs* that contains its objects precisely in the way that any state of affairs can be said to ''contain'' objects—in the way, for example, that the state of affairs of a page being in a book contains the page in question, as opposed to the way in which the book *itself* contains that page. That consciousness is indeed contrasted with its objects in virtue of being a special sort of state of affairs involving those objects seems evident from a number of things that Sartre says:

> All consciousness is positional in that it transcends itself in order to reach an object, and *it exhausts itself in this same positing*. (P. 11/18; emphasis added)

> For consciousness there is no being outside of that precise obligation to be a revealing intuition of something, i.e., of a transcendent being. (P. 23/29)

> Thus the very meaning of the for-itself is outside in being . . . it is nothing but the manner in which being is revealed as not being the for-itself, the manner in which *there is* being. (P. 251/230)

Consciousness, in other words, is not something that *is* aware of objects; it is the very *fact* of awareness of objects. It is not the subject of consciousness, but ''subjectivity itself'' (p. 17/23). This seems to be at least a good part of Sartre's meaning when he claims that the ''essence'' of consciousness lies in its ''existence'': consciousness is not a *particular* capable of entering into various states of affairs involving objects of consciousness; it is just a certain *sort* of state of affairs involving objects (pp. 15/21, 22–23/28–29).

Whereas consciousness, then, does not ''contain'' its object, on Sartre's view, in the way that one *thing* might be said to contain another, it does appear to contain its object in precisely the way that a *state of affairs* might be said to contain the objects involved in it:

> Consciousness has to be its own being, it is never sustained by being; it sustains being in the heart of subjectivity, which means once again that it is inhabited by being but that it is not being. . . . (P. 105/102)

> The for-itself is the in-itself losing itself as in-itself in order to found itself as consciousness. . . . It follows that this in-itself, engulfed and nihilated in the absolute event which is the appearance of the foundation or upsurge of the for-itself, remains at the heart of the for-itself as its original contingency. (P. 130/124–125)

But what *else* does such a state of affairs contain, in addition to the object of consciousness? Sartre's answer is plain: it contains nothing else. The motivation leading to this conclusion, furthermore, is precisely the same as what we have already encountered in Spinoza—the rejection of a thinking ''substance'' in which mental acts occur:

> The ontological error of Cartesian rationalism is not to have seen that if the absolute is defined by the primacy of existence over essence, it can not be conceived as a substance. Consciousness

has nothing substantial, it is pure "appearance." . . . But it is precisely because consciousness is pure appearance, because it is total emptiness (since the entire world is outside it)—it is because of this identity of appearance and existence within it that it can be considered as the absolute. (P. 17/23)

Consciousness, that is, is a state of affairs involving an object of awareness. But, contra Descartes, that state of affairs does not involve some special relation *between* that object and a subject that is aware of it. Having rejected the subject of consciousness, as something existing over and above its acts, Sartre can only conclude that consciousness is a state of affairs involving its objects and nothing else besides; hence that consciousness is, in a certain sense, "nothing" in addition to its objects (or, alternatively, *is* those very objects, regarded as consciousness of themselves).[34]

This much of an analysis of Sartre's view, together with our discussion of Spinoza, may at least serve to establish the basis for maintaining that Spinoza's claims about the identity of thought and object reflect the conflation of two independent lines of thought, one of them determined by his general conception of the attributes, the other by just the sort of consideration leading to Sartre's primary conclusions in *Being and Nothingness.*[35]

University of Tennessee

[34] This line of thought is perhaps more evident in Sartre's "La Transcendance de l'ego," *Recherches philosophiques* 6 (1936):65-123. For example, "The Transcendental Field, purified of all egological structure, recovers its primary transparency. In a sense, it is a nothing. . . . But this nothing is *all* since it is *consciousness* of all these objects" (trans. Forrest Williams and Robert Kirkpatrick [New York: Noonday Press, 1957], p. 93).

[35] Another point of comparison worth pursuing concerns Spinoza's and Sartre's conception of self-consciousness. Like Spinoza (2, 21, note; see n. 25, above), Sartre denies that consciousness of consciousness involves two distinct acts of consciousness: "This self-consciousness we ought to consider not as a new consciousness, but as *the only mode of existence which is possible for a consciousness of something*" (p. 14/20). Both thinkers regard this view as a consequence of their conception of consciousness as an "act" and not a "thing" (see Sartre, p. 14/20; Spinoza, 2, 43, note). As I have suggested above, this seems to reflect some confusion on Spinoza's part concerning the distinction between states of affairs and the entities contained in them; the same unclarity, as I have argued in "Two Problems of Being and Non-Being," permeates Sartre's thinking. Oddly, Sartre himself takes Spinoza to be denying his view of self-consciousness (p. 12/18), perhaps basing his interpretation on the *Improvement of the Understanding* rather than on the *Ethics.* (In both places, however, Spinoza and Sartre remain separated by Spinoza's commitment to an *infinite regress* of self-consciousness.) It is perhaps also worth observing that Hallett's defense of Spinoza in "On a Reputed Equivoque" bears a remarkable similarity to Sartre's arguments concerning the distinction between the body "for-itself" and "for-others" (pp. 404ff./368ff.).

[2]

HISTORY OF PHILOSOPHY QUARTERLY
Volume 8, Number 3, July 1991

CAUSATION AND SPINOZA'S CLAIM OF IDENTITY

Michael Della Rocca

SPINOZA expresses his position on the relation between the mental and the physical in his pronouncement: "a mode of extension and the idea of that mode are one and the same thing, but expressed in two ways" (2p7s).[1] At least part of what Spinoza means by this claim (which I will call the claim of identity) seems to be that there is a full-blown numerical identity between a mode of extension and the idea of that mode. What further import is carried by the qualification "but expressed in two ways" is a controversial and difficult issue which I will not address directly here (although some of my comments about Spinoza's notion of causation will have some bearing on this point). Instead, I want to focus on the question of whether we are in fact entitled to regard Spinoza as holding that a mode of extension and the idea of that mode are numerically identical. This interpretation, the numerical identity interpretation, is certainly appealing since it allows us to see Spinoza as, in some way or another, a precursor of contemporary identity theories of mind and body or mind and brain.[2]

But, however tempting it may be to interpret Spinoza in this way, there is an obstacle: it has been argued that the claim that a mode of extension and a mode of thought are numerically identical is incompatible with certain basic features of Spinoza's system. This line of argument is pursued in similar ways by R.J. Delahunty and Jonathan Bennett. In each case, I will argue, the particular contradiction which the numerical identity interpretation allegedly faces does not actually obtain. The numerical identity interpretation can avoid these inconsistencies if it can be shown that Spinoza holds a certain thesis about causation. I will present evidence that Spinoza does hold this thesis. But the fact that Spinoza adheres to this position on causation puts the numerical identity interpretation in danger of a different conflict with passages other than those which Delahunty and Bennett emphasize in this regard.

I

Delahunty's argument relies upon Spinoza's ban on causal relations between attributes. Spinoza introduces this ban in the following way. For

Spinoza, there are two different kinds of causes of modes. He says that the one substance God is the cause of each mode (1p18 and 1p25) and he says that each mode is the effect of another mode (1p23 and 1p28).[3] In terms of the distinction Spinoza draws in 1p18, the former kind of causal relation is one of immanent causation and the latter is one of transitive causation. (I follow here Curley's translation of "*immanens*" and "*transiens*.") It is not important here to see what the connection is between these two different kinds of causal relations. I simply want to note that in terms of each kind of cause of finite modes, Spinoza erects a causal barrier between different attributes.

For Spinoza, the cause of extended mode x cannot be God *qua* thinking, but can only be God *qua* extended. The modes of each attribute "have God for their cause only insofar as he is considered under the attribute of which they are modes, and not insofar as he is considered under any other" (2p6d). (See also 2p5 where Spinoza makes this claim for the attribute of thought in particular.)

Spinoza makes a similar point concerning causal relations between modes of different attributes:

> The formal being of the idea of the circle can be perceived only through another mode of thinking, as its proximate cause, and that mode again through another, and so on, to infinity. Hence so long as things are considered as modes of thinking, we must explain the order of the whole of nature, or the connection of causes through the attribute of Thought alone. And insofar as they are considered as modes of Extension, the order of the whole of nature must be explained through the attribute of Extension alone. I understand the same concerning the other attributes (2p7s).[4]

Delahunty claims that the numerical identity interpretation conflicts with Spinoza's denial of transitive causation between modes of different attributes. He argues:

> If he [Spinoza] accepted an identity theory..., then he can scarcely deny interactionism.... If a mode of Thought, X, simply *is* the corresponding mode of Extension, Y, then if X is the cause of mental mode Z, it seems to follow that Y is also the cause of Z; and if Y is the cause of extended mode Z, then so must X be. (This is connected with the fact that while ".. explains..." is not referentially transparent, ".. causes..." is arguably so....)[5]

This is certainly an important objection to the numerical identity interpretation. Any conflict between that interpretation and such a fundamental Spinozistic position as his denial of interactionism may be grounds for rejecting that interpretation.

Despite the importance of this objection, as far as I know, no one who holds a numerical identity interpretation explicitly tries to meet it. There are, however, at least two potential ways in which the numerical identity interpretation might be defended in the face of this objection. The first strategy would be to deny Delahunty's claim that there is a conflict between a numerical identity position and Spinoza's denial of causation between attributes. Alternatively, we might say that although Spinoza did hold a numerical identity position and although Delahunty is right that there is a conflict between such a position and Spinoza's denial of interactionism, Spinoza did not realize that there is such a conflict. The latter strategy is, obviously, less attractive than the former. It would be acceptable only after we've exhausted other possible interpretations of Spinoza's claim of identity in order to see if any of these can do a better job of presenting Spinoza's views as consistent. I will not undertake such a comparison here since a more straightforward defense of the numerical identity interpretation along the lines of the first strategy is available.

As the parenthetical comment in the quote from Delahunty suggests, this criticism depends upon the claim that causal contexts (where transitive causation is concerned) are referentially transparent. This is the claim that if one event, state, etc. causes another, then no matter how they are described, it is true to say that they are so related. Take a sentence of the form "x causes y" where "x" and "y" are singular terms each of which refers to a particular event, state, etc. The above claim says that if we substitute co-referring singular terms for either "x" or "y" or both, the result will necessarily be a sentence with the same truth-value as "x causes y." Delahunty's objection points out that a denial of causal interaction between the mental and the physical and the view that each mode of extension is identical with a mode of thought (and vice versa) are jointly incompatible with the claim that (transitive) causal contexts are referentially transparent. Thus Delahunty's objection says that the numerical identity interpretation must hold that Spinoza is committed to denying the transparency of (transitive) causal contexts. This would constitute an objection to the numerical identity interpretation only if one or both of the following were the case:

(1) It would be absurd to attribute a denial of transparency in this case to Spinoza since it is obviously true that (transitive) causal contexts are referentially transparent.

(2) Spinoza explicitly commits himself to the view that (transitive) causal contexts are transparent.

Neither of these, however, is the case. (1) is false for the following reason: Although the claim that (transitive) causal contexts are transparent has a great deal of plausibility and has many proponents in our day, it is not uncontroversially true. Recently, both Mackie and Anscombe have, in different ways, cast doubt on it.[6] Thus, if the numerical identity interpretation must say that Spinoza is committed to denying the transparency of causal contexts, it is not thereby attributing to him an obviously false philosophical position.

(2) is also false: Spinoza never commits himself to the view that (transitive) causal contexts are transparent. In fact, he says fairly clearly that contexts involving *immanent* causation are *not* transparent. This gives us some reason to hold that for him transitive causal contexts are also

referentially opaque. Consider 2p6: "The modes of each attribute have God for their cause only insofar as he is considered under the attribute of which they are modes, and not insofar as he is considered under any other attribute." This suggests that Spinoza would regard the following claims as true and false respectively:

(a) The thinking substance causes mode of thought x.

(b) The extended substance causes mode of thought x.

Notice that for Spinoza (a) is true and (b) is false despite the fact that, as he says in 2p7s, the thinking substance *is* the extended substance. If this is correct, then Spinoza denies that contexts involving immanent causation are referentially transparent.

We have reached this conclusion independently of any claim as to the identity of modes of thought and modes of extension. But now this conclusion can be used to provide evidence for claiming that Spinoza would also maintain that there is a similar failure of transparency in contexts involving transitive causation and thus to provide evidence for claiming that the numerical identity interpretation is compatible with Spinoza's denial of interactionism. Thus, consider:

(c) Extended mode x causes extended mode y.

(d) Thinking mode w causes extended mode y.

Spinoza could regard (c) as true, though (d) must be false. A proponent of the numerical identity interpretation would say that Spinoza could regard (c) and (d) in this way even if thinking mode w = extended mode x. Thus, it might be argued, just as Spinoza denies that immanent causal contexts are transparent, he denies that transitive causal contexts are transparent.

Thus, even though the numerical identity interpretation commits Spinoza to a denial of transparency in contexts of transitive causation, that is no reason to reject the numerical identity interpretation and, in fact, there is independent evidence that Spinoza would deny transparency in this case. So, there is a straightforward way in which the numerical identity interpretation can answer Delahunty's objection.

There is, in principle, another, more radical way for the numerical identity interpretation to avoid Delahunty's objection. The strategy here would be the following: A proponent of the numerical identity interpretation could agree with Delahunty that causal contexts are transparent and that a numerical identity view is incompatible with any denial of interactionism. But this proponent would go on to say that, contrary to appearances, Spinoza does not want to deny interactionism. He only wants to deny a closely related position, viz. that we can *explain* modes of extension in terms of modes of thought and vice versa. The idea would be that a mode considered as physical must be explained only in terms of other modes considered as physical; while that same mode considered as mental must be explained only in terms of other modes considered as mental. Such a view, it would be pointed out, does not by itself commit one to a denial of causal interaction between the mental and the physical. On this line of thought, since causal contexts are transparent, even if there are no explanatory relations between the mental and the physical there can still be causal relations.

Such a separation of causal relations and explanatory relations is a possible position and, indeed, a popular one in contemporary philosophy. For examples, see the references to Davidson, Follesdal, Searle and Strawson in note 6. But, while it may be tempting to view Spinoza along these contemporary lines, there is much evidence against doing so. Spinoza does sometimes state his causal ban in ways that involve the notion of explanation—see, e.g., the long passage from 2p7s quoted earlier. But even here Spinoza brings in the notion of causation. Notice the claim that one mode must be perceived through another *as its proximate cause*. Thus, we cannot be confident that Spinoza's strictures in this passage cover only a denial of explanatory relations between the mental and the physical. Further, Spinoza often states his causal ban in terms of causal relations alone, without bringing in the notion of explanation at all. In 3p2d, Spinoza says, "What determines the mind to thinking is a mode of thinking and not of extension." (Later on in the demonstration, he makes a similar claim about causation in the opposite direction.) In the preface to Part 5, while attacking Descartes's view that mind and body causally interact, Spinoza says, "the forces of the Body cannot in any way be determined by those of the mind." (The contexts of these passages indicate that Spinoza here means "cause" by "determine.") For these reasons, we cannot interpret Spinoza as denying explanatory relations, but not causal relations between the mental and the physical. Thus, for the numerical identity interpretation, the best response to Delahunty's objection is to claim, as I outlined above, that Spinoza simply denies the transparency of causal contexts.

II

Bennett offers a related objection to the claim that for Spinoza a mode of extension is identical with a mode of thought. In his book on Spinoza, Bennett claims that in the claim of identity Spinoza "cannot be saying that physical P_1=mental M_1; that is impossible because they belong to different attributes."[7] This point is elaborated in Bennett's "Eight Questions about Spinoza."[8] There Bennett points out that for Spinoza (1) "the modes of extension involve the concept of extension and the modes of thought involve the concept of thought" ("Eight Questions About Spinoza," p.18). Since (1) is true, he claims that if (2) each mode of extension is numerically identical with a mode of thought, then it follows that (3) "every mode involves every attribute." He holds that this conclusion "would bring large parts of the *Ethics* to ruin" (*ibid*., p.18). Certainly it seems to conflict with 2p6d: "The modes of each attribute involve the concept of their own attribute, but not of another one." And Bennett points out that (3) would remove "Spinoza's ground for saying that we cannot explain physical actions in terms of mental causes" (*ibid*., p. 9). Bennett concludes that therefore (2) must be rejected as an interpretation of Spinoza.

I agree that (1) should not be given up. But I think that there are two different ways to take (3). On one way, (3) is acceptable and does not do the damage to Spinoza's system that Bennett fears. On another way of taking (3), it *is* unacceptable for the reasons Bennett states. I think that (3) follows from (1) and (2) *only* when it is taken in the former, acceptable sense. When (3) is taken in its unacceptable sense, it does not follow from (1) and (2) and thus does not force a rejection of the numerical identity interpretation.

To elicit the different possible senses of (3), I need to show that Spinoza holds that x involves the concept of y only if x is caused by y. Since Spinoza equates the notions of x involving the concept of y and x being conceived through y (see 1ax5 and 2p6d), we can express the claim that I will attribute to Spinoza as:

(e) x is conceived through y only if x is caused by y.

Evidence that Spinoza holds (e) comes from 1p25:

> God is the efficient cause, not only of the existence of things, but also of their essence. *Demonstration*: If you deny this, then God is not the cause of the essence of things; and so (1ax4) the essence of things can be conceived without God. But (1p15) this is absurd. Therefore God is also the cause of the essence of things.

Here Spinoza says that if x is not caused by y, then x can be conceived without y.[9] This implies that if x is not caused by y then it is not the case that x must be conceived through y. Expressing this latter conditional in terms of its contrapositive, we get: If x must be conceived through y, then x is caused by y. Or, equivalently, x must be conceived through y only if x is caused by y. This is essentially what (e) says. The only difference between this claim and (e) is that (e) says "is conceived" and this claim says "must be conceived." This difference is not, however, significant in this context. When making points about conceiving through another and conceiving through oneself, Spinoza often glides between claims of these kinds. See, e.g., 1def3 and its restatement in 1p10d.

Further evidence that Spinoza holds (e) comes simply from the fact that the only concepts which Spinoza says that the concept of a mode of a particular attribute involves are the concepts of the causes of that mode. As we have seen, Spinoza claims that God as well as certain other modes cause a given mode. These other modes and God are also the *only* things the concepts of which, according to Spinoza, the concept of that given mode involves. (See 2p6 and 2p7s.)

The fact that Spinoza holds (e) enables us to rebut Bennett's objection to the numerical identity interpretation. On that interpretation, a particular mode of extension is identical with a mode of thought. The numerical identity interpretation must also hold, as we have seen, that causal contexts are referentially opaque. Thus, considered as a mode of extension, a given mode causally follows from God considered as extended, and so the concept of that mode *qua* mode of extension involves the concept of extension. But, given the opacity of causal contexts, that mode *qua* mode of extension has no causal relations with God considered as thinking. Thus, a proponent of the numerical identity interpretation could say that it follows from (e) that the concept of that mode *qua* mode of extension does not involve the concept of thought. Similarly, that mode *qua* mode of extension causally interacts with other modes insofar as they are modes of extension and not insofar as they are modes of thought. Thus, it also follows from (e) that the concept of the mode *qua* mode of extension does not involve the concept of any mode of thought (or, more accurately, does not involve the concept of any mode *qua* mode of thought).

Parallel arguments would show that the concept of the same mode *qua* mode of thought does not involve the concept of extension or of any mode *qua* mode of extension.

Thus, by virtue of (e), we can say that whether a particular mode is conceived through extension and other modes *qua* modes of extension or whether it is conceived through thought and other modes *qua* modes of thought depends on what causal relations it enters into. We have also seen that what causal relations a mode enters into depends on whether it is considered as a mode of thought or as a mode of extension. From these two claims, it follows that what attribute and what other modes a particular mode is conceived through depends on whether that mode is considered as a mode of thought or as a mode of extension. Bennett's objection assumes that sentences of the form "x is conceived through y" are referentially transparent. However, due to the the referential opacity which Spinoza attributes to causal contexts and due to the connection between conception through another and causation, sentences of that form are also referentially opaque. This implies that from the fact that mode of extension A is conceived through mode of extension B and the fact that mode of extension A = mode of thought 1, we cannot conclude that mode of thought 1 is conceived through mode of extension B.

Thus the most we can conclude from (1) and (2) is that

(3′) For each mode there is a way of considering it by means of which it involves the concept of thought and of certain other modes of thought and there is another way of considering it by means of which it involves the concept of extension and of certain other modes of extension.

This conclusion, however, does not entail that there is any illegitimate mixing of conceptual chains involving thought and conceptual chains involving extension. In particular, this conclusion would not entail, as

Bennett fears, that one could explain physical actions in terms of mental causes. It would only follow that each physical action is explainable through another mode of extension and that this mode of extension is identical with a mode of thought. This does not entail that the physical action is explainable through that mode *qua* mode of thought.[10] Thus, I see no reason why Spinoza would reject (3'). If (3) is interpreted in this way, therefore, it poses no threat to the numerical identity interpretation, even though it follows from (1) and (2).

What we *cannot* conclude from (1) and (2) is that each mode *qua* mode of extension involves the concept of thought and of other modes *qua* modes of thought and that each mode *qua* mode of thought involves the concept of extension and of other modes *qua* modes of extension. (On this view, sentences of the form "x is conceived through y" are referentially transparent.) This kind of conclusion would be unacceptable to Spinoza—he explicitly denies it in 2p6d. Bennett takes (3) in this sense and, as such, rightly rejects it. But he does not see that in this sense (3) does not follow from (1) and (2) and so cannot be used to undermine the numerical identity interpretation.

III

So by seeing Spinoza as denying the transparency of causal contexts we can neutralize both Delahunty's and Bennett's objections to the numerical identity interpretation. However, if we attribute this denial to Spinoza as well as the numerical identity view, then the numerical identity interpretation faces a textual difficulty distinct from the ones that Delahunty and Bennett envisage. This difficulty arises from Spinoza's claim in 2p7s that the claim of identity explains parallelism.

Parallelism is expressed in 2p7: "The order and connection of ideas is the same as the order and connection of things." This is, in part, the thesis that for each mode of extension there is a mode of thought which enters into causal relations that correspond to those that the mode of extension enters into (and vice versa).[11] Thus, if the following is a segment of the causal chain of extended modes (em's)

emA → emB → emC

then there are three thinking modes (tm's) related in the same way:

tm1 → tm2 → tm3.

In 2p7s Spinoza says that parallelism follows from the claim of identity:

> a mode of extension and the idea of that mode are one and the same thing, but expressed in two ways....For example, a circle existing in nature and the idea of the existing circle, which is also in God, are one and the same thing, which is explained through different attributes. *Therefore* [*ideo*], whether we conceive nature under the attribute of extension, or under the attribute of thought, or under any other attribute, we shall find one and the same order, or one and the same connection of causes, i.e., that the same things follow one another.[12]

However, if Spinoza does indeed deny transparency, then if he holds the numerical identity view, he would have no ground for saying that the claim of identity explains parallelism.

The claim of identity on the numerical identity interpretation (or, as I will call it, the claim of numerical identity) is certainly able to explain something that is presupposed by parallelism as I've stated it here, viz. that for each mode of extension there is a mode of thought and vice versa. This follows directly from the claim of numerical identity since that claim simply states that each mode of extension *is* a mode of thought and vice versa. But the claim of numerical identity is not able to explain why for each mode of extension there is a mode of thought *which enters into matching causal relations* (and vice versa). The reason the claim of numerical identity cannot explain this aspect of parallelism stems directly from the fact that the numerical identity interpretation is forced to hold that Spinoza regards causal contexts as referentially opaque. This point can be made by considering a relatively complex segment of the causal chain of extended modes, a segment that involves joint causes and joint effects. Spinoza allows that there are such causal chains—see, e.g., 2def7 and 5p8.

(I)

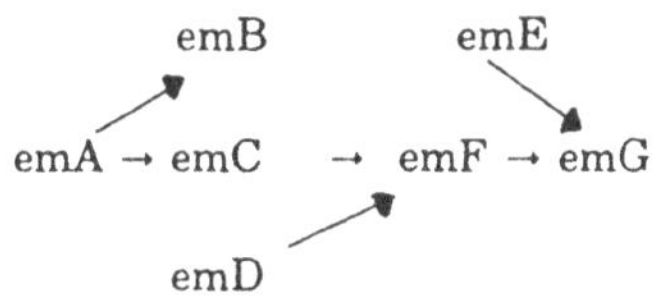

The claim of numerical identity does guarantee, as I noted, that there is a mode of thought for each of these modes of extension. Thus, without yet indicating their causal relations, we can represent this set of modes of thought as follows:

(II) tm2 tm5

tm1 tm3 tm6 tm7

tm4

The claim of numerical identity would allow us to say that tm1= emA, tm2=emB, etc. Parallelism would hold as a result of this numerical identity only if the causal links in (I) match those in (II), that is only if tm1 causes both tm2 and tm3, and if tm3 and tm4 jointly cause tm6, etc. We might expect that since tm1=emA and since emA causes both emB and emC and since emB=tm2 and emC=tm3, it follows that tm1 causes tm2 and tm3, etc. But such an inference would not be a valid one for a proponent of the numerical identity interpretation since, on that interpretation, Spinoza must be denying that causal contexts are referentially transparent. Given this denial, it is compatible with the claim of numerical identity that, although the previously noted identities hold and although the pattern of causes and effects among extended modes is as it is depicted above, nevertheless, the series of causes and effects among modes of thought has a different pattern. For example, the pattern might, for all the numerical identity interpretation says, have the following form:

(IIa)

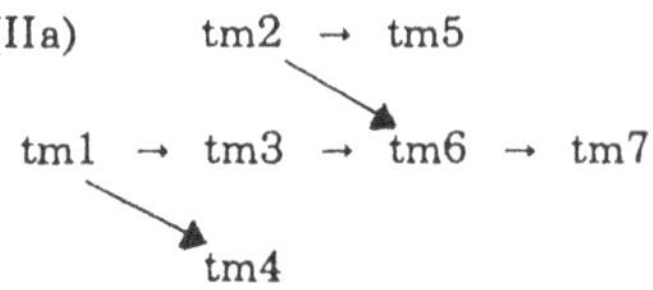

If the series of causes and effects in the two different realms were as it is depicted in (I) and (IIa), then parallelism obviously would not hold even though each thinking mode in (IIa) is identical with an extended mode in (I) and vice versa.[13]

Thus the numerical identity interpretation does not provide us with a way of showing how the claim of identity explains parallelism. This is a defect of this interpretation since Spinoza clearly sees the claim of identity as performing this function. This objection to the numerical identity interpretation is not, however, as serious as the (unsuccessful) ones Delahunty and Bennett raise. Those objections assert that the numerical identity interpretation conflicts with certain basic Spinozistic features, viz. the causal and conceptual barriers between thought and extension. Spinoza insists upon these positions often and in many different contexts. The objection I have raised points out a conflict between the numerical identity interpretation and Spinoza's statement that parallelism follows from the claim of identity. This statement is an important one, but it is certainly not one that Spinoza relies upon, or emphasizes, as much as his claims about the causal and conceptual barriers between thought and extension.

Further, one might be able to defend the numerical identity view from this objection by saying that although the claim of numerical identity does not explain parallelism, Spinoza thought that it did. This mistake on Spinoza's part was perhaps facilitated by the fact noted above that the claim of numerical identity does explain a claim presupposed by parallelism, viz. that for each mode of extension there is a mode of thought and vice versa.

This may indeed be a legitimate response to the objection I have raised against the numerical identity interpretation. However, as I noted before, we should avoid adopting an interpretation that attributes a mistake to Spinoza until we've examined other potential interpretations of Spinoza's view about the relation between a mode of extension and the idea of that mode to see if any of these other interpretations depicts Spinoza's view as more coherent.

A final evaluation of the numerical identity interpretation will thus have to await such a comparison with other interpretations. I have, however, shown that the numerical identity interpretation cannot be eliminated as summarily as the criticisms made by Delahunty and Bennett might lead us to believe. And, although the jury is still out on the numerical identity interpretation, we have in the meantime discovered an important and intriguing fact about Spinoza's notion of causation: there are strong indications that he denies that causal contexts are referentially transparent.[14]

University of California, Berkeley

Received May 21, 1990

NOTES

1. Spinoza's Latin is: *modus extensionis et idea illius modi una eademque est res, sed duobus modis expressa* . Spinoza repeats this theme of 2p7s in 2p21s and 3p2s. Here and throughout I rely on Gebhardt's edition of Spinoza's Latin text. See C. Gebhardt, ed., *Spinoza Opera* , vol. 2 (Heidelberg: Carl Winter, 1925). I follow Curley's translation and his system of numbering passages from the *Ethics* . See Edwin Curley, ed. and trans., *The Collected Works of Spinoza*, vol. 1 (Princeton: Princeton University Press, 1985). All references to works of Spinoza are to the *Ethics.*

2. When commentators attribute an identity view to Spinoza, they usually seem to have numerical identity in mind. See Henry E. Allison, *Benedict de Spinoza: An Introduction* (New Haven: Yale University Press, 1987), p. 86; Edwin Curley, *Behind the Geometrical Method*, (Princeton: Princeton University Press, 1988), pp. 68-9, 82; Douglas Odegard, "The Body Identical with the Human Mind: A Problem in Spinoza's Philosophy" in *Spinoza: Essays in Interpretation*, Maurice Mandelbaum and Eugene Freeman (eds.) (LaSalle, Illinois: Open Court, 1975), pp. 67-8. See also Stuart Hampshire, "A Kind of Materialism" in *Freedom of Mind and Other Essays* (Princeton: Princeton University Press, 1971), pp. 225-6. Hampshire, however, only claims that each mental thing is identical with a physical thing; he does not discuss the further claim that each physical thing is identical with a mental thing.

3. There is, however, one exception to the general claim that each mode is the effect of another mode. The exception concerns infinite modes. Spinoza says that many infinite modes "follow from" other such modes (1p23). This kind of relation may constitute a causal relation between infinite modes for Spinoza. But even if there are causal relations between some infinite modes and others, it is not the case for Spinoza that *each* infinite mode is the effect of (or follows from) another infinite mode. (And certainly, for Spinoza, no infinite mode is the effect of a finite mode.) In 1p23 Spinoza claims that some infinite modes follow not from other infinite modes, but from "the absolute nature of some attribute of God." Since the absolute nature of some attribute of God is, presumably, not a mode, there is then a special case in which a mode is not caused by another mode.

4. Spinoza repeats this claim with regard to the attribute of thought in 2p9. See also 3p2.

5. R. J. Delahunty, *Spinoza* (London: Routledge and Kegan Paul, 1985), p. 197.

6. See Anscombe, "Causality and Extensionality" in *The Collected Philosophical Papers of G.E.M. Anscombe* , vol. II (Minneapolis: University of Minnesota Press, 1981) and Mackie, *The Cement of the Universe* (Oxford: Clarendon Press, 1974), Chapter 10. See also Sorabji, *Necessity, Cause and Blame* (Ithaca, NY: Cornell University Press, 1980), pp. 13-16. Proponents of the claim that causal contexts are transparent include Davidson, "Actions, Reasons, and Causes," "Causal Relations," "Mental Events," in his *Essays on Actions and Events* (Oxford: Clarendon Press, 1980); Follesdal, "Causation and Explanation: A Problem in Davidson's view on Action and Mind," in LePore and McLaughlin (eds.), *Actions and Events : Perspectives on the Philosophy of Donald Davidson* (Oxford: Blackwell, 1984);

Searle, *Intentionality*, (Cambridge: Cambridge University Press, 1983), pp. 116-117; Strawson, "Causation and Explanation" in Vermazen and Hintikka (eds)., *Essays on Davidson: Actions and Events* (Oxford: Clarendon Press, 1985).

7. Jonathan Bennett, *A Study of Spinoza's Ethics* (Indianapolis: Hackett, 1984), p. 141.

8. A paper delivered at the 1989 Jerusalem Spinoza Conference and forthcoming in the proceedings of that conference to be published by E. J. Brill.

9. Spinoza seems to think that 1ax4 supports the inference from "x is not caused by y" to "x can be conceived without y." But 1ax4 says, "The knowledge of an effect depends on, and involves, the knowledge of its cause." This appears to claim that if x is caused by y then x is conceived through y. (Spinoza takes 1ax4 in this way in 1p3d. Represented in this way, 1ax4 is the converse of (e).) But from the fact that x is not caused by y and from the fact that if x is caused by y then x is conceived through y, it does *not* follow that x is not conceived through y (or that x can be conceived without y). Thus there may be a difficulty in Spinoza's proof of 1p25, but my concern here is not with the legitimacy of the particular way in which Spinoza supports the claim that if x is not caused by y then x can be conceived without y, but with the fact that he does make that claim in 1p25d. I am grateful to a referee of the *History of Philosophy Quarterly* for calling my attention to the relevance of Spinoza's reliance on 1ax4 here.

10. On this kind of point, see Douglas Odegard, "The Body Identical with the Human Mind: A Problem in Spinoza's Philosophy," pp. 67-8.

11. I use the qualification "in part" for two reasons here. First, parallelism, I believe, includes the claim that each idea *represents* its corresponding element in extension. This feature of parallelism does not play a role in the numerical identity interpretation, so we can omit this feature in what follows. However, the fact that Spinoza sees the relation between parallel modes as a representation relation *is* important in understanding the claim of identity though I do not have the space to explain why here. Second, parallelism, as stated in 2p7 (and the first half of 2p7s) does not speak only of a parallelism between ideas and modes of extension, but between ideas and *things*. Modes of extension are given as a particular example of the things which parallel ideas. For our purposes, however, I can treat the claim of parallelism made here as one concerning the causal chain of ideas and the causal chain of extended modes only.

12. My italics. See also 3p2s (beginning).

13. This claim needs to be restricted somewhat. Even if the causal chains are as portrayed in (I) and (IIa), parallelism could still hold since when we consider the causal chains of extended modes and of modes of thought *in their entirety*, there may be a way of mapping the segment of the causal chain of extended modes in (I) onto a segment (different from (IIa)) of the causal chain of modes of thought. Similarly, there may be a way of mapping the segment of the causal chain of modes of thought in (IIa) onto a segment (different from (I)) of the causal chain of extended modes. If this were the case, then parallelism could still hold. But, it is important to note that in this situation, the parallelism would *not* hold as a result of the claim of numerical identity, for in this situation, the modes parallel to one another would not be numerically identical. The numerical identity would then be irrelevant to the parallelism.

14. I would like to thank Janet Broughton and Wallace Matson for their helpful advice.

[3]

HISTORY OF PHILOSOPHY QUARTERLY
Volume 13, Number 3, July 1996

SPINOZA'S DEMONSTRATION OF MONISM: A NEW LINE OF DEFENSE

Mark A. Kulstad

P14: Except God, no substance can be or be conceived.

Dem.: Since God is an absolutely infinite being, of whom no attribute which expresses an essence of substance can be denied (by D6) and he necessarily exists (by P11), if there were any substance except God, it would have to be explained through some attribute of God, and so two substances of the same attribute would exist, which (by P5) is absurd. And so except God, no substance can be or, consequently, be conceived, q.e.d. (IP14D; Curley, p. 420)

Two of the best-known features of Spinoza's *Ethics* are its monism—the view that there is only one substance—and its geometrical method—the demonstration, *more geometrico*, of all the propositions of the *Ethics* from its definitions and axioms. A persistent concern about these two elements of the *Ethics* is whether Spinoza can maintain them both. Specifically, the concern—indeed, typically the criticism—is that Spinoza has not satisfied his own demand relating to these two items, namely, the demand for a geometrical demonstration of monism from the definitions and axioms of the *Ethics*. Many, indeed, have felt that Spinoza's demonstrations in this regard are hopelessly inadequate. But such a view is too harsh. Here an interpretation of Spinoza's demonstration will be presented which embodies a defense—admittedly not a complete defense—of the claim that Spinoza's monism does indeed follow validly from the definitions and axioms of Part I of the *Ethics*.

Of course, it is not enough to provide an interpretation that makes sense of the logic of proposed demonstrations—however much this side of the interpretive enterprise recommends itself in virtue of the principle of charity. So it will also be argued here that the proposed interpretation does a reasonable job of plausibly construing important passages in Spinoza's demonstration, sometimes in interesting new ways. In short, the present interpretation of Spinoza's demonstration is based on grounds of both validity and textual plausibility. In this way a defense of Spinoza's demonstration of monism from the definitions and axioms of the *Ethics* is provided. Along the way, there are suggestions of possible relevance for the interpretation of Spinoza's metaphysics more generally.

To be sure, only a limited defense of Spinoza's demonstration of monism is proposed. What are the limitations? One is implicit in what is said above. The assertion that Spinoza's demonstration of monism is sound—an assertion which would of course entail that Spinoza's monism is true—is not defended here. Rather, what is defended—and even here only partially—is the thesis that Spinoza's monism can be derived validly from the definitions and axioms of Part I of the *Ethics*.

Here a bit of explanation is in order. The passage quoted above, including IP14 and IP14D, is of course critical for our purposes. But, as with most of Spinoza's explicit demonstrations of particular propositions, IP14D does not explicitly take the derivation of Proposition 14, Spinoza's monism, all the way back to the definitions and axioms of the *Ethics*, that is, to the theses on which the demonstration, in theory, ultimately and exclusively rests. To be sure, IP14D explicitly invokes one definition, that of God, ID6. But the other two official premises of IP14D are not definitions or axioms, but rather earlier propositions of Part One, specifically, Propositions 5 and 11. With respect to Proposition 5, we are in luck. We are preceded by the brilliant work of Don Garrett on this subject, who has argued that proposition 5 does indeed follow validly from the definitions and axioms of Part I.[1] While a good deal of Garrett's fine work is accepted, this paper nonetheless departs from him on both the question of the proper reading of IP5 and on the proper response to what Garrett calls the Bennett-Leibniz objection to Spinoza's demonstration of IP5.[2] With respect to proposition 11, however, which is crucial in bringing an existence claim to Spinoza's metaphysics—the claim of the existence of God—, things are more problematic. There is no attempt, in this paper, to defend the demonstration of proposition 11 from the axioms and definitions of Part I, although some important features of its demonstration are considered. This is a key respect in which the present defense of the demonstration of monism must be termed limited.

There are, finally, two issues, one major, one less so, that are not addressed in this paper. The minor one is that the last three words of Spinoza's formulation of monism are not addressed: "Except for God no substance can be *or be conceived*" (IP14; Curley, p. 420; emphasis added). The major one concerns an issue that becomes central in Part II of the *Ethics*, namely, that, according to Spinoza, "the thinking substance and the extended substance are one and the same substance, which is now comprehended under this attribute, now under that" (IIP7S; Curley, p. 451). This notoriously difficult aspect of Part II of Spinoza's metaphysics cannot be dealt with here, although it is to be hoped that the interpretation presented in these pages may provide a new perspective from which to grapple with this important topic.

The paper has the following divisions. In the first section, a preliminary reconstruction of the explicit demonstration of monism, that is, of IP14, is developed. It turns out that there is a significant objection to this reconstruction, worrisome to friends of Spinoza. namely, that analysis of it reveals an implicit contradiction. This leads to reconsideration, in section two, of some crucial aspects of Spinoza's demonstration and the presentation of an interpretation that would not force Spinoza into this contradiction. Unfortunately, this interpretation is in turn subject to the objection that a contradiction can be derived from it also, as is outlined in section three. Finally, section four provides a revised interpretation that handles the problems of both contradictions and another leading interpretive problem as well, the Bennett-Leibniz objection to the demonstration of IP5. This solution, which leads also to a revision of Garrett's defense of the demonstration of IP5, contains as a subpart a second reconstruction of the Spinoza's demonstration of monism, which is claimed to be a valid deduction of monism—bracketing questions about the deduction of IP11—from the axioms and definitions of Part One of Spinoza's *Ethics*.

I. First Reconstruction of the Demonstration of Monism

The first reconstruction of the demonstration of monism is as follows:

(1) God is a substance having all attributes (Df. 6).

(2) God exists (IP11).

(3) There are no two substances having the same attribute (IP5).

(4) Every substance has at least one attribute (suggested in the course of Spinoza's demonstration of IP14).

(5) So, if God exists and some substance other than God exists, then there are two substances having the same attribute (1,4).

(6) So, it is not the case that God exists and some substance other than God exists (3,5).

(7) So, either God does not exist or no substance other than God exists (6).

(8) So, no substance other than God exists (2,7).

A few comments are in order here. First, premise (1) is not a simple statement of Definition 6. It is an interpretation of what that definition says or implies. Although the interpretation is plausible, it is not defended here beyond citing Jonathan Bennett's persuasive work in defending the critical move from the actual definition to the insertion of 'all attributes' above.[3] Secondly, premise (4) is not an official, explicitly-cited premise of Spinoza's argument, as ID6, IP5, and IP11 clearly are. In fact, it is not even explicitly stated in Spinoza's demonstration. More exactly, what Spinoza says there is that "if there were any substance except God, it would have to be explained through some attribute of God" (IP14D; Curley, p. 420). But this would seem to imply premise (4), an essential premise of the argument in the present reconstruction.[4] This still leaves the question whether premise (4) could in principle be derived from the axioms and definitions of the *Ethics*. It is not implausible that Spinoza would have thought that it could be, given the close links between substance, essence and attribute in his thought. (IP4 would obviously be critical here.) But this question will not be taken up in detail in what follows. This too, then, will be a limitation of the present defense.

Despite the initial plausibility of the reconstruction above, there seems to be a problem with the argument for monism reconstructed in this way. To get a different perspective on how the argument works, and hence to see how the problem arises, it may be helpful to consider just one non-monistic, that is, pluralistic, state of affairs, namely, a state of affairs in which God exists and some other substance, say a substance of exactly one attribute, the attribute of extension (call this Ed), also exists. How does Spinoza's demonstration of IP14, as reconstructed above, rule out this state of affairs? First, we apply Proposition 5 and Definition 6, the definition of God, to get the result that only one or the other of God and Ed can exist; they cannot both exist, for then there would be two substances having the same attribute, the attribute of extension, contra Proposition 5. We can put this by saying that if God necessarily exists, then Ed cannot exist; but we can also put this, equivalently but more worrisomely, by saying that if Ed necessarily exists, then God cannot exist. The first of these provides the most obvious entrée for Proposition 11, God necessarily exists. Putting the first together with Proposition 11, a simple application of modus ponens gives us the result that Ed cannot exist. And the same sort of argument would, given the implicit premise that every substance must have an attribute, yield the same result for every other substance not identical to God, so that, granting God's necessary existence, no other substance could exist. But the very power of this procedure—along with the simplicity of the first demonstration of Proposition 11—gives rise to a serious concern about the argument. Might not one use the logic of Proposition 11 to arrive at a very different result?

Let's look at the first demonstration of Proposition 11, that God necessarily exists.

> Dem.: If you deny this, conceive, if you can, that God does not exist. Therefore (by A7) his essence does not involve existence. But this (by P7) is absurd. Therefore God necessarily exists, q.e.d.

The most interesting thing about this demonstration is that its two official premises, A7 and P7, are perfectly general—they are about things and substances, not exclusively about God. This means that Spinoza's version of the ontological argument is quite different from more traditional versions, which clearly do turn on distinctive features of God (e.g., that God is the most perfect being, or a being than which none is greater, or a being having all perfections). In the demonstration of P11 just quoted, the only features of God that are required for the argument to proceed are that God is a thing and that God is a substance—both extremely non-distinctive features, at least before the demonstration of monism. In short, any substance—being both thing and substance—can be shown to exist necessarily by the logic of this argument.[5]

It may be granted that this is very different from the logic of traditional ontological arguments, but how does the difference relate to the demonstration of monism? Well, recall our finding, relative to that demonstration and the sample case of Ed and God, that if Ed necessarily exists, then God cannot exist. Now imagine a kind of anti-Spinoza turned loose on this sample case and given the ammunition of the demonstration of IP11 just considered. Our anti-Spinoza knows that the logic of the demonstration of IP11 yields the result that the substance, Ed, necessarily exists. Given this result, an application of modus ponens to the conditional above yields for our anti-Spinoza the result that God cannot exist.

This result would, obviously, be a problem for a Spinozist. But the real problem would not be that someone opposed Spinozist views. More strictly, the problem looming for Spinozists is rather that the theses of the *Ethics* up through Proposition 14 and its demonstration seem to provide the basis for a contradiction, that is, to imply a contradiction, that is, to be logically inconsistent. If the existence of a substance having exactly one attribute, that of extension, can be deduced by the logic of the demonstration of IP11, and the existence of God—who is such that no attribute can be denied him—can also be so deduced, then it appears that there will be a contradiction with one of the official premises of IP14D, namely, IP5, that there cannot be two or more substances of the same nature or attribute. God and Ed, both of whom (we are supposing) demonstrably exist, have the attribute of extension.

It should not be thought that this problem marks the end of the road for Spinoza's demonstration of monism. But the problem does force us to rethink the steps of reasoning leading to Spinoza's monism, and perhaps to see his thought in a new way. Let us consider some steps of dialectic proceeding from this problem.

II. A Resolution of the First Contradiction Problem, Based on the Guéroult-Loeb Interpretation

An interesting first step is provided by an unusual interpretation of Spinoza's views on the relation of substance and attribute. It comes from such interpreters as Martial Guéroult and Louis Loeb and will here be called the Guéroult-Loeb Interpretation.[6] There are two key theses involved, both of which are at least somewhat controversial. The first is that God is a compound substance, a substance consisting of multiple simple substances, each "having" exactly one attribute.[7] The second is that each simple substance, or a substance "having" exactly one attribute, just is that attribute; attributes and simple substances are identical. (Of course, we can still say that this simple substance "has" the attribute of extension, thus verbally suggesting a difference between the simple substance and the attribute. But in fact there is, on the Guéroult-Loeb Interpretation, no difference: the attribute of extension just is the simple substance in question.) Proceeding from these theses, and examining more closely the different shades of meaning possible in the interpretation of Spinoza's statement of monism and IP5, one can find an escape from the threatening contradiction.

Putting this differently—and more contentiously—the Contradiction Problem may arise on many interpretations of Spinoza's *Ethics*, but, granting a certain not unreasonable hypothesis, it does not go through on the Guéroult-Loeb interpretation. This may be taken as providing one reason for favoring that interpretation over many others.

To see how the Guéroult-Loeb interpretation might lead to a resolution of the contradiction problem, let us return to the test case of God and Ed (our substance having exactly one attribute, that of extension). An assumption of the argument for contradiction above is that God and Ed are "distinct" substances. For many this will be seen as a well-grounded assumption: God is a substance who has multiple attributes; Ed is a substance who, having exactly one attribute, does not have multiple attributes; therefore, God and Ed are "distinct" substances. But on the Guéroult-Loeb Interpretation, things take on a different look, suggesting a reconsideration of some central concepts.

Consider the two following cases, sketched in accordance with the Guéroult-Loeb Interpretation: (1) there is just one substance, the compound substance, God, consisting of infinitely many attributes, each of which is itself a (simple) substance and one of which is the simple substance consisting of (identical with) an attribute of extension; and (2) there are two substances, the compound substance, God, described just as above, and, over and above this, a simple substance consisting of an attribute of extension. Note that in case (2) there are *two* simple substances consisting of an attribute of extension; in case (1) there is only one.

Now consider IP5, that "in the universe there cannot be two or more substances of the same nature or attribute." The not unreasonable hypothesis mentioned above is that Spinoza intends this to apply only to something like case (2)—not to case (1), where just God, with his infinitely many attributes, exists. Let us bring in the case of God and Ed to explore this a bit more. Ed, it will be recalled, is a substance having exactly one attribute, the attribute of extension. On the Guéroult-Loeb Interpretation, of course, Ed is just a simple substance, not at all distinct from the attribute it "has." But as cases (1) and (2) reveal, this simple description of Ed leaves us with an important question unanswered. For all we know about Ed, Ed could be the simple substance with the attribute of extension that is, as in case (1), just one of the simple substances constituting the compound substance God, rather than being the simple substance with the attribute of extension that in case (2) is supposed to exist over and above God and all the constituent simple substances of God. The fact that there is an ontological argument for Ed, that is, for the existence of a substance having exactly one attribute, that of extension, as well as an ontological argument for the existence of God, in no way shows that case (2) obtains rather than case (1).[8] But by the not unreasonable hypothesis about IP5 presented above, it would only be case (2) that contradicts IP5.

Let me put this another way. The contradiction problem is supposed to be that IP11 can be used to prove not only the existence of God, but also the existence of a substance having exactly one attribute, that of extension, and that this leads to a contradiction of IP5. The question is, what state of affairs is supposed to be demonstrated by this double proof? Is it supposed to show that case (2) obtains? Not necessarily, since the double proof is perfectly consistent with case (1), with the existence of a simple substance having the attribute extension being proved twice over, once via the ontological proof of God, once via the ontological proof of a substance having exactly one attribute, that of extension. Nothing about the two ontological proofs shows that the existence of *two* simple substances having the single attribute extension has been demonstrated. Keep in mind that, given the hypothesis mentioned above, only case (2) contradicts IP5.

But, some will reply, is it not the case that God is nonetheless numerically distinct from Ed, even granting the points made above? For Ed is a simple substance, and God is a compound substance. They must be numerically distinct. Hence, the present line of thought, based on the Guéroult-Loeb interpretation, also contradicts Spinoza. Specifically, it contradicts his monism. There is no gain here via the Guéroult-Loeb interpretation. We get inconsistency either way.

Do we? The preceding discussion brings to the fore an important question about monism in relation to difference or distinctness. What precisely Spinoza is claiming in saying that "except for God no substance can be or be conceived" (Curley, p. 420), or, as another translator puts it, "no substance can be or be conceived *external to* God."[9] Is Spinoza saying that there is no substance numerically distinct from God? Or is Spinoza saying that there is no substance distinct from God in another sense, the sense appealed to above in saying that the proof of the existence of Ed proves the existence of nothing over and above—nothing in some second sense distinct from—what is already proved to exist by the ontological proof of the existence of God?

It turns out that very much the same idea involved in the hypothesis above about IP5 will help answer this question too. But at this point we need to refine the idea, spelling out explicitly the notion of distinctness involved there. Here then is a definition of a notion of distinctness different from numerical distinctness. (The label, 'distinct_1,' will sometimes be used to refer to numerical distinctness; the label, '$distinct_2$,' to refer to the second sense of distinctness.)

> x and y are $distinct_2$ =df. (i) x and y are $distinct_1$, i.e., numerically distinct, and (ii) neither x nor y is a constituent of the other.[10]

It will help in explaining this concept of distinctness if we apply it to cases (1) and (2) above. In case (1), we supposed the compound substance God, one of whose constituent substances was a simple substance having the attribute of extension. So as not to beg any questions about the relationship of this simple substance to the substance named earlier, Ed, let us name this simple substance Elvis. Again, this newly-named Elvis is a simple substance, a constituent of the compound substance God, and has the attribute of extension. We may grant that God and Elvis are $distinct_1$, that is, numerically distinct—since one is a compound substance and the other is simple—while nonetheless holding that they are not distinct in another sense, that of $distinctness_2$. For Elvis is a constituent of God (v. condition (ii) of the definition above), one of the simple substances that together constitute the compound substance God. More informally, Elvis is not a substance existing over and above what exists once the existence of God is granted, and so is not distinct from God in this sense. Or, again informally—and adopting the language of one of the translations presented above of Spinoza's statement of monism—we can say that Elvis, while not being numerically identical to God, is not "external" to God either (since Elvis is a constituent of God), so that the existence of Elvis does not contradict the monistic thesis that "no substance can be or be conceived external to God."[11]

Applying this now to the substance named Ed, we get the following results. If Ed is numerically identical to Elvis, then, obviously, Ed is not $distinct_2$ from God any more than Elvis is. And this is true even if we grant that Ed is numerically distinct from God. But if Ed is numerically distinct from Elvis, then if Ed and God (hence also Elvis) exist, we would have two substances, Ed and Elvis, being of the same attribute, the attribute of extension. And this, of course, would violate IP5. But in the case of just Elvis and God, or, if you wish, the case of Elvis (=Ed) and God, we are just in case (1), and, we have asserted, this constitutes no violation of IP5.

At this point some may be desirous of a fuller defense of the claim that the existence of the numerically distinct substances, God and Elvis, both in some sense "having" the attribute of extension, does not violate IP5. We are now in a position to give this defense. Again, the second sense of distinctness is central. IP5 asserts that "In nature there cannot be two or more substances of the same nature or attribute." The present suggestion is that this be taken as asserting, in essence, that there cannot be two $distinct_2$ substances of the same attribute. God and Elvis, while numerically distinct, are not $distinct_2$, since Elvis is a constituent of God, so the case of Elvis and God does not violate IP5 as it has just been interpreted.

And on the Guéroult-Loeb interpretation this reading is not altogether implausible. For, in the first place, on the Guéroult-Loeb interpretation, God's "having" the attribute of extension is really a quite different thing from Elvis's "having" that attribute of extension: the compound substance God "has" the attribute of extension solely in virtue of having Elvis as a constituent simple substance, which substance, in turn, "has" the attribute of extension in virtue of simply being numerically identical with it. One is tempted to say that in this case it is only verbally, but not actually, that the attribute is instantiated twice, and that this has some relevance to what Spinoza was getting at in IP5. And, in the second place, in a sense God and Elvis are not really two substances, any more than we would want to say, without qualification, that my body and my heart are two substances. (In a sense they are; in a sense they are not. Perhaps it was the second sense—the sense of distinctness$_2$—that Spinoza had in mind in talking about two substances in IP5.)

Even if it is granted that the preceding provides a strong response to the contradiction problem of the preceding section, some may be uneasy about the response's reliance on an interpretation, the Guéroult-Loeb Interpretation, that has been acknowledged to be controversial. Where, it may be asked, is the argument for this interpretation? It may surprise the reader that there is no direct argument for this interpretation in the present paper, despite the fact that the interpretation does in fact figure in the final version of the defense of the demonstration of monism proposed in this paper. Rather, the present paper offers an indirect argument, while also referring the reader to Guéroult and Loeb themselves for the direct arguments these two offer for the interpretation.[12] Briefly, the indirect argument will take this form: the Guéroult-Loeb interpretation makes possible a defense of Spinoza's demonstration of monism, a defense that frees it from some serious difficulties. If there is no equally plausible model that also makes possible a defense of that demonstration from these difficulties, then, by the Principle of Charity, we would seem to have a significant, albeit indirect, argument in support of the Guéroult-Loeb interpretation as a viable interpretation of Spinoza. Again, no claim is made that such an indirect argument would by itself be decisive. Not only would direct arguments have to be considered, but also counterarguments, including arguments in support of models inconsistent with the Guéroult-Loeb interpretation. Such a complete survey will not be attempted here. Rather, the argument is restricted to developing a particular line of defense, incorporating the Guéroult-Loeb interpretation, in support of Spinoza's demonstration of monism.

But it should not be thought that the procedure adopted here is uncritical, or that any claim has been made that the exposition of the intended line of defense is already essentially complete. In fact, even though the defense so far (which at base consists of the adoption of the Guéroult-Loeb interpretation, along with an interpretation of P14 and P5 in terms of distinctness$_2$ instead of distinctness$_1$) is a genuine answer to what will be called the First Contradiction Problem, it is not itself free of difficulties. In what follows, an important criticism will be raised and a response to it will be given. To be sure, the response requires a revision of what has been presented so far in this defense of Spinoza's demonstration of monism, but it retains the Guéroult-Loeb interpretation as a central element.

III. THE SECOND CONTRADICTION PROBLEM

We have seen above that the derivation of the original contradiction, that is, the contradiction allegedly brought about by the assertion of IP5 and the logic of IP11D—the latter implying, contra the former, both the existence of God and the existence of a substance of exactly one attribute, extension—, could be blocked by adopting the Guéroult-Loeb interpretation along with an interpretation of IP5 and IP14 in terms of distinctness$_2$. But of course the entailment of any contradiction by any valid route is sufficient to show the inconsistency of a set of propositions. And it might be argued that a second contradiction seems derivable from the newly interpreted IP5 and IP11.[13]

Consider the following two substances: S_1 is a compound substance constituted by two attributes (or simple substances), the first being the attribute of thought, the second, one of those attributes that surpass human comprehension—call it attribute *X*; and S_2 is a compound substance also constituted by two attributes, but this time by the attribute of extension, not thought, along with, again, mysterious attribute *X*. Pretty clearly, (a) the logic of the first demonstration of IP11 should apply to these two substances as well as it did to Ed and Elvis, so that both necessarily exist, and (b) S_1 and S_2 are two numerically distinct substances, since one has an attribute that the other lacks, and (c) S_1 and S_2 are of the same attribute, since they both have or involve attribute *X*.

A quick look at these results suggests that these two points already give rise to a contradiction: the existence of substances S_1 and S_2 contradicts IP5. But our work above with the case of God and Elvis reveals that we must proceed with care here before drawing such a conclusion. For God and Elvis met conditions (a), (b) and (c). And yet the existence of both of those substances could be handled on the Guéroult-Loeb interpretation, without contradiction, assuming that IP5 asserts only that no two distinct$_2$ substances are of the same attribute.

But the problem raised by this new case, of substances S_1 and S_2, is that a similar resolution will not work in this case. S_1 and S_2 *are* distinct$_2$, whereas God and Elvis were not: (i) S_1 and S_2 are numerically distinct and (ii) neither is a constituent of the other; in the case of God and Elvis only the first condition was satisfied. So, after all, it appears that the existence of S_1 and S_2 will lead to a contradiction, even on the assumption of the

Guéroult-Loeb interpretation and the interpretation of IP5 in terms of distinctness$_2$.

To some this may look like an overly technical difficulty. But it should be taken seriously. In what follows, a solution to the difficulty, via a modification of the Guéroult-Loeb approach adopted above, is proposed.[14]

IV. A Resolution of the Second Contradiction Problem and More

A key part of the approach mentioned above was the emphasis on a special sense of distinctness, one that we called distinctness$_2$, different from numerical distinctness. This sense was invoked in giving an interpretation of both IP14 and IP5. Of these two applications, the one to IP5 is probably the more controversial.[15] Most scholars have interpreted IP5 in terms of distinctness$_1$, that is, numerical identity, rather than in terms of distinctness$_2$. Of course, the interpretation of IP5 in terms of distinctness$_2$ was an essential part of our earlier solution to the first contradiction problem, and so is not something to be given up easily in constructing a defense of Spinoza's demonstration of monism from threatening contradictions. Nonetheless, we should indeed explore the option of giving up this interpretation of IP5, and set about constructing a reading in terms of this option, one that satisfies the following conditions, or desiderata, if you will, of a satisfactory interpretation of Spinoza's demonstration of monism: (i) the problem of the first contradiction must be resolved; (ii) the problem of the second contradiction must be resolved; (iii) the Leibniz-Bennett objection to IP5 must be avoided.

The third desideratum warrants further explanation. One of the oldest objections to Spinoza's system is that of Leibniz to IP5 and its demonstration. Leibniz felt that there was nothing in Spinoza's demonstration that ruled out the possibility—made pressing by Spinoza's own embrace of the decidedly uncartesian idea of a substance with multiple attributes—that two substances, each having two attributes, might share a single attribute but nonetheless be distinguished by other of their attributes. His example was of two substances, A and B, the former having attributes c and d; the latter, attributes d and e, with d being the shared attribute, and c and e being the attributes which distinguish substances A and B.[16] This simple example seemingly undercuts one of the key moves of Spinoza's demonstration. Details will not be presented here, but it will be assumed that this is a genuine difficulty for Spinoza's demonstration of IP5—as is assumed also by Garrett in the article referred to above.

But Garrett's attempted resolution of this problem will not be followed. Rather, in the present defense of Spinoza's demonstration of monism, a theory will be proposed that solves both of the contradiction problems of this paper, plus the problem the Leibniz-Bennett objection poses for the demonstration of IP5—with the final point being obviously critical for any defense of the full demonstration of monism. One reason why Garrett's attempted resolution is not followed here is that, while his ingenious defense of the demonstration of IP5 is successful against the problems he addresses, it does not provide a promising base for the broader defense of monism against the difficulties raised in the present paper. Again, an attempt is made to provide a defense that both resolves his difficulties and resolves the ones presented in the present paper.

What, then, is the proposed resolution? The key modifications from the resolution presented in section two have to do with IP5. First, as indicated above, we return to the more traditional idea that IP5 concerns numerical distinctness, rather than $distinctness_2$. Secondly, we adopt an idea that several commentators have mentioned at one point or another,[17] namely, that IP5 is in fact restricted to substances of one attribute. On this interpretation, the demonstration of IP5 seems to leave gaps in the case of substances of more than one attribute for the very good reason that it was never meant to be a demonstration that covers such a case. Putting these two ideas together, we get the following interpretation of IP5: there cannot exist two numerically distinct substances of one attribute that are of the same attribute. Here again, the argument for this interpretation will not be direct, but rather indirect: taken together with the Guéroult-Loeb interpretation and certain other features of the interpretation presented above, it allows for a reasonable defense of Spinoza's demonstration of monism. A challenge for rival interpretations would be to show that they too can provide the basis for a reasonable defense in the face of the kinds of difficulties we have raised.

To begin in mounting this indirect argument, let us examine whether these two changes in our interpretation of IP5 undercut our ability to deal successfully with the first contradiction problem, or whether they too allow for a resolution of that problem. Recall the contradiction involving God and Ed, with Ed being a substance having exactly one attribute, the attribute of extension. The existence of both God and Ed was proved via the logic of the first demonstration of IP11. But the existence of both of these substances appeared to contradict IP5.

Now one might be tempted by the view that the existence of these two substances poses even less of a problem on the new interpretation of IP5—restricted as it is, on this interpretation, to substances of one attribute—than it did on the previous interpretation of IP5 (in terms of $distinctness_2$). For God is obviously a substance of *multiple* attributes.

But such a view would be too hasty. For the newly interpreted IP5 still does apply to the case of God and Ed, albeit indirectly. To see this, one needs to follow out a bit the consequences of the proof of the existence of God. We know from the definition of God that God is a substance having infinite, that is (on Spinoza's unusual use of 'infinite'), *all* attributes, so we know—assuming the Guéroult-Loeb interpretation—that the proof of the existence of God brings with it a proof of the existence of the simple substances of one attribute which constitute God, including—since extension is an attribute—a simple substance having an attribute of extension, or a simple substance which just is identical to this attribute of extension. As before, we call this simple substance of the attribute of extension Elvis.

Once this is seen, it becomes clear that we can—indeed, must—apply the newly-interpreted Proposition 5 to the case of God and Ed. The question is, of course, whether this case violates the principle that there can be no two numerically distinct simple substances of the same attribute. And this in turn reduces to the question of whether Ed is numerically distinct from Elvis. Now it is no part of our purpose here to prove that Ed and Elvis are or are not numerically distinct. Rather, we argue only that nothing in the case so far presented forces the conclusion that they are numerically distinct. To put this differently, nothing in the "first contradiction problem" *forces* us to acknowledge a contradiction in Spinoza's metaphysics once we have adopted the interpretations that make up the present approach. The present approach allows for a reading that yields consistency, namely, that the proof of Ed just provides a second proof of the numerically identical simple substance, Elvis, implicitly proved in the proof of God. In sum, the modified interpretation does not undercut our ability to respond to the first contradiction problem.

What about the second contraction problem? As it turns out, there are some obvious analogues between the analysis of this case and the analysis of the case above. First, the newly-interpreted IP5 does not apply directly to the case of the two numerically distinct substances S_1 and S_2 (approximately in the way it did not apply directly to the case of God and Ed), since S_1 and S_2 are substances of two attributes each (with one being shared and one not); they are not substances of a single attribute. Secondly, the newly-interpreted IP5 does nonetheless apply indirectly to the case of S_1 and S_2, since—according to the Guéroult-Loeb interpretation—each is constituted by two simple substances, i.e., substances of exactly one attribute, and the newly-interpreted IP5 applies specifically to such substances. Thirdly, in considering whether any contradiction arises when applying the newly-interpreted IP5 in this indirect way to substances S_1 and S_2, we get a result similar to what we found above: no contradiction is forced. To be sure, there would be a contradiction *if* the simple substance of mystery attribute X that is a constituent of S_1 were numerically *distinct* from the simple substance of mystery attribute X that is a constituent of S_2. But nothing about the proposed ontological argument for substances S_1 and S_2 or about the newly-interpreted IP5 forces such a conclusion. Again, our present approach allows for a consistent reading, namely, that the proof of S_2 simply provides a second proof of the numerically identical simple substance of the mystery attribute X, already demonstrated in the proof of S_1.

What about desideratum (iii), that a proposed interpretation should provide an answer to the Leibniz-Bennett objection? In a way things are easy here; in a way hard. The easy response with respect to the Leibniz-Bennett objection is that on the newly-interpreted IP5 the Leibniz-Bennett objection simply disappears: the objection assumes that IP5 is intended to apply to substances of more than one attribute; but on the present interpretation it does not. Again, it applies (directly, one should add[18]) only to substances of just one attribute.

What is the hard part of the response to the Leibniz-Bennett objection? As it turns out, both Bennett and Garrett[19] address, more or less directly, the feature of the present interpretation of IP5 here under discussion, namely, that it applies only to substances of a single attribute. And both claim something that would be utterly devastating from the point of view of the present project: they claim that such an interpretation totally undercuts the demonstration of monism in IP14. It will not do to respond to the Leibniz-Bennett objection in such a way that there can be no defense of the demonstration of monism. Obviously, we must respond to this kind of objection.

To begin with, we note the two pieces of good news for our interpretation of IP5 that Bennett includes in his discussion: (i) he grants that the Leibniz-Bennett objection would indeed be blocked by this interpretation, and (ii) he acknowledges that there is some textual basis for adopting such an interpretation, given that Spinoza does not mention the possibility of a substance of multiple attributes until after the presentation of IP5, specifically, in IP9.[20] But Bennett proceeds immediately from these points to what he clearly views as the decisive objection against an interpretation of IP5 that would restrict its application to substances of one attribute:

> [A]re we to suppose that Spinoza intended his argument [for IP5] in that way, and simply forgot to review it in the light of his later thesis that substances can have more than one attribute? That is not credible, given how the pivotal p14d depends on combining 'There is a substance which has every attribute' with (p5) 'Two substances cannot share any attribute'. Surely in *this* context Spinoza could not just forget that p5d assumed that no substance has more than one attribute![21]

The point is clear: on Bennett's understanding of Spinoza's demonstration of monism, an interpretation of IP5 that restricted its application to substances of one attribute would completely undercut that demonstration.

How can such a powerful objection be handled? A first step is to say that while the objection may well be decisive within the framework of Bennett's own interpretation of Spinoza's metaphysics, it does not follow that it is decisive within the framework of other interpretations of Spinoza's metaphysics. So the real question for the purposes of the present project is whether the objection is decisive within the interpretative framework we are presently in. And with respect to that question, the answer is that the objection is not decisive.

To see this, let us revise our initial reconstruction of the demonstration of IP14 in accordance with our current interpretations. Naturally, a key element in this reconstruction will be the interpretation of IP5 as applying solely to substances of one attribute and being simply about numerical distinctness. But the Guéroult-Loeb interpretation and the interpretation of IP14 in terms of $distinctness_2$ will also be essentially involved. We shall see if such an interpretation leaves Spinoza's argument hopelessly unable to reach the conclusion of monism. The steps are:

(1) God is a substance having all attributes, that is, a compound substance constituted by simple substances of every attribute (Df. 6).
(2) God exists (IP11).
(3) There are no two numerically distinct substances of one attribute having the same attribute (IP5).
(4) Every substance has at least one attribute (introduced in the course of Spinoza's demonstration of IP14).
(5) So, if God exists and some substance $distinct_2$ from God exists, then there are two numerically distinct substances of one attribute having the same attribute (1,4).
(6) So, it is not the case that God exists and some substance $distinct_2$ from God exists (3,5).
(7) So, either God does not exist or no substance $distinct_2$ from God exists (6).
(8) So, no substance $distinct_2$ from God exists (2,7).

The steps of this second reconstruction have been intentionally kept in strict parallel with the steps of the first reconstruction, to make it easier to see precisely what has changed and what has stayed the same. One disadvantage of this approach is that it leaves obscure an additional, very significant, change in the final reconstruction. That change comes in the inference to line (5). This is the further point that needs to be explained.

Fortunately, the basic idea has already emerged in what has gone before: in some cases IP5 is applied indirectly rather than directly. Suppose, as Spinoza does in his demonstration, that there is a substance other than God. Of course, on the present interpretation of IP14, this means supposing that a substance exists that is $distinct_2$ from God. Keeping things simple, at least to start, we first suppose that this $distinct_2$ substance is a substance of a single attribute, say extension. For old times sake, call this substance Ed. Obviously, since God is constituted by simple substances of every attribute, and since, by P11, God exists, we know that there is a simple substance—that is, a substance of a single attribute—that is a constituent of God and is of the attribute of extension. Again, for old times sake, call this substance Elvis. Now, by our supposition, Ed cannot be numerically identical with Elvis. For the supposition was that there exists a substance $distinct_2$ from God; and if Ed were numerically identical with Elvis, Ed would of course be what Elvis is, namely, a constituent of the compound substance God, which would contradict Ed's being $distinct_2$ from God. But if Ed is not numerically identical with Elvis, then IP5 is contradicted. So the supposition must be rejected.

Now the same line of argument would obviously apply also for a simple substance of any attribute other than extension. And, though things would be more complicated, it is clear that it would also apply for any compound substance that was not a constituent of God.[22] So, on the hypothesis (more than plausible from the point of view of the Guéroult-Loeb interpretation) that all substances are simple or compound, the argument applies to all substances $distinct_2$ from God. Thus, the claim that interpreting IP5 as applying solely to substances of one attribute undermines the demonstration of monism is shown to be incorrect, at least within the framework of the interpretation we are here presenting.

But we are now in a position to say that all three of our desiderata have been attained in the context of giving a defense of Spinoza's demonstration of monism: (i) we are not forced to the conclusion that Spinoza's principles entail the first contradiction; (ii) we are not forced to the conclusion that Spinoza's principles entail the second contradiction; and (iii) we can avoid the Leibniz-Bennett objection to the demonstration of IP5. And we have done this, it bears repeating, on the assumption of the two central theses of the Guéroult-Loeb interpretation, on the assumption that Spinoza's monism is the thesis that there can be no substance that is $distinct_2$ from God, and on the assumption that Spinoza's IP5 is to be interpreted as stating that in nature there can be no two numerically distinct substances that are of the same nature or attribute. This result is evidence that the constellation of interpretative theses just stated deserves serious consideration as expressing Spinoza's metaphysical intent in presenting his demonstration of monism.

Rice University

Received January 3, 1996

NOTES

1. Don Garrett, "*Ethics* IP5: Shared Attributes and the Basis of Spinoza's Monism," in *Central Themes in Early Modern Philosophy: Essays Presented to Jonathan Bennett*, J. A. Cover and Mark Kulstad (eds.), (Indianapolis: Hackett, 1990), pp. 69-108.

2. What the Bennett-Leibniz objection is will be explained below.

3. What Definition 6 actually says is that God is "a substance consisting of an *infinity* of attributes . . . ," (*The Collected Works of Spinoza*, vol. 1, ed. and trans. by Edwin Curley, [Princeton: Princeton University Press, 1985] p. 409; emphasis added; hereinafter cited as Curley) not of all attributes. However, Bennett gives a nice argument that Spinoza's text warrants the move from the former to the latter. See his *A Study of Spinoza's* Ethics, (Indianapolis: Hackett, 1984), pp. 75-76.

4. Jonathan Bennett also seems to view this premise as necessary for the argument. See his *A Study of Spinoza's* Ethics, p. 70.

5. Don Garrett makes a similar point in "*Ethics* IP5," *op. cit.*, p. 85, as does Louis Loeb in *From Descartes to Hume: Continental Metaphysics and the Development of Modern Philosophy*, (Ithaca: Cornell, 1981), p. 165, n. 4.

6. For more on this interpretation see Martial Guéroult, *Spinoza I, Dieu (Ethique, I)* (Paris: Aubier-Montaigne, 1968), pp. 47ff, and Louis Loeb, *From Descartes to Hume*, pp. 161-66.

7. There are perhaps two main reasons for the controversial nature of this thesis. First, many view it as flatly contradicting Spinoza's monism, since it allows multiple substances. We shall address this question below. Secondly, it seems also to contradict Spinoza's insistence that substances cannot be divided, or have parts (v. IP12, IP12D, and IP13). Loeb responds to this second point in an interesting way. See Louis Loeb, *From Descartes to Hume*, pp. 170-71.

8. One might be tempted, à la Caterus, by other ontological arguments, e.g., by an argument starting from the following definition: SuperEd =df. a substance that (i) consists of exactly one attribute, extension, and (ii) is not a constituent of any compound substance of multiple attributes. Arguably, it is both the case that the logic of IP11D yields the existence of SuperEd as much as it yields the existence of God, and that the pair of ontological arguments, for God and SuperEd, does indeed force a situation of the type of case (2). In short, it might be argued that while the details of the section on the Contradiction Problem do not themselves prove that a contradiction can be derived from the resources of Spinoza's Part One, a refinement of those details does in fact lead to the derivation of a contradiction. To be sure, as Leibniz has taught us, one must not attempt an ontological argument from a definition that involves a contradiction. But it would appear that the challenge here would be to show that the definition of SuperEd does indeed involve a contradiction, while the definitions of God and Ed—or perhaps just of God—do not. Another route, this time following Descartes, in the Replies to Objections I, would be to argue that the concept of SuperEd does not involve a true and immutable nature, but only a fictitious one, due to a mental synthesis (*The Essential Descartes*, Margaret Wilson (ed.) [New York: Mentor, 1969], p. 229), while the definitions of God and Ed—or perhaps just of one of these two—do not. Interestingly, Spinoza seems to adopt something of this sort at one point. Bennett discusses it, but not sympathetically (v. *A Study of Spinoza's* Ethics, p. 75).

9. IP14D, in a translation by Samuel Shirley, *Baruch Spinoza: Ethics, Treatise on the Emendation of the Intellect, and Selected Letters*, edited with an introduction by Seymour Feldman (Indianapolis: Hackett, 1992), p. 39, emphasis added.

10. In turn one might request a definition of 'constituent'. While no formal definition will be offered here, some assistance may be given. A paradigm case of constituency for present purposes is the relation of simple substances to the compound substance that they constitute: each of these simple substances is a constituent of the compound substance. For compound substances, the key assertion is that (x)(y)(if x and y are compound substances, then x is a constituent of y just in case (z)(if z is a simple substance and z is a constituent of x, then z is a constituent of y)). (Perhaps some would qualify this a bit, saying that a constituent compound substance must lack a simple substance that the other compound substance has to rule out the verbally awkward result of a compound substance being a constituent of itself.) Finally, it is stipulated that one thing can be a constituent of another without being a part of the other. Thus, a simple substance could be a constituent of the compound substance God without being a part of it. This provides a terminological equivalent to Loeb's advocacy of what may be called the Guéroult-Loeb Interpretation of substances and attributes combined with Loeb's denial that the Guéroult-Loeb Interpretation conflicts with Spinoza's assertion that a substance cannot have parts into which it can be divided. (See IP12 and IP12D and also Loeb, *From Descartes to Hume*, pp. 165-66.)

11. Feldman, *Ethics*, *op. cit.*, p. 39.

12. See footnote 6 above.

13. The difficulty here is due to Darien Lynx, who has also provided useful comments on other points in this paper. I am indebted to him for his insights.

14. There are some interesting questions here that will be mentioned but, for reasons of space, not pursued. One is whether there could be such substances as S_1 and S_2 on Spinozistic premises. Another is whether the difficulty takes on a different color depending on whether S_1 and S_2 share a numerically identical simple substance (a simple substance of mystery attribute *X*)—as opposed to having in common that one of their constituent simple substances is "of the same attribute" as a constituent of the other, while not being numerically identical with a constituent of the other.

15. At least it can be said that others have supported a similar reading of IP14. See, for instance, Loeb, *From Descartes to Hume*, pp. 170-71.

16. See C. J. Gerhardt (ed.), *G. W. Leibniz: Die philosophischen Schriften*, 7 vol. (Berlin: Weidmann, 1875-90; reprint, Hildesheim: Olms, 1978), volume I, p. 142; translation from Leroy Loemker, ed., *G. W. Leibniz: Philosophical Papers and Letters*, 2nd ed. (Dordrecht: Reidel, 1969), p. 199. (For Jonathan Bennett's statement of the objection, see his *A Study of Spinoza's* Ethics, p. 69.) Notice the similarity between this and the case that gave rise to the second contradiction problem. One of the reasons for adding desideratum (iii) to the list above is that this similarity virtually forces recognition that there is a danger here of robbing Peter to save Paul, that is, as applied to this case, of solving the second contradiction problem while leaving unsolved the very similar problem of the Leibniz-Bennett objection. If the solution we offer satisfies all three conditions, we will have avoided this danger.

17. See, for instance, Jonathan Bennett, *A Study of Spinoza's Ethics*, p. 69. It should be noted that he mentions the idea only to reject it. Since the idea is retained in the present defense of Spinoza's demonstration of monism, it will be defended against Bennett's line of attack later in this paper.

18. One might think that there is still the problem of an indirect application. But here the Leibniz-Bennett objection differs from what we saw in the analysis of the second contradiction problem. To be sure, indirect application is possible in the case of the two two-attribute substances mentioned by Leibniz—and the results would be perfectly analogous to what we saw in the case of substances S_1 and S_2. But—and here is the difference—this is of no relevance to what is being argued in the Leibniz-Bennett objection. That objection is supposed to show that the demonstration of IP5 is invalid for substances of two or more attributes. But, on the present interpretation, Spinoza is *not* presenting a demonstration for such a conclusion: he is arguing only that there are no two numerically distinct substances *of one attribute* that are of the same nature or attribute.

19. See Bennett, *A Study of Spinoza's* Ethics, p. 69, and Don Garrett, "*Ethics* IP5," p. 94.

20. Bennett, *A Study of Spinoza's* Ethics, p. 69. On the same page Bennett adds that up until IP9, Spinoza "is thinking in terms of the concept of a one-attribute substance, a concept which explicitly appears in [I]p8d."

21. Bennett, *A Study of Spinoza's* Ethics, p. 69. Garrett makes a similar point in his "*Ethics* IP5," *op. cit.*, p. 94.

22. For reasons of space, this subargument is not given in full detail. The full version would make use of the explanation of a constituent given in an earlier note.

[4]

PHILOSOPHY RESEARCH ARCHIVES
VOL. XI, MARCH, 1986

SPINOZA'S CAUSAL LIKENESS PRINCIPLE[1]

HAROLD ZELLNER
KENT STATE UNIVERSITY

ABSTRACT. Axiom 4 of the *Ethics* of Spinoza runs:

> The knowledge (cognitio) of an effect depends upon and involves the knowledge of the cause.

Since this is in the ancestry of some of Spinoza's most important and characteristic claims, a clarification of its meaning would be highly desirable (in the literature it is left unhelpfully vague.) I argue that A4 is a causal likeness principle, according to which causal relationships always feature a property which in some sense is "passed" from the cause to the effect. This interpretation provides a key to understanding some darker passages.

Spinoza at least thinks that his doctrines follow from the axioms of the *Ethics*, so it would be desirable to be as straight about the latter as possible. I hope to do something here toward clarifying IA4,

> The Knowledge of the effect depends on, and involves, the knowledge of its cause

which is supposed to be in the ancestry of some of Spinoza's most characteristic claims. In the literature of which I am aware, A4 is taken to be a vague claim that if you know something about effects then you've got to know something about their causes: this sort of interpretation makes it mysterious how Spinoza might even have thought he was getting from it what he says he does.[2] It turns out that A4 is highly dubious, but it and the Spinozistic derivations employing it are reasonably intelligible.

One way of finding out what someone means is to look at what he thinks follows from it. Now by my count Spinoza appeals to IA4 in eight places in the *Ethics*. IP3Dem., IIP5Dem., IIP16Dem., and IIP45Dem. seem to fall in one group, which I will take up first. IP6Cor., VP22Dem., and IIP7Dem. constitute a different group and will be discussed separately. The basis for this division is that in the latter it is the "depends" half of A4 which is cited, whereas it is the "involves" part which is operative in the former.

IP6 is the familiar Spinozistic doctrine that one substance cannot be produced by another, because, we are told, two substances would have to have different attributes and thus would not have "something in common." Not having something in common, one cannot be the cause of the other, Spinoza concludes, appealing to P3. I take it what is at issue is non-relational properties; surely two substances could have the property of being-thought-of-by-me. If so, what we are being told is that if one thing causes another, they must share some nonrelational (and presumably, non-trivial) properties. The demonstration of P6, then, requires that P3 be a kind of causal likeness principle, albeit a weak one: causality implies at least one shared property. Now P3 is officially derived from A4 and A5, which is the only use of the latter in the *Ethics*:

> Things which have nothing in common with one another also cannot be understood through one another, or the concept of the one does not involve the concept of the other.

This looks like an explanation of the "involves" of A4, rather than a substantive principle in its own right, as the brevity of its appearance and its failure to appear in other derivations suggests. (I shall have more to say about this below.) If so, the concept of the cause "involving" the concept of the effect is a matter of cause and effect sharing properties in such a way that one can be "understood" through the other.

But in what way? Just what is A4 telling us from which a causal likeness principle is supposed to follow? And why all this talk about understanding one thing through another? My suggestion is that Spinoza holds what might be called a Transmission Theory of Causality. According to a Transmission Theory, for a to make b become F, a must already be F, in some sense. That is, causality is a "passing along" of some property from the cause to the effect. Some instances of causality do look like this, at least superficially. To move an ordinary billiard ball in ordinary circumstances one must transfer motion from something that is already in motion, such as a billiard cue or another billiard ball. To heat water in everyday circumstances we must transfer heat to it from something which is already hot. There are counterexamples, of course, such as moving metal balls with magnets or generating heat by mixing plaster of paris. But the counterexamples are unobtrusive enough to be overlooked, especially if one has powerful theoretical considerations pushing one towards such a theory. Recent authors have attributed Transmission Theories to both Descartes and Aquinas, though I shall not speculate here about influences.[3] A Transmission Theory is not quite the same as a causal likeness principle, though authors have often used the latter name for both. If a Transmission Theory is true, then causes and effects must at least resemble one another enough to be capable of having the same kinds of properties; in Spinozistic language, they must fall under the same attribute. But a causal likeness principle would not necessarily mean that causality was transmission.

Spinoza's most unambiguous statement of a Transmission Theory occurs in his book on Descartes's *Principles of Philosophy*, where he is explaining Descartes's version of it.

> Whatever reality or perfection there is in a thing exists either formally or eminently in its first and adequate cause. . . . This axiom depends on the preceding one, for on the supposition that there is nothing in the cause or less in the cause than in the effect, nothing in the cause would be the cause of the effect. But this is absurd (by preceding axiom); consequently, the cause of an effect cannot be anything one pleases, but precisely that which contains eminently, or at least formally, the entire perfection which exists in the effect.[4]

The "preceding axiom" says that things and their perfections cannot have "nothing" or a "nonexistent thing" as cause of their existence. It is tempting to think this is just a case of something looking self-evident at one time which looks obviously false or unintelligible at another, but I believe that would be a mistake. What we have in Spinoza's case is a man with a deep commitment to the idea that absolutely everything can be explained--that there are no "brute facts" about the universe. If you think this, and you also think that causal explanations must be in terms of one particular causing a change in another, you will be under pressure to say that a feature of an effect must be explained by some corresponding feature of the cause. Otherwise there will be no explanation of there being the feature: that is, it will spring "from nothing." Worse, you have the problem of why that feature rather than some other, so to make this intelligible you will be under pressure to picture all cases of causality as a "passing along" of a property from cause to effect. I take it the idea that the caused property might be "eminently" rather than "formally" contained in the cause is a bow to the fact that so many examples of causality don't fit the pattern. Rather than saying that a feature in the effect must also be in the cause it is supposed to be enough that the feature has some representative or stand-in.

It is notorious that in this early work Spinoza is not always giving his own views, but there is something similar in the correspondence. In a 1661 letter Oldenburg comments on an early version of the first part of the *Ethics*:

> . . . the fourth axiom, namely, things which have nothing in common cannot be one the cause of the other is not so obvious to my dull intelligence as to need no further light for its illumination. For God has nothing essentially in common with created things, yet He is held by almost all of us to be their cause.[5]

Spinoza's answer:

> Fourthly . . . it follows that things which have nothing in common between them cannot be one the cause of the other. For when the effect has nothing in common with its cause, then whatsoever it might have, it would have from nothing. As to your contention that God has nothing essentially in common with created things, etc., I stated the exact opposite in my definition.[6]

It looks as though a Transmission Theory--or at the very least a causal likeness principle--is being discussed in connection with the *Ethics* or a prototype thereof, and the principle is being identified as the "fourth axiom" and apparently confirmed as such by Spinoza himself. Of course, this does not mean that the *Ethics* as we have it contains such a theory, much less that our IA4 is the same as the one Oldenburg cites; in fact, the quote makes it clear they could not have been exactly the same. But it certainly should alert one to possibilities. My guess is that A4 is saying that the caused feature in the effect is "involved in"-that is, had-by the cause. If so, then A4-and A5-are much less epistemological than they sound. But perhaps that is all right, since Spinoza frequently seems more epistemically concerned than he really is, as when he speaks interchangeably of "ideas" and "knowledge". (It is noteworthy he speaks of "ideas or Knowledge" in the statement of A4 in a 1675 letter to Schuller.[7]) I have not forgotten that Spinoza does not just say that knowledge of the effect involves the knowledge of the cause; he also says that one depends on the other. The addition is important, and I shall try to show later that A4 is a conjunction of two claims.

Why does Spinoza appeal to A5 in the derivation of P3? If I'm right, A4 tells us that cause and effect will in some sense share the caused property. A5 is saying that things with no common properties are incapable of acquiring shared properties. Conclusion: no trans-attribute causality. That there is no trans-attribute causality is also the gist of IIP6Dem., IIP5Dem. and IIP45Dem., which I invite the reader to compare with IP3Dem.; nothing different in principle seems to occur. Yet all these cite IA4, and none mentions IA5. I think this confirms the suspicion that A5 is intended largely as an explanation of A4; it is the latter which is supposed to power the machinery.

Admittedly something weaker than a Transmission Theory would do for the inference of P3 and the cousins mentioned above. All that is required is that there be some conceptual overlap between cause and effect, not that they share the caused property. (And it might be that this is all Oldenburg and Spinoza are talking about.) But understanding A4 in terms of transmission has the advantage of explaining why Spinoza thinks there must be overlap. It also explains some passages which are otherwise difficult to fathom.

IIP16Dem. runs:

> . . . all the modes in which a body is affected follow from the nature of the affected body, and at the same time from the nature of the affecting body (by A1). So the idea of them (by IA4) will necessarily involve the nature of each body. And so the idea of each mode in which the human body is affected by an external body involves the nature of the human body and of the external body.

In the second corollary to this proposition Spinoza concludes that

> . . . the ideas which we have of external bodies indicate the condition of our own body more than the nature of the external bodies.

Before trying to interpret this, it is helpful to see what Spinoza does with it. Citings of P16 or the Corollaries appear in the demonstrations of Propositions 25-28, and these in turn are among the ancestors of P29. This whole sequence seems to reach a climax in P29Cor.:

> . . . so long as the human mind perceives things from the common order of nature, it does not have an adequate, but only a confused and mutilated knowledge of itself, of its body, and of external bodies.

One way of explaining all this would be to say that Spinoza is singing a very old song indeed. Perception is liable to error since it is causally dependent on bodily states which may be abnormal, and the strong language is meant to deprecate this kind of knowledge in comparison with the error-free a *priori* kind. Perhaps there's also the idea that this kind of knowledge is essentially incomplete since we can never know the whole causal story of what's determining our present perceptual states. On this view A4 has little to do with P16, whether interpreted along my lines or not.

This story does not seem to me to account adequately for Spinoza's strong language. After all, the mere fact that perception is causally dependent on the body tells us nothing about its reliability or content. The kinds of considerations cited above might give one a ground for downgrading perceptual knowledge so far as certainty is concerned, but Spinoza's complaints seem to be more far-reaching than that. Just because television reception is dependent on picture tubes we cannot conclude TV images are unreliable, or "involve" the nature of the picture tube as much as that of the object televised, whatever that would mean. Now suppose that Spinoza really is thinking of causality as transmission. Then when two objects cause a change in a third the last will be characterized by some sort of "blend" of the features being transmitted from the causes. Spinoza could be taken to mean that in perception "ideas" will combine features which occur separately in nature, blending properties of external objects with features of my own body. (Of course on Spinoza's Parallelism the ideas will not be caused by my body and the external object, but the ideata of those ideas will, and the ideas will mirror them.) With one special exception to be noted below, perception would then normally be like the case where I confuse my own high fever with the temperature of a room. Or it would be like a camera which could not take pictures without showing things as possessing some of the features of the camera. It may be this is what Spinoza means by "confused and mutilated" ideas. They blend (confuse) characteristics cut from different sources ("mutilation".) This really would mean that normally there would be something wrong with perception, not just that we couldn't tell when there was.

Now, there are two sections in the *Ethics* which make sense under this interpretation, and so far as I know, no one has a rival. IIP38 is the claim that that which is "common" to all physical objects and their parts can only be perceived adequately. The demonstration runs in part:

> Let A be something which is common. . . . I say that A can only be conceived adequately. For its idea . . . will necessarily be adequate in God, insofar as he has the ideas of its affection, which . . . involve in part both the nature of the human body and that of external bodies. . . . The Mind therefore . . . necessarily perceives A adequately, and does so both insofar as it perceives itself and insofar as it perceives its own or any external body.

The most striking things about this passage is the conflation of perception and conception. But it was typical of pre-Kantian modern philosophy to treat the sensory and the intellectual as being on the same continuum, rather than as differing in kind. It is likely that in this passage Spinoza is caught in the same muddle and supposing that without further ado you can say the same sorts of things about the one as you do about the other. (Of course, Spinoza officially distinguishes the two in IIDef.3. But as in this case he thereafter runs them together again.)

Above I made a suggestion about what "confused and mutilated" means, and with that would go an account of what inadequate ideas are. Adequate ideas are a difficult subject in their own right, but perhaps we already have enough to see our way through the proposition and its demonstration. I think what he is saying is something like this:

> A can only be conceived adequately, for its idea won't be a blend of properties which don't go together in nature. (Since we have supposed everything has A.) Whether we consider our body or external bodies neither can be misrepresented by the other as having A, since everything's got A. Thus a certain kind of illusion is ruled out. And that's why we can only perceive A adequately.

Properties which all bodies have in common could hardly be miscombined in the way envisioned above; if everything has them, ideas cannot misrepresent objects as having them, no matter what their ancestry.

IIP39Dem. and Corollary seem to require a similar treatment. P39Cor. also has P16 ancestry and is the dubious thesis that:

> . . . the Mind is the more capable of perceiving many things adequately as its body has many things in common with other bodies.

Notice that the ideas of things "common" will be "caused from without" just as much as inadequate ideas will, and in the cases now relevant these need not even be common to all bodies.

> Let it be posited now that the human body is affected by an external body through what it has in common with it, i.e., A; the idea of this affection will involve property A (by P16) and so (by P7C) the idea of this affection, insofar as it involves property A, will be adequate in God insofar as he

> is affected with the idea of the human body, i.e., (by P13) insofar as he constitutes the nature of the human mind.

As in the demonstration of P38 we have the claim that the commonality of properties possessed by the external object and the human body guarantees the adequacy of the corresponding ideas. This invites the question: what would be the difference if A were not common? Just what does the commonality of A rule out? Certainly not the causal dependence of perception on the body. On the other hand, commonality would mean a kind of distortion would be absent, which on my story Spinoza would think present in ordinary perception.

I must admit that I am hard put for examples, but then any interpretation of these passages is going to face that problem. Perhaps the following from I appendix is relevant:

> . . . if the motion the nerves receive from objects presented through the eyes is conducive to health, the objects by which it is caused are called beautiful; those which cause a contrary motion are called ugly. Those which move the sense through the nose they call pleasant-smelling or stinking; through the tongue, sweet or bitter, tasty or tasteless; through touch, hard or soft, rough or smooth, etc.; and finally those which move the ears are said to produce noise, sound, or harmony.

This list is certainly heterogenous. It was standard seventeenth century fare to treat taste, odor, and sound as secondary properties, and a little before this passage Spinoza mentions warmth and cold. It is surprising color is not mentioned; it's probably an oversight. The references to all these are slighting. Though he doesn't explicitly say so, Spinoza seems to think an objective, basic scientific account of the world would mention none of the items on the list above; it's shortly after this we are told the notions of ordinary people indicate only the "constitution of the imagination" and not "the nature of anything." Now my account of Spinoza's intentions can hardly run: human bodies do stink but nothing else does, and one of the confusions of perception is that due to A4 we sense other things as having a property they don't really have. That is really hopeless. Perhaps the story is this: the quality of stinking corresponds to a "motion" in our bodies, rather than in the external object. But since ideata of my idea of the external object is caused by "motions" in my body as well as in the external world, my idea of the object presents it as having a feature it does not really have.

If I'm on the right track, Spinoza's position is an interesting contrast to that of say, Locke. Locke also claims that some of the most distinctive features of perceptual experience are not instantiated in physical objects, and like Spinoza thinks there are counterparts of the secondary properties. But for Locke these are arrangements of particles on the surfaces of the physical objects themselves, which is perhaps why there is such a different flavor to what is otherwise a very similar position. By contrast, for Spinoza the counterparts of the same features are characteristics of our bodies. If so, it is small wonder Spinoza gives sense perception much lower marks than does Locke, at least in the case of those things which are not "common."

There is at least one other passage in the *Ethics* which seems to point in the right direction even though it is not directly related to adequate or inadequate ideas. IIP27 is the "imitation" doctrine which is important in the development of Spinoza's egoism and political theory; the demonstration runs in part:

> . . . if the nature of the external body is like the nature of our Body, then the idea of the external body we imagine will involve an affection of our Body like the affection of the external body. Consequently, if we imagine someone like us to be affected with some affect, this imagination will express an affection of our body like this affect. And so, from the fact that we imagine a thing like us to be affected with an affect, we are affected with a like affect.

I take it to be clear this is an account of perception rather than of imagination in the ordinary sense. So understood the sense seems to be that when F is a property which I am capable of acquiring, perceiving someone who is F will make me F. Presumably the qualification is supposed to handle counterexamples such as the male who sees a pregnant woman without becoming pregnant. Now, the basis for this reasoning is supposed to be IIP16. It should be clear that on anybody's interpretation of Spinoza this move should be invalid. Putting things as neutrally as possible, P16 says that perception will be influenced by the "nature" of the human body as well as by the "nature" of the object perceived; the above is saying that if something is part of the nature of an external object we are perceiving that feature will tend to pop up in us. Leaving aside any other problems about this, what he's got in P16 is

> If it's influenced us, then it's from both

whereas what he uses is something like

> If it's from either, then it influences us.

That is, it looks as though he's muddled on which way a conditional runs. So any rescue can only be partial. But the derivation is intelligible if IA4 and P16 are to be understood in terms of the Transmission Theory. If you muddle which way *that* runs, you get that if it's got it then it passes it along. And that's exactly what he needs to get what he thinks he can.

I shall now turn to the "depends" part of A4 and the second group of descendants. Let us first look at IIP7:

> The order and connection of ideas is the same as the order and connection of things. Dem.: This is clear from IA4. For the idea of each thing caused depends on the knowledge of the cause of which it is the effect.

Down to this point I have felt no need to bring in spinozistic "ideas", simply because nothing about them seemed to be operating in the passages I was trying to explain. One could "read through" ideas to things without going far wrong. Here that is not possible. It is fairly clear that Spinoza is assuming the existence of his thought-realm counterparts of things, and A4 is being read as:

> the idea of each thing caused depends on the idea of the cause of which the thing is the effect.

Presumably the "depends" is causal, so what A4 is saying is that "ideas" of effects, whatever they are, are caused by ideas of causes, just as the effects are caused by the causes. If you add that to determinism and the thesis that for everything there is a thought realm counterpart you get IIP7.

Now I think there is a connection between the Transmission Theory and the idea-dependence claim. If the former is true, then we can always infer the existence of F-ish type causes from the existence of F-ish type effects. And I take it that's another way of saying the idea of the cause can be inferred from the idea of the effect. If you then conflate inferrability and causal dependence, isn't the result the "depends" reading of A4? I am suggesting, then, that A4 is a Transmission Principle-caused properties are transmitted from causes which have the same properties-together with the supposed consequence about the causal/logical relations between the thought-realm counterparts of the cause and effect. A4 would thus be telling us that causal connections are logical connections also. (This is a standard claim about Spinoza, of course, but I hope to have elucidated just what it means, and especially how Spinoza might have thought it was true.)

On anybody's interpretation the "causal, so logical" move is going to be important in the "depends" group. An especially well-known and interesting use is in IP6Cor.:

> . . . if a substance could be produced by something else, the knowledge of it would have to depend on the knowledge of its cause (by A4). And so (by D3) it would not be substance.

I shall not consider here just why Spinoza thinks the concept of substance must be independent of anything else. But he does, and this is taken to mean substance must be causally independent also. I see the Transmission Theory as the bridge between causal and logical connection here and in the other members of the "depends" group. What I would appreciate from critics is any other interpretation which would explain as much about the *Ethics* as the Transmission Theory does.

ENDNOTES

[1] This paper was written under a National Endowment for the Humanities Summer Seminar for College Teachers grant. Quotations from the *Ethics* are from E.M Curley (tr. & ed.), The Collected Works of Spinoza, vol I, Princeton University Press, Princeton, 1988.

[2] See, e.g., William Charlton, "Spinoza's Monism", *Philosophical Review* XC No. 4, October 1981, 515. I mention this example because of its high quality.

[3] See Kenneth C. Clatterbaugh, "Descartes's Causal Likeness Principle", *Philosophical Review* LXXXIX, No. 3, July, 1980, 379-402; see also Linwood Urban, "Understanding St. Thomas's Fourth Way", *History of Philosophy Quarterly*, Volume 1, No. 3, 284-85.

[4] *Earlier Philosophical Writings*, Frank A. Hayes, trans., Bobbs-Merrill Co., Indianapolis, 1963, 28-29.

[5] *The Correspondence of Spinoza*, A. Wolf, trans., Russell and Russell, Inc., 1966, 79.

[6] Wolf, 83.

[7] Wolf, 341.

[5]

Emilia Giancotti

On the Problem of Infinite Modes

The theory of infinite modes is not only one of the most controversial in Spinozist philosophy but also a kind of intersection of concepts upon whose clarification or interpretation the ultimate meaning of this philosophy depends. Interpreted in a certain way, it can present an insurmountable barrier to the secular reading of Spinozist thought which seems to me to be closest to its letter and spirit.

The point of view from which I shall examine the argument is that of an immanent criticism. This approach has been dictated not by the conviction that it is useless to seek its sources — many of which have, in fact, been identified — but rather by the realization that such reconstruction — in the historical form which it has taken[1] — is not sufficient to resolve the internal problems: it neither explains the function of the infinite modes within the system nor determines the possible significance which Spinoza attributed to them. Although I do not claim to have been successful in explaining the first or in having determined the second, this has been the direction of my research.

The function and significance of the infinite modes are basic to the system of Spinozist thought since, whatever their historic antecedents, they only come to light in relation to other elements of the theory to which they belong. Therefore I shall not evaluate the plausibility of the Platonic, Neo-Platonic (which appears most probable), cabalistic, scholastic, Renaissance or Cartesian origins of the theory of modes. Nor will I speculate as to other hypothetical or probable origins of the theory, as this has been sufficiently worked over. What I propose to attempt here is an internal reconstruction of the theory — through a precise analysis of the relevant texts — in order to establish the extent of the difficulties which many have already pointed out, and to try to resolve them or to confirm their irreducible persistence. A great problem in Spinozism seems to be the lack of the "deduction" of the finite from the infinite. From this perspective, the definition of the nature and function of the infinite modes takes on a particular significance. Although

Y. Yovel, ed., God and Nature: Spinoza's Metaphysics, 97-118.

the salient terms are familiar, it is worthwhile reproducing them here synthetically to provide a stable point of reference for my discussion.

The first point concerns terminology: the commonly used expressions "immediate infinite modes" and "mediate infinite modes" are not, in this specific form, original Spinozist terms; they seem to have been constructed from the canonical texts of this theory (in particular from KV I:9 and EIp23d). But they are indicative of Spinozist concepts.

The texts on which my analysis is based can be divided into two groups: those in which the theory is explicitly stated, and those which, in one way or another, contribute to the determination of its conceptual field. The first group includes: KV I:2 (Dialogue 1); KV I:8; KV I:9; KV II:22/4 note; EIp21–23; EIp28s; EIp29s; EIp30; EIp31; EIp32c2; EIIp4,d; EIIp11c; EIIp43s; EIVp68s; EVp40s; Letter 9; Letter 54; Letter 64; TTP 7 (G102/21–24). The second group includes: KV I:2; KV I:3; KV IIpref, §7 of note (G52/4–6); KV II:4; KV II:5/5 (G62/27ff.), 9, 10; KV II:19/6–8, 15 note; KV II:20; KV II:26; KV app1ax3; KV app1p4,d; KV app2/11,13,14; TIE 98–103; CM II:7 (G263–264); EIp16d,c1; EIp28; EIapp (G80/16–17, 83/26–32); EIIp7s; Letter 6; Letter 32; Letter 73; Letter 81; Letter 83.[2] These two groups will be examined either in succession or in relation to each other. Although not every passage indicated is subjected to direct analysis, their content is presupposed.

The first formulation of the theory occurs in the *Short Treatise on God, Man, and His Well-Being.* It is preceded by the discussion between Erasmus and Theophilus on divine causality (KV I:2, Dialogue 2; Curley, pp. 76ff.). The argument is confronted from three different points of view: the compatibility of immanent and remote causality; the possibility that God is the immanent cause of things without there being any change in his essence; and, finally, the possibility that God is the immanent cause of things which perish although the cause continues to exist. Dealing with this last question one distinguishes between "effects he has produced immediately, through his attributes alone without any further circumstances" (Curley, p. 78) and "effects whose existence does not depend immediately on him but which have come to be from some other thing" (ibid.). Of the first it is said that God is "properly a cause." And of the second, it is said that God is "an internal cause ... (except insofar as their causes neither do nor can act without God or outside him)." Since the problem which arises here is that of the relationship between the immanent causality of God and effects which can be destroyed, the distinction between effects which God has immediately produced and effects whose existence does not immediately depend on him but have their origin in other things anticipates the twofold distinction between immediate infinite modes and finite modes, and between proximate and remote causality (cf. KV I:3/8; EIp28s). Of the mediate infinite modes there is no mention.

The canonical text of the theory, however, is that made up of two passages from KV I:8–9. In KV I:8, after having divided Nature into *Natura naturans* and *Natura naturata*, the latter is further divided into universal and particular *Natura naturata*. Of the universal it is said that it "consists of all those modes which depend on God immediately" (Curley, p. 91). And of the particular it is said that it "consists in all those singular things which are produced by the universal modes" (ibid.). It should be noted that, even here, as in Dialogue 2 quoted above, there is no reference to the mediate infinite modes: universal *Natura naturata* is made up only of the "modes which depend on God immediately." In the second passage (KV I:9), after having reconfirmed the identification of the universal *Natura naturata* with "those modes or creatures which immediately depend on, or have been created by God," their content is specified by saying: "we know only two of these: Motion in matter and Intellect in the thinking thing, ... these have been from all eternity, and will remain to all eternity, immutable, a work truly as great as the greatness of the workman" (Curley, p. 91). Apart from the rhetorical emphasis of this last expression, the place attributed by Spinoza to the "modes or creatures which immediately depend on, or have been created by God" in the comprehensive architecture of reality is of primary importance, because it is primarily in them that infinite power expresses itself.[3]

Regarding motion, Spinoza specifies:

> With regard particularly to Motion, it belongs more properly to a treatise on Natural sciences than here, [to show] that it has been from all eternity, and will remain to all eternity, immutable, that it is infinite in its kind, that it can neither exist nor be understood through itself, but only through Extension. So we shall not treat any of these things here, but shall say only that it is a Son, product or effect, created immediately by God (Curley, pp. 91–92).[4]

What is said in this passage is completed by a note to the passage quoted above: "What is said here of Motion in matter is not said seriously. For the Author still intends to discover its cause,[5] as he has already done, to some extent, a posteriori. But it can stand as it is here, because nothing is built on it, or depends on it" (Curley, p. 91). Even though he indicates the ontological status of movement, Spinoza delegates the task of explaining the cause of this status to the science of nature and, exercising caution, specifies that what is said here (that is, throughout the *Short Treatise*) about motion is not said with certainty since it is only a result of an a posteriori knowledge.

Of the Intellect[6] Spinoza says: "this, too, is a Son, product or immediate creature of God, also created by him from all eternity and remaining immutable to all eternity. Its sole property is to understand everything clearly and distinctly at all times. From this arises immutably a satisfaction infinite, or most perfect. ..." (Curley, p. 92). This "son of God" — here defined as "infinite intellect" — is referred to again in a note to KV II:22, dedicated to a treatment "Of true Knowledge, Rebirth, etc.," where, in dealing with the immediate character of the knowledge of God, the union with God and its effects, with reference to the concept of the mind as an idea of the body, he says: "This explains what we said in the first part, viz. that the infinite intellect must exist in Nature from all eternity and why we called it the son of God. For since God has existed from eternity, so also must his Idea in the thinking thing, i.e., exist in itself from eternity; this Idea agrees objectively with him."[7]

Taking these passages together we observe that:

- modes or creatures exist which depend on or are created immediately by God (immediate infinite modes);
- these modes are eternal because they are created by or dependent on God from all eternity. (Infinity is not mentioned here, but it is, evidently, implicit, and in any case is anticipated by the passage in KV I:3 in which it is affirmed: "God is the proximate cause of those things that are infinite and immutable, and which we say that he has created immediately; but he is, in a sense, the remote cause of all particular things." [Curley, p. 81]);
- God is the proximate cause of the infinite modes, whereas for those things which do not immediately depend on him he is the remote cause;
- the infinite modes are movement in Extension, and the understanding or infinite intellect/idea of God in Thought (KV II:22/4 note).

What has been said about movement must be integrated with what follows from the passages in which rest, in addition to motion, is described as a mode of Extension, and the complementary nature of both is emphasized. The passages to which I refer are:

> KV I:2/26 (preceding the passages quoted thus far), where it is affirmed that "it is clear that there is motion and rest in Nature" (Curley, p. 72).

The Preface to KV II (no. 7 in note; Curley, p. 95), which ascribes the constitution of bodies to the action of motion and rest: "Each and every particular thing that comes to exist becomes such through motion and rest. The same is true of all modes in the substantial extension we call body."

KV II: 19/6, which considers motion and rest, on the one hand, as "effects which we see depend necessarily on extension," and KV II: 19/8, which, on the other hand, considers them as being themselves the cause of all effects which derive from extension: "When we consider extension alone, we perceive nothing else in it except motion and rest, from which we find that all its effects derive" (p. 131).

KV app2/14–16, where motion and rest are designated the sole modifications of Extension ("there is no other mode in extension than motion and rest" [p. 155]); from their proportions the constitution of every single corporeal thing depends, so that "the human body ... is nothing but a certain proportion of motion and rest," (ibid.), whose corresponding objective essence in the thinking attribute is the mind of the body in which variations occur in correspondence with the variations produced in it by the changing proportion of motion and rest.

KV II: 19/15 note and KV II: 20/2 note 2, which point to motion and rest as the cause of sensations.

In concluding our examination of these passages from the *Short Treatise,* we can confirm that at this stage Spinoza places only one kind of infinite mode between God-attributes and particular things, the immediate ones: infinite intellect or the idea of God in thought, Motion and Rest in extension. This is usually neglected. Curley notes it in passing (p. 72).[8]

The mediate infinite modes first appear in the *Ethics.* Here, more precisely than in the *Short Treatise,* the formal determinations of the derivation (the modes *sequuntur*) from God of both the immediate infinite modes and the mediate infinite modes (EIp21–23) are put forward.[9] However, we do not find any explicit specification of their content, i.e., there is no identification of these modes with the infinite intellect, movement and rest. As in the *Short Treatise,* the infinite modes are referred to the *Natura naturata* (EIp31) and equated (EIp32c2) in their relationship with the nature of God. In comparison with the *Short Treatise,* the property of infinity is made explicit:

"All the things which follow from the absolute nature of any of God's attributes have always had to exist and be infinite, *or* are, through the same attribute, eternal and infinite." (EIp21); "Whatever follows from some attribute of God insofar as it is modified by a modification which, through the same attribute, exists necessarily and is infinite, must also exist necessarily and be infinite" (EIp22).

A suggestion as to what these modes are — which is missing in the *Ethics* — is present, as is generally known, in Letter 64. Here, in response to Schuller (who is also speaking on behalf of Tschirnhaus), Spinoza states: "Lastly, the examples for which you ask are, of the first kind, in Thought, absolutely infinite understanding, but in Extension, motion and rest; of the second kind, the face of the whole Universe, which, although it varies in infinite modes, yet remains always the same; on this subject see the Scholium to lemma 7 before Proposition XIV, Part II."[10] It should be noted that this is the only case in which that something which "follows from some attribute of God insofar as it is modified by a modification which, through the same attribute, exists necessarily and is infinite" (EIp22), is given a precise content. As already observed, in the *Short Treatise,* only that which depends on God or is immediately produced by him is mentioned, i.e., immediate infinite modes; whereas in the *Ethics,* the scholium to which Spinoza's answer refers speaks of "the whole of nature" as "an Individual, whose parts, i.e., all bodies, vary in infinite ways, without any change of the whole Individual," without identifying the mediate infinite mode.

The question Spinoza is answering in Letter 64 had been posed by Schuller as follows: "Fourthly, I should like examples of those things which are immediately produced by God, and of those which are produced by some infinite mediate modification" (Letter 63). The presence in this formulation of concepts like "immediate production" and "mediation" through an "infinite modification" shows that the texts of the *Ethics* which Schuller has in mind here are EIp22–23 and EIp28s. The interpretation of the latter text (the subject of which is similar to that of KV I:2 Dialogue 2 and of KV I:3), as is well-known, is controversial[11] even though Gueroult has judged it to be "très simple et très clair" (Gueroult, p. 342). It deserves to be quoted in its entirety:

> Since certain things had to be produced by God immediately, viz. those which follow necessarily from his absolute nature, and others (which nevertheless can neither be nor be conceived without God) had to be produced by the mediation of these first things, it follows: I. That God is absolutely the proximate cause of things produced immediately by him, and not [a proximate cause] in his own kind, as they say. For God's effects can neither be nor be conceived without their cause (by P15 and P24C). II. That God cannot

> properly be called the remote cause of singular things, except perhaps so that we may distinguish them from those things that he has produced immediately, or rather, that follow from his absolute nature. For by a remote cause we understand one which is not conjoined in any way with its effect. But all things that are, are in God, and so depend on God that they can neither be nor be conceived without him (EIp28s).

The doubt in its interpretation concerns two points: (1) are the *alia mediantibus his primis* (others [that had to be produced] by the mediation of these first things) "mediate infinite modes" or finite modes? and (2) the kind of causality through which God produces the single categories of the modes: Is God absolutely the proximate cause of all infinite modes (both immediate and mediate) and the proximate cause in his kind of the finite modes (Gueroult, p. 343); or is he absolutely the proximate cause of infinite immediate modes, proximate cause in his kind of the mediate infinite modes (Wolfson, p. 390, note 2) and, in this case, remote cause of the finite modes? Or even proximate cause of all infinite modes and remote cause of finite modes (Robinson, Curley)? The text seems to offer justification for each of these interpretations.

It is, in fact, true that that *alia mediantibus his primis* recalls the *mediante alia modificatione* of the conclusion in EIp23d, which, formulated as it is, could give rise to the misunderstanding which Gueroult fell into, that all infinite modes should follow from the absolute nature of an attribute of God (Gueroult, p. 342). The final passage of the demonstration in fact states that "Therefore, the mode, which exists necessarily and is infinite, has had to follow from the absolute nature of some attribute of God — either immediately (see P21), *or* by some mediating modification, which follows from its absolute nature, i.e. (by P22), which exists necessarily and is infinite, q.e.d." (my emphasis). The reference to "the absolute nature of some attribute of God," which precedes the distinction between that which follows immediately and that which follows with modifications, could, in fact, lead one to think that all infinite modes follow from the absolute nature of an attribute of God. But this conclusion cannot be isolated from the proposition which it demonstrates, EIp23, which distinguishes, without a shadow of doubt, between two types of modes: "Every mode which exists necessarily and is infinite has necessarily had to follow either from the absolute nature of some attribute of God, *or* from some attribute, modified by a modification which exists necessarily and is infinite" (my emphasis). The position of the disjunctive in the demonstration is different—and this creates the misunderstanding which Gueroult fell into — but this only seems to change the meaning. The meaning remains that of the proposition, the text of which logically follows from the two preceding propositions: in the first (EIp21), Spinoza deals with that which follows from the absolute nature of an attribute of God, whereas in the second (EIp22), he deals with that which follows from an attribute of God inasmuch as it is modified by a modification which, in virtue of the attribute, necessarily exists and is infinite. There is no mention here of the absolute nature of the attribute of God. Nor could there be.

Moving on to the problem of the different kinds of divine causality, the distinction between proximate causality in ***absolute proxima*** and ***proxima in suo genere*** could be referred, with a little effort, to two different effects: the first to all infinite modes, the second to finite modes; or the first to immediate infinite modes, the second to mediate infinite modes. But it can — and in fact I think it should — also be understood in the sense of exclusion, i.e., as a specification, with respect to the scholastic tradition,[12] that the causality under discussion here is absolutely proximate and not in its kind.

I now put forward my own point of view. The passage as a whole has the clear nature of a summary of what has been said beginning with EIp21. It seems that Spinoza anticipated the difficulties his readers would have when attempting to follow him from EIp21 (the *derivation* of immediate infinite modes from the absolute nature of an attribute of God), to EIp28 (determination of every singular thing from another singular thing), through EIp26–27 (necessary *determination* of all things on the part of God), and that he hoped to make things easier by providing a description, apparently free of any problematic points, of the productivity of God: productivity immediate of some things (*quaedam*) and mediate of others (*et alia*), both of which, however, are in God, because without God they can neither be nor be conceived, all being modes of God even though only some are infinite. (Incidentally — with respect to the question of the deduction of the finite from the infinite — it is not insignificant that here, as in other places, Spinoza uses the verbs *produci* and *sequi,* which indicate different kinds of relations between God and modes, indifferently. These relations are, however, assimilated to and placed in apposition to the one indicated by the *in esse*).[13] The effects of immediate productivity, as in the *Short Treatise,* are listed as being of only one kind (*quaedam*), with the difference that in the *Short Treatise* the second kind of infinite modes, the mediated ones, had not yet been introduced.[14] Here, however, Spinoza — having introduced mediate infinite modes at EIp22 — should have been more precise. How can we explain his lack of precision? If we reread propositions 21 and 22, we confirm — as I have stressed above — that only immediate infinite modes are derived from the absolute nature of an attribute of God, whereas the mediate ones derive from an attribute of God inasmuch as they are derived from the immediate ones ("quatenus modificatum est tali modificatione, quae et necessario, et infinita per idem existit"). Otherwise, what could the difference between the two kinds of modes consist of? The "ex absoluta [Dei] natura" of EIp28s and the "ex absoluta natura alicujus Dei attributi" of EIp23d cannot, then, refer to the two kinds of infinite modes,[15] but must instead refer to the immediate infinite modes.

If this is true, the fact that the mediate infinite modes are part of the *alia* produced *mediantibus his primis* is only a hypothesis, whereas on my view it is certain that the finite modes are contained in them, and that none of them — though not immediately produced by God — can either be or be conceived without God. But from this premise, which seems clear, Spinoza draws a consequence (*hinc sequitur*) which again creates obscurity. He distinguishes two kinds of causality, the absolutely proximate and the remote, the first referring to things which are immediately produced by God (thus, to immediate infinite modes), and the second to singular things. He specifies that one speaks of remote causality only relatively, i.e., to distinguish the causality of singular things from that of things which God has produced immediately, because if it were intended absolutely (i.e., as a cause which is not conjoined with its effect) it could not refer to singular things, which — like everything that exists — are in God and depend on God in such a way that they can neither be nor be conceived without him. The mediate infinite modes are not included in this classification. Nor does it seem to me possible for proximate causality in its kind to refer to them: the specification *non vero in suo genera*, as we have seen, signifies exclusion and is made as further clarification of the concept *causa absolute proxima.* Moreover, the identification of things which are immediately produced by God and things which follow from his absolute nature is reconfirmed, excluding the possibility that absolutely proximate causality could produce as effects the mediate infinite modes as well. Therefore, the conclusion which the texts, considered in themselves without conjecture, allows us to draw is that in EIp28s the so-called mediate infinite modes are not mentioned at all, just as they are not mentioned in the *Short Treatise.*

This observation will be taken up again after I examine the problem of identifying the mediate infinite modes which derive from the attribute of Thought. But, for the time being, let us put it aside and move on to an examination of what these infinite modes are.

As we have seen, Letter 64 is the text which contains most of the evidence on this point, because to the infinite intellect and Motion and Rest it adds the *facies totius universi* as a unique example of an infinite mode of the second kind (i.e., mediate). To clarify: as we have seen (p. 102, above), Spinoza refers to lemma 7, which concludes the short treatise on physics (in which, however, only the physical universe is discussed). This has given rise to a problem with which interpreters are still struggling: that of finding within the attribute of Thought the mediate infinite mode corresponding to the mediate infinite mode within the attribute of the Extension. I too have dedicated a great deal of time to analyzing the various solutions proposed, and have come to the conclusion that this is a useless effort. The question is whether it is possible that Spinoza left it out. It seems improbable. But if he was aware of this lacuna there must be a reason. If one excludes, as I think one should, a breach in the so-called parallelism, the reason why Spinoza gives only one example for the mediate infinite modes is because he felt that this was sufficient. In fact, since "The order and connection of ideas is the same as the order and connection of things" (EIIp7) and "God's ... power of thinking is equal to his actual power of acting" (EIIp7c), one can deduce that Spinoza intended the *facies totius universi* to include the mediate infinite modifications of Extension and Thought.[16]

The fact that Letter 64 refers back to lemma 7, which deals only with physical nature, is not proof to the contrary, as might appear at first glance. The reference to lemma 7 serves to explain, by means of an example drawn from the physical world, the composite and unitary structure of these modes, whose infinity is the result, the product, the effect of the unfolding (regulated by the necessary laws) of infinite divine power in the infinite series of his finite expressions. But we know that Spinoza's universe is not simply the corporeal universe but includes all the forms of the infinite power of Nature.[17] Therefore, the expression *facies totius universi* can legitimately be considered to include the immediate infinite mode of the attribute of Thought as well. And since *facies* has the meaning of "exterior form," "aspect," "image," and thus designates that which one sees and which is the object of experience, these mediate infinite modes are identified with the infinite and external totality of finite bodies and the ideas/minds which correspond to them — infinite and eternal whole because its cause is infinite and eternal.

If the mediate infinite modes are made up of the totality of finite bodies (as the explanation of the expression *facies totius universi* with its reference to lemma 7 attests), and of the totality of the ideas/minds of these bodies (as the theory of the identity of the order of ideas and things authorizes us to hold), perhaps it is not wrong to draw from it the following conclusion for the interpretation of EIp28s. As we have seen, this text poses the twofold problem of providing a content for the *alia mediantibus his primis* and identifying the effects of God's proximate causality, assuming the existence of two kinds of infinite modes, immediate and mediate. The analysis I have suggested herein has led to the observation that the mediate infinite modes do not appear in that text, just as they do not appear in the *Short Treatise.* But, if mediate infinite modes are nothing but the infinite whole of finite modes, it is also possible to hold that: (1) the *alia mediantibus his primis* of EIp28s include the mediate infinite modes and the finite modes, because the former are identified with the latter as a whole, and (2) God is *causa absolute proxima* of things which are immediately produced by him, that is, of immediate infinite modes and — assuming a distinction between these effects of immediate divine productivity and other effects — remote cause of singular things. If it is true that mediate infinite modes are made up of finite modes as a whole, it is not necessary to isolate a particular aspect of divine causality to produce them.

Two related objections to my hypothesis could be raised. The first is that these mediate infinite modes are spoken of as being something which necessarily exists and is infinite, whereas the finite modes are necessarily determined to exist but do not necessarily exist and are, in fact, finite. That necessary existence could be the result of a sum of existences necessarily determined (and hence compulsory, but not necessary), and that the infinite could be the result of the sum of an infinite series of finite modes, are theoretical hypotheses which seem paradoxical. The second objection is that, from the way the formal determination of the derivation of mediate infinite modes from an attribute of God is formulated, it seems possible to deduce that they are individuals and not a group of individuals. For the sake of simplicity, I shall reply to this hypothetical objection by referring the reader back to the *facies totius universi* and to its explanation in terms of the image of corporeal Nature as a compound individual made up of an infinity of bodies.

The first objection leads us to the problem of the deduction of the finite from the infinite.

First of all, it could — and has been — said that, when related to the modes, eternity or necessary existence and infinity cannot have the same meaning as when related to substance and its attributes. The attributes, as the substance which they constitute, are by their very nature infinite and eternal. In the case of modes, however, both eternity or necessary existence (Spinoza ascribes the former to immediate infinite modes and the latter to mediate infinite modes), as well as infinity, are only the effects of the eternity and infinity of the attributes of substance. This is true in both kinds of infinite modes, which — though infinite — belong to *Natura naturata* and are effects of the causality of *Natura naturans*. They are eternal or exist necessarily and are infinite by virtue of the attributes of which they are modes. But, if the necessary existence of infinite modes is not the existence, implicit in the essence, of substance and of its attributes, but rather the necessary existence which the infinite modes enjoy because of their being effects of the necessary productivity of substance, from the point of view of quality, there is no difference between the necessary existence of the infinite modes and the necessary existence of the finite modes. In both cases, it is an *ab alio* and not an *a se* necessity. The difference is between the existence of substance, which is implicit in its essence, and the existence of the modes, which is necessary in virtue of their cause: for infinite modes, in virtue of the attributes of God, and for finite modes, in virtue of, or rather through the action of, a finite mode of the infinite series. But, even in this case, as EIp26–27 tell us, the cause which determines the existence of the finite modes is also God himself. Thus, even finite modes exist necessarily (*contingens non datur*), are necessary effects of divine causality, just as infinite modes are.

It is impossible to clarify this without mentioning the concept of essence, even if the importance of this subject demands wider treatment. As is well known, this concept is defined by EIIdef2, which establishes a reciprocal relationship between the essence of a thing and the fact that the thing, of which it is the essence, exists. This definition — the motivations of which are made clear by EIIp10s2 — seems to conflict with EIIax1, which denies the implication of man's existence by his essence, and EIIp8, which establishes a correspondence between the way in which "the ideas of singular things, *or* of modes, that do not exist [are] comprehended in God's infinite idea" and the way in which "the formal essences of the singular things, *or* modes, are contained in God's attributes." There is an anticipation of this correspondence in EIp8s2, where the possibility of having "true ideas of modifications which do not exist" is stated, "for though they do not actually exist outside the intellect, nevertheless these essences are comprehended in another in such a way that they can be conceived through it." These texts clearly refer to the essences of non-existent modifications, and this fact, at least apparently, contradicts EIIdef2. The solution lies in the distinction — which is discussed in EIIp8c, EIIp45s and EVp29s — between two forms of existence or actuality of singular things. One form of existence is that whereby they "do not exist, except insofar as they are comprehended in God's attributes, [and] their objective being, *or* ideas, do not exist except insofar as God's infinite idea exists" (EIIp8c), in which case, "we conceive them to be contained in God and to follow from the necessity of the divine nature," and therefore "we conceive them under a species of eternity" (EVp29s), "insofar as they are in God," and "the force by which each one perseveres in existing follows from the eternal necessity of God's nature" (EIIp45s).

Another form of existence or actuality of singular things is that whereby "they are said to have duration, [and] their ideas also involve the existence through which they are said to have duration" (EIIp8c), in which case, "we conceive them to exist in relation to a certain time and place" (EVp29s), because "each one is determined by another singular thing to exist in a certain way" (EIIp45s).

In light of this distinction — maintained consistently from the first through the fifth parts of the *Ethics* — we obtain two results. The first is that the contradiction which seemed to arise between EIIdef2 and EIIax1 or EIIp8 disappears, because the reciprocity between the essence and the existence of the singular things in EIIdef2 remains both in the case that things exist inasmuch as they have a duration in time and are located in one place, and in the case that things exist inasmuch as they are in God before any temporal or spatial determination. The second result is that the difference between the necessary existence of substance and the existence of modes (infinite or finite) — also necessary but as an effect of the necessity of substance — reproduces the distinction between *Natura naturans* and *Natura naturata,* between substance as cause and substance as effect, and therefore reveals itself as a distinction within the sole necessity of substance, which, as intrinsic necessity (implication of essence and existence) produces itself and pluralizes itself in an infinity of infinite and finite modes, which are the modes of its very being.[18]

The problem, then, shifts to the plane of the relationship between finite and infinite. Thus it is not a question of explaining how an infinite mode which exists necessarily can be made up of an infinite number of modes whose existence is limited, and hence finite, but also necessary. Rather, it is a matter of explaining how an infinite mode can be identified with the infinite series of the finite modes. The question can be considered from the point of view of the finite or from that of the infinite. In the former case, it would be necessary to explain how, by summing finite and finite, one could arrive at the infinite. Given that the Spinozist infinite is positive and has a qualitative value, one can only conclude that the passage from the finite to the infinite is impossible to achieve. And not only that; it would not even help to explain the deduction of the finite from the infinite. Thus, the infinite mode cannot be thought of as a "result."

If we begin from the infinite, we must explain the passage from the infinite to the finite, i.e., we come up against the central problem of the theory of modes. It has been said that the deduction of the finite from the infinite and of the many from the one is impossible in Spinoza's system.[19] The theory of infinite modes, elaborated in order to join the infinity of the substance with the finiteness of the modes, i.e., singular things, would seem to have failed. It appears that Spinoza himself justified these conclusions by asserting (in KV I:9 note) that he still had to find the cause of motion, by referring (in EII lemma 7 to the opportunity for further explanation and demonstration of the structure of the corporeal universe, and by declaring (Letter 83) — a year before his death — the impossibility of providing an a priori demonstration of the variety of things starting only from the concept of extension.

I shall begin with an examination of these passages to see if Spinoza was right in seeking a cause of motion and in expressing the need for an a priori demonstration of bodily multiplicity. My negative response to these questions may come as a surprise. But I believe there is evidence that, while he was right in considering a wider explanation of physics necessary, he was wrong in believing[20] that he had to look for a cause of motion or an a priori demonstration of bodily multiplicity. In fact, Spinoza already had the cause of motion in the attribute of Extension-matter, which, unlike Descartes, he conceived not as a mass in a state of rest (Letter 81) — from which, therefore, it was impossible to demonstrate the existence of bodies — but rather as an attribute, i.e., as the substance's own power to act. In the attribute of Extension-matter, Spinoza already had the principle from which to deduce the variety of bodies; Letter 83 seems to me to be very clear on this. After having repeated what he said in Letter 81 — i.e., that it is not possible to demonstrate the variety of things simply from the (Cartesian) concept of extension — he concludes (Letter 83) that "[matter] must necessarily be defined by an attribute which expresses eternal and infinite essence." His reference to the concept of Extension-matter as an attribute has precise significance. The attribute expresses the essence of the substance. But essence equals power: "God's power is nothing except God's active essence [*actuosa essentia*]. And so it is as impossible for us to conceive that God does not act as it is to conceive that he does not exist" (EIIp3s). God acts simply because he is. Substance is a dynamic principle which, starting from the constituent forms of its being (the attributes), transmits and continues itself in an infinity of forms which together make up the universe. Multiplicity is at the heart of substance, because it includes infinite qualitative determinations. The problem of the deduction of the many from the one does not exist. The process of "finitization" (excuse the neologism) takes place already within the essence of the substance, because each of the attributes of which substance *consists, expresses* its essence in a certain form, it is *infinitum in suo genere,* not *absolute.* The passage from the attributes to the modes does not entail the deterioration or loss of power of being, but rather the articulation and infinite pluralization of the modes of the being, as a spontaneous explication of its power. Within this process — which is an expression of the essence/power of the substance — all the modes occur: immediate infinite, mediate infinite, and finite.

Spinoza's insistence on the concept "being in God" (the concept with which EIIp45s indicates the existence of things insofar as they follow from the eternal necessity of the nature of God) cannot be without significance. Although it is said that some things are produced immediately by God and others through these, i.e., that God is the absolutely proximate cause of some things and the remote cause of others, it must be stressed that all are in God because God is the immanent cause of all. More precisely, "God must be called the cause of all things in the same sense in which he is called the cause of himself" (EIp25s). Hence "particular things are nothing but affections of God's attributes, *or* modes by which God's attributes are expressed in a certain and determinate way" (EIp25c). It is only an error of the imagination to conceive the modes of substance, i.e., the many, as separate from substance itself (Letter 12). But the intellect comprehends them in their concrete unity. The concept of the indivisibility of substance, even of corporeal substance, is a sign of the presence of the multiplicity of finite things in the infinite unity of substance.[21]

Some clarification of "what" these infinite modes are is still needed. The task is made easier by the clues Spinoza gives us as to their identity: as noted above, motion in matter, understanding in the thinking thing, according to the text in which they are spoken of for the first time as "modes or creatures which immediately depend on, or have been created by God" (KV I:9).

The contents of Letter 6 and the *Theologico-Political Treatise* must be added to what has already been said about movement. Both these texts are important. Letter 6 — one of Spinoza's few texts dealing with a scientific argument — contains (as is well known and as its subtitle indicates) annotations to Boyle's book *De Nitro, Fluiditate, et Firmitate.* In section 1 of the second part ("On Fluidity"), Spinoza distinguishes between the "notions which are derived from popular usage, or which explain Nature not as it is in itself, but as it appears to human sense," which must not be counted among the highest generic terms, or "chief kinds," and those which are defined with a remarkably evocative expression as "pure" notions, "which explain Nature as it is in itself." Motion, rest and their laws are cited as notions of this latter kind. This means that the "pure" notions (those not altered by random experience?) of motion, rest and the laws according to which motion and rest develop, provide scientific knowledge of physical nature, allowing us to recognize it as it is in itself, and express its essence. Analogously, in TTP 7, Spinoza states that: "In the examination of natural phenomena we try first to investigate what is most universal and common to all nature — such, for instance, as motion and rest, and their laws and rules, which nature always observes, and through which she continually works — and then we proceed to what is less universal." Motion and rest are brought forward here as real entities, universally present in physical nature; their laws and rules brought forward as laws followed by physical nature without any exception, hence, as laws according to which all natural phenomena occur.[22] The origin of the *facies totius universi* lies in the development of motion and rest according to particular laws, and therefore appears as a system of finite modes (bodies and ideas) related to one another.

To understand the true meaning of the notion "infinite intellect" one must fulfill two conditions: (1) one must not lose sight of Spinoza's polemic against the attribution of intellect and will to the nature of God (EIp17s, EIp33s2, Letter 54[23]), their reduction to the sphere of *Natura naturata* (Letter 9, EIp31, EIp32c2),[24] and their having the same relationship to God as do all natural things, and (2) one must refuse every hypothesis as to the permanence of traces of positive Christian religion. With respect to this hypothesis, suggested by the use of the expression "son of God" and by the identification of Christ, son of God, with the *sapienta Dei,* it must be noted[25] that the extension of the expression "son of God" to movement indicates an attempt to bring the sacred back to the sphere of the natural, while the identification of Christ with the *sapientia Dei* attests not only to the esteem Spinoza had for Christ, but also to his intention to humanize him. This intention is demonstrated by the conviction, expressed in Letter 73 to Oldenburg, that the eternal wisdom of God manifested itself to the maximum level in Christ, but manifests itself in all things and above all in the human mind.[26] This interpretation is supported by the theory that the human mind — insofar as it understands and perceives things according to the truth — is part of the infinite intellect of God (EIIp11c, EIIp43s, EVp40s; Letter 32),[27] or, conversely, that the infinite intellect of God expresses itself in human minds and is none other than the true knowledge of everything that is. Spinoza's theory of infinite intellect is a direct confutation of the theory of a divine creative intellect.

Even the analogies with concepts from the Neo-Platonic tradition — which certainly hold great interest from the point of view of reconstructing a possible historical picture to refer to — are to be considered deviant insofar as understanding the relationship between substance and modes, infinite and finite. As noted above, in addition to terms and concepts such as *sequi* = "to derive from" or "to follow," terms and concepts such as *producere, determinare, in esse* are also used frequently. The derivation of the modes (both finite and infinite) from the infinite substance is the same as their being produced and determined by the substance and their "being within" the substance.[28] The causality with which the substance produces its modes and expresses itself in them is immanent causality. None of these, then, are comparable to the process of Neo-Platonic emanation.[29]

I think the elements which have emerged so far allow us to draw the following conclusions. The problem of the identification of the mediate infinite modes in the attribute of Thought does not exist. The explanation (in Letter 64) of the *facies totius universi* with its reference to EII lemma7, and the principle of the identity of the order of ideas with the order of things, authorize the hypothesis that, in response to Schuller, Spinoza intended infinite modes of the "second kind" to indicate the infinite totality of all the finite bodies which make up the physical universe and the infinite totality of the minds of these bodies. I believe, therefore, that this idea is not a conjecture made by us in reading and trying to interpret Spinoza's text, but rather that it was already present in Spinoza's reply to Schuller.

However, due to their constitution, mediate infinite modes represent the most advanced point of the process of pluralization of the modes of being, with respect to which the immediate infinite modes serve as regulating and propelling principles.[30] This explains why Spinoza says so little about mediate infinite modes.

The problem of the deduction of the finite from the infinite is a spurious problem, which Spinoza himself contributed to with his statements on motion and on the need to study the argument further. But this problem is partly due to our misunderstanding of the text of Letter 83. By opposing the Cartesian concept of extension to his own, Spinoza confirms in this text his original conception of substance and its attributes as a dynamic principle which, though it remains one, is realized by pluralizing itself to infinity. This structure makes the moment of mediation superfluous, since it is already within substance itself.

114 *Emilia Giancotti*

Notes

1. I limit myself to mentioning a few of the classical texts and a selection of contemporary literature on the subject:
E. Schmitt, "Die unendlichen Modi bei Spinoza," *Zeitschrift für Philosophie und philosophische Kritik,* 140 (Leipzig, 1910); G. T. Richter, *Spinozas philosophische Terminologie* (Leipzig: Barth, 1913): 90–111; G. Huan, *Le dieu de Spinoza* (Arras: Schoutheer, 1913); C. Gebhardt, "Spinoza und der Platonismus," *Chronicon Spinozanum* 1 (1921): 178–234; E. Schmitt, "Zur Problematik der unendlichen Modi," *Chronicon Spinozanum,* 2 (1922), pp. 155–173; J. Freudenthal, *Spinoza, Leben und Lehre,* Zweiter Teile: *Die Lehre Spinozas,* (Heidelberg: Curis Societatis Spinozanae, 1927), particularly *Ontologie und Metaphysik,* pp. 108–145; L. Robinson, *Kommentar zu Spinoza Ethik* (Leipzig: Verlag von Felix Meiner, 1928), particularly pp. 196–207, 213–215, 217–219, 308–312; H. A. Wolfson, *The Philosophy of Spinoza,* 2 vols. (New York: Meridian Books, 1960), pp. 370 ff.
M. Gueroult, *Spinoza, Dieu* (Paris: Aubier-Montaigne, 1968), particularly pp. 308 ff.; A. Matheron, *Individu et communauté chez Spinoza* (Paris: Les Editions de Minuit, 1969); E. M. Curley, *Spinoza's Metaphysics* (Cambridge, Mass.: Harvard University Press, 1969), particularly pp. 58–74; J. G. Lennox, "The Causality of Finite Modes in Spinoza's Ethics," *Canadian Journal of Philosophy,* 4 (1976), pp. 479–500; C. E. Jarrett, "The Concepts of Substance and Mode in Spinoza," *Philosophia* 7 (1977), pp. 83–105; P. Macherey, *Hegel ou Spinoza* (Paris: Maspero, 1979), particularly pp. 180–198.

2. I have used the following editions: Spinoza, *Opera,* ed. C. Gebhardt, 4 vols. (Heidelberg: Winter, 1924); *The Collected Works of Spinoza,* ed. and transl. Edwin Curley, vol. I (Princeton: Princeton University Press, 1985; henceforth cited as Curley, and page number); *The Correspondence of Spinoza,* translated and edited by A. Wolf (London: George Allen and Unwin, 1928).

3. I deliberately use the concept "expression," which I consider as much part of Spinoza's conceptual inheritance as the concepts "dependence," "derivation," "determination" and "inherence." The concept "creation," still present in the *Short Treatise,* disappears in the *Ethics.*

4. On the use of the expression "son of God," cf. the observation of M. Francès in note 67 of her translation of the text: "The choice of traditional religious terms, here, is more than a simple concession to the language habits of the listeners. In fact, it is nothing less than the desacralization of an idea recognized by several Christian sects." (Spinoza, *Oeuvres complètes* [Paris: Bibliothèque de la Pléiade, 1954], pp. 1378–1379; transl. by D. Maisel). With respect to this interpretative hypothesis — with which I tend to agree — F. Mignini's observations are not necessarily to be considered alternatives (Spinoza, *Korte Verhandeling, Breve Trattato,* [L'Aquila: Japadre Editore, 1986], n. 20, pp. 564–565). Adoption of terminology from the old- and neo-testamentary literature does not imply — as the passage from TTP quoted by Mignini shows — acceptance of the significance linked to it; nor does it exclude desecrating intent on the part of Spinoza with respect to Christian dogmatics. It is surprising that Gebhardt (I, p. 471), should refer the expression to the attribute rather than to the infinite modes of God.

5. Mignini adds here, in square brackets, *"a priori."*

6. The Dutch text, though, has *Verstaan* here and not *Verstand* as the English "intellect" would lead us to expect, as Mignini observes in n. 17 of p. 564 of his commentary (and as we can see from the *Lexicon Spinozanum*, vol. 2, s.v. *Verstaan*, p. 1302).

7. In KV I:2 ("What God is"), the infinite intellect of God is mentioned several times without including the concept within the theory of infinite modes or, indeed, specifying its relationship with God's nature. In this text Spinoza discusses the concept of God, engaging in polemics with the traditional theory of the pluralism of substances and of creationism, using the concept of God's infinite intellect to sustain his argumentation.

8. According to a line of interpretation which Yovel propounds as well, the *fixae atque aeternae res* which are treated in the *Treatise on the Emendation of the Intellect* (Curley, pp. 39–42) are also intended as infinite immediate modes. I would like to put forward some doubts as to the solidity of this hypothesis, even if in a problematic and synthetic manner, by drawing attention to a specific passage.

Although we are given to understand that sufficient knowledge of the *res fixae et aeternae* has not yet been attained (Curley, p. 42), it is said of them that: (1) without them singular, changeable things "can neither be nor be conceived"; (2) "although these fixed and eternal things are singular, nevertheless, because of their presence everywhere and most extensive power, they will be to us like universals or genera of the definitions of singular, changeable things, and the proximate causes of all things" (Curley, p. 41). The first aspect is typical of the relationship between substance and modes. Therefore it does not seem to be a possible relationship between immediate infinite modes and changeable, singular things, that is, finite modes. It seems that even omnipresence (*ubique praesentia*) and omnipower (*latissima potentia*) cannot be ascribed to infinite modes, whereas — also taking into account the fact that these *res fixae et aeternae* are designated "singular" — they could perhaps be ascribed to the attributes of substance. The function Spinoza ascribes to them, as "genera of the definitions of singular changeable things, and the proximate causes of all things," might appear more convincing. But is that sufficient for identifying the *res fixae et aeternae* with the immediate infinite modes as a stable point of the theory of infinite modes? I think this is doubtful because — although a passage of the *Short Treatise* (KV I: 19/8) points to motion and rest as that from which all the effects of extension derive (cf. p. 101, above) — the concept of proximate causality is used by Spinoza to indicate the relationship between substance and immediate infinite modes rather than to indicate the relationship between the latter and changeable, singular things, finite modes.

9. See Richter, pp. 90–91.

10. The kabbalistic origins of the expression *facies totius universi* have been stressed by Wolfson, vol. 1, pp. 244–247. S. Zac also drew attention to these origins during the discussion which followed the presentation of my paper.

11. For the various positions see: Robinson, p. 215; Wolfson, p. 390, n. 2; Gueroult, p. 342; Curley, *Spinoza's Metaphysics*, p. 71.

12. On the definition of *causa absolute proxima, causa proxima in suo genere* and *causa remota*, see Heereboord (also quoted by Robinson, pp. 213–214) in *Meletemata philosophica* II, d.22 and in the *Syntagma logicae*, I, c.17.

13. I believe Richter is right when he calls attention to the category of inherence as

well in his examination of the relationship between substance and modes.
14. Curley's reference to the *Short Treatise* (*Spinoza's Metaphysics*, pp. 71–72) is correct as long as one does not forget this.
15. Gueroult (pp. 312, 342) — with whom it seems to me that Curley agrees — takes this hypothesis for granted, completely annulling the very clear distinction which Spinoza makes in EIp21 and EIp22, and repeats in EIp23. On this point see also the observations of Gebhardt (II, p. 353).
16. On this particular point, I believe Freudenthal was right; see Freudenthal, pp. 138–140.
17. Everyone is familiar with the passage in the *Theologico-Political Treatise:* "Nothing, then, comes to pass in nature in contravention to her universal laws (N.B. I do not mean here by "nature," merely matter and its modifications, but infinite other things besides matter") (TTP 6).
18. I consider what is said here about the concept of essence a preliminary treatment of the argument, which merits a thorough examination.
19. The most authoritative supporter of this impossibility was Hegel, with his interpretation of Spinozism as acosmism. An eminent representative of Italian neo-idealism, G. Gentile — author, among other things, of an important set of notes appended to the Italian edition of the *Ethics* (Florence: Sansoni, 1963) — shows in a note to EIp13c that he agrees with this interpretation. Gentile affirms that "in Spinoza's system there is no place for multiplicity; as there is not, in general, in the Neo-Platonic intuition which Spinoza shares" (pp. 695–696).
20. If, that is, the note is actually his; see in this connection Mignini, p. 563, n. 13.
21. An analogous concept is found in the following passage from Macherey's *Hegel ou Spinoza:* "He must conclude that there is an absolute identity between the infinite and the finite. These are not like two separate orders, between which there can be only a relationship of correspondence or submission; one can say that one is nothing without the other, and that one is nothing outside the other, if not from the abstract perspective of the imagination which divides them" (p. 198; transl. by D. Maisel).
22. The universality of these laws is limited to the physical world; therefore, strictly speaking, they cannot be identified with the "universal laws of nature," which are equated "with decrees of God following from the necessity and perfection of the Divine nature," dealt with in TTP 6 (G 82–83). The latter are the laws of the nature of God itself, which as a whole constitute the fixed and unchangeable order which regulates the unfolding of substance in all the modes of its infinite attributes; thus, they include not just the laws which are the object of natural sciences, but also those according to which the *infinita cogitandi potentia* unfolds through the infinite series of its finite modes and according to which all the infinite attributes of God must unfold in the infinite series of their finite modes.

Yovel suggests that we consider the infinite modes the metaphysical place of natural laws, and he tends to consider the two notions interchangeable. I believe that, on a logical and analytical plane, both the distinction between *leges naturae universales* = *decreta Dei* (which regulate all reality) and the laws which regulate motion (i.e., the laws of physical nature), and that between infinite modes and natural laws, must be firmly maintained. The infinite (immediate) modes are metaphysical or real entities — in which the attributes of substance immediately express themselves — which unfold according to laws. For Motion and Rest, these laws are natural laws, and for the Intellect, they can even be found, as Yovel suggests, in the laws of association.

23. "I shall show later ... that neither intellect nor will pertain to God's nature" (EIp17s); "I too, in order not to confuse the Divine with human nature, do not assign to God human attributes, such as Will, Understanding, attention, hearing, etc. Therefore, I say, as I have said just now, that the world is a necessary effect of the Divine Nature, and was not made by chance" (Letter 54).

24. "I think I have demonstrated clearly and evidently enough that the intellect, though infinite, pertains to *Natura naturata,* not to *Natura naturans*" (Letter 9; Curley, p. 195). "The actual intellect, whether finite or infinite, like will, desire, love, etc., must be referred to *Natura naturata,* not to *Natura naturans*" (EIp31); "It follows, secondly, that will and intellect are related to God's nature as motion and rest are, and as are absolutely all natural things, which (by P29) must be determined by God to exist and produce an effect in a certain way" (EIp32c2).

25. See M. Francès' observations referred to in n. 4, above.

26. "Lastly, to open my mind more clearly on the third head, I say, that it is not entirely necessary to salvation to know Christ according to the flesh; but we must think far otherwise of the eternal son of God, that is, the eternal wisdom of God, which has manifested itself in all things, more especially in the human mind, and most of all in Christ Jesus" (Letter 73).

On Christ, see Zac's analyses in his *Spinoza et l'interprétation de l'écriture* (Paris: PUF, 1965) and A. Matheron, *Le Christ et le salut des ignorants chez Spinoza* (Paris: Editions Aubier Montaigne, 1971).

27. "From this it follows that the human Mind is a part of the infinite intellect of God. Therefore, when we say that the human Mind perceives this or that, we are saying nothing but that God, not insofar as he is infinite, but insofar as he is explained [*explicatur*] through the nature of the human Mind, *or* insofar as he constitutes the essence of the human Mind, has this or that idea" (EIIp11c); "Add to this that our Mind, insofar as it perceives things truly, is part of the infinite intellect of God (by P11C); hence, it is as necessary that the mind's clear and distinct ideas are true as that God's ideas are" (EIIp43s); "These are the things I have decided to show concerning the Mind, insofar as it is considered without relation to the Body's existence. From them — and at the same time from IP21 and other things — it is clear that our Mind, insofar as it understands, is an eternal mode of thinking, which is determined by another eternal mode of thinking, and this again by another, and so on, to infinity; so that together, they all constitute God's eternal and infinite intellect" (EVp40s); "You see, then, in what way and why I think that the human Body is a part of Nature. As regards the human Mind I think it too is a part of Nature: since I state that there exists in Nature an infinite power of thought, which in so far as it is infinite, contains in itself subjectively the whole of Nature, and its thoughts proceed in the same way as nature, which, to be sure, is its ideatum. Then I declare that the human mind is this same power, not in so far as it is infinite, and perceives the whole of nature, but in so far as it is finite and perceives only the human Body, and in this way I declare that the human Mind is a part of a certain infinite intellect" (Letter 32).

28. In KV I:3, on divine causality, Spinoza also uses the term *uytvloejende oorzaak* = "emanative cause," but he equates it to the *daarstellende oorzaak* = "productive cause."

29. "The main property of Plotinus' God is transcendence. It is a serious mistake, in interpreting Spinoza, to attribute this property to Spinoza's substance ... Whereas Plotinus' One is beyond the psychic and the physical, beyond all qualities, Spinoza's substance is entirely psychic, entirely physical, and has all — i.e., infinite — qualities. Although the actions of the One are tightly attached to it, they are outside it and far away from it, whereas all the actions of the substance always remain within it. Whereas generation and degeneration do not touch upon the perfect being of the One, Spinoza's substance lives through what happens in Nature. Thus, any variation in the regularity of Nature means a variation in the substance-God. Were it possible to eliminate the smallest part of a body, the entire substance would be taken away with it. Whereas Plotinus' One is beyond all knowledge, knowledge of Spinoza's substance grows with each finite thing known. For, not only is every finite thing in the substance, but the substance is also in every finite action; every single thing is substance itself under a defined condition. This is why Spinoza can say '*Deus sive Natura,*' while this idea would be unthinkable for Plotinus" (Richter, pp. 108–109; transl. from the German).

30. As *principia individuationis*. See E. Schmitt, *Zur Problematik der unendlichen Modi*, p. 157.

[6]

PHILOSOPHY RESEARCH ARCHIVES
VOL. XIV, 1988-89

SPINOZA'S TEMPORAL ARGUMENT FOR ACTUALISM

HAROLD ZELLNER
Kent State University

ABSTRACT. In three places Spinoza presents an argument from (a) determinism and (b) God's "eternity" to (c) "actualism", i.e., the doctrine that this is (in some sense) the only possible world. That he does so shows that he distinguishes (a) from (c), which he has been thought to conflate. On one reading of 'eternal', he is claiming that an infinite past entails no other world was a "real" possibility. As might be expected, the argument is a failure, but it may help explain why Spinoza holds that there are no contingencies.

Spinoza's temporal argument for actualism is an attempt to get from determinism, together with the claim that "God" is eternal, to the conclusion that this is the only possible world. As might be expected the argument is not persuasive, but it does seem to show Spinoza was clear about a distinction he has often been thought to ignore. Moreover, it may well help us understand Spinoza's attitude towards contingencies.

A version of the argument appears in the *Ethics* at Part I, P33 Scholium 2:

> . . . even if it is conceded that will pertains to God's essence, it still follows from his perfection that things could have been created by God in no other way or order. It will be easy to show this if we consider, first, what they themselves concede, viz., that it depends on God's decree and will alone that each thing is what it is. For otherwise God would not be the cause of all things. Next, that all God's decrees have been established by God himself from all eternity. For otherwise he would be convicted of imperfection and inconstancy. But since in eternity there is neither when, nor before, nor after, it follows, from God's perfection alone, that he can never decree anything different, and never could have, or that God was not before his decrees,. and cannot be without them.[1]

It is clear that this is supposed to be an *ad hominem* argument against traditional theism, and thus cannot be safely used as a guide to Spinoza's own views. Some commentators explain the passage in terms of the Aristotelian identification of the omnitemporal with the necessary, and this is not an implausible reading of the text.[2] However, there is a similar argument in the *Short Treatise* which suggests a much more interesting and Spinozistic line of thought.

> But now the dispute arises again as to whether God can omit doing everything that is in his Idea and that consequently he can produce perfectly? and whether such an omission is a perfection in him? We say that since everything that happens is done by God, it must be predetermined by him. Otherwise he would be changeable, and that would be a great imperfection in him. And since this predetermination by him must be from eternity, and since in eternity there is neither before nor after, it follows inevitably from this that God was not able before to predetermine things in a way different from that in which they are now determined from eternity, and that before or without these determinations God could not have been.[3]

A quick check of the text will show that in this case Spinoza is speaking for himself, rather than drawing conclusions from someone else's views. This strongly suggests the argument in the *Ethics* does not turn primarily on the views of medieval Aristotelians, though of course Spinoza thinks he and his opponent have enough in common that he too should accept the Spinozistic conclusion. What will the argument look like purely from Spinoza's point of view?

Before turning to that question, it is worth making a point based on the general structure of these passages. Using 'actualism' to mean the (obscure) doctrine that this is the only possible world, the arguments have the form:

Determinism

+

"Eternity"

So

Actualism.

Here, at least, Spinoza clearly distinguishes between determinism and actualism, a distinction some scholars have thought Spinoza was muddled about.[4] If Spinoza conflated the first item with the third, he would not have thought to bring in the second. Of course, it is happy that one of the passages is from the early *Short Treatise*; if he had the distinction earlier on, presumably he had it later.

There has been considerable debate about what Spinoza means by 'eternity'.[5] There is a tradition according to which temporal predicates are not ascribable to God, and in that sense he is not supposed to be "in" time. On this interpretation 'eternal' would mean 'timeless'. The other school of thought is that an "eternal" object would have infinite durations, together with necessary existence. This would have it that, when Spinoza says that there is no "before or after" in eternity, what he means is that, since an eternal thing exists at all times, there are no times before which it existed or after which it does not exist. I am going to assume that this latter view is correct, though I am not going to argue for that here. In the present case the issue is especially complicated, since in one passage Spinoza is not speaking wholly for himself, while the other quote is from an early work where his views may not be the same as they are in the *Ethics*. In any case the most interesting argument uses 'eternal' in the way I suggest.

Everybody knows that Spinoza is a pantheist, so when he says God is eternal what he must mean is the universe (or some aspect of it sufficient to cause the rest) is eternal. Though Spinoza explains eternity in terms of necessary existence, a careful reading of the texts shows this is not what is operative in the move to actualism. It is something about God's filling all times that is supposed to bring that off (the "God was not able before . . ." in the *Short Treatise* argument). I would suggest the following as a reasonable reconstruction of the argument:

1. (Eternity Premiss) The actual world exists at all times throughout an infinite past and future.

2. (Determinism Premiss) All the facts obtaining at a time in the actual world can be causally explained by the facts at earlier times together with the laws of nature.

3. If a presently non-actual world could have been actual then there is a time at which God could have caused such a world to exist. (Assumed)

4. If there is a time at which God could have caused a presently non-actual world to exist then that time was before a time at which the presently actual world began to exist. (From the Determinism Premiss)

5. If 1., then it is not the case that there is a time before a time at which the presently actual world began to exist (since there is no time at which it began to exist).

6. It is not the case that there is a time before a time at which the presently actual world began to exist. (1,5)

7. It is not the case that there is a time at which God could have caused a presently non-actual world to exist. (4,6)

8. It is not the case that a presently non-actual world could have been actual. (From 3,7)

Several unconnected comments are in order before we turn to what all this means and why Spinoza might have found this convincing. I should perhaps hasten to point out that I am not claiming the above is clear or coherent, much less true; at the moment I am interested in what seems to be behind a text. Using the possible worlds terminology is not entirely happy, but it is the easiest and most familiar way of restating the argument. The price is that as in so many such discussions, God and a temporal metric are being thought of as "outside" possible worlds in a peculiar way. I have lumped together in 1. the infinite duration claim with the idea that the universe exists at all times; these are really distinct, but I think Spinoza would conflate them, or at least infer the latter from the former.[6] In the above reconstruction it is the actual world, rather than God, which is "eternal", but that should be all right on a Spinozistic reading of the texts. It will be remembered Spinoza sometimes speaks of God as the one substance and the attributes as distinct from the system of modes, so having some sort of distinction between the world and God does not necessarily conflict with the pantheism. There is a large amount of metaphysical stuffing in this argument that will fall outside the scope of this paper, e.g., what he means by 'God' and 'God's causality', etc. It is 4. that is distinctive, that is, the claim that God's causing a different world would have had to occur before the origin of whatever world actually obtains. What this comes to is that only a world with a finite history would have had alternatives; since he thinks our world does not have a finite history, it has no alternatives.

There is an immediate and obvious objection to the above which goes:

> All that this argument could show is that no other world could have existed, given the one we've got; whatever the details, what happens is that something about the actual world conflicts with any other world's being actual.

The conclusion is thus trivial, and it hardly requires an elaborate argument to show what is essentially the claim that no two possible worlds could be actual. This objection is conclusive if what the argument is supposed to do is show that only one world is consistently describable or something of the sort; but I shall try to show that there is a weaker but still interesting reading of the conclusion. Suppose that determinism is true, and we want to say that an event could have occurred which did not occur. And we do not mean just that it was logically possible or self-consistent. Presumably we mean (at least) that there is a true counterfactual conditional which describes that event in its consequent. Of course, such a conditional analysis of 'could have done otherwise' has been important in compatibilist theory about human action given determinism. I think it is this kind of 'could have' which may (who knows?) be leading Spinoza's thought in the quoted texts. Using this kind of 'could have' to claim a different world could have been actual would mean that, if some state of affairs had obtained which did not in fact obtain, then that world would have become actual. Consider the way (as it is often said) Leibniz introduces contingency, through the free choice of God.[7] For Leibniz, to say that a different world could have been actual means that it would have if God so chose at the time of the creation. Whether Leibniz himself thought of this choice as being in time is not a question which need concern us here; the important point is that this is the sense in which Spinoza seems to be saying that a different world is not possible, since there is no time at which things might have been different.

Without worrying about the historical Spinoza for a moment, consider present-day accounts of how counterfactuals are to be understood under determinism. According to one theory, the truth of a counterfactual should be understood in terms of what would have happened given that each in a series of events in the whole past of the antecedent and consequent would have been different. According to the other theory, a true counterfactual under determinism would state what would happen given a "divergence miracle" just prior to the events described in the antecedent and consequent.[8] Now will either account allow for it to be true that, if certain states of affairs had obtained which did not, then a different world would have been actual? With a finite past, a different world might have existed if there had been a different start-up, i.e., initial facts. Or it might be that a divergence miracle just after the start-up would have produced a different series. Does the supposition of an infinite past mean that no such counterfactual could be true?

Now some counterfactuals about when different worlds would have existed will be trivially true, even given an infinite past. Trivially, if a series of events in the past—infinite or not—were different from what it was, a different possible world would exist. Trivially, if an event currently ruled out by the laws of nature were to occur, a different possible world would exist. That is why even if determinism is true and past time is infinite, every time is still a time at which some other world could have been actual (in one sense), for of each time it will be true that, if something had been different at that time, a different world would have existed. But these counterfactuals will not relate a whole possible world to states of affairs which would explain how such a world might have been caused. I think Spinoza believes that to explain how a different world might have obtained it is necessary to have a true counterfactual in which the entire world is treated as an outsized event which would have occurred after, and been caused by, events described in the antecedent. Moreover, he would claim that the antecedent would have to explain not only why a different world would have obtained, but also why this one would not. Now consider the counterfactual corresponding to an analysis (altered past or divergent miracle) of "our world might not have existed" under determinism:

> If p at t, then, after t, not-our world (rigid designator) and a different world instead

where p is thought of as somehow causing or producing a state of affairs in which no sequence in our world ever exists and the whole of some other world exists instead. (The reading must be stronger, that is, than the trivial 'If something happened which did not happen, then that would be a different world'.) Given a beginningless past there will be no time at which t will precede our world and thus there could be no explanation of why our world would not have existed or why some other world would have existed in its place. In that sense, Spinoza can claim, our world could not have failed to exist, and no other world could have existed. I take it that this is what he means in the texts above when he says the "decrees" and "ways" could not have been different.

I do not see how to make the argument look more interesting from a contemporary—and non-Spinozist—point of view. If we take out the dubious notion of whole possible worlds being caused, what would be left is:

> 3'. If a presently non-actual world could have been actual, then there is a time at which that world might have existed.
>
> 4'. If there is a time at which a presently non-actual world might have existed, then that time was before a time at which the presently actual world began to exist.

with 1., 2., 5., and the conclusion as before. But 4'. looks obviously false. As pointed out above, even assuming determinism and an infinite past, every time is a time at which a different possible world might have been actual, in the sense that, if anything had been different, a different world would have been actual. Perhaps Spinoza would respond that the only time at which a different world could have been actual would be before the existence of this one, since in our world nothing can be other than what it is. But this move will be question-begging in what appears to be an argument that there are no contingencies.

Even if the argument could be repaired in some way, there would be grave problems about getting from its conclusion to what Spinoza wants. At best the argument would show that there was a sense in which the totality of things could not have been other than what it is. But it would be doubtful at best to move from that to the claim that each thing cannot be other than what it is. Even if there were no true counterfactuals about under what conditions other possible worlds would have obtained, that would not mean that there were none about objects or events in worlds. However, the fact that the move will not come off does not mean that Spinoza did not make it; and I submit that the argument above may be one explanation of Spinoza's believing that there are no contingencies.[9] □

ENDNOTES

1 Edwin Curley, trans., *The Collected Works of Spinoza*, Volume I (Princeton: Princeton University Press, 1985), 437.

2 See R.J. Delahunty, *Spinoza: The Arguments of the Philosophers* (London: Routledge and Kegan Paul, 1985), 148-50.

3 Curley, 81-2.

4 E.g., Jonathan Bennett, *A Study of Spinoza's Ethics* (Indianapolis: Hackett Publishing Co., 1984), 123.

5 Martha Kneale, "Eternity and Sempiternity", *Proceedings of the Aristotelian Society*, Vol. 69 (1968-69).

6 Spinoza often says 'infinite' when he means 'all'. See Bennett, 75-9.

7 See, e.g., Nicholas Rescher, *The Philosophy of Leibniz* (Englewood Cliffs, NJ: Prentice-Hall, Inc., 1967), 32.

8 David Lewis, "Counterfactual Dependence and Time's Arrow", *Nous* 13, 1979, 455-76, and Jonathan Bennett, "Counterfactuals and Temporal Direction", *Philosophical Review* 93, 1984, 57-91.

9 How does the argument above relate to an *ad hominem* argument against someone who believes in a personal God? I suspect that the crucial changes are:

1'. God is eternal (timeless).
5'. If God is eternal, then it is not the case that there is a time at which God could have caused a non-actual world to exist.

with the conclusion as before, using 3. I should suppose the argument for the crucial 5'. is that, since facts about God are not datable, it could not be a fact that at some time God caused one world in preference to another; perhaps the idea is that, since God is not in time, he could not decide at a time on one world rather than another. The role of determinism here is to make it plausible that worlds must be chosen whole and once and for all. Presumably, this is the argument in Chapter III of the Appendix to *Descartes' Principles of Philosophy*; see Curley, 309. However, the latter work must be used with caution as a guide to Spinoza's own views and arguments. This argument depends in part on theological considerations, and I shall not consider it further here.

Part II
Knowledge, Truth and Error

[7]

Reprinted from the Philosophical Review
Vol. LXXX - No. 3 - July 1971

SPINOZA'S THEORY OF IDEAS

DAISIE RADNER

THE THEORY of ideas found in Part II of the *Ethics* has been severely attacked. The theory may be rescued from at least some of the objections to it if it is interpreted as an attempt to overcome certain difficulties raised by the Cartesian form of representationalism. The first section of this paper gives a brief statement of Spinoza's theory of ideas and an account of some of the criticisms that have been directed against it. The second section presents two basic problems in the Cartesian theory of ideas, which are taken to be the motivations of Spinoza's theory. The third section shows how Spinoza attempts to solve one of these problems by introducing a distinction between the object of an idea and the thing represented by an idea. The fourth section is an examination of Spinoza's theory of the adequacy of ideas as an attempt to solve the other problem. The final section concerns Spinoza's classification of adequate and inadequate ideas.

I

In Part II of the *Ethics* Spinoza considers the nature of the human mind and its relation to the human body. He argues that the relation between the mind and the body is the same as that between an idea and its object, since the mind is nothing but that idea of the body which exists in the infinite intellect of God. Just as the human body is a highly complex individual, consisting of many individuals of different kinds, so the mind is a highly complex idea, consisting of many ideas. Each of the ideas of which the human mind is composed is the idea of some affection of the human body; and for each affection of the human body, there is an idea of it in the mind. All that the mind perceives it perceives by means of the ideas of the affections of its own body. For example, when the mind perceives the sun, the idea by means of which it perceives the sun is the idea of the human body as affected by the sun. The mind knows external bodies, itself, and the human body only by means of its ideas of the affections of the human body. Thus it has inadequate knowledge or ideas of these things. The things of which the mind has adequate knowledge or ideas include that which is common to all bodies and "the eternal and infinite essence of God." One can determine the adequacy of these ideas without reference to the things of which they are the ideas.

Among critics' objections to this theory are the following. First of all, the word "idea" is used ambiguously. H. Barker criticizes Spinoza for his "use of the one and the same word *idea* to denote sometimes *conceptus*, sometimes *mens*."[1] Similarly, A. E. Taylor objects to "the standing and apparently unconscious Spinozistic equivocation by which 'the idea of Peter' may mean either 'the mental complex which corresponds to Peter's brain and nervous system, the mind of Peter,' or 'the mental complex which exists when Paul thinks of Peter, Paul's "idea" of Peter.' "[2] Celestine Sullivan's version of the criticism is that "Spinoza employs the term 'idea' to mean simply the form or nature of a mode of extension, while yet he also means ambiguously by this same term 'idea' an element in the conscious life of man."[3] In the latter two statements, an idea in the first sense is something whose *ideatum* is the human body, or else something in the mind whose *ideatum* is some aspect of the human body; while an idea in the second sense is something in the mind whose *ideatum* may be an external body. To illustrate by means of the previous example of the mind perceiving the sun, the idea in the second sense is the idea of the sun; while the idea in the first sense is the idea of the human body as affected by the sun.

Spinoza himself points out this ambiguity in the Scholium to Proposition 17, but he is not always careful to take it into account. In particular, his demonstration of Proposition 13, that "The object of the idea constituting the human mind is a body . . .

[1] Barker, "Notes on the Second Part of Spinoza's *Ethics*," *Mind*, XLVII (1938), 418.

[2] Taylor, "Some Incoherencies in Spinozism," *Mind*, XLVI (1937), 153.

[3] Sullivan, "Critical and Historical Reflections on Spinoza's 'Ethics,' " *University of California Publications in Philosophy*, XXXII (1958), 3.

actually existing, and nothing else,"[4] is based upon a confusion of the two senses of the word "idea." In order to support the final step of the demonstration, that the object of the idea constituting the human mind is an existing body, *and nothing else*, Spinoza appeals to Axiom 5, which says in effect that we have no other ideas except those of bodies and of modes of thought. Our ideas of bodies and of modes of thought are *ideas* in the second sense of the word, however, while the mind as the idea of a particular body is an *idea* in the first sense of the word. The fact that we have no other ideas (in the second sense) besides those of bodies and modes of thought does not serve to demonstrate that the object of the idea (in the first sense) which constitutes the mind is nothing else than a particular existing body.

A further criticism of Spinoza's theory of ideas concerns the claim that the mind knows external bodies only by means of the ideas of its own bodily affections. According to Barker and James Martineau, this claim involves a confusion of the process by which we come to have ideas with the relation that an idea has with its object. Spinoza accepts the scientific doctrine that we perceive external bodies only in so far as they affect our own body. He appeals to this doctrine in support of his contention that the ideas by which we perceive external bodies are ideas whose objects are the affections of our own body. This approach reveals his failure to distinguish between the conditions for having an idea and the object of the idea. In order that we may perceive an external body, our own body must be affected in a certain way; but the object of our idea is not our body thus affected.[5]

Critics have also attacked Spinoza for his notion of the adequacy of ideas. First of all, they object to his claim that it is possible to determine whether an idea is adequate, without referring the idea to that of which it is the idea. They insist that Spinoza, in his attempt to provide an internal mark of adequacy,

[4] Quotations are from the W. H. White translation of the *Ethics*. For the sake of brevity, Spinoza's citation of proposition and axiom numbers will be deleted, unless required for purposes of discussion.

[5] Barker, *op. cit.*, pp. 295-300; Martineau, *A Study of Spinoza* (London, 1882), pp. 138-139.

overlooks the inescapable fact that ideas are ideas of things other than themselves. Barker puts the objection as follows:

> Spinoza . . . introduced the definition of an adequate idea in II, because, in view of the complete independence of the attributes he wished to insist that there is a wholly internal criterion of the truth of an idea . . . . But . . . he does not really escape the external reference of ideas, for (1) they know their ideata, (2) they correspond to their ideata.[6]

Likewise, Taylor insists that any attempt, such as Spinoza's, to "treat of knowing without ever introducing the reference to anything non-mental which is a known object" will invariably fail, because "no account can be given of knowing with the least vestige of plausibility which ignores the most patent characteristic of knowing, *viz.* that it is always the knowing of an object other than itself."[7]

Second, critics object to Spinoza's view that ideas may be inadequate in one context and adequate in another. Spinoza allows that some ideas are inadequate in the human mind (*E.* II, 24–31); yet he insists that "All ideas . . . in so far as they are related to God are true and adequate" (*E.* II, 36 Dem.). Apparently, then, he believes that those same ideas which are inadequate in the human mind are adequate when they are related to God. Barker finds this position untenable, for the following reason. In order that an inadequate idea may be adequate in relation to God, there must be some alteration in the content of the idea. Thus the idea which is inadequate in the human mind and the idea which is adequate when related to God are not one and the same. Barker writes:

> The ideas which are inadequate in man must surely undergo a change in order to become adequate in the *intellectus infinitus* . . . . Or, conversely, ideas which are adequate in the *intellectus infinitus* must undergo a change in order to have a place as inadequate ideas in the mind of man, a change which implies not merely a diminution but a distortion.

[6] Barker, *op. cit.*, p. 433.
[7] Taylor, *op. cit.*, p. 150.

In other words, the ideas which are adequate in one reference and inadequate in another are not the same ideas.[8]

A final difficulty with regard to Spinoza's notion of adequacy is this. Given that there is an exact correspondence between ideas and the things of which they are the ideas (*E.* II, 7), how can there be inadequate ideas at all? Barker challenges Spinoza with the following dilemma:

If ideas agree with their ideata and are true, they cannot be confused and inadequate. If they are confused and inadequate, they cannot agree with their ideata and be true.[9]

Why does Spinoza put forth a theory of ideas that is open to such obvious and serious criticisms? Most philosophical theories are developed to solve particular philosophical problems. A theory which solves no problems and which is beset with difficulties in its own right deserves little consideration. A theory which is put forth to solve certain philosophical problems at least deserves an evaluation in terms of whether it solves the problems it is meant to solve.

Spinoza's theory of ideas may be taken as an attempt to improve upon the Cartesian account of the relation of ideas to the things they represent. When the theory is taken in this light, the objections noted above are not so insurmountable as they first appear. In particular, the two senses of "idea" constitute, not an unconscious equivocation, but a deliberate attempt to give a more satisfactory account of the representation relation than Descartes was able to offer. The alleged confusion of the object of the idea with the conditions for having the idea disappears, as the idea takes on two distinct relations to the things external to it. The difficulties concerning the adequacy of ideas are overcome when the notion of adequacy is assigned the proper role in Spinoza's representationalism. But before we examine how Spinoza attempts to overcome the difficulties raised by the Cartesian theory of ideas, let us first give brief consideration to the Cartesian theory itself and the problems to which it gives rise.

[8] Barker, *op. cit.*, p. 438.
[9] *Ibid.*

SPINOZA'S THEORY OF IDEAS

II

According to Descartes, when the mind thinks of something external to it, the situation may be analyzed as follows. First of all, there is the mind or the substance which thinks. Second, there is the act of thinking, which is a modification of the mind. Third, there is the content of thought or that of which the mind is directly aware in the act of thinking. This content of thought, or *idea*,[10] is something mental and is representative of something else which is nonmental. Finally, there is the nonmental thing which is represented in the idea. This nonmental thing is called the *object* of the idea. That which is in the object of the idea *formally* is in the idea itself *objectively* or "by representation."[11]

Why does the cognitive situation require the presence of an idea, which is distinct from and representative of the object? The reason is this. All that the mind knows, it knows by means of that of which it is directly aware. The only things of which the mind is directly aware are things which are immediately present to the mind. The only things which are immediately present to the mind are mental things or things which are in some way *in* the mind. Thus all that the mind knows, it knows by means of that which is in the mind. Now the mind is commonly said to know nonmental things. Its knowledge of nonmental things must take place by means of its direct awareness of things in the mind. Thus the mind knows nonmental things by means of its ideas, which are representations in the mind of nonmental things.

If the mind knows the objects of its ideas only by means of its ideas, how can it distinguish between ideas which represent their objects as they really are and ideas which do not? For example, suppose I have the idea of a material body as something capable of motion and rest. How do I know whether the body really is capable of motion and rest? Again, suppose I have the idea of a material body as something colored. How do I know

[10] I overlook the ambiguity in Descartes's use of the word "idea," whereby he uses it to refer both to the act of thinking and to the content of thought. My concern here is not so much to give an accurate statement of Descartes's theory, as to outline it as Spinoza seems to have understood it.

[11] Haldane and Ross (trans.), *Philosophical Works of Descartes*, I, 162 (hereafter cited as *H.R.*).

whether color really belongs to the body? Since I am not directly acquainted with the material body itself, but know it only by means of my idea of it, I cannot compare my idea of the body with the body itself in order to see whether my idea represents the body accurately. Descartes's response to this problem is to appeal to the principle that whatever is clearly and distinctly conceived is true. Those ideas which are clear and distinct give accurate representations of their objects, while those which are obscure and confused do not. Thus it is possible to determine which ideas represent things as they really are, without any need of direct comparison of the ideas with their objects. Clearness and distinctness are defined independently of the relation which the idea has to its object. Once it is established that an idea has the internal characteristics of clearness and distinctness, one may be assured that the idea gives an accurate representation of its object or, in Descartes's terms, that the idea is true.

The way in which Descartes separates the clearness and distinctness of ideas from the accuracy with which they represent their objects raises one set of difficulties for the Cartesian theory of ideas. (*a*) How does one determine which ideas are clear and distinct and which are obscure and confused? The question of whether or not an idea is clear and distinct must be answered before one can answer the question of whether or not the idea is true. Hence in ascertaining the clearness and distinctness of an idea, one cannot presuppose that the idea is true. (*b*) Given that it is established that an idea is clear and distinct, how does one know that the idea is true? The principle that clear and distinct ideas are true is not itself a self-evident truth. Descartes supports this principle by an appeal to the veracity of God. Thus, having inserted a skeptical wedge between the clearness and distinctness of ideas and their truth, Descartes must supply a method of ascertaining clearness and distinctness which can be implemented without reference to truth;[12] and he is then obliged to bridge the gap between the clearness and distinctness of ideas and their truth by means of a divine guarantee.

[12] For an examination of Descartes's response to this difficulty, see Alan Gewirth's article, "Clearness and Distinctness in Descartes," *Philosophy*, XVIII (1943), 17-36.

SPINOZA'S THEORY OF IDEAS

A further difficulty for the Cartesian theory concerns the nature of the representation relation. How do ideas represent their objects? Descartes seems to consider representation as a kind of resemblance or likeness between an idea and its object. An idea represents its object, by resembling it in some way. Thus in the *Third Meditation* he speaks of ideas as being "like [pictures or] images" of their objects.[13] Similarly, in the *Reply to the Second Objections*, he says that what is in the idea objectively, by virtue of being represented by the idea, is in the object itself formally, if "the way in which it exists in the object is exactly like what we know of it when aware of it."[14] In order for there to be a resemblance or likeness between two distinct things, however, the two things must have something in common. What is there in common between an idea and a material object? Ideas belong to the realm of thought, material things to the realm of extension; and "there is nothing at all common to thought and extension."[15] If there is nothing in common between ideas and their objects, then there can be no resemblance between them. But if there is no resemblance between ideas and their objects, how do ideas represent their objects? Having introduced a total disparity between the mental and the material, Descartes must provide an explanation of how an idea can represent a material object without resembling it.[16]

Spinoza attempts to rescue the representative theory of ideas from difficulties such as these. In answer to the question of how ideas represent their objects, he offers an explanation of representation which does not make it a kind of resemblance between the mental and the material. As for the problems arising from the separation of clearness and distinctness from truth, Spinoza attempts to undercut them by establishing a necessary connection between the internal characteristics of ideas and the accuracy with which they represent their objects.

---

[13] *H.R.* I, 163.

[14] *H.R.* II, 53.

[15] *H.R.* II, 212.

[16] For an examination of other Cartesians' attempts to solve this problem, see Richard A. Watson's *The Downfall of Cartesianism: 1673-1712* (The Hague, 1966).

III

Spinoza agrees with Descartes that the mental and the material are totally distinct realms; that we are directly aware only of things in the mental realm; and that we can have knowledge of material things by means of ideas which represent them. How do ideas represent material things? The key to Spinoza's theory of the nature of representation is his distinction between the object of the idea and that which the idea represents. The term "the object of the idea" is not synonymous with the term "that which is represented by the idea," although in some cases the two terms have the same reference. The object and the thing represented stand in two different relations to the idea. The relation between the idea and its object is explicated in terms of the distinction between objective and formal reality. The relation between the idea and what it represents is explicated in terms of the resemblance of the thing represented to the object of the idea.

The object of the idea and the idea are related as formal reality to objective reality. A thing has formal reality in so far as it exists in itself, and objective reality in so far as it is thought of. This version of the distinction between objective and formal reality is found in the following passage from Descartes's *Reply to the First Objections*:

> Hence the idea of the sun will be the sun itself existing in the mind, not indeed formally, as it exists in the sky, but objectively, i.e. in the way in which objects are wont to exist in the mind.[17]

Spinoza understands the distinction between objective and formal reality as it is expressed in this passage. According to Spinoza, all individual things are modes of the one substance, and they are all thought of by the one substance. For everything that exists, there is an idea of it in the infinite intellect. In Spinoza's words, "whatever follows *formally* from the infinite nature of God, follows from the idea of God [*idea Dei*] in the same order and in the same connection *objectively* in God" (*E.* II, 7 Cor.). And for every idea in the infinite intellect, there is an existing thing

[17] *H.R.* II, 10.

which is its object. In Spinoza's words, "that which is objectively contained in the intellect must necessarily exist in nature" (*E.* I, 30 Dem.). The realm of ideas and the realm of objects are coextensive, because it is one and the same thing which exists both objectively and formally. Spinoza writes:

> Thus, also, a mode of extension and the idea of that mode are one and the same thing, expressed in two different ways . . . . For example, the circle existing in nature and the idea that is in God of an existing circle are one and the same thing, which is manifested through different attributes [*E.* II, 7 Schol.].

In so far as the individual thing exists formally, it is considered "under the attribute of extension." In so far as it exists objectively, it is considered "under the attribute of thought."

The human body is a finite mode of the attribute of extension. There is an idea of it in the infinite intellect of God. This idea is the human mind. Since a mode of extension and the idea of that mode are one and the same thing, the human body and the human mind are one and the same thing, viewed on the one hand under the attribute of extension and on the other hand under the attribute of thought. The human body consists of many parts, which are affected in many ways. The mind, as the objective reality of the body, contains ideas of all the ways in which the body is affected. In Spinoza's words, "the ideas of the affections of the body are in God in so far as He forms the nature of the human mind"(*E.* II, 19 Dem.).

Just as there are in God ideas of the human body and its affections, so there are also in God ideas of all other material bodies. These ideas bear the same relation to their objects as the mind bears to the body.

> For of everything there necessarily exists in God an idea of which He is the cause, in the same way as the idea of the human body exists in Him; and therefore everything that we have said of the idea of the human body is necessarily true of the idea of any other thing [*E.* II, 13 Schol.].

For example, there is in God the idea of the sun, and this idea constitutes the objective reality of the sun.

Descartes would say that when we perceive an external body, we have an idea of that body—that is, an idea of that body is present in our mind. Spinoza sometimes adopts this usage. For example, in *Ethics* II, Proposition 16, Corollary 2, he speaks of "the ideas we have of external bodies." Similarly, in the Scholium to Proposition 17 he speaks of "the idea of Peter himself which is in another man" when the other man perceives Peter. For Spinoza, as for Descartes, our idea of an external body is that idea in our mind which represents the external body to us. But Spinoza, unlike Descartes, distinguishes between the idea which represents the external body to us and the idea whose object is the external body. In Spinoza's system we do not perceive external bodies by means of ideas whose objects are external bodies, since such ideas are not in our mind. The only ideas in the human mind are ideas whose objects are the affections of the human body, and external bodies are not affections of the human body. Since we perceive external bodies only by means of ideas which are in our mind, we perceive them by means of the ideas of our affections. Thus Spinoza says that the ideas of our affections "represent to us external bodies" (*E*. II, 17 Schol.; III, 27 Dem.).

Representation for Spinoza is a matter of *making known*, and it is always with respect to a particular knower. Idea *X* represents object *Y* to knower *K*, just in case *X* makes *Y* known to *K*. How can the idea of our bodily affections make external bodies known to us? Spinoza's answer is that since external bodies are causes of our affections, our affections have something in common with external bodies, and thus the ideas of our affections "involve the nature" of external bodies.

Spinoza adopts the Cartesian principle that there is nothing in the effect which did not first exist in the cause. Whatever is in the effect must have been present in the cause; otherwise "whatsoever it might have, it would have from nothing."[18] Thus, given that one knows that the effect has property *P*, one may infer that the

[18] A. Wolf (trans. and ed.), *The Correspondence of Spinoza*, Letter IV, 83. Cf. *H.R.* I, 162; II, 34-35.

cause also has property *P*. In this way "the knowledge (*cognitio*) of an effect . . . involves the knowledge of the cause" (*E*. I, Ax. 4).

The affections of the human body are produced by the action of external bodies upon the parts of the human body. In *Ethics* II, Postulate 3, Spinoza writes:

> The individuals composing the human body, and consequently the human body itself, are affected by external bodies in many ways.

Since that which is in the effect must first be present in the cause, and since the affections of the human body have external bodies as their cause, the affections have something in common with external bodies. Thus the idea of each affection is the idea of something which is present, not only in the human body, but also in external bodies. Spinoza writes:

> The idea of every way in which the human body is affected by external bodies must involve the nature of the human body, and at the same time the nature of the external body [*E*. II, 16].

Because the ideas of the affections of our body involve the nature of external bodies, they can serve to represent external bodies to us. In Spinoza's words, "the human mind perceives the nature of many bodies together with that of its own body" (*E*. II, 16 Cor. 1). Nevertheless, the ideas which represent external bodies to us have as their objects the affections of the human body. In Spinoza's words, "the ideas we have of external bodies indicate the constitution of our own body rather than the nature of external bodies" (*E*. II, 16 Cor. 2).

The ideas of our bodily affections do not represent external bodies by being mental pictures or images of them. These ideas represent external bodies to us, by containing objectively something which the external body, as cause of the affection, contains formally. For example, suppose I have an idea which represents the sun to me. This idea is not a mental picture of the sun. It is the objective reality of an affection of my body, an affection which is produced by the action of the sun upon the parts of my body. Since the sun is cause of this bodily affection, and since there must be something in common between cause and effect

(*E.* I, 3), there is something in common between the sun and my bodily affection. My idea represents the sun to me, by virtue of the fact that its object is an affection which has something in common with the sun. Thus the resemblance or likeness is not between my idea and the sun. It is between my bodily affection and the sun; or, since "The order and connection of ideas is the same as the order and connection of things," it is between the idea whose object is my bodily affection and the idea whose object is the sun.

Thus, by separating the relation between an idea and its object from the relation between an idea and the thing it represents, Spinoza suggests an interpretation of the representation relation which does not make it a relation of likeness between two things which are essentially unlike. It seems, however, that the problem of resemblance between two unlike things is merely transferred from the representation relation to the relation between an idea and its object. If it is difficult to see how an idea and a material body can belong to totally distinct realms and yet have something in common, it is even more difficult to see how they can belong to totally distinct realms and yet be one and the same thing. The problem which arises for Spinoza at this point is basically the same as the one which arises from his claim that thought and extension, though totally distinct and independent attributes, yet constitute the essence of one and the same substance (*E.* I, 10 Schol.).[19] The problem on the level of finite modes derives from the problem on the level of attributes; for the idea and the object are distinguishable only in terms of the attributes under which they are conceived. The problem of how an idea and its object can be one and the same thing jeopardizes, not only Spinoza's account of the nature of representation, but also his attempt to bridge the Cartesian gap between the internal characteristics of ideas and the accuracy of their representation.

[19] The problem with regard to the attributes is raised by Martineau, *op. cit.*, p. 185, as well as by Reginald Jackson in his article, "The Doctrine of Substance in Descartes and Spinoza," *Australasian Journal of Philosophy*, IV (1926), 208.

SPINOZA'S THEORY OF IDEAS

IV

How does one know which ideas in the human mind represent things as they really are and which do not? In answer to this question, Spinoza appeals to the principle that whatever is adequately conceived is true. Thus the notion of adequacy plays the same role in Spinoza's system as the notion of clearness and distinctness plays in Descartes's. That is, it enables one to determine which ideas give accurate representations, without directly comparing the ideas with the things they represent. Spinoza, however, attempts to characterize adequacy in such a way that there is a necessary connection between the adequacy of ideas and their truth.

At the beginning of Part I of the *Ethics* Spinoza characterizes a true idea as follows:

> A true idea must agree with that of which it is the idea [Ax. 6].

At the beginning of Part II he gives the following definition of an adequate idea:

> By adequate idea, I understand an idea which, in so far as it is considered in itself, without reference to the object, has all the properties or internal signs (*denominationes intrinsecas*) of a true idea.
> *Explanation*: I say internal, so as to exclude that which is external, the agreement, namely, of the idea with its object [Def. 4].

In a letter to Tschirnhaus Spinoza makes explicit the relation between adequacy and truth:

> I recognize no other difference between a true and an adequate idea than that the word true refers only to the agreement of the idea with its ideatum, while the word adequate refers to the nature of the idea in itself; so that there is really no difference between a true and an adequate idea except this extrinsic relation.[20]

From these passages it is apparent that adequacy and inadequacy have to do with certain internal characteristics of the idea, while

[20] Wolf, *op. cit.*, Letter LX, 300.

truth and falsity have to do with the relation that the idea has to that of which it is the idea.

The phrase "that of which it is the idea" has acquired an ambiguity in Spinoza's system. It may refer either to the object of the idea or to that which is represented by the idea. To which of these does Spinoza refer when he characterizes the truth as the agreement of the idea with that of which it is the idea? If a true idea were one which agrees with its object, then all ideas would be true; for all ideas are themselves the objective reality of their objects and so "agree" with their objects. Spinoza allows that all of God's ideas are true (*E.* II, 32); but he maintains that some of our ideas are false (*E.* II, 35 and Schol.). Close examination of the passages in which he speaks of the falsity of our ideas (for example, the falsity of our idea of the sun's distance from us) reveals that he considers these ideas as false, not with respect to their objects, but with respect to what they represent. The truth or falsity of ideas has to do, not with whether or not the ideas agree with their objects, but with whether or not they represent things as they are in themselves. A true idea of *X* is one which represents *X* as it is in itself. A false idea of *X* is one which represents *X* but not as it is in itself.

An adequate idea is one which, considered in itself, has all the internal signs of a true idea. Since a true idea is one which represents a thing as it really is, an adequate idea is one which, considered in itself, has all the internal signs of an idea which represents a thing as it really is. An idea is adequate or inadequate, not with respect to its object, but with respect to that which it represents.

Thus it is possible to overcome Barker's objection that the existence of inadequate ideas cannot be reconciled with the exact correspondence of ideas and things. Although all ideas agree with their objects, not all ideas agree with the things they represent. Since ideas are adequate or inadequate with respect to that which they represent, the existence of inadequate ideas does not conflict with the exact correspondence of ideas and their objects. An idea may be inadequate with respect to that which it represents and yet agree with its object.

We have an adequate idea of *X* if the idea which represents *X* to us is the idea which represents *X* to God—that is, if the idea by which God knows *X* is in God in so far as he forms the nature of the human mind (*E.* II, 34 Dem.). We have an inadequate idea of *X* if the idea which represents *X* to us is not the idea which represents *X* to God—that is, if the idea by which God knows *X* is not in God in so far as he forms the nature of the human mind. Our idea of *X* is inadequate, even though God's idea of *X* is adequate, since our idea of *X* is not God's idea of *X*.

Given this interpretation of adequacy, it is possible to overcome Barker's objection to Spinoza's claim that all of God's ideas are adequate, while some of our ideas are inadequate. Barker believes that this claim implies that there are ideas which are inadequate when related to the human mind and adequate when related to God. He argues that since there cannot be such ideas, Spinoza's claim must be false. This argument rests upon a failure to recognize that ideas are adequate or inadequate in so far as they represent things, and that there is not a one-to-one correspondence between ideas and the things they represent, as there is between ideas and their objects. One and the same idea may represent two different things; and it may be adequate in so far as it represents one, but inadequate in so far as it represents the other. Furthermore, two different ideas may represent one and the same thing, though one idea is adequate and the other is inadequate. While it is true that one and the same idea cannot be adequate when it represents *X* to God and inadequate when it represents *X* to the human mind, it does not follow that it cannot be the case that God's idea of *X* is adequate and our idea of *X* is inadequate. For the idea which represents *X* to God and the idea which represents *X* to the human mind need not be one and the same. When Spinoza claims that all of God's ideas are adequate, while some of our ideas are inadequate, all he means is this. For any *X*, the idea which represents *X* to God is adequate; while for some *X*, the idea which represents *X* to the human mind is inadequate. From this it does not follow that there is an idea which is adequate in so far as it represents *X* to God and inadequate in so far as it represents *X* to the human mind.

How does one determine which of the mind's ideas are adequate and which are inadequate? Spinoza offers a method of ascertaining the adequacy of ideas which is based upon the kinds of things the ideas represent. Certain things are such that our knowledge of them can only be adequate. If our ideas represent things of this kind, they are adequate.

In *Ethics* II, Proposition 38, Spinoza says that the things of which we can only have adequate knowledge are "those things which are common to everything, and which are equally in the part and in the whole." In the demonstration he explains that things of this kind can only be adequately known by us, because the human mind knows them by means of the ideas by which God knows them. Since God's ideas of them are adequate, and since our ideas of them are the same as God's, our ideas of them are adequate.

> Let there be something *A*, which is common to all bodies, and which is equally in the part of each body and in the whole. I say that *A* can only be adequately conceived. For the idea of *A* will necessarily be adequate in God . . . in so far as He has ideas which are in the human mind [*E.* II, 38 Dem.].

When is the idea by which the human mind knows *A* the same as the idea by which God knows *A*? The idea by which God knows *A* is the idea whose object is *A*.

> A knowledge of everything which happens in the individual object of any idea exists in God in so far only as He possesses the idea of that object [*E.* II, 9 Cor.].

If the idea by which the human mind knows *A* is the idea whose object is *A*, then the idea by which the human mind knows *A* is the idea by which God knows *A*. Thus Spinoza's method of determining the adequacy of an idea, on the basis of the kind of thing it represents, amounts to this. One determines whether an idea is adequate by considering whether it represents its object or some aspect of its object. If the idea which represents *A* to the human mind is the idea whose object is or includes *A*, then the idea which represents *A* to the human mind is adequate.

Spinoza's demonstration of Proposition 39 may be seen as an

application of this method. Spinoza shows that the human mind has an adequate idea of that which is common to the human body and the external bodies which affect it, and which is equally in the part and in the whole of these bodies, by showing that the idea which represents this property to us is an idea whose object has this property. Let *A* be a property of the kind just described. Spinoza argues:

> Let it be supposed that the human body is affected by an external body through that which it has in common with the external body, that is to say, by *A*. The idea of this affection will involve the property of *A*, and therefore the idea of this affection, in so far as it involves the property of *A*, . . . is . . . adequate in the human mind [*E*. II, 39 Dem.].

Since the external body affects the human body through *A*, *A* will be present in the affection of the human body, as it is present in the cause of the affection. Thus the idea whose object is the affection is an idea which includes *A* in its object. When the idea of the affection represents *A* to the human mind, it represents something in its object, and thus it is an adequate idea of *A*.

Since an idea is adequate in so far as it represents its object or something included in its object, there is a necessary connection between the adequacy of an idea and its truth. For if the idea which represents *A* is the idea whose object is *A*, then the idea is itself the objective reality of *A*. The idea and that which it represents are one and the same thing, considered objectively on the one hand and formally on the other. If the idea is the objective reality of the thing it represents, then it represents the thing as it is in itself. And an idea which represents a thing as it is in itself is a true idea of that thing (*E*. II, 43 Schol.).

V

Spinoza claims that the ideas whose objects are the affections of the human body are inadequate when they represent the parts of the human body (*E*. II, 24), external bodies (*E*. II, 25), the human body (*E*. II, 27), the affections of the human body (*E*. II,

28), and the human mind (*E.* II, 29). These ideas are adequate when they represent those things in which all bodies agree (*E.* II, 38 Cor.) and when they represent "the eternal and infinite essence of God" (*E.* II, 45–47).

Our ideas represent external bodies by virtue of the fact that the objects of our ideas have something in common with the external bodies. When our ideas represent to us that which their objects have in common with external bodies, they are adequate ideas. When our ideas represent to us the external bodies themselves, however, they are inadequate; for external bodies are not the objects of our ideas, nor are they wholly included in the objects of our ideas.

Why does Spinoza claim that our ideas are inadequate when they represent the affections of our body, but adequate when they represent "the eternal and infinite essence of God"? The situation seems to be the other way around. If we have an adequate idea of *A* when the idea which represents *A* to us is the idea whose object is *A*, then our ideas, whose objects are the affections of our body, are adequate when they represent those affections. Also, since the idea which represents God's essence to us (*E.* II, 45) is not the idea which represents God's essence to God (*E.* II, 3), our idea of God's essence is not an adequate idea.

Spinoza's claim that we do not have adequate ideas of our bodily affections is based upon a confusion between "adequate" as applied to ideas and "adequate" as applied to causes. We have already observed that Spinoza speaks of ideas as being adequate or inadequate. He also speaks of causes as being adequate or inadequate. At the beginning of Part III of the *Ethics* he defines an adequate cause and an inadequate cause as follows:

> I call that an adequate cause whose effect can be clearly and distinctly perceived by means of the cause. I call that an inadequate or partial cause whose effect cannot be understood by means of the cause alone [Def. 1].

A particular thing *X* is the adequate cause of another thing *Y*, just in case *Y* follows from the nature of *X* alone. *X* is the inadequate or partial cause of *Y*, just in case *Y* follows, not from the nature of *X* alone, but from the nature of *X* together with that of another thing *W*.

The idea representing *A* is an *adequate idea*, just in case it has the internal characteristics of a true idea—namely, that which it represents is or is included in its object. On the other hand, the cause of the idea representing *A* is an *adequate cause*, just in case the idea follows from the nature of this cause alone, without the concurrent influence of anything else. Spinoza confuses these two notions of adequacy, when he argues as follows. Since the human body is not the adequate cause of its affections, the idea of the human body, or the human mind, is not the adequate cause of the ideas of these affections. Thus the human mind does not have adequate ideas of the affections of the human body (*E.* II, 28 Dem.). The principle underlying this argument is that if the mind has an adequate idea, then the mind is the adequate cause of this idea. In *Ethics* III Spinoza proves that if an idea is adequate in the human mind, then the mind is adequate cause of what follows from the idea (*E.* III, 1 Dem.). But he nowhere proves that if an idea is adequate in the human mind, the mind is adequate cause of the idea itself. On the contrary, from what has been said of the adequacy of ideas and of causes, it seems perfectly possible that the mind may have an adequate idea and yet not be the adequate cause of this idea.

Spinoza's claim that we have an adequate idea of "the eternal and infinite essence of God" is based upon a misapplication of the principle that what is common to everything and equally in the part and in the whole can only be adequately conceived. Spinoza reasons as follows. Since God is the cause of all things, "in so far as He is considered under that attribute of which they are modes," the idea of every existing individual involves "the eternal and infinite essence of God" (*E.* II, 45 Dem.). Since every idea involves God's essence, God's essence can only be adequately conceived.

Whether a thing be considered as a part or as a whole, its idea . . . will involve the eternal and infinite essence of God. Therefore that which gives a knowledge of the eternal and infinite essence of God is common to all, and is equally in the part and in the whole. This knowledge therefore (Prop. 38, pt. 2) will be adequate [*E.* II, 46 Dem.].

Since the human mind possesses ideas which involve God's essence, it "possesses an adequate knowledge of the eternal and infinite essence of God" (*E.* II, 47).

The principle introduced in Proposition 38 is that what is common to every individual can only be adequately conceived. Spinoza, however, does not say that God's essence is common to every individual. He says only that the idea of every individual involves God's essence. From what he says in Part I of the *Ethics*, it is apparent that the idea of each individual involves a different aspect of God's essence. For the idea of each individual involves God's essence in so far as God is cause of the individual; and God is cause of individual things "in so far as the attribute is modified by a modification which is finite" (*E.* I, 28 Dem.). Although every individual has God as its cause, the ideas of different individuals involve God's essence in so far as it is modified by different modifications. What is common to every individual is that it has God as cause in so far as he is modified by something finite. God's essence per se is not common to every individual. Thus Spinoza cannot support his claim that we have adequate knowledge of it by appeal to Proposition 38.

If an idea is adequate in so far as it represents its object or something included in its object, then some of the ideas which Spinoza says are inadequate turn out to be adequate, and some of the ideas which he says are adequate turn out to be inadequate. Spinoza attempts to make the class of adequate ideas coincide with Descartes's class of clear and distinct ideas. According to Descartes, our ideas of the affections of our body are not clear and distinct, while our idea of God as a perfect being is clear and distinct. In like manner, Spinoza claims that the former ideas are not adequate, while the latter idea is adequate. Given the way in which Spinoza characterizes the adequacy of ideas, however, the class of adequate ideas does not coincide with Descartes's class of clear and distinct ideas. Spinoza succeeds in dispensing with the Cartesian gap between the internal characteristics of ideas and the accuracy with which they represent things external; but the way in which he does so prevents him from drawing the distinction between adequate and inadequate ideas in the same way as Descartes draws the distinction between ideas which are clear and distinct and those which are not.

DAISIE RADNER

644 Main Street West, Apt. 1512
Hamilton, Canada

[8]

Alexandre Matheron

Ideas of Ideas and Certainty in the *Tractatus de Intellectus Emendatione* and in the *Ethics*

When we compare what Spinoza tells us about how ideas relate to ideas of ideas in the *Treatise on the Emendation of the Intellect* and what he tells us about this in Part II of the *Ethics*, our immediate impression is that there are several contradictions between these two texts. Thus:

1. In TIE 33 Spinoza declares that an idea *can* be the object of another idea, whereas in EIIp20,22 he demonstrates that there is *necessarily* an idea of every idea. But this does not really amount to an outright contradiction. For, after all, what is necessary is a fortiori possible. And to prove conversely that there not only can be but must be an idea corresponding to each idea, one has to presuppose Spinoza's whole ontology: the definition and existence of God; the unity of substance; the parallelism of the attributes, etc. None of that is available to the intended readers of TIE; on the contrary, that work aims to pick its readers up where it finds them and lead them gradually to the discovery of the *premises* of the Spinozist ontology, which therefore cannot be relied upon at the outset. Quite generally, the mere absence from TIE of a doctrine of the *Ethics* proves nothing.

2. Here is something more serious. In TIE 33,34 Spinoza insists that the idea of an idea is *other than* the idea of which it is an idea. In EIIp21s, on the other hand, he demonstrates that an idea and the idea of it are really *one and the same thing*. Yet here again there may be no formal contradiction, for it can happen that two things constitute only one in a certain respect while remaining distinct in another respect. And just that account of the matter is what emerges from Spinoza's demonstration of the identity of the idea with the idea of the idea: thus, EIIp21s explicitly relies on EIIp7s, where it has been established that a body and the idea of it are one and the same thing under two different attributes — which we can generalize to say that an idea and *its ideatum*, each

Y. Yovel, ed., Spinoza on Knowledge and the Human Mind, 83-91.

under its own attribute, are the same thing. The *ideatum* of the idea of an idea is just the idea, but it is the idea only considered in a certain respect, namely, in its formal essence and not in its objective reality. Accordingly, the idea of the idea, as EIIp21s itself explains, is identical (this time under the same attribute) with "the form of the idea insofar as this is considered as a mode of thinking without relation to the object"; but it is not false to say that there is at least a distinction of reason between it and the idea considered in its representative content, or the idea insofar as it is the "objective essence" of something. Indeed, how far that is from being false is shown by the fact that the doctrine of the idea's identity with the idea of it (and furthermore of the idea's identity with its *ideatum*) is expounded only in scholia which bring us to a level of "intuitive knowledge" higher than that of the rest of the *Ethics*: if we attend only to the propositions of Part II and their demonstrations, we get no help whatsoever in deciding whether the series of ideas and that of ideas of ideas (or the series of ideas and that of bodies) are one and the same series or two distinct series that merely run parallel. So even supposing (and it is not proved) that at the time of the TIE Spinoza had fully arrived at his definitive doctrine on this matter — a doctrine which, even in the context of the *Ethics*, is rather esoteric from the standpoint of the demonstrative apparatus strictly so-called — it is understandable that he does not bring it into the earlier work.

3. There is, however, something more serious still. In TIE Spinoza seems to say that *there is not even a parallelism* between the idea and the idea of the idea. Thus, in TIE 34 he winds up his analysis of relations between the idea and the idea of it by announcing two theses:

Thesis 1: "In order for me to know, it is not necessary to know that I know."
Thesis 2: "To know that I know, I must first [*prius*] know."

So Spinoza seems to say that it is possible to know without knowing that one knows: he appears to think that one knows first (*prius*), and then later one does or doesn't, as the case may be, become conscious that one knows; one first has a true idea, and then later the true idea of this true idea does or doesn't, as the case may be, make its appearance. But in EIIp43 he is quite clear about it: someone who has a true idea *knows at the same time* that he has a true idea; so it is impossible to know without knowing that one knows, and the knowledge of the knowledge is strictly synchronous with the knowledge itself. This time, don't we have an outright contradiction?

Perhaps not, after all. So far as Thesis 1 is concerned, it may be remarked that Spinoza does not say outright that it is possible to know without knowing that one knows. That thesis, at least when taken in context, means only that our knowledge of a thing's nature in no way depends, so far as its content is concerned, on premises taken from our knowledge of the nature of our knowledge of that thing: taking Spinoza's example in TIE 34, all I shall ever understand of Peter's essence will come to me purely from my true idea of that essence, and my true idea of that true idea will never be able to add any information about the essence of Peter; for the idea of a thing and the idea of that idea have two different *ideata*, and quite generally an idea can never give us knowledge of anything but its own *ideatum*. In other words, epistemology is not a part of science. But that does not necessarily mean that it could happen that one knows without knowing that one knows: the thesis is that, if that is impossible, it is not because of a supposed logical dependence of the idea on the idea of the idea; but it may be impossible for other reasons. In brief: in TIE Spinoza does not answer this question and does not even ask it, whereas in the *Ethics* he answers it in the negative.

Looking at Thesis 2 against that background, one might very well hold that in it the meaning of the adverb *prius* is purely logical and in no way chronological. Taken strictly, the thesis could be saying: "To know that I know what a thing is, I must know what that thing is, and I must know this by means of a knowledge that does not depend in any way on knowledge of knowledge." And of course that is a *part* of what it is saying. But reducing it to that would be a bit artificial, forcing the sense of the words. Anyway, it is ruled out by what comes next.

In TIE 35 Spinoza applies what he has just said to the problem of certainty. Here is what he says about this: "From this it is clear that certainty is nothing but the objective essence itself" (in other words, the idea); and he explains, "i.e., the mode by which we are aware of the formal essence [*modus, quo sentimus essentiam formalem*] is certainty itself." Spinoza does not explicitly demonstrate this; he merely says that it is clear from what has gone before. But if the idea is admitted to be necessarily contemporaneous with the idea of it, it is not at all evident what makes that conclusion immediately "clear": all that can be immediately inferred from what has gone before is that in our mind there are ideas, ideas of ideas, ideas of ideas of ideas, etc. ..., and that certainty is to be found at one or another of these levels. But there would seem to be no reason to identify certainty in particular with the idea rather than with the idea of the idea — or, if there is a reason, it does not spring to the eye. On the other hand, if we acknowledge that the idea comes chronologically before the idea of it, then indeed Theses 1 and 2 directly and immediately entail that certainty is located at the level of the idea. Here is why.

1. Let us suppose that certainty is found at the level of the idea of the idea and not at that of the idea. And let us admit the possibility of having a true idea before having a true idea of it. In that case we could, for an initial period, know

(for example) that the sum of the angles of a triangle is equal to two right angles, without yet being certain of it; we could, that is, represent the triangle to ourselves as necessarily having angles whose sum equals two right angles, the demonstration being complete and perfectly understood, while still wondering whether our inability to think the contrary really proves that the contrary is impossible. It would only be later on, after the idea of the idea put in an appearance, i.e., after an epistemological scrutiny of our idea of the triangle had shown us that it possessed all the characteristics of a true idea, that we would be certain that the triangle really has — and could not possibly lack — angles equal to two right angles. But, obviously, if that were how things stood, the transition from the earlier time to the later would have given us new knowledge *about the triangle*: learning that a thing fits our idea of it is getting information about the thing itself and not about the idea of it. But that would contradict Thesis 1: my true idea of my true idea of the triangle could not possibly tell me anything about the triangle, because the triangle is not its *ideatum*; all it can tell me is the nature of my true idea of the triangle — the nature it had before I knew what it consisted in. So, really, the only new knowledge that my idea of my idea gave me is that my idea of the triangle, all by itself, made me understand that the triangle really cannot fail to have angles equal to two right angles. But, then:

2. Since our true idea of the triangle has not changed between the earlier time and the later, there is no denying that, from the earlier time and before the idea of the idea put in an appearance, the idea of the triangle — unaided, all by itself — made us understand that the angles of the triangle have a sum equal to two right angles. For if that were not so, it would contradict Thesis 2: we would have a true idea of a true idea which informed us that we knew at the earlier moment something which in reality we did not then yet know; this idea of idea would have misinformed us, and would therefore not be true after all. So, indeed, we have to conclude that our true idea, right from the beginning, really involved certainty, all by itself, at its own level.

But this quasi-direct proof, if I have reconstructed it correctly, obviously rests on the possibility of distinguishing between two successive times, and thus on the possibility that the idea should chronologically precede the idea of it. Now, do we really have a contradiction between TIE and the *Ethics*, which asserts that such temporal precedence is impossible? No, there is not really a contradiction here. What shows that there is not is the fact that in the *Ethics* itself, *immediately after* having demonstrated in EIIp43d, that the idea and the idea of it are strictly contemporaneous, Spinoza gives us a new demonstration of the identity between the idea and certainty in EIIp43s, and in this demonstration, which is just like the one in TIE, the word *prius* appears right after an explanation whose purpose is to justify the use of that word in its *chronological* sense. If TIE were in conflict with the *Ethics* on this matter, there would also be a contradiction between the two adjacent passages in the *Ethics* — which is hardly likely. In fact, neither contradiction exists, as we shall see when we analyze these two texts.

First, EIIp43 informs us that "He who has a true idea at the same time knows that he has a true idea, and cannot doubt the truth of the thing." And the demonstration is strictly parallelist: it comes down to saying that when we have an adequate idea we necessarily have at the same time an adequate idea of that adequate idea, that is, an idea that enables us to know the truth of our true idea, and that therefore — "as is manifest through itself" — we are certain of having the truth. This demonstration is perfectly convincing, and it establishes exactly what it is supposed to establish. Still, it is not the best possible demonstration, just because it risks leading us to think that certainty exists *only* at the level of the idea of the idea. Spinoza does not say this, because it would be false: he simply says that if there is a true idea of a true idea, it is evident that there is also certainty (we could take him to mean: then a fortiori there is certainty), without presenting this sufficient condition as being also necessary. But nor does he say explicitly that the true idea involves certainty *by itself*, just as a first-level idea, and that it is not made certain by the idea of the idea. That is why, in the ensuing scholium, he faces up to the need to correct this false impression.

In the first sentence of EIIp43s, Spinoza points us in the right direction: "In P21S," he tells us, "I have explained what an idea of an idea is." As that scholium had established that the idea of the idea is nothing but the form of the idea, it obviously follows that, even if certainty is identified with the idea of the idea, we should infer that it is also identical with the idea, as that is the same thing. But that is not enough; for if we left it at that we might think that certainty is identical with the idea only considered in its role as "idea of itself" or reflexive idea, and not in its role of "idea of something." All this is straightened out in what follows.

Further on in EIIp43s Spinoza tells us that, without relying on parallelism or even on a distinction of reason between idea and idea of idea, the preceding proposition is "manifest through itself": whoever has a true idea is, ipso facto, certain of having the truth. And although this is self-evident, Spinoza gives us a two-part demonstration of it — a demonstration which is enormously like the one that can be extracted from TIE, with just two additions. The first part of this demonstration is devoted to showing, as in TIE, that insofar as there is certainty at the level of the idea of the idea, that certainty can only consist in knowledge of the fact that the idea, by itself and considered as a first-level idea, already involves certainty at its own level. All Spinoza adds are two details which had not appeared in TIE: (1) he points out to us that we always have this knowledge, at least implicitly, just because we have a true idea; and (2) he explains to us how we can have the psychological illusion that we don't have this knowledge or that we haven't always had it. In TIE, on the other hand, he did not explicitly rule out the hypothesis that, at the earlier time, we really did not have that knowledge because the idea of the idea had not yet put in an appearance. However, those two details aside, this first part of the demonstration is based on a principle which is equivalent to Thesis 1 of TIE. As for the second part of the demonstration, it is absolutely identical with the one that can be found in TIE, and is explicitly based on Thesis 2. Let us, then, examine these two parts.

The first part consists in the statement and demonstration of the following thesis: "[N]o one who has a true idea is unaware that a true idea involves the highest certainty." This is demonstrated in two stages.

In the first stage, Spinoza tells us: "[T]o have a true idea means nothing other than knowing a thing perfectly, *or* in the best way." Notice that he does not yet tell us that the true idea *is* perfect knowledge of the thing; he simply tells us that the concept of "true idea" means the same thing, or has the same content, as that of "perfect knowledge": when we think that an idea is true, we thereby think that it gives us knowledge of its *ideatum*, knowledge from which nothing is missing and which has no need to be completed by anything else — above all, not by any knowledge of knowledge (which is equivalent to Thesis 1). Now, knowing something perfectly means knowing that the real nature of this thing cannot be other than we conceive it to be; for if we did not know that, our knowledge of the thing would not be genuinely "perfect" — something essential would be missing from it. From that we can indeed infer that the concept of perfect knowledge implies that of *certainty*, and that therefore the concept of true idea implies it also.

That conclusion, of course, is true if and only if the concept of true idea is equivalent to that of perfect knowledge. Yet we can still ask whether this equivalence is really justified. The answer is that, despite appearances, it is. And that is what the second stage of the argument shows.

In this second stage, Spinoza tells us: "And of course no one can doubt this" (that is, the equivalence between the concepts of true idea and of perfect knowledge) "unless he thinks that an idea is something mute, like a picture on a tablet, and not a mode of thinking, vis. the very [act of] understanding [*ipsum intelligere*]." This is a fairly complex argument; let us analyze it.

1. It is impossible to doubt that, quite generally, "having a true idea" means

the same thing as "knowing perfectly," at least if it is granted that "having a true idea" means "understanding"; for it is obvious, and everyone knows, that "understand" means "know perfectly." Therefore:

2. Whoever has a given particular true idea cannot doubt that he has perfect knowledge of the corresponding thing unless he doesn't recognize this idea for what it is, namely, an act of the intellect. For as soon as he thinks "I understand this thing" he will necessarily think "I know this thing perfectly." But:

3. In fact, someone who has a true idea cannot be completely ignorant of the fact that it is an intellectual act (he knows it implicitly, at least), for one cannot understand without at that very instant being to some degree conscious of understanding. So someone who has a true idea cannot really doubt that he knows the thing perfectly and thus is certain. However:

4. There are people who *doubt psychologically*, or *think they doubt*, that they know perfectly the thing of which they have a true idea. When that happens, it is because, although they know that their true idea is an intellectual act, this knowledge is darkened, or veiled, or pushed to the back of their mind by an imaginative preconception concerning the nature of ideas in general. Their true idea is made known to them as an intellectual act by the true idea that they have of it; but at the same time they have an imaginative idea of an idea, which represents ideas in general to them as being analogous to pictures on a tablet. And this is because in their mind the imagining of the word "idea" is associated with the imagining of a tablet, and because, in their body, the auditory image of the word "idea" is associated with the visual image of a tablet. As these two ideas of ideas come together, their owners doubt psychologically: they waver between two contradictory assertions, according to which of the two ideas is uppermost in their mind. And this is why it can happen that, at the earlier time, these people ask if the tablet they have in their mind really conforms to the original: it is just the sort of question to be asked about a tablet. But, really, even at that earlier time, they never ask this question directly about their true idea; they ask it about the tablet which they more-or-less vaguely imagine and which they call "idea," and they apply it in an external way to their true idea when they do not have it truly present to their mind. However, when they *truly* think of their true idea, attending to it, they stop doubting. Consequently:

5. It is quite evident that if, at a later time, an epistemological reflection on their true idea allows them to eliminate their preconception (by detaching the word "idea" from the representation of a tablet), these people will come to know explicitly what they already know implicitly. They can do this without acquiring any new knowledge: the mere disappearance of their preconception (in their body as well as in their mind) will raise what they already know ("I understand this") to a higher level of consciousness.

So, at last, the original thesis — that whoever has a true idea knows that this idea involves certainty at its own level — is well established. But this is only the case if it is made clear that this knowledge can be psychologically doubted — even pushed to the margins of our consciousness — by a conflict in our minds between it and certain imaginative preconceptions, and that it can be turned into explicit knowledge only through the disappearance of these preconceptions. Well, then, can it be immediately inferred that the true idea, by itself, makes us really certain? In the case where the preconceptions have vanished, yes indeed. In the case where they have not yet disappeared, though, it is less clear what the answer is. For if we know only implicitly that we are certain, doubting this psychologically, can it truly be said that we have already achieved certainty? Can one be certain when one is still psychologically unsure that one is certain? Yes, after all; and that is what the second part of the demonstration will show.

Here is how Spinoza puts the second part of the demonstration in EIIp43s: "And I ask, who can know that he understands some thing unless he first [*prius*] understands it? I.e., who can know that he is certain about some thing unless he is first [*prius*] certain about it?" Now, that is a literal reproduction of Thesis 2 of TIE, including the two occurrences of *prius*: to know that one knows, one must first know. The very same reasoning, then, is used in TIE and here. Suppose, says Spinoza, that at first I had preconceptions about the nature of ideas and that I am now free of them. Right now I know explicitly that my true idea is really an intellectual act which, as such, really involves certainty. This knowledge that I now have is clearly true knowledge. But it can be so only if my true idea is really an intellectual act which really involves certainty. Now, my true idea has not intrinsically changed between the time when I was a victim of my preconceptions and the time when I was freed from them. Nothing has changed except the imaginative context, which changed with the disappearance of the associative tie between the word "idea" and the representation of a tablet. Since the idea in itself has stayed unchanged, it has to be admitted that my true idea, in itself, involved certainty at its own level even before I knew explicitly that it did so. The upshot is that it naturally involves certainty, just as a first-level idea, independently of what else it can teach us as a reflexive idea.

So, indeed, I could not know explicitly that I am certain unless I had — literally — *first* been certain. This is how we are to take the word *prius*, as having a chronological sense. Of course *prius* also has a logical sense ("independently of any idea of an idea"); but at the same time it means: "Even before I lost the imaginative preconception which gave me an illusory psychological impression of not knowing whether or not I was certain, and the dispelling of which has given me the illusory psychological impression of having just begun to be certain."

I conclude that there is no contradiction between TIE and the *Ethics*. In TIE Spinoza stays with immediate appearances: he does not explicitly eliminate the hypothesis that knowledge of knowledge could come later than knowledge. In the *Ethics*, on the other hand, Spinoza uses his doctrine of parallelism (which was entirely unavailable to him in TIE) to demonstrate that when one knows, one knows that one knows. But immediately after that, in EIIp43s, he shows how it is possible to provide for the immediate appearances, with which he was dealing in TIE, without going against parallelism: he explains *why* one can thoroughly imagine that one has first had a true idea, and only later a true idea of it. In fact, one has had the idea of the idea from the outset, since it is nothing but the idea itself insofar as it necessarily reflects on its own formal reality. But it presented itself to consciousness more weakly as an idea of an idea than as a first-level idea, because it ran into imaginative preconceptions concerning the nature of ideas in general. However, when these imaginative preconceptions are dispelled, it is — or tends to be — as fully present to consciousness *qua* idea of an idea as it is *qua* first-level idea. In other words, the idea and the idea of the idea are indeed contemporaneous (since they are one and the same thing), but there can be a temporal gap between the time when the idea reaches its highest degree of perfection and the time when the idea of the idea does so; for the true idea of the true idea reaches its highest degree of perfection only after a change in the imaginative context of the true idea itself. And that is exactly why methodology always comes *after* science, even if it then in turn helps science to make further progress.*

* This paper was translated by Jonathan Bennett.

[9]

Australasian Journal of Philosophy
Vol. 72, No. 1; March 1994

SPINOZA ON TRUTH[1]

Edwin Curley

I.

In *Why Does Language Matter to Philosophy?*,[2] Ian Hacking observes that

> Many scholars have stated, with intense conviction, that Spinoza held a [coherence] theory of truth. Others have urged, with comparable passion, that he held a correspondence theory.

And Hacking goes on to suggest that this dispute arises because neither coherence nor correspondence is a clear enough notion to permit us to decide which Spinoza would have preferred to use in the analysis of truth. Similarly Thomas Mark's book, *Spinoza's Theory of Truth*,[3] starts out from the same division among the commentators, argues that neither conception of truth is adequate to describe Spinoza's

[1] I originally wrote the paper which follows for a conference on Spinoza held at the Jewish Theological Seminary in New York in 1978. It was to have been published with the proceedings of that conference. Those proceedings never appeared and it is clear that they never will. Deciding how much to revise the paper for publication now (= April 1993) presented an interesting problem. I do not accept now all the assumptions I made then, but it seemed to me that there was a good reason not simply to rewrite the paper in accordance with my present views.

The paper's project is to trace the development of Spinoza's theory of truth from his earliest works to his most mature works. In 1978 I believed (as I think most other Spinoza scholars did) that the *Short Treatise on God, Man and his Well-Being* was Spinoza's earliest work, followed by the *Treatise on the Emendation of the Intellect*. In 1982 I became aware of Filippo Mignini's theory that the *Treatise on the Emendation of the Intellect* was in fact earlier than the *Short Treatise*. [See Filippo Mignini, 'Per la datazione e l'interpretazione del *Tractatus de Intellectus Emendatione*', *La Cultura* 18 (1980) pp.223-273, and his edition of the *Short Treatise, Korte Verhandeling van God, de Mensch, en deszelvs Welstand* (L'Aquila: L.U. Japadre, 1986). I discussed Mignini's theory in my edition of Spinoza's *Collected Works*, vol. I (Princeton: Princeton University Press, 1985).] Although I am not sure that I would accept all of Mignini's arguments for that theory, I do find the difficulties I encountered in writing this paper on the more traditional assumption to be persuasive evidence for Mignini's theory. On the usual assumption about order, we find Spinoza first adopting a correspondence theory, then rejecting it, and finally returning to it. On Mignini's theory, the development of his ideas would proceed more smoothly. Because my paper, as originally written, illustrates the difficulty of proceeding on the usual assumption, I have felt that I should not rewrite it, but let it stand as an argument in favour of Mignini's theory.

Some of the material in the paper (e.g., fn. 12 and the paragraph it is attached to) comes from a revision done in the early 1980s, at a time when I still thought the proceedings of the conference would be published. Any material added subsequently will be in brackets and dated.

[2] Ian Hacking, *Why Does Language Matter to Philosophy?* (Cambridge: Cambridge University Press, 1975) p.131. Where I have supplied the term 'coherence' in brackets in the first sentence, Hacking's text has 'correspondence'. Clearly there is something amiss with the proofreading, probably in the first sentence, possibly in the second.

[3] Thomas Mark, *Spinoza's Theory of Truth* (New York: Columbia University Press, 1972). The ontological theory of truth holds that in addition to being a property of ideas and statements,

position, and presses on us a third alternative, which he calls 'the ontological conception of truth'.

In *Spinoza's Metaphysics* [4] I enlisted in the ranks of those who ascribe to Spinoza a correspondence theory — not passionately, or with intense conviction, but as I now think, altogether too casually. What I should like to do here is to reconsider the question. For it seems to me that none of the recent discussions of it, including my own, are very satisfactory. Perhaps this is partly for the reasons that Hacking and Mark suggest, but surely one respect in which the controversy is unsatisfactory is methodological: it has fed on an unbalanced diet of texts, drawn primarily from the *Ethics*, with a little help from the *Treatise on the Emendation of the Intellect* and the *Metaphysical Thoughts*, but without much attention to the *Short Treatise* or the correspondence or to the possibility that Spinoza's position may be significantly different in different works. My aim is to remedy this by focussing attention on some of the texts I think have been neglected, beginning with the *Short Treatise* and continuing through the earlier works, up to the *Ethics*, which I regard as having been substantially completed by 1665. My hope in undertaking this developmental approach is that it will tell us more about the latent processes of thought which lay behind the geometrical method than we would learn by rummaging among the works of Spinoza's medieval predecessors. But I will not claim that the story I tell shows that Spinoza had a clearcut allegiance to any of these theories of truth as they are now understood.

II.

First, then, the *Short Treatise*. Spinoza takes up the topic of truth and falsity in chapter 15 of Part II, beginning with definitions of truth and falsity:

> Truth, then, is an affirmation (or denial) concerning a thing, which agrees with the thing itself.
> Falsity is an affirmation (or denial) about a thing which does not agree with the thing itself.[5]

Taken by itself this text certainly sounds as though it expresses a correspondence theory of truth and it has the advantage over the similar sounding statement in the *Ethics* (IA6) in that it gives us explicit definitions rather than an axiom.

We might, of course, define a correspondence theory very narrowly, so that someone couldn't be said to have a correspondence theory of truth unless he were

[3] *continued...*

truth is also a property of actions and physical objects. Cf. pp.85-89. According to Mark this conception of truth 'need not be a rival of the correspondence theory; it can take the form of a more general theory which includes the correspondence theory as a special case'.

[4] Edwin Curley, *Spinoza's Metaphysics* (Cambridge, MA: Harvard University Press, 1969), see particularly pp.56, 122-126.

[5] Spinoza, *Opera* (ed.) C. Gebhardt (Heidelberg: C. Winter, 1925) 1/78/20 (i.e. volume I, page 78, circa line 20). Henceforth all references to Spinoza's works will be given in parentheses to the volume, page and line numbers of the Gebhardt edition. Translations are my own.

operating with a very explicit metaphysic of propositions and facts, and accounted for truth in terms of some well-defined relation between these entities, or unless he formulated his theory of truth in a philosophic context in which alternative theories were well-developed and well-known. But the more usual view is that Aristotle expressed the heart of the correspondence theory when he wrote that:

> To say of what is that it is not, or of what is not that it is, is false, while to say of what is that it is, and of what is not that it is not, is true [*Metaphysics* 1077b26],

as did Aquinas, when he adopted Isaac Israeli's formula:

> Truth is the adequation of thing and understanding,[6]

or Descartes, when he wrote that:

> One can certainly explain, to those who don't understand the language, what the word truth means by telling them that in its proper signification it denotes the conformity of thought with the object . . .[7]

If these can all properly be counted as various ways of expressing the same central conception, then it looks as though there are as good grounds for ascribing a correspondence theory of truth to Spinoza as there are for ascribing it to any historical figure.

A word of caution, however, is necessary. Consider the following three quotations, this time from 20th Century writers. The first is an attack on pragmatic theories of truth:

(1) When you ask what it is which makes each idea right or wrong, you cannot exclude its agreement or its discord with fact other than my will . . . In selecting my means I am forced to consider their relation to the facts, and if my idea works, it is because of this relation, which is not made by my idea. And it is in this relation that we have to seek the distinctive nature of truth.

The second passage lays down a condition any adequate theory of truth must satisfy:

(2) The truth or falsehood of a belief always depends on something which lies outside the belief itself. If I believe that Charles I died on the scaffold, I believe truly, not because of any intrinsic quality of my belief . . . but because of an historical event which happened two and a half centuries ago . . . truth and falsehood are properties of beliefs . . . dependent on the relations of the beliefs to other things . . .

[6] Aquinas, *Disputed Questions on Truth*, qu. I, art. 1.

[7] Letter to Mersenne, 16 October, 1639. Of course, some interpreters of Descartes want to ascribe a coherence theory of truth to *him* also. E.g., H. Frankfurt, *Demons, Dreamers and Madmen* (Indianapolis: Bobbs-Merrill, 1970). I have argued against this interpretation in *Descartes Against the Skeptics* (Cambridge, MA: Harvard University Press, 1978).

The third runs:

(3) A judgment e.g., is true, if the thoughts whose union is the judgement 'correspond' to the facts whose union is the 'real' situation which is to be expressed. *My* judgment is true if *my* ideas, as asserted by me in my judgment, correspond to the facts.

One of these passages comes from that paradigmatic correspondence theorist, Bertrand Russell. But the other two come from paradigmatic coherence theorists, F.H. Bradley and H.H. Joachim. And while differences of literary style, technical vocabulary and punctuation may help you to decide which was which, you should find it difficult to decide on grounds of the philosophic content alone of these passages, taken in isolation from the contexts in which they occur.[8] We should not, then, get too excited if Spinoza sometimes says things which *sound* very like standard formulations of the correspondence theory. There is grist here for Hacking's mill. Differences in what Spinoza goes on to say may put him very far from Russell, Moore and the younger Wittgenstein.

Well, what does Spinoza go on to say in this passage from the *Short Treatise*? His next step is to draw one apparent consequence of his definitions of truth and falsity:

> Either there is no distinction between the false and the true idea, or there is no real distinction between them, but only a distinction of reason, because affirming or denying this or that are only modes of thinking, and have no other distinction between them than that the one agrees with the thing and the other does not. [I/78/25]

This seems to mean that on Spinoza's definitions the only difference between a true idea of an object and a false idea of that object will be in their relation to the thing they make an affirmation or denial about, and that this entails that there will be no *intrinsic* difference between, say, my true idea that the tower is round (when in fact it is round) and my false idea that it is round (when in fact it is square). If this is what Spinoza means, then it seems right (though perhaps it commits Spinoza to denying one form of the idealist doctrine that all relations are internal). But Spinoza does not commit himself to this consequence. He continues in a hypothetical vein:

> And if that should be so, one could rightly ask what advantage one man has with his truth, and what harm the other has through his falsity? And how will the one know that his conception or idea agrees more with the thing than the other's does? Finally, how is it that the one errs and the other does not? [I/78/30]

[8] (1) comes from Bradley's *Essays on Truth and Reality* (Oxford: Clarendon Press, 1914) p.77; (2) from Russell's *Problems of Philosophy* (New York: Oxford University Press, 1959) p.121; (3) from Joachim's *The Nature of Truth* (New York: Greenwood, 1969) p.19.

Spinoza responds to the second of these three questions, first, by announcing a doctrine familiar to readers of the *Treatise on the Intellect*:

> The things that are clearest of all make known both themselves and also falsity . . . For because they are clearest of all, there can never be any other clarity through which they could be explained. So it follows that truth manifests both itself and falsity . . . but falsity is never manifested or indicated through itself. [I/79/5]

And he goes on to draw the consequence that someone who has the truth cannot doubt that he has it.

Though familiar, this response is nonetheless puzzling. *Prima facie* it can only provide an answer to the question posed if we identify possession of a true idea with possession of an idea which is clearest of all (*aller klaarste*). And there does not seem to be any reason to do that, so long as we define truth as Spinoza did in chapter 15.

We might try to alleviate the puzzle by harking back to earlier chapters of the *Short Treatise* and recalling that Spinoza's project at this point is to discover the effects of 'true belief', which is, in fact, the *Short Treatise*'s analogue of *ratio*, and is defined at one point as

> a strong proof, based on reasons, by which I am convinced that the thing truly is, outside my intellect, such as I am convinced in my intellect that it is. [I/29/25]

If Spinoza were to mean only that true belief *in this sense* manifests itself, then we should not be so surprised at his confidence that someone who has the truth knows that he has it. We would not be dealing with just any true idea. No doubt this expedient removes the paradox only at the expense of having Spinoza switch definitions in mid-stream. But I observe that Spinoza appears to be guilty of a similar equivocation in the *Ethics* (IIP43D), where he treats IIP34 as if it established, not merely that adequacy entails truth, but also that truth entails adequacy.

We might note in passing that the very form of the question Spinoza raises — how will the one know that his idea agrees *more* with the thing than the other's does? — suggests that agreement, and hence, truth, will be a matter of degree, a doctrine which certainly brings to mind the subsequent teachings of the coherence theorists.

And this seems congruent with Spinoza's explanation of error here. Pursuing the question, 'Why the one [who has a true idea] is more aware of his truth than the other?', Spinoza replies that this happens

> Because the idea of affirmation (or denial) in the first agrees completely with the nature of the thing, and consequently has more essence . . . . note that the intellect . . . is a pure passion, i.e. that our soul is changed in such a way that it acquires other modes of thinking it did not have before. Now if someone, because the whole object has acted in him, acquires such forms or modes of thinking, it is clear that he acquires a completely different sensation of the form or quality of

> the object than another who has not had so many causes, and so is moved to affirm or deny by a different, and slighter action (since he becomes aware of it in himself by a few, or lesser, affections). [I/79/25-30]

Now it looks as though *some* of what is said here is to be dismissed as expressing a doctrine Spinoza abandoned. Certainly it is hard to imagine the Spinoza of the *Ethics* saying that the intellect is a pure passion.[9] On the other hand, it also looks as though we have here the germ of Spinoza's later conception of an adequate idea.[10] That is to say, we have implicitly an anticipation of the claim made in the *Ethics*, that falsity consists in the privation of knowledge which inadequate ideas involve (*E* IIP35). At least one difference (if not the only difference) between the man who has a true idea and the man who has a false idea is that the one has a complete idea of the thing and the other has only a partial idea of it. The completeness envisaged here does not appear to imply that the man with a true idea is aware of all of the relations of the object to other objects. Rather, the completeness of his idea is a function of the extent to which the object acts on him via sensation. The possession of a true idea does not require knowledge of the whole of reality.

One result of this is that Spinoza is not so much concerned with the question 'Whether we, being finite, ever have a true idea?' as he is with the question 'Whether we ever have false ideas?'.[11] This comes out clearly in the subsequent chapter of the *Short Treatise* (II, 16), when he explores some further consequences of the doctrine that understanding is a pure passion:

> We have said that understanding is a pure passion, i.e., a perception in the soul of the essence and existence of things. So it is never we who affirm or deny something of the thing; it is the thing itself which affirms or denies something of itself in us. [I/83/15]

Spinoza conjectures that Cartesians will reject this, because they confuse mental affirmations or denials with linguistic utterances, in which we can say anything we please. But once that confusion is cleared up, he must still deal with a further objection:

> If it is not we, but only the thing, which affirms or denies [something] of itself in us, then nothing can be affirmed or denied except what agrees with the thing. So there is no falsity. For we have said that falsity is affirming (or denying) something of a thing which does not agree with the thing, i.e. that the thing does not affirm or deny that of itself. [I/83/35]

[9] Cf. *E* IID3Exp. Wolf, in his edition of the *Short Treatise* (New York: Russell and Russell, 1963) p.221, says we should not take this expression too literally, and perhaps he is right, though I find it difficult to be certain.

[10] I say 'later' because the Dutch term (*evenmatig*) which translates *adaequatus* in contemporary Dutch translations of Spinoza's Latin works does not seem to occur in the *Short Treatise*. Nor is there any other term which seems to require translation into English by the term *adequate*.

[11] Cf. Mark's criticism (relying on different texts) of the coherence theorists' interpretation of Spinoza, *op. cit.*, pp.46-50.

It is interesting to note that in presenting this objection Spinoza reaffirms the formula that earlier sounded so much like an expression of the correspondence theory, but goes on to gloss it in a way which presupposes his doctrine that a mode of extension and the idea of that mode are one and the same thing, expressed in two different ways (*E* IIP75). The object, which in this context seems to be a physical object of sense perception, can be considered as affirming or denying something of itself, and our thought about it is true if our thought affirms of the thing what the thing affirms of itself.

By the way, I still think, as I did when I wrote *Spinoza's Metaphysics*,[12] that this does not, for Spinoza, entail that ordinary physical objects, like stones, are 'thinking things' with any degree of consciousness. And I still think (partly because of the passage under discussion here) that the general line of interpretation of the relation between thought and extension that I suggested then has something in it. But I cannot argue that here.

Now I suppose that what the thing 'affirms' of itself necessarily agrees with itself (cf. *KV* II, 20, I/97) and that this is what Spinoza has in mind when he writes in the *Ethics* that 'all ideas, insofar as they are referred to God, are true' (IIP32). But this kind of agreement does not seem to be what Spinoza has in mind here when he considers the objection that his theory denies the existence of falsity. The point of the objection seems to be that the only way I can have a thought about an object is to be acted on by that object, and that if the object does cause my thought, there must be an agreement between my thought and the object, conceived as affirming something about itself. In any case, I do not find Spinoza's answer very clear:

> If we attend properly to what we have already said about truth and falsity, we shall see this objection satisfactorily answered. For we have said that the object *is* the cause of our affirmation or denial about it, whether [that affirmation] is true or false, and that falsity arises from this: perceiving only something or a part of the object, we imagine that the object affirms or denies this of itself as a whole. [I/84/5; I/581]

Spinoza goes on to say that this is most common in 'weak souls' which

> . . . very easily receive a mode or idea through a very slight action of the object on them . . .

[12] Though I now think I have better grounds for this conclusion than I had then. Then I invoked Letter 56 (*Spinoza's Metaphysics*, op. cit., p.126). Now I would appeal primarily to *E* VP39S, which seems to be Spinoza's most explicit statement about consciousness. There Spinoza makes consciousness a function of the aptitude of the body to do many things and its independence of external causes. But I still think that he also associates consciousness with having an idea of one's ideas. Margaret Wilson has challenged this (in a paper in *The Philosophy of Baruch Spinoza* (ed.) Richard Kennington (Washington: Catholic University of America Press, 1980) and in *Spinoza's Metaphysics* (p.128) I offered no direct textual evidence for it. Now I would contend that the association is presupposed by the demonstration of *E* IVP8. As yet I have no reply to her other criticisms of that chapter. [I am still in the same position in April 1993.]

but that

> . . . apart from this there is no other affirming or denying in them.

My main difficulty with this passage is that it looks as though my imagining that the object affirms something of itself as a whole ought to be an affirmation distinct from anything the object does affirm of itself, whereas Spinoza seems to say that the mind does not contain any affirmation about the object except those that the object affirms of itself in us. But then I encountered essentially the same difficulty years ago, relying primarily on passages from the *Ethics*, and I'm still inclined to think this indicative of a problem in Spinoza's system rather than in my way of understanding it.[13]

Let me now try to sum up the results of our investigation so far. Spinoza does, I think, begin in the *Short Treatise* with a correspondence theory of truth. The version of that theory expounded there may be expressed in a somewhat idiosyncratic way, insofar as Spinoza glosses the agreement of idea and object in terms which presuppose his peculiar metaphysical scheme, but it still bears a recognizable family resemblance to other pre-20th Century versions of the theory. However, Spinoza does seem to be drawn very quickly, by his desire to avoid skepticism, into positions which I would take to be incompatible with the correspondence theory as conceived, say, by Bertrand Russell. Hence, he wants to be able to say (though he does not say explicitly) that there is an intrinsic as well as an extrinsic difference between true and false ideas, that the difference between them is never *merely* a difference in their relation to their object. This I take to be the import of his doctrine that the truth manifests both itself and falsity, and of his burgeoning doctrine of adequate ideas. And this development does, I think, lead him to adopt a doctrine of degrees of truth which is similar to, though not identical with, that of later coherence theorists. But so far, at any rate, that agreement does not spring from any tendency to locate the essence of truth in a relation that ideas have to one another.

III.

I now turn to the *Treatise on the Intellect*, which I propose to deal with more briefly, not because it is less rich as food for our thought on this topic, nor because it is easier, but because I find this work far too mysterious to say much about it with confidence.

One thing, however, is clear: in §69 Spinoza does say explicitly what was only implicit in the *Short Treatise*, that there is an intrinsic difference between true and false ideas:

> As for what constitutes the form of the true, it is certain that a true thought is distinguished from a false one not only by the extrinsic, but chiefly by the intrinsic denomination. For if some architect conceives a building in an orderly fashion,

[13] Cf. *Spinoza's Metaphysics*, pp.129-137.

> then although such a building never existed, and even never will exist, still the thought of it is true, and the thought is the same, whether the building exists or not. On the other hand, if someone says, for example, that Peter exists, and nevertheless, does not know that Peter exists, that thought, in respect to him, is false, or if you prefer, is not true, even though Peter may exist. [I/26/15-25; §69]

So far as I can see there is no way in which this passage can be fitted into an interpretation of Spinoza as a correspondence theorist regarding truth.

The opening sentence offers some reason to think that Spinoza has not completely abandoned the initial position of the *Short Treatise*, insofar as it seems to allow that there will also be an extrinsic difference between true and false ideas, even if it is not so important as the intrinsic difference between them. 'Not only *p*, but chiefly *q*' seems, at least, to entail *p*. So far the compromise position implicit in the *Short Treatise* is available. All and only true ideas have the relational property of agreeing with an object, but fortunately for the epistemologist they all also have an intrinsic property which enables us to distinguish them from false ideas, But the continuation of the passage undermines that compromise. The examples indicate, ever so plainly, that the existence of a corresponding object is neither necessary nor sufficient for the truth of a thought.[14]

Why is there this apparent shift? It is not sufficient here, I think, to invoke a concern with skepticism, for the compromise position would be sufficient to deal with skeptical anxieties (or would be, at least, if there were some plausible reason for supposing that there really was an intrinsic difference between true and false ideas). I can only suggest the possibility that reflection on the nature of mathematical truth may have prompted Spinoza to abandon the correspondence theory altogether. Certainly Descartes, in expounding the ontological argument, had wanted to emphasize that mathematics provided him with countless ideas of things which, even if they had never existed outside him, had 'true and immutable natures' which did not depend on his mind.[15] So he would apparently allow that an idea might be true even if it had no corresponding object actually existing at any point of space or time.

One question this raises is: 'What definition of truth would make sense of this possibility?'. And perhaps the Leibnizian-sounding definition of Descartes' Geometrical Exposition is intended as an answer to this question:

> When we say that something is contained in the nature or concept of a thing that is the same as if we said that it is true of this thing.[16]

[14] So perhaps Elwes was on the right track when he rendered the first line: '. . . a true idea is distinguished from a false one *not so much* by its extrinsic object as by its intrinsic nature'. Appuhn and Caillois, however, both have something equivalent to the translation offered in the text. Paul Eisenberg, in a translation published in *Philosophy Research Archives* 3, no. 1170 (July 1977), has '. . . not merely . . . but especially . . .' which is perhaps the best rendering. 'Not merely *p* but especially *q*' does not so clearly entail *p*.

[15] Fifth Meditation, AT VII, 64.

[16] AT VII, 162. Spinoza, of course, reproduces this definition in *his* geometrical exposition of Descartes (I/150-151), but neither he nor Descartes makes it clear whether *truth* is the definiens or the definiendum.

This might explain why, in the passages immediately following §69, Spinoza is concerned with mathematical objects, why he emphasizes that mathematical truth is independent of the existence of its objects, and why in §72 he says that

> falsity consists only in this: that something is affirmed of a thing which is not contained in the concept we have formed of it. [II/27/25]

I hasten to add that there is a good deal in the detail of these passages which I find difficult to understand, in these or any terms.

In any case, it would appear that Spinoza is moving very far now from a conception of truth in terms of correspondence. And in some passages it seems that he may be moving towards a conception of truth in which the notion of membership in a coherent system would play an important part. Consider the thesis that truth manifests both itself and falsity. This thesis does reappear in the *Treatise on the Intellect*, sometimes, I think, without being greatly illuminated (e.g. in §§35 and 36). But there is one passage which I am inclined to read as providing us with a helpful gloss on it:

> When the mind attends to a fictitious thing which is false by its very nature, so that it considers it carefully, and understands it, and deduces from it in good order the things to be deduced, it will easily bring its falsity to light.
> And if the fictitious thing is true by its nature, then when the mind attends to it, so that it understands it, and begins to deduce from it in good order the things which follow from it, it will proceed successfully, without any interruption — just as we have seen that, from the false fiction just mentioned, the intellect immediately applies itself to show its absurdity . . . [II/23/25ff; §61]

I take this to mean that it is of the nature of truth that if you start from a true idea, you will be able to deduce consequences from it indefinitely without encountering absurdity, whereas this is not possible with a false idea. I suppose that that is a plausible view if you restrict yourself to mathematical truth and that it comes close to what the coherence theorists had in mind. Of course, '. . . is a possible member of a coherent system of ideas' does not sound like an *intrinsic* property of ideas; but if you take the view that a proposition's logical consequences are 'contained' in it, then this property would have its basis in what is internal to the idea.

So far, then, I conclude that the dispute between those who ascribe a correspondence theory to Spinoza and those who ascribe a coherence theory to him does not arise simply from the unclarity of these categories. Each interpretation has support in Spinoza's texts, depending on which texts you choose.

IV.

At this point chronology dictates that we take up two texts from the period between the *Treatise on the Intellect* and the *Ethics*: one in the *Metaphysical Thoughts* and another in the letter written a few months before the *Metaphysical Thoughts* were published, but probably not long after that work's composition. I shall treat them as contemporaneous.

The first is interesting because in it Spinoza seems to revert, unequivocally, to a correspondence theory of truth:

> To perceive these two, the *true* and the *false*, rightly, we shall begin with the signification of the words, from which it will be plain that these are only extrinsic denominations of things and are not attributed to things except metaphorically. But since ordinary people first invent words, which afterwards are used by the philosophers, it seems desirable for one seeking the original signification of some term to ask what it first denoted among ordinary people — particularly where other causes which would be drawn from the nature of the language are lacking. [*CM* I, 6; I/246]

So Spinoza is engaged in a bit of conceptual analysis, one with a beguilingly modern methodology: ask first what a term means in ordinary language and then consider how that usage may have been extended by the philosophers.

> The first signification of the *true* and *false* seems to have had its origin in stories: a story was called true when it was of a deed which had really happened, and false when it was of a deed which had never happened. Afterwards philosophers used this signification to denote the agreement of an idea with its object and the contrary. So an idea is called true when it shows us the thing as it is in itself, and false when it shows us the thing otherwise than it really is. For ideas are nothing but narratives, or mental histories of nature. [*Ibid.*]

Spinoza goes on to say that the (common medieval) attribution of truth to things other than ideas and stories is a confused and illegitimate metaphorical extension of the philosophical usage. In so doing he rejects — so far as I can see — Professor Mark's ontological theory of truth, the theory that in addition to being a property of ideas and statements, truth is also a property of actions and physical objects.[17]

Interesting as this passage may be, however, it is deeply puzzling. Why does Spinoza seem to go back to an account of the nature of truth which he had apparently embraced in the *Short Treatise* and rejected in the *Treatise on the Intellect*? Is he, in the *Metaphysical Thoughts*, merely expounding the views of someone else (Descartes? the scholastics? his own self at an earlier date?), views he did not hold himself, or at least, did not hold when he published the *Metaphysical Thoughts*? Or is the *Treatise on the Intellect*, in this respect, an aberration from which Spinoza recovered?

The correspondence suggests that the doctrine of the *Treatise on the Intellect* is aberrant. Recall the problem set for Spinoza in Letter 8 by his friend, Simon de

[17] It is true, of course, that Spinoza does adopt the Augustinian formula that God is truth (I/63/1-2). But since he also seems to think of that identification as being explained by the argument of *KV* II, 15 (cf. I/79/15), it needs some argument to show that the similarity between Spinoza and the Christian Neoplatonists is more than superficial.

Vries: should we accept Borrelli's view of definitions (that the definition of a subject must consist of a property or construction which is first, essential, best known to us, and true) or that of Clavius (that it does not matter whether a definition is first, or best known, or true, so long as the definition given is not said to agree with a thing unless it has first been shown to agree with that thing (IV/40/9-15)). Spinoza undertakes to go between the horns of the dilemma by drawing a distinction between a definition

> which serves to explain a thing whose essence only is sought, as the only thing there is doubt about, and one which is proposed only to be examined. For because the former has a determinate object, it ought to be true. The latter does not require this. [IV/42-43]

These two kinds of definition he subsequently characterizes as those which 'explain a thing just as it is [NS: in itself] outside the intellect' and those which explain it 'as we conceive it, or can conceive it'. As an example of the first, Spinoza cites a description of the Temple of Solomon, which ought to be 'a true description of the temple as it was'. With this is contrasted a description of a temple I have conjured up in my mind, as one I want to build.

This passage is usually attended to for what it says about Spinoza's concept of definition. But it is relevant to our present concerns more for what it says about his concept of truth. The requirement that a definition of the first kind be true is clearly treated as equivalent to requiring that it bear a certain relation to external reality, and this fits in well with the account in the *Metaphysical Thoughts*. Moreover, Spinoza seems here to retreat from the position adopted in §69 of the *Treatise of the Intellect* regarding ideal objects. There the thought of a non-existent building was true, whether the building existed or not, so long as it was conceived in an orderly fashion. Here the description of a non-existent building is again subject only to constraints internal to the world of thought — it must at least be of a conceivable object — but it is not conceived *sub ratione veri*, which I take to mean that the concept of truth is not applicable to it. Spinoza's reason for exempting definitions of this second kind from the requirement of truth is that they lack 'a determinate object'. So Spinoza's thoughts about truth and ideal objects appear to have undergone some development from the *Treatise on the Intellect*. And the passage which caused the greatest difficulty for the interpretation of Spinoza as a correspondence theorist is at least partly repudiated.

V.

I pass now to the *Ethics*, which I propose to treat briefly, partly because I don't, at this stage, see any reason to retract what I have already said in my book about the main passages.[18] But I would like to comment on two passages which I did not discuss then, and which I think have generally been neglected in discussion of Spinoza's theory of truth.

[18] [Still true in April 1993.]

The first is a gloss on IA6 which Spinoza provides us with when he deploys that axiom in the demonstration of IP30:

> A true idea must agree with its object (by A6), i.e. (as is known through itself) what is contained objectively in the intellect must necessarily exist in nature.

Spinoza uses this axiom (fallaciously, I think)[19] to demonstrate that the intellect, whether finite or infinite, must comprehend God's attributes and affections, but nothing else. What interests us here, however, is his explanation of what the axiom means. And that explanation seems to confirm that Spinoza is thinking of truth in terms of a correspondence between thought and reality.

On the face of things, however, the other passage I want to call attention to contradicts this. I refer to IP8S2, where Spinoza is explaining why men would regard P7 ('That it pertains to the nature of substance to exist') as axiomatic if they attended to the nature of substance. He begins by paraphrasing his definitions of substance and mode:

> By substance they would understand what is in itself and is conceived through itself, i.e. that the knowledge of which does not require the knowledge of another thing. By modifications they would understand what is in another, and those things whose concept is formed from the concept of the thing in which they are.

So far there is nothing to greatly surprise us.[20] But as Spinoza proceeds, we seem to have a reversion to the doctrine which I found most puzzling in the *Treatise on the Intellect*:

> That is why we can have true ideas of non-existing modes, since, although they do not actually exist outside the intellect, their essence is nevertheless so comprehended in another that they can be conceived through it. But the truth of substance is not outside the intellect unless it is in the substances themselves, because they are conceived through themselves.

Spinoza then proceeds to explain that it would be absurd for us to claim to have a true idea of substance and at the same time to express doubt as to whether a substance corresponding to our idea existed. But what interests me here is the clear implication that an idea of a mode might be true without such correspondence.[21] Why does Spinoza think this? I suppose that he thinks the following mathematical possibility has its analogues in nature. We can, in a weak sense of the term, conceive of geometrical figures which we have never experienced, and which, for all

[19] The reference to truth in this premise of the demonstration drops out in the conclusion. So far as I can see, the conclusion should read: 'The intellect, insofar as it has true ideas, must etc.'

[20] Though, as I've observed in *Spinoza's Metaphysics* (p.15), I believe the gloss on the definition of substance is significant.

[21] I observe, in passing, that our text is closely analogous to IIP8 and its Scholium. Since the latter passage no doubt contains the key to Spinoza's doctrine of the eternity of the mind, we might hope to make headway with that notoriously obscure doctrine if we could understand IP8S2.

we know, never have been or will be instantiated in nature, i.e., we can construct logically consistent definitions of such figures. Our knowledge of the first principles of geometry (axioms, definitions and postulates) enables us to demonstrate various properties of these figures in spite of their non-existence. In so doing we come to conceive them in the stronger sense which Spinoza deploys in his definitions of substance and mode and which he warns us of here, i.e. we come to know certain truths about them. The truth of these truths consists simply in their deducibility from the relevant first principles and definitions. I take it that something like this philosophy of mathematics and mathematical truth is implicit in Descartes and is accepted by Spinoza as early as the *Treatise on the Intellect*, though not without some apparent vacillation in subsequent writings. And I suppose that he thinks that the fundamental laws of physical nature permit us to make similar truth claims about non-actual, but possible, physical objects.

If this is right, then Spinoza's rejection of the apparent requirements of a correspondence theory in certain passages of the *Ethics* is accompanied by a conception of truth which certainly does have affinities with the idealist notion that the truth of an idea is, in some fundamental way, connected with its membership in a system of ideas. Note that I say only that there is a certain affinity here. In an interesting article Jonathan Cohen has argued that the classical idealist form of the coherence theory of truth embodied a much stronger notion of coherence than has been typical among positivist adherents of the theory.[22] The weak notion of coherence characteristic of positivist coherence theories allows a set of propositions to be coherent if its members are mutually consistent and as comprehensive as possible. The strong notion of coherence characteristic of idealist coherence theories requires not merely mutual consistency and comprehensiveness, but mutual entailment. This is most clearly stated by Blanshard, who writes that

> Fully coherent knowledge would be knowledge in which every judgment entailed, and was entailed by, the rest of the system.[23]

It should be evident that, while Spinoza would have thought that the ideal system of knowledge involved something more than just a set of consistent and comprehensive propositions, he could not have accepted so strong a notion of coherence as Blanshard embraces. The truths constituting an ideal system of knowledge would have to be so ordered that each of them was involved in some entailment relations with other members of the system, either as ground or as consequence. But if every truth entailed every other, it would follow, for Spinoza, that each element of reality depended on every other. And that is a manifestly unacceptable consequence. Spinoza's vision of reality is hierarchic, not democratic, as Blanshard's is. Modes depend on substance (or its attributes). They exist in it and are conceived through it. But the converse relation does not hold. Substance (or its attributes) is in itself and is conceived through itself.

[22] L. Jonathan Cohen, 'The Coherence Theory of Truth', *Philosophical Studies* 34 (1978) pp.351-360.

[23] Brand Blanshard, *The Nature of Thought* (New York: Humanities Press, 1939) vol. II, p.264.

[So truths about substance or its attributes should constitute the ultimate premises of the ideal system of knowledge, which are not derived from any other truths, though all other truths are in some way derived from them. If the account of Spinozistic causation and necessity which I offered in chapters 2-3 of *Spinoza's Metaphysics* is correct, not every truth will follow from the truths about substance or its attributes alone; truths about finite things will follow from truths about the attributes only in conjunction with other truths about finite things. (Contrast IPP21-23 with IP28.) But every truth will be linked by entailment relations with every other truth. Added April 1993.]

One final note about the *Ethics*: I spoke cautiously above about Spinoza's 'rejection of the apparent requirements of the correspondence theory'. Since Spinoza is capable of combining *prima facie* statements of the correspondence theory (not only IA6, but also IP30D) with *prima facie* rejections of it (in IP8S2) in the same part of the same work, it looks as though he saw no inconsistency in this conjunction of positions. Why this should be so is a puzzle to which I have no answer.[24]

What are we to conclude from all of this? The difficulty, it seems to me, is not primarily that the notions of correspondence and coherence are too unclear for us to say which Spinoza would have preferred to use in the analysis of truth, or that it is anachronistic to raise this question. No doubt both notions could be much clearer than they are, and the customary contrast between the two theories is weakened by the disposition of some coherence theorists to grant the partial truth of the correspondence theory. And no doubt there is some anachronism involved in supposing that, when Spinoza says things that sound like expressions of a correspondence theory, he is aware of taking sides in a debate. As Mark has noted, the coherence theory, as we know it is essentially a creature of the 19th Century. Nevertheless, both notions do seem clear enough to allow the following conclusions:

1. That Spinoza does persistently, from his earliest work up to and including the *Ethics*, embrace a version of the correspondence theory.
2. That he also frequently, from his earliest work up to and including the *Ethics*, expresses himself in a way that is very difficult to reconcile with any version of the correspondence theory.
3. That the philosophic problems which apparently lead him away from a correspondence theory apparently also lead him in the direction of a version of the coherence theory, though a version quite different in character from the main contemporary versions of that theory.
4. That Spinoza does not appear to have been aware of these conflicting tendencies in his thought.

ADDENDUM

I wrote this paper partly because it seemed to me that historians of philosophy often did not consider sufficiently seriously the possibility that the figures they studied

[24] [Still true in April 1993.]

might have changed their minds over time. In the interim I have worked extensively on Hobbes, who provides another illustration of the phenomenon. Hobbes scholars often write as if there were no importance to be attached to the dates of his works, e.g., as if *Leviathan* (1651) did not differ significantly from *De Cive* (1642). It is ironic that this paper itself is one more example.

I have discussed the relation of Leibniz's theory of truth to those of Descartes and Spinoza in 'Der Ursprung der Leibnizschen Wahrheitstheorie', *Studia Leibnitiana* 20(1988) pp.160-174. I intend to publish an English language version of that paper as soon as I can resolve certain difficulties in it which trouble me.

University of Michigan

Received August 1992
Revised April 1993

[10]

Spinoza on error

Jonathan Bennett
Syracuse University

J.L. Austin once said to me, of a certain philosophical opinion of Leibniz's, that "It is a very great mistake, and only a very great philosopher could have made it." In that paradoxical remark there is at least this much truth: a philosopher can be led into error by the very power of his thought, making serious mistakes that he might not have made if he had seen less and probed less deeply. In this paper I shall illustrate that with an example, namely Spinoza's profoundly wrong view that what we ordinarily call error is really a species of ignorance.

My account of this will involve Spinoza's doctrine of systematic parallelism between the mental and the physical. I had better play safe by declaring my understanding of that doctrine — what it is and why Spinoza held it. That will take two sections.

Naturalism and rationalism

Spinoza was a naturalist about mankind: he didn't think that we are radically special in any way at all. The whole story about people, he held, can be told with concepts that are needed, anyway, to describe other parts of Nature. Not just other organisms; the concept of life itself has no *basic* place in the true story of the universe, and the difference between organic and inorganic is like that between complicated and simple, or orderly and jumbled — a smooth difference of degree with no ultimate significance.

What about minds? Spinoza followed Descartes in holding that there is a mental side to each of us, and that it is genuinely additional to the physical side. He also held — as Descartes sometimes did — that there could not be a causal flow between physical and mental. If we look at how mentality seems to be distributed on our planet, it seems reasonable to suppose that

> Physical systems with the right kind of degree of complexity cause mental events to occur; less complex physical systems, or ones that are complex only in other ways, don't have any mental effects.

But this was anathema to Spinoza because he thought that all causal connections must make sense — it must somehow *stand to reason* that this thing has this effect and not that — and he thought that mental items are so unrelated, conceptually, to physical ones that nothing could make it reasonable that they should be linked in one way rather than another. In his view, if some physical systems had mental effects and other didn't, that would have to be a brute fact, an unadulterated case of "that's just the way it is, for no reason": and Spinoza is sure that there are no brute facts, that if something is the case then there is a reason why.

So, faced with the seeming fact that pebbles and chisels and rivers and many other physical things have no mental side, Spinoza denies that it is a fact. He offers the bold hypothesis of panpsychism, "mind everywhere", according to which there is a mental item matched with every physical item. My mind, for example, is a mental complex corresponding to the physical complex which is my body, and this is just one case of a general correspondence between the physical realm and the mental, a correspondence such that any physical simple will be matched by a mental simple, any physical complex by a mental complex. Just as my body is the complex upshot of putting a lot of atoms together in the right way, so my mind is the complex upshot of putting a lot of mental atoms (so to speak) together in the corresponding way; and each of these elementary constituents of my mind corresponds to an elementary constituent of my body. That does not mean that each physical atom in my body has a mind. An atom — or even a cell — is not complex enough for its mental side to be complex enough to count as a "mind", ordinarily so-called. But an atom or a cell has a mental aspect which differs from having-a-mind only in that it lacks the right kind and degree of complexity.[1]

Parallelism

As well as having to confront the appearance that mentality is distributed unevenly throughout the physical world, Spinoza must also face the fact that within systems that undeniably do have mental as well as physical aspects —systems such as you and me — there seems to be causal interaction between the two. It seems obvious that if you are jabbed with a pin a cry out in pain this is because of a causal chain of the form

Jab → Pain → Cry,

but Spinoza must deny this, because it involves causal flow in both directions between the body and the mind. At a minimum, he owes us an explanation of why it *seems* to us that such interaction occurs. His explanation is as follows.

(See *Ethics* 2, props 5-7,9.)

His panpsychism is a *parallelism:* it says that there is a mental realm that corresponds to the physical realm, *matching it detail for detail.* The mental item corresponding to any physical item *x* is what Spinoza calls "the idea of *x*", and his doctrine of parallelism says that there is an idea of each physical thing or fact or event such that if *x* is like *y* then the idea of *x* is like the idea of *y*, and if *x* caused *y* then the idea of *x* caused the idea of *y*. In short, there is a mapping of physical similarities onto mental ones and of physical causal chains onto mental ones. In Spinoza's words: "The order and connection of ideas is the same as the order and connection of things" (*Ethics* 2, prop 7).

And so the appearance of a causal chain from Jab to Pain to Cry comes from the fact that there are two causal chains going on:

idea of Jab → Pain → idea of Cry

Jab → brain event → Cry.

Of these six items, there are three that we know little about, namely the ideas of the jab and of the cry, and the brain event corresponding to the pain. So we fasten onto the other three and out of them concoct a spurious causal chain that runs from physical jab to the mental pain and back to the physical cry.

I am not much inclined to believe in Spinoza's parallelism. But I think that it is a possible solution to the question of how the mental and physical realms are interrelated and it's no worse than any of the other solutions so far propounded.[2] So I think it deserves respect, even if not credence.

I should mention that although the mental and physical realms are *metaphysically* on a par, Spinoza in his *explanations* nearly always gives precedence to the physical side. At *Ethics* 2, prop 13 (corollary and scholium), for example, he says that to understand what is excellent about the human mind you need to grasp what is excellent about the human body, and then embarks on some physics and biology to introduce his psychology.

The two "idea of" relations

Spinoza thought, as presumably we all do, that at least some of our mental states have representative features — i.e. they are in some way about items other than themselves. In his terminology, representation comes in through the language of what an "idea" is "of"; and "ideas" include not only perceptions and images but also beliefs, so that an idea of something can be a belief about it.

There are two distinct "of" relations at work in the *Ethics*. I have already mentioned one: if a mental item *I(x)* has physical item *x* as its counterpart under the parallelism, then *I(x)* is "the idea of" *x*. I shall express this by saying that *I(x)* is the idea *directly of x*. This is obviously not much like what we ordinarily think of as mental representation. Each belief of mine is the idea directly of a state of my body and especially of my brain[3], but few of my beliefs are in any ordinary sense *about* my brain. If Spinoza is to provide for what we ordinarily take to be thoughts about things — things other than our own bodies — it will have to be through something other than the directly-of relation.

He does also introduce a different way for an idea to represent something. He says that an idea of mine can be not only of my body but of your body; and I'll put this by saying that an idea directly of my body may also be *indirectly of* your body. Although Spinoza is absurdly brief in what he says about this, one gets the general idea: when I see, touch, hear you, I come to be in a mental state that represents you because it corresponds to a condition that my body was caused to be in by your body. (See *Ethics* 2, prop 16 (corollaries 1,2) prop 17 (scholium).)

This indirectly-of relation embodies a causal theory of perception, but that is not all. The "ideas" that Spinoza speaks of include not only perceptions but also beliefs: my idea directly of my body — he seems to hold — is a vast, complex, mostly unconscious belief that my body is thus and so, and my idea indirectly of you is a belief that there is a body of such and such a kind in front of me. And so the indirectly-of relation provides for beliefs that are about things other than the believer's own body.

What makes a belief true?

What makes a belief true? Spinoza says that a true belief is an idea which "agrees with" its "object".[4] Well, what relations are expressed by "agrees with" and "is the object of"? From Spinoza's handling of these expressions in his arguments, we learn this much about agreement:

If *x* and *y* are counterparts under the parallelism, *x* agrees with *y*.

And we learn this much about the "object" relation:

If *x* is *y*'s object, *x* and *y* are counterparts under the parallelism.

From these two it follows by trivial logic that every idea agrees with its object, from which it follows, by Spinoza's account of truth, that *every idea is true*, i.e. that *there cannot be any false beliefs*. Spinoza accepts that argument to that astonishing conclusion (*Ethics* 2, props 32, 33), and works hard to convince us that what we call error is really a species of ignorance — that it doesn't involve false "ideas" but merely a certain kind of lack of "ideas".[5] I shan't discuss those arguments.[6] My concern is with what pushed Spinoza into them, namely his doctrine that error cannot consist in outright false belief.

A better account of truth?

Why did Spinoza stick with an account of truth that had such an absurd consequence? It is of course initially plausible to say "A belief is true if it agrees with its object"; but that is plausible only because we are thinking of a belief's "object" as the item that the belief is, in the ordinary sense, *about*, and that is not what Spinoza means by the "object" of a belief. Our ordinary notion of a belief's object — or, anyway, our ordinary notion of the source of a belief's truth-value — is indicated by how we refer to the belief in a phrase of the form "John's belief that..." with a sentence in the gap. For instance, *my belief that there is a keyboard on my desk* is true or not depending on whether *there is a keyboard on my desk*— its truth value doesn't depend on the state of my brain!

Spinoza's indirectly-of relation is something like our ordinary notion of what a belief is about. Let us use the form "*y* is an indirect object of *x*" as the converse of "*x* is an idea indirectly of *y*".[7] Then we can try the conjecture that the truth-value of a belief depends upon how it relates to its *indirect* object(s). Isn't that better than what Spinoza offers? If so, why did he not at least explore it?

Here is a possible explanation:- "Spinoza deals sketchily with indirectly-of because he thinks of it as a superficial, unimportant, dispensable part of our conceptual armoury. He would say of the mental item that I have called

my belief that there is a keyboard on my desk

that if we knew enough we could replace that phrase by a much better one, namely one of the form

my belief that....

with the gap filled by a description of the corresponding brain state. In his account of truth, then, Spinoza is speaking of the best possible notion of truth — the one that is geared to the best possible notion of the content of a belief, the notion that brings in correlated brains states (direct) rather than associated external states of affairs (indirect)."

If that were Spinoza's position, he would be flagrantly abusing the concepts of "belief that P" and "true belief", under the guise of offering superior versions of them. Some of our contemporaries look forward to the day when we shall have what they see as a properly mature and scientific treatment of cognitive states, one that attends solely to their neural correlates, ignoring their relations to the outer world; and Spinoza would have sympathy with that project — this is suggested by his emphasis on the directly — of relation and by other indications as well.[8] But that is a project that involves throwing out the concept of belief — one of the relevant books is subtitled "The Case Against Belief" — or at least throwing out any distinction between true and false beliefs. Spinoza might say: "Just so! We now cannot have that distinction, so we should count all beliefs as true!". But that would be absurd. By moving to a purely internalist way of looking at cognition, we don't find a basis on which to call all beliefs true; rather, we find a basis for dropping the very notion of belief, or at least the notion of a belief as a cognitive state that has content and a truth-value.

In fact, however, I don't think that that is Spinoza's position; and even if it is, it certainly doesn't bring us to the bottom of our problem about why he lets himself be committed to saying that there could not be a false belief. I now explain why.

Trouble with the indirectly-of relation

Although Spinoza uses the directly-of relation in arguing that "there is nothing positive in ideas by virtue of which they can be called false", when he discusses cases of error, trying to convince us that they don't really involve false beliefs, *all his examples involve the indirectly-of relation*. For example, he discusses the child's naive belief that the sun is a few hundred feet away, and the belief that men are free; these beliefs or ideas are clearly being thought of as erroneous because of how they relate to something outside the believer's body, and Spinoza is arguing that their erroneousness lies in their involving a kind of ignorance — rather than a positive misrepresentation — of *that outer reality*.

That shows clearly enough that he does not propose a purely internalist handling of the concept of belief, in which the question of a belief's relation to anything outside the believer's brain is not even raised. But it also suggests that something extremely peculiar is going on: there is an enormous split through his procedure, with the directly-of relation being used in his argument against false "ideas", the indirectly-of relation dominating his defence of his conclusion through the examination of examples. It looks as though Spinoza thought that the notion of false belief — except when construed as a species of ignorance — is unsupportable even when what is in question is falsity with regard to the indirect rather than the direct "object" of the belief.

And so there is. Despite my hints that Spinoza could have allowed for beliefs to be false if he had defined truth in terms of the indirectly-of relation, really that is not so, *given his account of the latter*. If we ask Spinoza what entitles a belief of yours to count as being about an item other than your body, his answer is that this can be so because the belief is the idea directly of a state of your body that results from your being "affected by" that external item (*Ethics* 2, prop 16). And that is the only provision he makes for a belief or idea to be of or about something other than the believer's body; it is his entire theory of the indirectly-of relation. Spinoza has committed himself to saying that a belief can have *x* as its indirect object only by being the mental counterpart of a brain state that is *caused by x,* and it can be the belief that *P* (where *P* is not about the believer's body) only by being the mental counterpart of a brain state that is *caused by the fact that P*.

This implies that a belief of mine can count as my belief that there is a keyboard on my desk only if its cerebral counterpart — its direct object — is caused by the presence of a keyboard on my desk; it can count as my belief that the desktop is black only if its cerebral counterpart is caused by the blackness of the desktop; and so on. But obviously a state of my brain cannot be caused by something that does not exist: I can't be in a cerebral state caused by a keyboard that isn't there, or by the blackness of a desktop that is really brown. So even if we define truth and falsity through the indirectly-or relation, we still get the devastating result that there are no false ideas.

There is one exception to this, pointed out by Spinoza himself in *Ethics* 2, prop 17. My idea indirectly of *x* will remain unchanged, even after *x* has altered or moved, just so long as my body does not relevantly alter. So Spinoza does have room for positive misrepresentation by an idea of its indirect object, but only in the special case where a belief which was true is now false. But this kind of falsehood, which consists in merely being out of date, is only a miniscule fraction of what we normally think of as false belief.

A suggested explanation

The fact that Spinoza cannot make liberal provision for false (indirect) belief, any more than he can for false (direct) belief, suggests that although he demonstrates his thesis in terms of the directly-of relation and defends it in terms of the indirectly-of relation, he really means (or half means) to be talking, throughout, about both at once. He may have realized (or half realized) that he ought to regard falsehood indirectly about things as being almost as hard to come by as falsehood directly about them.

It would not be out of character for him to demonstrate in terms of one relation a doctrine that he meant to apply also to another. There is plenty of evidence that his official demonstrations in the *Ethics* do not always give his real or his best reasons for the conclusion. I think that this is because he liked his demonstrations to be short, and to depend upon his technical terminology; and sometimes his real or best reasons couldn 't be forced into that mold. That applies to our present case. To attack the notion of false belief in terms of the indirectly-of relation would have required more words than he liked; and it would also have been more informal than he liked, for the indirectly-of relation is not part of the technical apparatus — it occurs mainly in one of the informal discussions, playing virtually no part in the official deductive structure.

That suggested explanation of Spinoza's strange performance was the best I could find for many years. (I didn't find the secondary literature helpful, because no-one else seems to have noticed the problem.) But I became dissatisfied with it, because I am sceptical about philosophical coincidences. It is not likely — I came to think — that a philosopher should provide bases for two utterly different ways a belief can have content, one relating it to the brain, the other to the external world, and that it should just *happen* that each of them is unfriendly to the concept of false belief. And when it is as tough and deep a philosopher as Spinoza, I thought, one should suspect that there is a common cause for both halves of this story, presumably in the form of a positive, deep, general reason for thinking that there is something problematic about the notion of a false belief — a reason that does not arise from the details of any theory.

A better explanation

That suspicion was right: there is a profound reason why Spinoza should deny that any belief can be downright false — a reason that he did have, in the sense that it arises out of views that he did hold. He doesn't explicitly present it anywhere, but I am pretty sure that it had actual force in his thinking, even if only subliminally, and that it deeply explains his holding that false belief is impossible.

In one of his discussions of error, Spinoza says of certain mental contents which might generate so-called error that "they are not contrary to the true" (*Ethics* 4, prop 1 (scholium)). That phrase is suggestive. I think that *Spinoza was grappling with the problem of how something that is real can be contrary to the true.* Suppose that there were a genuinely false belief — that is, for some value of *P* it is the case both that *P* and that *x* believes that not-*P*. Both of those reports must be part of the total true story about the universe. On the one hand, the speed of light is finite, and on the other John believes that the speed of light is infinite. John's mind is a natural object, it's a small chunk of the real, just as his body is or as the pebble in my shoe is. How can a natural object contain something false? Or, in other words, how can a false proposition occur in a true account of reality?

An aside: Falsehood compared with pathology

Spinoza's denial that anything real can be untrue is similar in spirit to his denial that anything real can be wrong or defective. Because he thinks that Nature (or God) has no purposes and is subject to no external standards, he firmly rejects the idea that there is any pathology of Nature: he pours scorn on certain common attitudes by saying that they imply that "Nature has gone wrong". We tend to think there is something intrinsically wrong or bad or substandard about a child with leukemia; but Spinoza would say that a child that has leukemia is a perfect specimen of one kind of natural object, and is not evidence that Nature has made a mistake.

A single frame of mind can encourage one to think that nothing real is intrinsically bad and that nothing real is false. I accept the former doctrine, while strenuously rejecting the latter; but I agree with Spinoza that the notions of real pathology and real falsehood are *problematic*, differing from him only in thinking that in the case of falsehood the problem can be solved. Let us now consider how.

"A queer kind of medium, the mind"

How could a false proposition have a use in a true description of some part of the universe? A philosopher who was asked that question, out of the blue, might well say:

> A falsehood can be part of a truth when the truth reports the existence of a false belief. We can get "the speed of light is infinite" into an account of the real world if we can find someone who believes that the speed of light is infinite. There is no problem about how something real can be false, or about how a falsehood can be needed in telling the whole truth. The appearance of mystery vanishes as soon as we remember that some parts of reality are mental representations such as thoughts and beliefs, for we know that such mental items have that feature of being-about-something which enables them to be at once real and false.

Spinoza would refuse to say anything like that unless he could explain why it was true. He would want some account of how the trick is worked, of what is conceptually going on when a falsehood is nested inside a true report of a belief. And he would refuse to be fobbed off with the pseudo-explanation that this just is a fundamental property of the mind. We have been warned against such "explanations" by Wittgenstein:

> Understanding, meaning, interpreting, thinking...seem to take place in a queer kind of medium, the mind; and the mechanism of the mind, the nature of which, it seems, we don't quite understand, can bring about effects which no material mechanism could. Thus e.g. a thought (which is such a mental process) can agree or disagree with reality; I am able to think of a man who isn't present; I am able to imagine him, "mean him" in a remark which I make about him, even if he is thousands of miles away or dead.[9]

The attitude behind Wittgenstein's irony in this passage is the attitude Spinoza takes to all such matters; he is never willing to regard any aspect of the mind as simply "queer" and to be taken on faith.

Indeed, Spinoza must think that every truth about how the mind functions is isomorphic to some truth about how the body functions; this is required by the parallelism doctrine (*Ethics* 2, prop 7). It follows that if any belief is false, some corresponding sort of physical item must also be false or at least have some analogue of falsity. That, for him, must always be the acid test of whether some mental operation has been properly understood.

So he does have a genuine problem: the question of how something real can be "contrary to the true" is a superb one. It is not answered just by insisting that mental representations obviously can be contrary to the true, if that is left unexplained; and we should not be put off by the plea that it can't be explained because it's mental and therefore occult or queer or inherently mysterious.

The functionalist account of belief

A decent explanation of how beliefs can be false must be embedded within a general account of what kind of state a belief is. Spinoza offers no such account. Instead of trying to analyse the concept of belief, he merely introduces the term "idea", says that ideas are beliefs, and leaves it at that. So he has no theoretical framework within which to answer his wonderful question about how a belief could be false. How can we fill this gap in what he offers us?

The functionalist kind of theory that is most popular these days looks just right for the purpose. It associates *the belief that P* with a certain kind of function from sensory inputs to behavioral outputs. Stated roughly and abstractly, it says that to believe that *P* is to be disposed when one has such and such perceptions to engage in such and such actions. When the details are filled in,[10] the result is an explanation of *how* a proposition about the external world — such as "There is a keyboard on my desk" — gets into a description of someone's mind, including an explanation of what makes a particular doxastic state count as the belief that *P* rather than the belief that *Q*. Functionalism splendidly does *not* take refuge in the plea that it is a fundamental unanalysable property of the mental that mental states can be representative.

Like Spinoza's account of the indirectly-of relation, functionalism does relate beliefs to their causes. But whereas Spinoza implies that if I believe there is a keyboard on my desk then

> The cause of my cognitive state is *the fact that there is a keyboard on my desk*,

which implies that there cannot be false beliefs, functionalism only says something of the form: if I believe there is a keyboard on my desk then

> The cause of my cognitive state is *something that has relation R to states of affairs in which there is a keyboard on my desk,*

Where *R* is some kind of similarity relation. That allows that a cognitive state can count as the belief that *P* even if it is not caused by the fact that *P*; and so it places no obstacle in the way of allowing that some beliefs are false.

So functionalism has foundations that one might expect Spinoza to approve, yet does not end up with his catastrophic conclusion that there cannot be false beliefs. Why didn't he think of it? There doesn't have to be any answer to this; but I think there are two answers — that is, that two features of Spinoza's thought would have impeded him even if he had come somewhere near to thinking of the functionalist account of belief.

One obstacle: counterfactual conditionals

Functionalism understands beliefs as dispositions to act in certain ways in such and such circumstances. It analyses "John believes that *P*" as a counterfactual conditional, that is, a proposition saying that if such and such *were* the case John *would* behave thus and so.

That is acceptable to me, because I think counterfactual conditionals are all right: we use them all the time, and have some idea of how to sort out the true from the false ones. But there is evidence that Spinoza did not think that counterfactuals are all right. As a strict determinist, he thought that whatever happens is inevitable, and this seems to have persuaded him that there is something fishy about the notion of "...if the course of events had been different from what it actually was...". Even today some philosophers hold that if determinism is true then counterfactuals are out of business; I don't agree, but their view is not an absurd one.[11] The question of how counterfactuals can be legitimate in a deterministic world is quite tough.[12]

Anyway, for whatever reason, Spinoza was visibly hostile to counterfactual conditionals. The most graphic example is in a letter where he is replying to a correspondent who has asked, among other things, "whether by our precaution we can prevent what would otherwise happen to us". This cries out for the obviously true counterfactual answer, which goes like this: Yes; for it can happen that I step back onto the kerb and survive, whereas *if I hadn't taken that precaution I would have been run over*. But here is what Spinoza says instead:

> Since one could ask a hundred such things in an hour without arriving at any conclusion about anything, and since you yourself do not press for an answer, I shall leave your question unanswered.

He was often rude in his letters, and this correspondent had proved to be especially annoying. Still, this surly rebuff needs to be explained; and I think it shows Spinoza's unwillingness to use a counterfactual conditional if he could talk his way out of it.

That's one reason why Spinoza was not well placed to accept the functionalist account of belief.

The second obstacle: teleology

There is another obstacle as well, involving a feature of functionalism that I have so far kept out of sight. It is that the functionalist account of *belief* also uses the concept of *desire* or *want*. Any attempt to get at what someone thinks through how he behaves, or how he would behave if..., must appeal to what he wants.

If you are not sure that that is right, I offer you a challenge. Take any belief *P* you like, any kind of action *A* you like, and describe circumstances *C* (not involving desires) in which you think that someone who believes *P* is bound to do *A*; and I undertake to describe to you someone who believes *P* and is in circumstances *C* but does not do *A*, because his desires are different from those of the person you were thinking of. For example, you might think that the belief that (*P*) the liquid in this glass is poisonous would certainly lead, in

(*C*) normal circumstances where there are no threats or duress, to (*A*) the person's not drinking the liquid in the glass. But that assumes that the person does not want to be ill or die! Replace that person by one who is suicidal, and the likely behavioral expression of the belief that *P* is quite different. So, as I said, functionalism has to relate belief not only with input-output but also with desire. Its basic doctrine about belief is that it is first and foremost a cognitive state which combines with a conative state — a wish or want or hope or desire — to produce behaviour.

Spinoza would have rejected this, because he had declared war on the ordinary concept of desire. Really, there is nothing wrong with this concept, but he thought there was, for an honourable reason: he had noticed a real difficulty in the concept, which I shall now explain.

We use the concept of desire or want or intention in an explanatory way. We say things like

> He raised his hand so as to deflect the stone
> He drank because he wanted to slake his thirst

and so on, as answers to the question of why he raised his hand, why he drank. Spinoza noticed that in doing this we explain a certain event by referring to something — the deflection of the stone, the unthirsty state — that comes later than it and is indeed caused by it. This is absurd, he said; it "reverses the order of nature", treating an effect as though it were a cause, e.g. treating the deflection as though it produced the arm-raising instead of vice versa.[13] He concluded that it can never be right to explain an action in terms of a want or purpose or desire or intention; and this attempt to keep teleology out of his system dominates much of the later parts of the *Ethics* — a reader who doesn't grasp that much has no chance of properly understanding those parts.[14]

Now, Spinoza is wrong about that. We may grant him his point about kinds of causation: there are only efficient causes, pushes from earlier events; there are no final causes, pulls from later ones. But a possible future state or event can come into the explanation of an action without helping to cause it, because an action can be caused by a *thought about* a possible future — my raising my hand may be caused by my believing that if I raise my hand the stone will be deflected.

In fact, this is a topic that functionalism is also helpful with. One way of saying what functionalism offers is this: It analyses the concept of belief in terms of behaviour and desire, and analyses the concept of desire in terms of behaviour and belief. But that makes it sound circular, which it is not. Really, functionalism presents belief and desire as distinct but collaborating functions from perceptions to actions, or as two kinds of mental state that jointly produce overt behaviour; and it does this in such a way that you cannot understand its account of one of them without at the same time coming to understand its account of the other. You might suspect that a theory which thus runs the two concepts in a single harness is in danger of fusing them together, smudging the line between the cognitive and the conative. In fact, however, for reasons that I cannot go into here,[15] that does not happen.

I am not contemptuous of Spinoza for his failure to see how to legitimize the concept of desire. So far as I know, nobody had much understanding of it until quite recently, and I cannot find a glimmering of an understanding of it in any of his predecessors or contemporaries. When he rejected the ordinary concept of desire, then, he was not being stupid and narrow; rather, he had seen something which everyone else had overlooked.

Conclusion

So there we have it. Spinoza ends up asserting the patently false thesis that what we call error is not outright false belief, but rather a kind of ignorance; and that fact, taken on its own, might suggest that the man is simply incompetent. If we add his definition of truth in terms of agreement between idea and object, he looks not just incompetent but stubborn: why did he retain a doctrine of truth that had that consequence? Throw in the material about the indirectly-of relation and the waters get muddier still, without making Spinoza's reflection in them any less unattractive.

But then see him asking how anything real can be contrary to the true. The one reply that seems right and relevant is that a mental representation such as belief can be real in itself and yet false in what it represents or says. But it's a sign of Spinoza's greatness that he would never accept this in the spirit of "It's just a fact about beliefs that they can be false", and that before accepting the notion of a false belief he would insist upon knowing what was going on, having an explanation of how false belief is possible. We can explain this, but only with help from counterfactuals (which Spinoza rejected for sober reasons) and from the ordinary concept of desire (which he also rejected for sober reasons). And so he was trapped.

In this matter, I see Spinoza as up to his neck in error, and yet as gloriously shining with insight and intellectual integrity. And I have learned a lot about belief from thinking about his failures with it. This is the only way I know to learn from Spinoza — trying to pin down what he thought, and wrestling with him to find out his deepest reasons for thinking it.[16]

NOTES

1. If you want to explore this further, the best place to start is Thomas Nagel's "Panpsychism", in his *Mortal Questions* (New York, 1979).
2. In that remark I am agreeing with Nagel, *op. cit.*
3. I don't insist on *states* of the brain; it could be events or processes in the brain.
4. In *Ethics* 1, axiom 6 he says only that if an idea is true it agrees with its object, but he clearly means this as a biconditional.
5. Spinoza sometimes writes as though there were "false ideas", but that is not his considered view, as he pretty well admits in saying that "As regards the difference between a true and a false idea, it is clear from *Ethics* 2, prop 35 that the former is to the latter as being is to non-being". *Ethics* 2 prop 35 says that "falsity consists in the privation of knowledge".
6. See *Ethics* 2, prop 17 (scholium), prop 35 (scholium), and *Ethics* 4, prop 1 (scholium). These passages are examined in G.H.R. Parkinson, *Spinoza's Theory of Knowledge* (Oxford University Press: Oxford, 1954), pp. 123-126; and in Jonathan Bennett, *A Study of Spinoza's Ethics* (Hackett: 1984), section 40.
7. I use the indefinite article each time because the indirectly-of relation is many-many; the directly-of relation is one-one.
8. Actually, the contemporary enemies of mental representation, most of whom are outright materialists, hold that beliefs *are* neural states rather merely being *correlated* with them. I have used the latter formulation so as to conform with Spinoza's property dualism. It does no harm in the present context, for the issue about mental representation cut right across the issues between outright materialism and Spinoza's kind of parallelism.
9. L. Wittgenstein, *The Blue and Brown Books* (Basil Blackwell: Oxford, 1958), p. 3f.
10. As they are, more fully than elsewhere in the functionalist literature, in Jonathan Bennett, *Linguistic Behaviour* (Cambridge University Press: Cambridge, 1976), chapters 2-4.
11. See for example Richmond Thomason and Anil Gupta, "A Theory of Conditionals in the Context of Branching Time", *Philosophical Review 89* (1980), 65-90, at p.75f.
12. For two contrasting views about it, see David Lewis, "Counterfactual Dependence and Time's Arrow", *Nous* 13 (1979), 455-476; Jonathan Bennett, "Counterfactuals and Temporal Direction", *Philosophical Review* 93 (1984), 57-91.
13. *Ethics* 1, Appendix.
14. For a fairly full discussion of this much neglected matter, see Jonathan Bennett, *A Study of Spinoza's Ethics*, ch. 9 and the pages referred to under "teleology" in the index.
15. They are expounded in my *Linguistic Behaviour*, p.51f.
16. A French translation of this paper is appearing in the 1986 volume of *Studia Spinozana*.

[11]

Yirmiyahu Yovel

The Second Kind of Knowledge and the Removal of Error

Writers on Spinoza (the present one not excepted[1]) frequently complain of his parsimonious account of the third kind of knowledge. We tend to think that we know far more about the second kind, *ratio* — and in a rather scattered way, indeed we do. But, as Bennett has pointed out,[2] Spinoza's formal account of the cognitive use of *ratio* is hardly more generous than that of *scientia intuitiva.*[3]

My aim in this paper is to begin a reinterpretation of the second kind of knowledge, with a special emphasis on how it is related to the first and the third kind. I shall rely primarily on *Ethics* II, with occasional reference to other of Spinoza's works, as well as to his often tacit polemics with Descartes, since a good working clue to what Spinoza had in mind is what he objected to.

I. Error as Primordial

The first thing to notice about the second kind of knowledge is that, indeed, it must always come second. Knowledge never starts with adequate ideas; Man's original cognitive stance is necessarily sense-perception, and hence, according to Spinoza, error. Therefore, *ratio* is essentially the correction of error, its removal and replacement by adequate ideas.

Why is this so? Why doesn't the human mind begin immediately with adequate

---

[1] In my "The Third Kind of Knowledge as Alternative Salvation," in: E. Curley and P.-F. Moreau, eds., *Spinoza —Issues and Directions; Proceedings of the Chicago Spinoza Conference* (Leiden: E.J. Brill 1990), pp. 157–175; and, in revised form, in chapter 6 of my *Spinoza and Other Heretics*, Volume I, *The Marrano of Reason* (Princeton: Princeton University Press, 1989).

[2] J. Bennett, *A Study of Spinoza's* Ethics (Cambridge: Cambridge University Press 1984), p. 182.

[3] The concept makes its first appearance in EIIp40s2 and is discussed over four more propositions, before reappearing in Part V (props. 10, 12–15, 28–29).

Y. Yovel, ed., Spinoza on Knowledge and the Human Mind, 93-110.

ideas — especially as the mind, according to Spinoza, is eventually capable of forming these ideas out of itself? In other words, why aren't humans free, self-determining beings by nature, instead of having, at best, to undergo a process of ascent and emancipation? This query sounds like a variation of the broader metaphysical complaint known as theodicy: Why should evil, sin, error, suffering, etc. exist at all? Descartes tried to explain error by attributing its origin to human free will, as if it were a cognitive "sin" and even a kind of negative miracle intervening in nature. Spinoza, of course, denies free will as a transcendent fiction. Error, like everything else in the world, must be explained causally, from within nature itself. What makes humans begin in error is that our minds are inseparable from our bodies, and our bodies are causally linked with, and therefore resonant of, the rest of extended nature. Hence the mind, as the body's idea, must begin its cognitive career by reflecting the effects of external causes on the body[4] in the random order in which they occur — that is, confusedly.

At first glance it might seem that in making error the primordial human condition, Spinoza, too, offers a semi-theological version of original sin; but in fact, Spinoza bypasses theology and expresses human finitude in strictly naturalistic terms and on immanent grounds. Precisely because man is ontologically on a par with the rest of nature, he cannot escape being constantly affected by the environment; this produces sensory imprints on his body (*imagines*) and their reflective counterparts in the mind (*ideae*); and the latter, because they correspond neither specifically to the external object nor to the bodily system which the object affects, but to their conflation, are necessarily confused ideas.

But why should the mere having of confused ideas be ipso facto error? Don't we have to *assert* these ideas in order for error to arise? Yes, says Spinoza, but we do so anyway, all the time. Descartes required a special act of assent or denial to turn perception into judgement and thereby produce truth or error. Spinoza understands ideation itself as an act of assertive judgement. To have an idea is to assert that its object exists and its content is true. Since the mind, as a

---

[4] Or, under a broader interpretation: on the *organism*. Substituting "organism" for "body" will suit a strictly parallelist reading of Spinoza (which the logic of his system may require, although the relevant text of *Ethics* II does not favor). On the strictly parallelist reading, the organism as a whole, existing both as body and as mind, is affected by external causation in two parallel avenues; as a result, two types of ideas are generated, those inadequately representing the body and the external physical objects, and those inadequately representing the mind and the external mental items affecting it. This reading is less faithful to the text, more faithful to the principle of parallelism, and changes nothing essential in our analysis, although it complicates the description.

purely natural being, becomes aware of the world's existence and of the objects within it by reflecting in ideas their ongoing impact on the body (the organism) (EIIp26), the mind also automatically asserts these ideas to be the true expressions of external objects, which of course they are not.

Even of *itself* the mind can be aware, at first, only on the basis of its reflecting the body, and through it the external world that causally affects it (EIIp23). Self-awareness is the confused idea of bodily modifications. It is not — *pace* Descartes — a monadic mental event that can occur even without a body.

The first kind of knowledge is therefore based on my mind echoing the world-network from the particular spot that my body happens to occupy within it. This is what sense perception means, in both its sharp and hazy forms. The sharp form puts a single object in focus, and the hazy form echoes an accumulated, partly distant and gradually fading impact of the environment; but we must be careful not to confuse the "sharp" perception with a "distinct" idea. What is visually sharp may well constitute an idea that misses its true object or represents a conflation of several objects, and is hence a confused idea.

Sense perception represents all things in a temporal dimension, and as nominologically unrelated particulars. Compounded by the mechanism of natural association (the supreme psychological law, analogous in status, I think, to the Law of Motion and Rest in physics), sense perception produces memory and imagination and, through them, generates *experientia vaga*, the uncritical form of experience. *Experientia vaga* unwittingly interprets the world on the basis of "the common order of nature," a rather inept term Spinoza uses to denote the random order in which our bodies encounter external things. The "common order of nature" is very different from the "order and connection of things" which actually structures the world, and hence the purely subjective, associative nexus of ideas it generates is to be distinguished from the "order and connection of ideas," i.e., the system of all true ideas about the universe.

Error, we might therefore say, is the incongruity of the subjective nexus of ideas in my mind with the relevant segment in the "order and connection of ideas." Thus error is neither some "positive negation" in being, so to speak, nor a "black hole" in the continuous order of ideas, but a mismatch between the order and connection of ideas in my mind and in the infinite intellect. This mismatch can be set aright by reassigning ideas to their proper objects and, thereby, withdrawing the improper commitments which our initial judgements involved.

This is the role of *ratio*; but before analyzing the way it works in removing error, we must inquire into its origin and claim to validity. Where does it come from? How closely linked is the second kind of knowledge to the first?

II. Common Notions: The Naturalist Foundations of Reason

The foundations of reason (*fundamenta rationis*) are common notions (EIIp44c2d), and these, Spinoza assures us, are formed a priori. They are the axioms and first principles of all branches of rational knowledge,[5] most clearly of mathematics and physics, but also of logic, metaphysics, and almost certainly of psycholog} (the law of association [6]). All adequate ideas are either common notions or derived from them according to rules that are also common notions.

As such, common notions are clear and distinct ideas in a sense similar (though not identical) to that which Descartes has given to the term. In forming common notions the mind uses its inborn power, which grows and becomes more adept through being exercised in increasingly complex tasks; but, at all stages, the mind uses ideas that are generated by itself. This self-sufficiency is constitutive of the meaning of adequate ideas,[7] and is the measure, even the primary definition, of human freedom in Spinoza.

But doesn't spontaneity evade, or even break, the immanent causal nexus of nature and thereby involve a kind of transcendent miracle? Other so-called rationalists (Descartes, Leibniz, later also Kant) were not disturbed by this implication of rational spontaneity, not being committed to naturalism as Spinoza was. Spinoza, however, goes a long way in attempting to reconcile the spontaneity of *ratio* with his immanent, naturalistic view of man. In essence, his doctrine of common notions tries to draw the foundations of reason from the same metaphysical inherence of man in nature that accounts for error as well.

Spinoza realizes that to maintain his immanent metaphysics he cannot allow for an external intellect that perceives the truths of reason without that intellect itself being part of the immanent system of nature. Therefore, although *ratio* is spontaneous, it must not be construed as springing out of nothingness; nor should *ratio* seem possible (as it is in Descartes) without there being a world and a body. In order to clearly perceive the "foundations of reason," the mind must have a body, and the organism as a whole[8] must have had a history of residence in the world.

---

[5] Or, in the words of the TTP (end of Chap. 14), the "foundations of philosophy" (in the broad sense, which includes the natural sciences).

[6] That this law appears as prop. 18 of *Ethics* II is no objection — quite a number of propositions in the *Ethics* could have figured as axioms, and vice versa. That the law of association is the most general psychological pattern (after ideation itself) is witnessed by its further use in the system (including its affective varieties in Part III).

[7] Bennett (ibid., p. 178) even sees it as exhaustive of that meaning.

[8] I shall use "organism" in the sense of the body/mind unity, i.e., the whole individual.

Herein lies Spinoza's modification of the Cartesian "natural light," which Descartes opposed to the *super*-natural light of Revelation and mystical experience, but which Spinoza continues to view as transcendent miracle. To be really natural, the "light" of intuitive reason must be anchored within nature and have its counterpart in extension. More precisely, intellectual intuition is natural if and only if it belongs to an embodied being that exists within nature and is interwoven with its causal network. A free-floating *cogito*, disembodied and worldless, would not be able to perceive rational truths, because it would have no object in which to perceive them and no *natural* medium through which to do it.

It should be remembered that, although intellectual intuition is spontaneous, it does not create the truths it knows; its spontaneity lies in its reflecting rational truths that exist in nature objectively, i.e., are inscribed in the laws and infinite modes by which the universe works. These truths do not inhere in a Platonic world of ideas or in the mind of a transcendent God which is internally bound by these truths (Leibniz) or which chooses and guarantees them to the human mind (Descartes).[9] The rational truths must therefore be reflected from *within* nature, that is, read off the infinite modes by a mind whose organism shares in the general causality of nature, yet is not determined by anything particular in it.

To meet all these concerns, Spinoza constructs a difficult theory of the genesis of common notions, which makes him the most empirically oriented thinker among the rationalists. According to this theory, the autonomy of reason is

---

[9] In Plato, the mind observes rational ideas existing in themselves outside nature. In Descartes and Leibniz, the human mind draws the rational truths by rationally intuiting itself (or, in Leibniz's case, the divine intellect). But in Spinoza, the mind draws the common notions by observing *nature* — though not through the senses, but through intellectual intuition, which Spinoza construes as the mind's capacity to grasp law-like uniformities when observing itself, not as a being outside the world of objects, but rather as representing it in the organism's accumulated experiences.

A possible way, therefore, to characterize the difference between Leibniz and Descartes (within the non-Spinozistic camp of rationalists) is the following. In all three cases, a basic realism is assumed: the mind does not produce the rational truths but perceives and records them. Wherefrom? In Descartes, merely from itself, with God guaranteeing that its ideas correspond to the laws of reality. The mind does not perceive the outside reality of these laws (which lack logical necessity), but only reflects their reflection within itself, and hence the query, the doubt: do they represent the actual laws of logic and mathematics? In Leibniz, the mind, in perceiving its innate ideas, perceives something real outside itself — i.e., the inner object of God's own intellect — so no further divine guarantee is required. The philosopher's mind, which he observes a priori, serves as a kind of medium to the inner object of God's mind, which binds a priori his creation. (Cf. Hegel metaphorically equating logic with God's mind prior to the creation of the finite world.) In Spinoza the rational mind perceives the fundamental laws of reason *within* nature itself and through its mediation.

exercised by "*drawing* its ideas from nature" (Spinoza's terms; my emphasis[10]); though this does not require passing through particulars but is accomplished directly, by grasping the law-like patterns inscribed in the universe. And the medium in which the mind perceives these laws is the dimly felt and unindividuated bulk of experiences that the organism, by virtue of its residence in nature, has accumulated in its previous career and maturation.

This unusual theory diverges from the Cartesian doctrine of innate ideas while also rejecting the empiricist model of abstraction from particulars. It claims that deductive relations and nomological universals can be grasped directly from an experiential substrate. Among the questions that it leaves unanswered, is the ambiguity about whether this substrate — and, more specifically, the body's exposure to the causality of nature — is a *condition* of *ratio* or merely its counterpart under the body–mind parallelism.

Even without being able to solve all the questions, I think it is at least possible to grasp Spinoza's aims and to obtain a sense of his train of thought. So let me elaborate my reading of "common notions."

Perceiving Uniformities

I take common notions to be the epistemological counterparts of fundamental laws of nature, i.e., their expressions as mental representations. As such, common notions have their metaphysical basis in the infinite modes (or the "fixed and eternal things"[11]), that is, in the law-like uniformities which structure the actual world and serve as legitimate universals within it.[12] Spinoza believes that *ratio* has (or rather, is) the inborn power (*vis nativa*)[13] to perceive these uniformities directly, without passing through particulars and, consequently, without relying on a process of induction or abstraction.

---

[10] "Common notions are the foundation of Philosophy and should be *drawn from nature alone*" (as against the foundations of Faith which should be drawn from the Bible alone; TTP, end of Ch. 14, my translation and emphasis).

[11] TIE 101. I discussed the infinite mode in my paper at the former Jerusalem Conference on *Ethics* I. See Y. Yovel, "The Infinite Mode and Natural Laws in Spinoza," in Y. Yovel, ed., *God and Nature: Spinoza's Metaphysics (Spinoza by 2000 — The Jerusalem Conferences, vol. I*; Leiden: E.J. Brill, 1991), pp. 79–96.

[12] As distinguished from illegitimate universals, like genera, or transcendentals. Here I agree with Bennett, Haserot and others, that Spinoza accepts uniform law-like patterns as real aspects of the universe; see also the paper cited in note 1, above.

[13] See TIE 31, where an analogy with inborn bodily tools is made. Spinoza, there too, defines "inborn" as "what is not caused by external causes."

The natural medium by which the foundations of *ratio* are gained is the general causality of nature, which affects our organism in law-like patterns that are the same "in the part and in the whole" (EIIp37–39), i.e., are uniform. By virtue of this uniformity, there is something in my experience which the mind will always reflect in the same way, and with respect to which it makes no difference that my body echoes the world from a particular angle, or that my ideas conflate the properties of external objects with those of my own body — in all these cases, the same law-like properties (e.g., the laws of motion, or of logic) are involved.

Spinoza, I think, implies that the causal impact of the world provides us with two kinds of input, particular and universal, overt and tacit. When bodily events affect my organism, they furnish it with latent, unconscious information about the uniform laws of nature. At first, the mind is not even aware of having this information (or rather, pre-information); it reflects it vaguely and rather automatically by virtue of merely existing in the world and having a body. But if, again by natural causes, the mind attains pure intellection and, concentrated, is free of external influence, it can grasp this information with clarity and shape it in adequate ideas: now the mind actually knows the universal laws, and knows that it knows them.

What makes Spinoza believe that his account of common notions fulfills the difficult condition mentioned above? In perceiving those immutable modes of the universe which remain the same from any possible perspective ("are equally in the part and in the whole"; EIIp37), the mind is observing a *natural reality* and yet is free of determination by anything *in particular*. What determines the mind in this case is the infinite mode directly, i.e., God as spelled out in this immutable mode.

Natural laws spell out God's essence and transmit his necessity to all there is, in the "horizontal" sequence laid down in EIp16 and its derivatives. Normally, natural laws determine particular entities and events by determining how *other* particulars — their proximate causes — exist and act upon them.[14] But, in the case of common notions, Spinoza seems to say that the mind is not determined by any other particular, only by the universal law as such — which means that it is immanently and directly determined by an infinite mode of God. And this is also the basis for human freedom — the only kind of freedom that a finite mode, as distinguished from a substance, can possibly enjoy.

[14] I discussed this at some length in the text cited in note 1.

Deductive Relations

In addition to universal laws, reason's *vis nativa* is also capable of directly grasping deductive relations that obtain in nature, both in human discourse and in the way things work.[15] In EIIp29s, Spinoza equates the mind's being "determined externally" with its perceiving things by randomly encountering them "from the common order of nature"; and he equates the mind's being "determined internally" with the "fact that it regards a number of things at once, to understand their agreements, differences, and oppositions." I take the latter to refer to deductive (rather than inductive)[16] relations, where the mind, in a single overview, intuits the consequence's derivation from its antecedents as flowing from what is mutually common and different in them. However, because in grasping specific deductive relations we may be said to also grasp their underlying law (at least implicitly), I shall continue to discuss *ratio* from the standpoint of laws.

All Sciences Refer to the World

The theory of common notions contains another crucial implication: since common notions are the "foundations of *ratio*" in general — in logic and geometry no less than in physics — and since all are "drawn from nature" in the way discussed, Spinoza implicitly denies that a fundamental distinction separates "formal" sciences from "substantive" or "empirical" sciences. Rather, all sciences are equally *substantive* (and this in the dual sense that they all refer to the same real substance and are all materially significant sciences). Logic, too, is not abstract, but describes uniform patterns in which all physical and mental things in the universe act and exist. No science is "formal" in the sense of "not being about the world," since all true ideas have reality as their *ideata*. For these and other reasons, the distinction between "synthetic" and "analytic" truths is also basically foreign to Spinoza's thinking.

Spinoza's Brand of Rationalism

Spinoza's "rationalism" thus lies partly in his belief that the mind has the inborn capacity to intuit universal laws directly, without recourse to induction

---

[15] I say "also" because Spinoza, in the passage to be quoted, goes on to mention "this or *another* way" (my emphasis) in which reason is determined from the inside. So grasping deductive relations is only one example; the other (and more comprehensive) is grasping laws.

[16] Although Spinoza's words might fit also an act of abstraction, the logic of his position excludes this reading.

or abstraction from particulars. However, he tries to anchor the second kind of knowledge as much as possible in the domain of the first kind, which is important for "situating" reason within nature and, again, for avoiding abstraction. Abstraction is invalid, first, because it is based on blurring out differences and thus produces a confused idea (EIIp40s1); and, second, because its typical products are genera and species, which are fictitious entities (ibid.). "Common notions," on the other hand, are not species but law-like uniformities, which do have real metaphysical counterparts in nature (through the "infinite modes"), and as such constitute the only valid universals in Spinoza's world. Yet the mind's special capacity to grasp the common notions requires a natural substrate, something that the mind possesses by its very being in the world and within which it intuits the rational laws.

At this point, Spinoza dissents from the ordinary rationalist view of innate ideas. In forming common notions, the mind does not perceive itself in the abstract but as having had some previous cognitive career in which the substrate for intellectual intuition has been accrued. Thus the inherence of man in nature that inevitably produces error is the very same terrain from which reason arises.

Here an important question arises to which I referred earlier. What is the precise role of the bodily counterparts of *ratio* (or, speaking for a strict parallelist, of the mental stimuli as well)? Do they *engender* the common notions, or merely serve as their natural substrate and (in the case of bodily items) as their counterpart under the parallelism? In other words, are these accrued imprints the *pre*-conditions or the *co*-conditions of *ratio*? While I think that neither of these answers finds solid support in Spinoza's text, or in its somewhat ambiguous logic, discussing their merits and weaknesses here would sidetrack my present purpose.

Nomological Universals and Explaining Individuals

We have found common notions to be the ideas of nomological universals which replace the abstract universals (genus and species) in serving as the valid tool of science. Nomological universals are real aspects of the universe, they are quantifiable and free of teleological bias, whereas generic forms are qualitative, abstract and alleged paradigms of perfection (and hence teleological). Spinoza (as Haserot and Bennett, among others, rightly claim) is not a nominalist, but an opponent of abstraction. Law-like universals are metaphysically real and indispensable to philosophy and modern science.

But, one might wonder, indispensable for what? A few statements in the *Ethics* (e.g. EIIp37; EIIp40s2, §II; EIIp44c2d) may seem to suggest that *ratio* is limited solely to grasping universal patterns. Looking more closely, however, we realize that the subject of EIIp40s1,s2 (the "kinds of knowledge" scholia) is not knowledge in general, but the ways in which universals are formed: concerning the formation of universals, which are the basis of knowledge for him, too, Spinoza tells us in these famous scholia that *imaginatio* fabricates empty universals (abstractions), whereas *ratio* forms valid ones (common notions).

As for EIIp44, this proposition invokes the super-temporal element that attaches to the objects of rational knowledge by virtue of their universal aspect; thus *ratio* enables us to understand individuals, too, as enjoying a certain aspect of eternity. But rational knowledge is not knowledge only of universals. If that were so, if no individuals could be known prior to the third kind of knowledge, then we might wonder what happens when an engineer explains why my car radio broke down or a meteorologist predicts tomorrow's weather — whether they are practicing Spinoza's third kind of knowledge. This is clearly absurd. These scientists are characteristically engaged in knowledge of the second kind, that is, in applying valid universals (laws) to individual objects, in order to explain them by their true causes and reproduce their adequate ideas in the mind.

There are, however, two kinds — or levels — of applying universal patterns to particular cases. Watching the sun in the mode of uninformed sense-perception, I see it as a small balloon, the size of a football, suspended next to the top of that tree. When *ratio* intervenes to correct my initial error, it may lead me to two different levels of knowledge:[17]

1. *The concrete level*: I know the sun's actual diameter and its exact distance from me.
2. *The level of explanatory schemes*: I know that the sun is really at a very great distance and is several times bigger than the earth, and I also know how to further investigate and explain this; but I do not know the exact figures that express the sun's size and its distance from me.

In the latter case I am launched on the correct cognitive track, without having particular knowledge. In other words, here reason supplies a general *explanatory scheme*, though it does not provide this scheme with the corresponding filling.[18]

---

[17] As indicated, I see *ratio* as getting down to scientific detail. Certainly one does not need the extraordinary mode of *scientia intuitiva* to acquire such ordinary results (nor does their acquisition contribute to any unusual mental experience, as *scientia intuitiva* is supposed to do).

[18] I was pleased to find a similar idea applied to Spinoza by Donald Davidson, in his "Spinoza's Causal Theory of the Affects" (The Jerusalem Spinoza Lectures 1993), forthcoming in Y. Yovel, ed., *Desire and Affect: Spinoza as Psychologist (Spinoza by 2000 — The Jerusalem Conferences, vol. III).*

This may be the level of an enlightened person who is not an astronomer, or it may be an interim level in the astronomy student's way to precise scientific knowledge. The explanatory scheme is in any case the work of *ratio*, since it applies the laws of movement, gravitation and optical perspective to the kind of problem involved in my seeing the sun. And although this scheme does not contain fully particularized (=precise) knowledge, we have left the abstract level of mere natural laws and created a more definite pattern, one whose universality is adapted to the kind of problem at hand and therefore provides a specifically oriented explanatory scheme for it.

It is crucial to Spinoza's project to hold that the explanatory scheme, although still general, is a full-fledged idea, in that it has the power to annul the assent attached to the former sensuous idea and eject it from the mind. Thus the occurrence of a true explanatory scheme, even one that is unaccompanied by much detail, can help provoke a radical change in a person's perception of the cosmos and his/her position in it.

The two-level structure will become clearer when we examine how the removal of error works in Spinoza.

III. The Removal of Error

Error presents a serious problem for Spinoza. According to the fundamental proposition EIIp7, the order and connection of ideas is the same as that of things; hence, every idea has *some* real object for its *ideatum*. Spinoza also insists that there is nothing negative in an idea as such by virtue of which it is false. However, if all ideas are positive and refer to something real, how can false ideas exist?

Spinoza answers that false ideas occur when true ideas are confused or "mutilated," that is, are taken to be pointing to something other than their actual object; or are fragmentary and lacking in context; or are made of ingredient-ideas that have been torn out of their proper setting and placed in arbitrary combinations. Correcting a false idea therefore involves an analysis (through *ratio*) in which complex ideas are dismantled and their ingredients are reassigned to their true objects and linked in different combinations, which express the order and connection of things.

So, in essence, error is due: (1) to a mismatch between true positive ideas and the things alleged to be their objects/ideata; or (2) to fragmentary, partial contexts in which true ideas are placed. In the second case, error is a special kind of ignorance, not blank ignorance but incomplete knowledge — knowledge that is partial or fragmentary or that lacks sufficient context.

The most common form of error is the mismatch between ideas and their objects. While all ideas are true in some respect (i.e., with respect to their actual *ideata*), in the usual course of life they appear in confused combinations and are taken to point to other objects than their own. This is due to the contingent, merely temporal order in which our body happens to encounter events in the world, and to the purely associative links which *imaginatio* establishes between the ideas of these events. This "common order of nature" (the subjective order of experience) deviates from the order and connection of ideas as they are in God's infinite intellect, that is, as they constitute the objective map of being.

Even when ideas are adequately matched with their objects, error may exist in the second form, when the matching idea is too hazy or indefinite to provide a sufficient account of the object to which it points correctly. The deficiency in this second kind of error lies not in the idea's intentionality but in its poverty — in its loose substance, imprecise detail, lack of saturation or fragmented context. These are specific forms of ignorance which can be removed by fleshing out the idea and giving it substance, context and detail. Another, less demanding though quite efficient remedy is to provide a specific context for it — namely, its proper explanatory scheme (which is, as I will argue later, a meta-context as it were, more methodological than substantive).

Before a further examination of the removal of error, we should recall that Spinoza rejects Descartes' theory that false beliefs can be retracted at will, by denying assent to the ideas that express them. To Spinoza, as mentioned above, ideas carry their own assent in themselves, and can be removed only within a natural, causal cognitive process. More specifically, only a new positive idea can retract a false belief and remove error. Negation — also the negation of error — is actually the affirmation of a new idea which ejects the former false idea and takes its place.

The new corrective idea is of two kinds, or works in at least two ways, corresponding to the two main kinds of error named above. First, a new idea — the actual idea of the object *x* — ejects the former one, which was not at all the idea of the intended object *x*, but of some other, unintended object *y*, or of a haphazard combination of ingredients from *x*, *y*, *z*, etc. The former idea is not annulled, however, but is reassigned to some other location in the map of being (in the *ordo et connexio rerum*), to which it actually belongs. For example, a sense-idea that fails to adequately express either its intended object or the perceiver's physiology, will be reasigned to a complex of events in the universe to which it does refer — namely, the interplay of a sense-organ with some external physical entity. In no case will the idea be abolished or its positive character turned negative.[19]

---

[19] The presupposition, of course, is that it is a true idea (with some *ideatum* in the world),

The Removal of Error — a Graphic Illustration

Using the example of sense perception, the following Figures 1-4 illustrate how a typical error is created and removed.

Objective State of Affairs

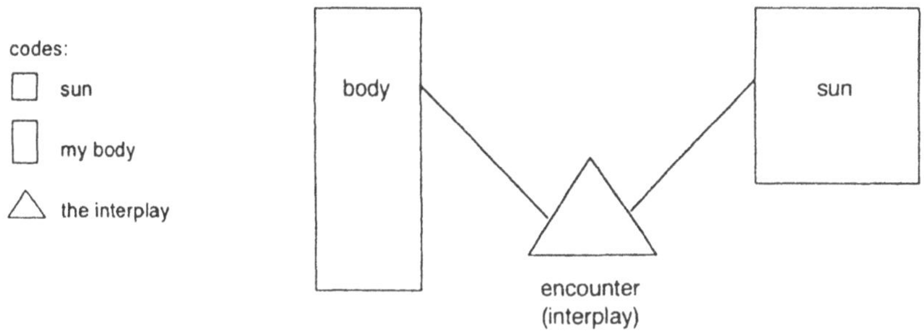

Fig. 1.

Fig. 1 shows what takes place when I look at the sun. Two distinct physical entities, my body and an external body (the sun), come into a causal relation. This leads to an encounter (interplay[20]) between the sun's radiation and my body's sensory system. Parallel to this, an idea is formed in my mind, expressing the interplay in the attribute of Thought. This idea purports to be the idea of the sun when, indeed, it is the idea of the interplay; thus it misses its true object. According to Spinoza (though he does not fully explain why), this kind of intentional bias is built into sense-perception generally. All sense-ideas[21] fail to adequately refer to their true objects. They are not empty ideas, since they (or rather, their main ingredients) do have something real in nature (in the

---

and not a fiction or an empty word. Even fictions and empty words, however, like "free will" or the names of alleged universals, gain some hold on reality whenever someone *believes* they are not empty; for *that* mistaken belief is a real event in the world, for which some true idea (=explanation) exists in God's infinite intellect.

[20] I say "interplay" because, in producing the sense-perception, the body receives and modifies the effects of the light according with its own features. Of course my body does not affect the sun in return, but the light reaching my body then works conjointly with the sensory system to produce the phenomenon of vision; and within that context their features modify each other to produce something else.

[21] I use "sense-idea" to indicate that we are analyzing actual ideas and not mere "images." An image (*imago*) is a physiological entity, a configuration created in the sense-organs by external stimulation. Sense-perception, however, as I understand it, also has a mental aspect proper to itself, which I designate by the term "sense-idea."

"order and connection of things") as their object. Yet that object is quite different from what the idea purportedly refers to.

Thus error is generated, in the sense of a mismatch between the idea and its object. At the same time, error also arises in the sense of confusion and incompleteness. The content of the sense-idea is a confluence of ideas expressing elements that are drawn partly from my body's nature and partly from the sun's nature. These ingredients have been torn out of their natural contexts (they are "mutilated" ideas) and reassembled to new clusters, which are imaginary (or erroneous) in the following senses. First, the cluster is a confused idea, i.e., is made up of items that belong in nature to different and sometimes incongruous contexts. And second, there is nothing in nature that corresponds to the cluster as such. The ingredients of the cluster have real objects scattered in various places in the map of being.[22] But the cluster itself, as a single complex idea, has no such object. Its error lies in its principle of organization — the imagination — which mirrors the temporal "common order of nature" rather than the atemporal "order and connection of things."

Notice that the sense-idea fails to be an adequate idea even of its own object, namely, the interplay, because it presents it in the manner of *imaginatio*. To be adequate, the idea would have to provide, through *ratio*, a scientific account of the natural laws and causal conditions under which the interplay occurs. But the sense-idea does not of course do this. Also, the essence of the sense-idea lies in the very sensual quality — such as "yellow" — which *ratio* dismisses as an imaginary quality, because it cannot be predicated on anything real in the world.

To clarify the latter point: my experiencing the quality "yellow" is a real event in the world, for which there is a scientific explanation — a true idea — in God's intellect; yet attributing that quality to anything real in the world would be a fallacy. In other words, my seeing yellow is a real thing, though no real thing is yellow. And that is what the imagination characteristically does — present something in the mind that isn't real, or that cannot be predicated on anything real.

In consequence, we may say that there is absolutely nothing in nature of which the sense-idea is the adequate idea; for it misses its purported object and misrepresents its actual object. Thus it manifests error in both its forms, as a mismatch (of idea and object) and as a confused idea.

---

[22] This may indicate an inner link between "mutilation" and "confusion" (and even with "mismatch"). A confused idea puts together ingredients that belong to different contexts in God's intellect; and a mutilated idea is torn out of its natural context, thus making a confused idea possible. In both cases, the tearing away and the re-assembly are characteristically the work of the imagination.

Now a philosophical observer enters the scene. (Figures 2, 3 and 4 describe the observer's action in three stages. The broken line represents an idea.) At first the observer ascertains the actual state of affairs. This stage (Fig. 2) is still

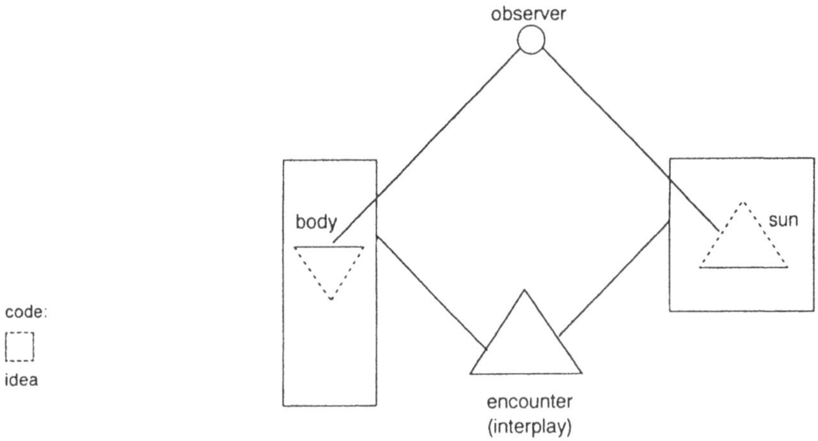

Fig. 2

only critical. The philosopher rejects the sense-idea's claim to be the adequate idea of the sun, because he knows it to be a confluence of elements expressing the sun and elements expressing his own body. Thereby, however, the observer also recognizes there is something in the sun and something in his body which the idea of their interplay (the sense-idea) represents partially or in "mutilated" form. This is depicted in Fig. 2 by the two inverted triangles (drawn in broken lines), embedded in the square and the rectangle. Yet none of these ingredients corresponds either to the square or to the triangle (i.e., either to the sun or to my body).

The next stage (Fig. 3) is the most important. Now the philosopher-observer proceeds to correct the error by reassigning each idea to its proper object. Thus, the sense-idea is re-assigned to the interplay, while both the sun and the observer's body have each a distinct idea expressing its nature and place in being. In order to accomplish the reassignment, the philosopher must already understand, through *ratio*, what the respective natures of these objects are, what sciences and laws of nature pertain to them and, most important, what explanatory scheme should serve as a basis in their investigation. (An explanatory scheme applies a specific crossing of relevant scientific laws, data and pertinent methodology to the needs or features of a given domain of investigation; as such it is both general and specific). At this stage, for example, I already know — through the general but also specific explanatory scheme drawn from astronomy, optics, etc., as applied to this kind of situation — that the sun is not a small ball suspended next to the tree but a huge heavenly body millions of kilometers away, and that I see it as small because of the nature of optical perspective. No less important, I also know in what way and direction to seek further, more precise information of the matter at hand. This knowledge derives from the explanatory scheme, and is part of the *meaning* of having a matching idea.

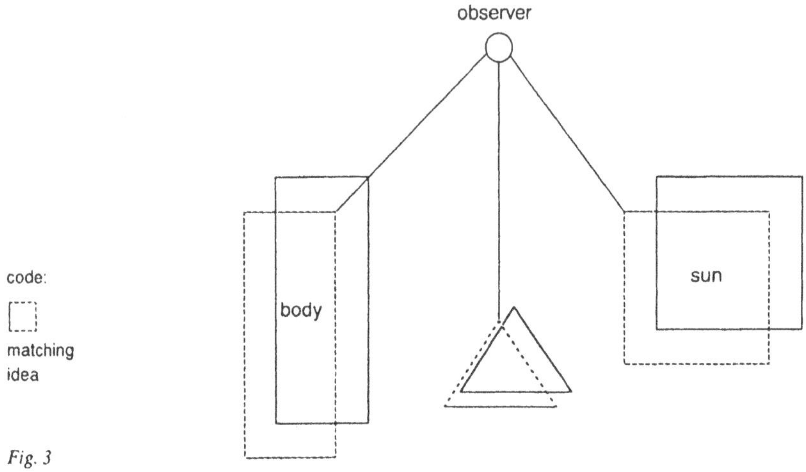

Fig. 3

Equipped with a matching idea and explanatory scheme, the observer may, however, still be lacking a great deal of knowledge about the object. In other words, his idea is rather thin and under-informed. But his epistemological situation has radically changed. Now his ideas match their objects, and *he is placed on the right cognitive track*, which he can pursue to the degree of detail he prefers. Moreover, the explanatory scheme has provoked a deep change in his view of the world. For example, now he knows the true order of magnitude of the earth with respect to the sun, and therefore knows his place — and that of all human life and civilization — in the cosmos, even while ignoring the exact figures or many of the specific data; or (in a different domain) he knows that everything in the world is necessary, so alleged miracles and free will must be reexplained causally, even if he does not yet possess the precise explanation for a given case; or again, he knows the actual nature of sensation and the intentional bias built into it, even though he still ignores the exact mechanism by which neurons transmit the effects of light to the brain. To Spinoza, such cognitions have the full-fledged status and power of true ideas that are capable of ejecting former, false ideas from the mind, thus revoking a person's naïve beliefs and superstitions even though they are still poor in detail and not spelled out much farther than the general level of an explanatory scheme.[23]

Yet error also has the meaning of incomplete knowledge, or *specific* ignorance, ignorance of something we need to know or have a special reason or interest in knowing. Such a need or interest may emanate from the *conatus* in its shape as *conatus intelligendi*, the desire to be liberated through rational enlightenment; or, more narrowly (and confined to epistemology), it may arise from our already having posed a specific question and delineated the explanatory scheme for answering it, so that we now have a natural reason to wish to complete our knowledge of the matter.

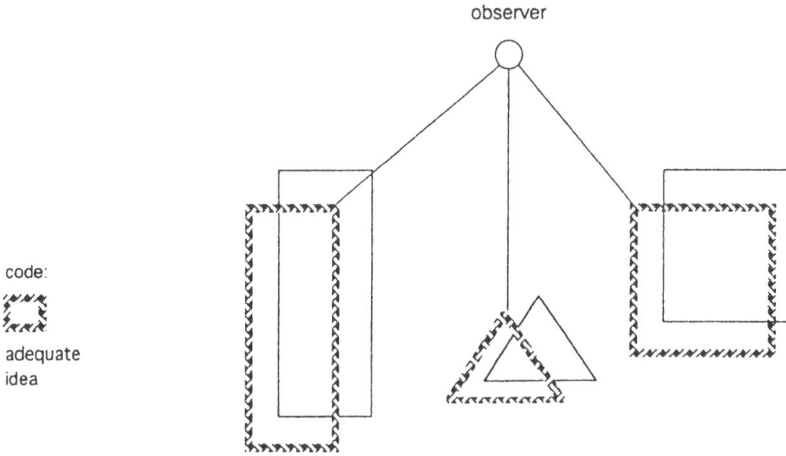

Fig. 4

This is depicted in Fig. 4. The matching ideas are fleshed out by detailed, exact, and substantive research. Now the observer knows exactly what the sun is and how it works, what the constitution of his nervous system is and how it reacts to electromagnetic radiation, producing the brain movements that are the physiological aspect of vision. In consequence, as the diagram shows, the three matching ideas are now also filled in, though not saturated. They provide substance and detail as well as much of the relevant context and background.

Strictly speaking, the complete context of any singular idea is of course the

---

[23] It would be misleading to say that explanatory schemes are "methodological guidelines" as distinguished from substantive "ideas." Spinoza's first move in philosophy was to deny the validity of such a distinction (in the TIE, where he argues that method does not precede substantive knowledge and is not gained independently of it). The explanatory scheme is an idea all right, substantive and informative, although placed at a certain level of generality which is specific to the domain at hand. In that respect it occupies an intermediary zone between a common notion (or a general law of nature) and the idea of an individual. Moreover, within that zone it may occupy various positions, that is, become more or less specific.

totality of ideas in God's intellect. There is a sense in which the famous Hegelian dictum, "the true is the whole," applies to Spinoza as well. However, this is an impossible condition to fulfill and insisting on it is pedantic and unnecessary. Even from a purely cognitive point of view, the extent to which one must provide context and detail largely depends on the kind of question that has been asked. Limiting ourselves to that scope is rational (even in Spinozistic terms), because it is compatible with a system that recognizes that humans, as finite modes, can reach infinity only in a qualitative and not in a quantitative manner.

This coincides with the way I have interpreted error-as-ignorance. If we distinguish between ignorance at large and specific ignorance, and agree that ignorance is specific to the kind of interest projected or question asked, then in order to remove error-as-ignorance it will be sufficient to flesh out the idea to the extent that meets this need or interest. Of course, the limit must be understood as dynamic, since new answers provoke further questions and interests. But the goal is to flesh out the idea, not to saturate it.

Knowledge is subsidiary to an ethical goal in Spinoza. Its goal is liberation through enlightenment, and this does not always require a very detailed degree of knowledge. Enlightened people will often know important things about themselves and the universe without having to delve into encyclopedic omniscience. We need not all be astronomers, physiologists or biochemists to have adequate ideas or to benefit from their enlightening effect. Having true explanatory schemes will be sufficient to eject former superstitious ideas and produce the liberating ethical effects Spinoza is seeking, if only to a limited extent, without actually fleshing the idea out in detail.[24]

Our discussion of error and its removal has shown the various levels on which *ratio* is supposed to apply common notions and natural laws. While *ratio* does apply universal laws and explanatory schemes to particular instances, it can perform this work on various levels, depending on the question and interest in the background. There are cases where *ratio* will remain on a fairly general level of explanation, not because it is incapable of descending lower toward concrete particulars, but because, in these cases, such concreteness is not deemed necessary for the ethical goal underlying the philosopher's enterprise. Detailed particular knowledge is more important in the *third* kind of knowledge, and Spinoza does say that the more we know particular things, the more we know God. Yet this "more" is desirable, not conditional — certainly not for ordinary *ratio* and its kind of enlightenment.

---

[24] A possible exception may be psychology, where he seems to require relatively more detailed and subtle knowledge.

[12]

The Philosophical Review, LXXXVII, No. 2 (April 1978).

DISCUSSIONS

SPINOZA'S DISTINCTION BETWEEN RATIONAL AND INTUITIVE KNOWLEDGE

Spencer Carr

I

There seems to be a growing consensus about how Spinoza's distinction between rational knowledge (*ratio*) and intuitive knowledge (*scientia intuitiva*) is to be understood. Central to this body of agreement is the view that the distinction, as it is drawn in the *Ethics*, is one between the knowledge of general truths and the knowledge of individual things; the former is to be associated with reason, the latter with intuition.[1] We will soon have occasion to note other positions and consequences associated with this interpretation of Spinoza's distinction. But perhaps the most striking of these is the notion that the *Ethics* version of the distinction differs radically from the version Spinoza provides in the *Treatise on the Improvement of the Understanding*.[2] I find the evidence for both the central claim and its corollary unconvincing; acceptance of them leads to un-Spinozistic ideas and to the undermining of what I take to be Spinoza's fundamental insight. In what follows I will defend the constancy and integrity of Spinoza's distinction by arguing for an alternative reading of his texts, one which is plausible in its own right and which does not entail that he changed his mind between the *Treatise* and the *Ethics*.

It will be necessary to work fairly close to Spinoza's texts, so I begin by setting out the crucial passages from, respectively, the *Treatise* and the *Ethics*.

---

[1] See, for example, G.H.R. Parkinson, *Spinoza's Theory of Knowledge* (London, 1954), pp. 182-5, and Errol E. Harris, *Salvation from Despair* (The Hague, 1973), pp. 107ff. More sustained defense of this view may be found in E. M. Curley, "Experience in Spinoza's Theory of Knowledge," in M. Grene, ed., *Spinoza* (Garden City, N.Y., 1973) and in J. J. MacIntosh, "Spinoza's Epistemological Views," in G.N.A. Vesey, ed., *Reason and Reality*: Royal Institute of Philosophy Lectures, Volume 5 (London, 1972).

[2] This is explicitly claimed by both Curley and MacIntosh. Because the views I am considering are held and defended most explicitly in these two papers, I will attend exclusively to them.

SPENCER CARR

Reflection shows that all modes of perception or knowledge may be reduced to four:

I. Perception arising from hearsay or from some sign which everyone may name as he pleases.

II. Perception arising from mere experience—that is, from experience not yet classified by the intellect, and only so called because the given event has happened to take place, and we have no contradictory fact to set against it, so that it therefore remains unassailed in our mind.

III. Perception arising when the essence of one thing is inferred from another thing, but not adequately; this comes when from some effect we gather its cause, or when it is inferred from some general proposition that some property is always present.

IV. Lastly, there is the perception arising when a thing is perceived solely through its essence, or through the knowledge of its proximate cause.[3]

From what has already been said, it clearly appears that we perceive many things and form universal ideas:

1. From individual things, represented by the senses to us in a mutilated and confused manner, and without order to the intellect [Corol. Prop. 29, pt. 2]. These perceptions I have therefore been in the habit of calling knowledge from vague experience.

2. From signs; as for example when we hear or read certain words, we recollect things and form certain ideas of them similar to them, through which ideas we imagine things [Schol. Prop. 18, pt. 2]. These two ways of looking at things I shall hereafter call knowledge of the first kind, opinion or imagination.

3. From our possessing common notions and adequate ideas of the properties of things [Corol. Prop. 38, Prop. 39, with Corol. and Prop. 40, pt. 2]. This I shall call reason and knowledge of the second kind.

Besides these two kinds of knowledge, there is a third, as I shall hereafter show, which we shall call intuitive science. This kind of knowing advances from an adequate idea of the formal essence of certain attributes of God to the adequate knowledge of the essence of things.[4]

The differences in these formulations lead Curley to conclude that the distinction between reason and intuition has altered radically.

In the *Treatise* the primary contrast between reason and intuition seemed to be that, whereas reason involved an inadequate, because inferential, knowledge of the essences of things, intuition involved an adequate and immediate knowledge of their essences. And there were two species of intuition—one exemplified by knowledge of the essence or definition of an attribute, the other exemplified by knowledge of the essence or definition of a mode.

[3] *Tractatus de Intellectus Emendatione. Spinoza Opera.* Herausgegeben von Carl Gebhardt. Heidelberg: Carl Winters. V. II. p. 10. (Translation by W.H. White) This passage appears on p. 7 of John Wild's *Spinoza Selections* (New York, 1930).

[4] *Ethics* in Gebhardt, op. cit. (Translation by Elwes, Bohn) IIP40S2. I.e. Part two, proposition 40, scholium two. In other references, A stands for "axiom," C for "corollary," L for "lemma."

RATIONAL AND INTUITIVE KNOWLEDGE

In the *Ethics* intuition seems to be conceived more narrowly. It includes adequate knowledge of the essences of singular things, i.e., finite modes, but it does not include adequate knowledge of the essences of divine attributes. . . . In the *Ethics*, though not in the *Treatise*, this kind of knowledge is classified under the heading of reason. . . . Reason is knowledge of the essences of those things that in the *Treatise* are described as fixed and eternal things—the attributes and infinite modes of the *Ethics*. Intuition is knowledge of the essences of those things that in the *Treatise* are described as singular mutable things—the finite modes of the *Ethics*. [Curley, pp. 56-7]

So Curley contends that the crux of the distinction has changed from one of inferential versus direct knowledge to one of knowledge of universals versus knowledge of the essences of finite things. This characterization of the latter distinction is based to a great extent upon an aside made by Spinoza at V36CS.

I thought it worthwhile for me to notice this here, in order that I might show, by this example, what that knowledge of individual objects which I have called intuitive or of the third kind [Schol. 2, Prop. 40, pt. 2] is able to do, and how much more potent it is than the universal knowledge, which I have called knowledge of the second kind.

Referring to this same passage, MacIntosh reaches the same conclusion. He offers "truths of a general nature vs. insights into the essence of individual things" as a slogan to catch the crucial distinction (MacIntosh, p. 47).

There are two crucial elements of this new consensus: (1) That the distinction in the *Treatise* is that between inferential and direct knowledge, and (2) That in the *Ethics* the distinction is that between knowledge of universal or general truths and knowledge of individual objects. These two elements lead to a third: (3) That whereas in the *Treatise* there are two kinds of intuition (intuition of an *attribute* involves grasping its essence; intuition of a finite *mode* involves grasping the essence of its proximate cause), in the *Ethics* intuition is only of modes. I am primarily concerned to contest these three positions.

There is no denying that Spinoza's language can suggest something like the consensus view. But there are a number of problems for this interpretation. Consider, for example, Spinoza's favorite illustration of the different kinds of knowledge:

Let there be three numbers given through which it is required to discover a fourth which shall be to the third as the second is to the first. A merchant does not hesitate to multiply the second and third together and divide the product by the first, either because he has not yet forgotten the things which he heard without any demonstration from his schoolmaster, or because he has seen the truth of the rule with the more simple numbers, or because from the 19th Prop. in the 7th book of Euclid he understands the common property of all proportionals.

> But with the simplest numbers there is no need of all this. If the numbers 1, 2, 3, for instance, be given, every one can see that the fourth proportional is 6 much more clearly than by any demonstration, because from the ratio in which we see by one intuition that the first stands to the second we conclude the fourth. [IIP40S2]

This example raises two questions. In the first place, unless Spinoza holds that numbers are finite modes, it is unclear how any mathematical knowledge can be intuitive if rational knowledge is of universals and intuitive knowledge is of particular things. And there is the related problem that Spinoza's use of the same example for all the types of knowledge implies that the same thing can be known in each of the ways of knowing. His classification is presented as one of different ways of knowing the same thing and not as one of different objects of knowledge. Now it is possible that an analysis of different ways of knowing can lead to the consequence that only certain objects are appropriate to certain ways, but according to the interpretation we are considering, a difference in object is at the heart of the distinction. The example of finding the fourth proportion strongly suggests that Spinoza did not see it in this way.

There are other difficulties for the interpretation. When Spinoza celebrates the intellectual love of God in Part V of the *Ethics*, intuitive knowledge is called the highest and most perfect form of knowledge. On the consensus interpretation this refers to knowledge of the essences of singular mutable things, that is, the finite modes. But this is surprising since the knowledge of fixed and immutable things is the likelier candidate for improving one's relationship with Spinoza's God. Another surprising consequence, one noted by both Curley and MacIntosh, is that intuition, the highest form of knowledge, necessarily depends upon rational knowledge, the lower form.[5]

Much of this is prima facie puzzling, but there is another, more general, problem to face. If we assume that Spinoza's epistemological views underwent a substantial change between the *Treatise* and the *Ethics*, we should also assume that he had some good reason for changing his mind. Even if we accept that the *Treatise* version is unclear and not completely thought out, it is difficult to believe that Spinoza would undergo a radical change in his beliefs unless he were under the influence of some newly perceived insight or problem. Neither Curley nor MacIntosh provide any such motivation.

[5] For arguments to this conclusion see Curley, p. 58, and MacIntosh, p. 47.

RATIONAL AND INTUITIVE KNOWLEDGE

II

I turn now to the task of providing an alternative interpretation of Spinoza's two formulations and their relationship. I shall first discuss the *Treatise* distinction between reason and intuition and argue that it is *not* a distinction between inferential and direct knowledge. I shall then try to show that the *Ethics* formulation of the distinction is consistent with a proper understanding of the *Treatise*. Then I shall consider, one by one, those passages from the *Ethics* which have seemed to require the interpretation of the consensus. If I am right, a close scrutiny of all these passages, with special attention to their contexts, will allow us to deal with them in a way that will not involve us with the difficulties attending to the consensus position.

Turning, then, to the *Treatise*, I suggest that the crux here is not that rational knowledge is inferential and that intuitive knowledge is not. In the first place, what Spinoza says is that rational knowledge arises "when the essence of one thing is inferred from another thing, but not adequately." This suggests that the essence of a thing might be inferred from another thing *adequately*, in which case we would not have an instance of rational knowledge. What would we have? Certainly not imagination or opinion. The only plausible candidate is intuitive knowledge. If so, then intuitive knowledge can be inferential. The same conclusion is suggested by the very formulation of intuitive knowledge itself, for intuitive knowledge arises "when a thing is perceived through its essence, or through the knowledge of its proximate cause." Perceiving a thing through the knowledge of its proximate cause amounts to perceiving it as following as a consequence from its proximate cause. So it looks as though inferential knowledge need not be rational and that it can be intuitive.[6]

What is essential to the distinction Spinoza is drawing is not the presence or absence of inference. Spinoza's first example of an inadequate inference is that of inferring from an effect to its cause, and the generality of his example clearly implies that any such inference will be inadequate. This contrasts with the case of intuitive knowledge in which we perceive a thing through its proximate cause, that is, in which we infer from cause to effect. Spinoza's point has to do with the proper *ordering* of our thoughts.

That this is the proper view to take of the distinction is supported by the overall structure and argument of the *Treatise*. The question is which

[6] Of those I have cited, Parkinson argues that intuitive knowledge may be inferential. See p. 183. Curiously enough, he argues this from the text of the *Ethics*.

form of knowledge is best, and Spinoza's answer, unequivocally, is that based on intuition. That this "best knowledge" is a matter of the direction of thought comes out many times in the *Treatise*:

> [D]oubt always proceeds from want of due order in investigation. [Gebhardt, V. II, p. 30; Wild, p. 32]

> [T]rue science proceeds from cause to effect; [Gebhardt, V. II, p. 32, Wild, p. 34]

> [T]he knowledge of an effect is nothing less than the acquisition of more perfect knowledge of its cause. [Gebhardt, V. II, p. 34; Wild, p. 36]

And, from the *Ethics*, "The knowledge of an effect depends upon and involves the knowledge of its cause" (IA4). Of course, everything that is not self-caused is an effect. So, on what I shall call the "ordering" interpretation, what is important about knowledge is that it is rooted in causes and moves toward effects. Since the ultimate cause of everything is to be found in substance, or God, intuitive knowledge must have its source in attributes of God. Rational knowledge arises when we infer from the nature of dependent things and not from the nature of substance. It starts at the wrong place and it moves in the wrong direction.

What has happened to this distinction in the *Ethics*? Spinoza says that knowledge of the second kind (rational knowledge) comes "from our possessing common notions and adequate ideas of the properties of things." Intuitive knowledge, however, "advances from an adequate idea of the formal essence of certain attributes of God to the adequate knowledge of the essence of things." What is of special interest here is the same notion of a certain source and direction of thought that we saw in the *Treatise*. Intuitive knowledge is an *advance from* the formal essence of an attribute *to* the adequate knowledge of the essence of things. Rational knowledge, on the other hand, arises from our ideas of things. Spinoza, I claim, is making the same methodological point in both places. He is willing to grant, at least in the *Ethics*, that there is adequate knowledge arising from our ideas of things; but this knowledge is less than the ideal, according to which "true science" will always move from cause to effect, from the independent to the dependent, from substance to mode, and from God to finite things.[7]

Why, then, if the two accounts are not fundamentally different, does Spinoza list two forms of intuitive knowledge in the *Treatise* and only one in the *Ethics*? This is not an inconsistency or a change of mind on Spinoza's part, but only a reflection of the different contexts in which the two classifications arise. In the *Treatise* Spinoza is talking about intuitive knowledge in general. It arises either when a thing is perceived

[7] Spinoza's attitude toward the adequacy of rational knowledge may be perceived as ambivalent. I will consider this question below in section III.

solely through its essence (this can apply only to attributes) or when it is perceived through the knowledge of its proximate cause (this applies to modes). In the *Ethics* Spinoza gives only one case because he is in the midst of a discussion of the relationship between knowledge and objects, that is, individual things or finite modes. This being so, it would not be relevant to mention the intuitive knowledge of attributes. That this *is* so is clear from the context, and it may be useful to spell this out.

In propositions 37-40 of Part II Spinoza explains that we can form common notions from our perceptions of objects and that these notions are necessarily adequate. His idea is, roughly, that all bodies, by virtue of involving the conception of the same attribute, agree in certain respects; every body and every part of every body has something in common. Spinoza's view of the relationship between mind and body and his doctrine of perception lead him to hold that any such common notion can only be perceived adequately. (At IIP13L2 Spinoza gives "being capable of motion and rest" as an example of what is common to all bodies.) The first scholium to proposition 40, the scholium just preceding the classification, is concerned with distinguishing these common notions from universal ideas drawn from experience which we might confuse with common notions. Throughout, Spinoza is talking about how we do and how we should handle our experience of natural objects. He sums up the two main ways in which we have knowledge of things from experience (imagination or opinion and reason); then he adds that there is another way of knowing things, one which gets to their essence. He describes this intuitive knowledge of the essences of things and, naturally enough, he does not mention intuitive knowledge of attributes. But it certainly cannot be inferred from this that Spinoza now believes that knowledge of attributes would not be intuitive. This would be an unmotivated change of a radical sort and one contrary to the spirit of Spinoza's thought. We should be hesitant to attribute such a change to Spinoza, and I shall now try to show that there is no real pressure to do so.

So far, then, there seems to be no clear reason to think that the two formulations are inconsistent. They can both be seen as attempts to articulate the point about the proper ordering of our thoughts. It remains to examine in some detail those aspects of the *Ethics* that have led others to the more complex view of the consensus.

(1) In the first place, why does Spinoza's description of rational knowledge insist that it arises out of *common* notions and adequate ideas of the *properties* of things? The consensus view takes these expressions to contrast with the *individuals* mentioned in the description of intuitive knowledge. In fact, though, Spinoza's arrangement of the types of

knowledge strongly suggests that the description of rational knowledge is meant to contrast more tellingly with that of imagination or opinion. Both reason and imagination or opinion deal with our experience of objects; the demand that reason involve common notions and adequate ideas of properties is a demand that in working from experience we take care to stick to what may be adequate. This is the demand that our experience be ordered and corrected by reason. It is not a demand that rational knowledge be knowledge of universals as opposed to knowledge of individuals. As we shall see, rational knowledge may arise out of common notions without being knowledge of general truths or universals.

(2) It may be argued that in introducing in the *Ethics* definitions of the types of knowledge as a way in which we form "universal ideas," Spinoza implies that rational knowledge is of universals. And if intuitive knowledge is of individuals, this would explain why Spinoza curiously tacks on intuitive knowledge without numbering it as a fourth way in which we perceive many things and form universal ideas. But this procedure, if curious, does not tell against an "ordering" interpretation. For one thing, if it follows from this procedure that rational knowledge is not of individuals, it would also follow that opinion or imagination is not of individuals. This is surely not Spinoza's view. In the second place, from the fact that Spinoza is here talking about deriving universal ideas from experience it actually follows that rational knowledge of individuals is possible. So suppose that one forms a common notion from the proper application of reason to one's experience of things. According to Spinoza (*Ethics* IIP49CS), every idea involves an affirmation or negation. The affirmation associated with a common notion will be the thought that everything following under the relevant attribute has the given property. But it follows from this that, for any such particular individual, *it* has the property. I take this to be rational knowledge of an individual. It is certainly *adequate*, and it cannot be intuitive since common notions are no part of the essence of a thing (*Ethics* IIP37). In a similar way, adequate ideas of properties should give rise to rational knowledge of individuals. And such knowledge of an individual is fairly described both as being "inferred from some general proposition that some property is always present" (from the *Treatise* version) and as arising "from our possessing common notions and adequate ideas of the properties of things" (from the *Ethics*).

(3) But what of Spinoza's remarks in Part V? Despite the appeal made to this passage by both Curley and MacIntosh, it does not support the consensus view. Spinoza does *not* say that intuitive knowledge is necessarily of individuals. The crucial line here is the relative clause "which I have called intuitive," and the issue turns on whether the clause is

interpreted as restrictive or nonrestrictive. A nonrestrictive reading of the clause would support Curley and MacIntosh; but the restrictive reading is, I claim, more plausible. It has Spinoza saying that of the various ways of knowing individual objects, it is intuitive knowledge of them that is more powerful. There is no suggestion that intuitive knowledge is of individuals only. He goes on to contrast this way of knowing individual objects with the universal knowledge of them, which he has called knowledge of the second kind. (This last "of them" is not in Spinoza; I have added it to make clear what I take to be his intentions in this passage.) Although I claim that my reading of V36CS is more plausible than the alternative, I need not do so to make my case. Just pointing out that the restrictive reading is possible suffices to show that we need not take V36CS to support the consensus.[8]

It remains to explain why Spinoza would call knowledge of individuals *universal*. Two possibilities come to mind. One is that such knowledge can be universal in the sense that all persons share in it. Spinoza believes that there is such knowledge, and it may be no accident that its possibility depends upon the common notions (*Ethics* IIP38C), the doctrine of which is central to the *Ethics* account of rational knowledge. More likely, I believe, it is that the knowledge can be said to be universal in that it *rests upon* our grasping properties that are in all things. This connects rational knowledge to (some) individuals, but it would not follow that the knowledge is *of* universals or *of* general truths. It is still reasonable to insist that the knowledge can be knowledge of particular things.

(4) Now the fact that in this same passage Spinoza contrasts the power of intuitive knowledge of objects with his demonstrations in Part I of the *Ethics* might be thought awkward, since it suggests that this knowledge is not intuitive, whereas on my view that the matter of direction of inference is crucial, all these proofs should produce intuitive knowledge. The answer is, again, that Spinoza is talking about knowledge of individual things. The knowledge of Part I *is* intuitive, but it is not intuitive knowledge *of individual things*. Intuitive knowledge of individual things, of course, must end in our knowledge of the essence of the thing; what we know about things from Part I is only what is common to them all, and what is common to them all

[8] Those familiar with the vagaries of 17th century texts will recognize that this issue cannot be settled by an appeal to Gebhardt. It is my understanding that, in fact, a restrictive reading of the clause is the more natural view to take of the Latin. But this is not decisive. Of more weight are considerations of immediate context and of consistency with one's general understanding of Spinoza's thought. Thus, arguments made elsewhere in this paper will weigh in on the side of the restrictive reading.

can be no part of the essence of any. In a nutshell, in Part V Spinoza is not contrasting intuitive and rational knowledge; he is contrasting intuitive knowledge of individual things with rational knowledge of them.

Notice that on the *Ethics* description of intuitive knowledge of objects there are two ways in which adequate knowledge of a thing can fall short of intuitive knowledge and, hence, be merely rational. Such knowledge may not be grounded in the formal essence of an attribute, or, being properly grounded, it may not advance to the essence of the thing. What Spinoza proves of objects in Part I fails to be intuitive knowledge because it fails the second requirement. It is significant that the second example of rational knowledge given in the *Treatise* is "knowledge inferred from some general proposition that some property is always present." This is a fair description of what is shown of objects in Part I, and it is further confirmation that the two accounts of intuition and reason are not fundamentally different.

(5) It would be unfair to finish this survey without taking note of the largest embarrassment to what I believe to be the correct way of seeing Spinoza's distinction. I refer to his example of our knowledge of the fourth member of a proportion. For when he speaks of knowing the fourth member intuitively rather than through a Euclidean proof, it does seem likely that he is thinking of a distinction between direct and inferred knowledge. However, it should be noted that it is notoriously difficult to reconcile any account of the ways of knowledge with the examples Spinoza provides. Furthermore, precisely the same strong suggestion of the inferential/direct knowledge distinction is present in the example Spinoza gives in the *Ethics*, in which, on the view I am criticizing, the distinction is supposed to have changed altogether. So in the *Ethics* this example is an embarrassment for both views. On the strength of the evidence provided above I suggest that it makes most sense simply to concede that Spinoza's example is misleading on this point. Probably he believed that once knowledge was well ordered and rooted in essences, one could "directly intuit" it in a way in which previously one could not. But this is not the distinction Spinoza describes or uses when he actually deals with the distinction between reason and intuition.

I conclude from all this that, considering just the relevant texts by themselves, my "ordering" interpretation comes off better than does the consensus view. It is consistent with what Spinoza says in these passages and makes better sense of them than does the alternative. Furthermore, the interpretation of Spinoza offered here avoids the problems facing what I have called the consensus. (1) If my reading is correct, then, with the proviso just noted, it is appropriate for Spinoza to cite mathematical examples as instances of intuitive as well as of rational knowledge. It is also appropriate in that it makes clear that the same things can be known by reason and by intuition. On the consensus view, Spinoza's example is wildly misleading on this point. (2) My reading does not have the consequence that knowledge of individual things is most blessed for Spinoza. It places knowledge of substance where it belongs–at the heart of Spinoza's doctrine of blessedness. The blessedness of the knowledge of things is derived from that of the knowledge of God. (3) Is intuition necessarily dependent upon reason, as it must be for the consensus? No, for intuitive knowledge of substance, or God, is not so dependent, and God's own knowledge will always be intuitive and not rational. It is only our knowledge of individual things which may, by reason of our finiteness, require reason. And even then, as the passage from Part V suggests, we may later come to know the same thing by intuition. That form of knowledge which constitutes human salvation does not *necessarily* rest upon an inferior form of knowledge. (4) Finally, the interpretation presented here does not have the consequence that Spinoza's view of knowledge underwent a drastic change in the *Ethics*. The distinction he draws is consistent, well-motivated, and tied to his fundamental insight about the universe and the methods we use to obtain knowledge of it: that what is independent and uncaused is ontologically prior to what is dependent and caused and that the highest form of knowledge must be rooted in the former.

III

Apart from the main issue of the reason/intuition distinction, both Curley and MacIntosh argue that Spinoza changed his mind about another matter. They claim that in the *Treatise* Spinoza restricts adequate knowledge to the highest form of knowing and that in the *Ethics* he reverses himself, allowing that both intuitive and rational knowledge are adequate. Actually, in the *Treatise* Spinoza is ambivalent about the status of rational knowledge. He says of it that "we may say that it gives us the idea of the thing sought and that it enables us to draw conclusions without risk of error; yet it is not by itself sufficient to put us in possession of the perfection we aim at" (Gebhardt, V. II, p. 13; Wild, p. 10). Spinoza is not going to endorse such knowledge; the point of the *Treatise* is to find the best form of knowledge, and that is clearly intuitive knowledge. But neither is he willing to go so far as to say that it may lead to error. (Perhaps it should be pointed out that the appearance of the phrase "not adequately" in Spinoza's definition of rational knowledge in no way commits him to the view that the knowledge itself is not adequate.) In the *Ethics*, of course, Spinoza is quite forthright in holding both that rational knowledge is adequate and that it is less desirable than intuition.

If this does constitute a shift of emphasis, there are at least two possible explanations for it. First, in working out the metaphysical details of the *Ethics* Spinoza may well have become more aware of just how deeply committed he was to the adequacy of forms of nonintuitive knowledge. This commitment grows out of his theory of the parallelism of the attributes of thought and extension and out of the doctrine of common notions. Second, Spinoza had to face the problem that confronts any philosopher with a special path to knowledge. What is to be said of what passes for knowledge among those who are not privileged to the special method? Do "underprivileged" scientists and mathematicians *really* know anything? It is hard to answer in the negative, but a positive answer seems to undermine the claims of the special method. Descartes was willing to take the heroic course of denying full epistemic credit to atheists since, without divine guarantee, they were vulnerable to the overthrow of reason. Spinoza is more tolerant. He allows that rational knowledge is necessarily true, at least so far as it goes. But his ultimate interest in the ethical rather than the epistemic allows him to grant knowledge claims to non-Spinozists without undermining the claims of his method. The method leads to joy and salvation; he can allow rational thinkers their claims to knowledge.[9]

Worcester, Massachusetts

[9] I wish to express appreciation to Martha Brandt Bolton for comments on an earlier paper that forced me to face the issues taken up here and to Robert M. Adams for assistance in dealing with Spinoza's Latin.

[13]

Infinite Understanding, *Scientia Intuitiva*, and *Ethics* I.16

MARGARET D WILSON

I

Spinoza defines 'substance' partly in terms of the way in which substance is conceived: "By substance I understand that which is in itself, and is conceived through itself: that is, that the concept of which does not need the concept of another thing, from which it must be formed" (E I.Def. 3).[1] But the following definition of 'attribute' suggests an even more central connection between *this* concept and that of a certain way of knowing: "By attribute I understand that which understanding perceives of substance, as constituting its essence" (E I.Def. 4). And the link between attributes and understanding is emphasized by Spinoza even more strongly in an early letter to de Vries (Letter IX). He writes:

> . . . [B]y substance I understand that which is in itself and is conceived through itself: that is, the concept of which does not involve another thing. I understand the same by attribute. Except that it is called attribute with respect to understanding. Attributing to substance a certain such nature.[2]

Some scholars have held that this definition of 'attribute' in terms of the perception of understanding indicates that the attributes in Spinoza's system are merely subjective or ideal.[3] Others have argued very persuasively against this reading.[4] Perceptions of understanding—and particularly understanding's ideas of the divine attributes—are held by Spinoza to be intrinsically true and adequate (II.37, 38, 41). But a true idea by definition has an agreeing ideatum. It follows that there are (formally) attributes that correspond to understanding's perceptions of attributes. And I take it further to follow that the attributes really do constitute and "express" God's essence, for otherwise they could hardly be said to "agree" with understanding's perception of them. Indeed, *Ethics* I.15, scholium, contrasts extension or quantity as conceived *abstractly* or *superficially* ("as we imagine it"), with the same conceived by understanding ("as substance"). This contrast suggests that the stress on understanding in Spinoza's account of attributes is particularly *meant* to carry the implications of truth and adequacy.

The subjective interpretation did, of course, have a reasonable motive: it was intended to resolve the problem of reconciling the unity of substance with the plurality of attributes. This remains a very difficult problem in interpreting Spinoza —one that I will not address here. Rather, I will argue that the definition of 'attribute' in terms of understanding ties in with conspicuous but initially mysterious references to understanding and "infinite understanding" in a key proposition concerning the causality of God, *Ethics* I.16. (Though conspicuous, these references to understanding have received relatively little notice in the literature on Spinoza.[5]) My approach involves the claim that *Ethics* I.16 involves the exemplification on the level of infinity of *scientia intuitiva* or the third (and highest) "kind of knowledge." It is *scientia intuitiva* that gives rise to the intellectual love of God and our highest possible peace of mind; it is also involved in the eternity of our minds. Hence, one consequence of my argument will be the demonstration of important continuity between the latter notions, which dominate the discussion of freedom and salvation in Part V, and the more broadly "metaphysical" definitions and propositions of Parts I and II. My reading also has significant implications for the interpretation of God's causality, as asserted in *Ethics* I.16.

II

Spinoza alludes to "understanding" in several of the early propositions of Part I, as well as in the definition of attribute. I want to focus though on I.16, where the expression "infinite understanding" also appears (and for the first time). The proposition reads as follows:

> From the necessity of the divine nature, infinite [things] in infinite ways (that is, all [things] which can fall under infinite understanding) must follow.
>
> *Ex necessitate divinae naturae, infinita infinitis modis (hoc, est, omnia, quae sub intellectum infinitum cadere possunt) sequi debent.*

The demonstration of this proposition consists of just two sentences. In the first Spinoza asserts that the proposition "ought to be evident to anyone" who only considers that:

> from the given definition of anything understanding infers (*concludit*) a number of properties, which indeed necessarily follow from it (that is, from the essence itself of the thing), and the more, the more reality that the definition of the thing expresses, that is, the more reality that the essence of the thing involves.

But, Spinoza concludes:

> Since the divine nature has absolutely infinite attributes, each of which also expresses infinite essence in its kind [all this follows from the definition

> Spinoza has given of God], from its necessity therefore infinite [things] in infinite ways (that is, all [things] which can fall under infinite understanding) must necessarily follow.

On initial reading, at any rate, *understanding* and *inference* certainly seem to be playing some important role in this proof. Comparing the first with the second sentence, we might even suppose that infinite understanding's inferential power plays some kind of *accessory* role in the derivation of things from the divine nature. But *this* cannot be right, for surely, on Spinoza's view, things just *do* follow with necessity from the divine nature: they do not need something different from that nature to bring them about. And Spinoza's remarks elsewhere clearly indicate that infinite understanding is *not* part of the divine nature: it is merely the *idea Dei* or an infinite mode under the attribute of thought.[6] So perhaps we should, taking heed of the parentheses, construe the references to infinite understanding in this proposition and proof as merely some sort of unimportant *obiter dictum*. But then why are the references there at all? Why would Spinoza muddle one of his key propositions with such distractions?

A more satisfying but (I am going to suggest) still inadequate explanation is the following. Spinoza clearly wishes to oppose the idea that God's creative activity involves a voluntary selection among the things existing in his understanding (as in the Leibnizian picture). He therefore mentions infinite understanding in the statement and demonstration of Proposition I.16 just to underline his view that God's understanding of things does not exceed in scope the things he actually brings into existence.

This explanation is inadequate because—I hope to show—the references to understanding in I.16 have a broader significance for the overall interpretation of the *Ethics*. They signal, namely, the involvement of this proposition, together with the definition of attribute, with *scientia intuitiva*.

In *Ethics* II.40 Scholium 2 Spinoza explains *scientia intuitiva* in the following terms: "This kind of knowing proceeds from an adequate idea of the formal essence of some attributes of God to an adequate knowledge of the essence of things."[7] I will argue that this definition fits exactly the role ascribed to infinite understanding in I.16—with several interesting implications for the interpretation of the content of the proposition.

As we noted in connection with the definition of 'attribute', understanding perceives the attributes adequately, and indeed can only perceive them adequately. Further, I think it is clear enough from what has been said that in perceiving the essence of substance, understanding perceives it "under" an attribute—or "attributes to substance a certain such nature." (See also E I.10S.) But the notion of inference or conclusion introduced in Proposition I.16 certainly suggests a "proceeding from . . . to."[8] So, to show that Proposition I.16 involves an exemplification, on the level of infinity, of *scientia intuitiva*, I think that we need to do only one thing. We need only show that Spinoza conceives of infinite understanding as proceeding (inferring) from its adequate ideas of the formal essences of the attributes of God *to an adequate knowledge of the essence of things.*

Now, Spinoza makes quite clear in the subsequent propositions and demonstrations that the "*infinita infinitis modis*" do at least *include* the essences of things. For example, in Proposition 25 of Part I (scholium), he observes that "God is the efficient cause not only of the existence of things but also of their essence," citing I.16. But does infinite understanding form *adequate* ideas of the essences of things that it infers from the essence of God's attributes (as the definition of *scientia intuitiva* would further require)? There are strong grounds for saying it does. According to Proposition 40 of Part II, "Those ideas are . . . adequate which follow in the mind from ideas which are adequate in it." And in I.16 *infinita infinitis modis* are precisely said to "follow" (*sequi*) from the ideas of God's attributes in infinite understanding—which, as we have noted, must be adequate. Also, Spinoza asserts in the proof of V.17 that "all ideas, as they are related to God, are true; that is to say, are adequate." I will take this as sufficient evidence at the present stage of argument. (There are some difficulties about the point, though, which I will mention later.)

It seems then that I.16 does portray infinite understanding as proceeding from the "adequate idea . . . of some attributes of God to an adequate knowledge of the essence of things"—and hence as complying with the definition of *scientia intuitiva*. But, one might object, I.16 still does not conform *exactly* to the definition. For, as we have just seen, Spinoza relates this proposition to the claim that the essences *and existences* of things follow from the divine nature. But *scientia intuitiva* seems to have to do with essences in contrast to existences.

This is a reasonable objection but one that can be answered. And the answer bears in an interesting way on the interpretation of I.16 itself.

Later in the *Ethics*, when Spinoza refers back to this proposition, he at least twice indicates that the sort of existence or reality that it asserts to follow from the divine nature must *not* be confused with "existence at a certain time and place." On the contrary, it has to do with a type of existence that Spinoza gives us rather clear license to construe in terms of "essence." Consider the scholium to Proposition 45 of Part II. The proposition itself reads: "Any idea of any body, or singular thing, existing in act, necessarily involves the eternal and infinite essence of God." The scholium refers us back to I.16.

> Here by existence I do not understand duration, that is, existence as it is conceived abstractly, and as a certain kind of quantity. For I speak of the very nature of existence, which is attributed to singular things, because [*propterea quod*] infinite [things] in infinite ways follow from the eternal necessity of the nature of God (see Prop. 16, p. I). I speak, I say, of the existence itself of singular things, as they are in God. For, even though any singular thing is determined by another to a certain mode of existing, nevertheless the power, by which any thing perseveres in existence, follows from the eternal necessity of the nature of God.[9]

But the power, by which a thing perseveres in existence, is characterized by Spinoza as the "actual essence" of the thing itself (E III.7). This characterization suggests that "the existence itself of singular things, insofar as they are in God" may be identified with the essences of things (in a certain special sense). And this is just the sort of "existence of things" that I.16 asserts to follow from the attributes of God. Hence, infinite understanding's inference even of "existences" in Proposition I.16 actually is subsumable under Spinoza's account of *scientia intuitiva*.

Ethics V provides further support for the proposed interpretation of I.16. For example, Proposition 29 of Part V reads as follows:

> Whatever the mind understands under the aspect of eternity, it does not understand from the fact that it conceives the present actual existence of the body, but from the fact that it conceives the essence of the body under the aspect of eternity.

The scholium to this proposition refers specifically to II.45, which I cited just above. The scholium reads:

> Things are conceived by us as actual in two ways, either as they exist with relation to a certain time and place, or as we conceive them to be contained in God, and to follow from the necessity of the divine nature. But those which are conceived in this second way as true or real, we conceive them under the aspect of eternity, and their ideas involve the eternal and infinite essence of God, as we showed in Proposition 45 of Part II.[10]

Taken together these two passages—and others from Part V could be cited—seem clearly to indicate that Spinoza distinguishes two senses of 'existence' or 'being-actual', only one of which relates to what infinite understanding infers according to *Ethics* I.16. The "existence" of things that infinite understanding infers from the divine attributes must be distinguished from their duration through time or at a certain place. Proposition I.16 specifies the inference of the essences of things from the essence of God: it tells us that (in the words of V.22, which also makes use of I.16) "in God there necessarily exists an idea which expresses the essence of this or that body under the aspect of eternity."

If I am right, then, Proposition I.16 has considerably broader significance and implication than is usually noticed. It does not merely tell us that all things follow "from the necessity of the divine nature" (or by necessity from the divine nature)[11] and that *everything* comprehended by infinite understanding does so "follow." It also relates this fundamental statement about the origin of beings in God to the concept of the third kind of knowledge, which is only explicitly developed later. Both Proposition I.16 and *scientia intuitiva* play important roles in Spinoza's development of his ethical and (loosely speaking) eschatological views in Part V. I hope that my interpretation, by showing the connection between this key proposition of Part I and the concept of knowledge explained in Part II, will be suggestive also as to the interpretation of Part V. More particularly, I hope it will advance efforts to interpret the *Ethics* as a truly unified work, in which metaphysical, epistemological, ethical, and eschatological themes are quite rationally and purposively intermingled.

I will not attempt here to apply my interpretation of I.16 to the detailed interpretation of the propositions of Part V. However, the proposed reading of I.16 as relating to the concept of *scientia intuitiva* does have one particular implication that seems worth pointing out. I mentioned above that the references to infinite understanding in the proposition might be construed as Spinoza's way of underscoring his opposition to the voluntarist view that God *selects* the things to be brought into existence from *among* the things he understands. But, at least in its Leibnizian version,[12] the voluntarist position has to do with creation in the sense of bringing into existence at a time and place. And, according to my reading of I.16, nothing about existence at a certain time or place (or about "actual" being in the corresponding sense) is supposed to be established by I.16. If this reading is correct, then, the proposition does not establish quite as complete and direct an opposition between Spinoza's metaphysics and, say, Leibniz's as might be supposed. Spinoza need not—indeed cannot—be construed as saying that everything that falls under the divine understanding exists in the sense of being instantiated at a time and place or of "having duration." Rather, he holds that everything that falls under the infinite understanding has *some sort of being* in the divine attributes (such as extension). (This result also accords well with Spinoza's statement that the "ideas of non-existent modes" have their ideata "in the attributes of God"—a claim crucial to his proof [in Propositions V.21-40] that the human mind is eternal.) Of course, even on this reading of I.16, Spinoza still differs from Leibniz in holding that all of God's ideas have *ideata* or objects that are *in some sense* actual. However, at least as far as Proposition I.16 goes, he need not be differing from Leibniz to the extent of denying that in *one* sense of 'existent' or 'actual' not everything in God's understanding is existent or actual.[13] (What is most peculiar is that Spinoza regards the notion of existence or being-actual involved in this second, commonsensically more obvious sense as an "abstraction."[14])

Now to consider some problems.

III

The reading of I.16 that I am proposing leads us into several difficulties. I believe that only one—the first to be considered—is a problem for the interpretation itself. The other difficulties I will mention seem to arise within Spinoza's system independently of the connection I am suggesting between I.16 and *scientia intuitiva*. They are relevant to my interpretation without being generated by it.

The first problem I want to consider has both a philosophical and a textual aspect. Philosophically, one may argue that my interpretation implausibly leaves the determination of existence *in the sense of duration* (or existence at a time and place) outside the scope of Spinoza's major statement on the origin of things in God. In fact, my reading of I.16 leaves the significance of I.16 rather unclear in general, for *what is* the "very nature of existence" in contrast to "existence at a time and place" or duration? Textually, the interpretation of I.16 that I am proposing runs up against the fact that some later propositions of Part I (and Part II) seem to suggest that the "causality of God" delineated in *Ethics* I.16 does after all include the determination of things' duration. In particular, in the corollary to Proposition 24 of Part I—"The essence of things produced by God does not involve existence" —Spinoza comments: ". . . [T]he essence [of things] cannot be the cause either of their existence *or of their duration*, but God only, to whose nature alone it pertains to exist" (italics added). This statement certainly suggests that Spinoza is including the duration of things in the scope of God's causality. And subsequent propositions explicitly presuppose that I.16 has established that God is the cause of all things, *both* of their essence and of their existence.

Now, I believe that the *textual* aspect of this objection to my interpretation can be met, though admittedly not in a philosophically satisfying manner. Spinoza does indicate that God is the cause of things' duration in I.24, but he does *not* there refer us back to I.16. And the subsequent passages, which do refer back to I.16, contrast existence and essence under the divine causality but make no mention of duration. It is, therefore, possible after all to render these propositions consistent with II.45 and V.29, which I have relied on in defending the view that I.16 exclusively exemplifies the third kind of knowledge. For one can, consistent with the former texts, construe Spinoza's position in the following way: (1) Though Spinoza may assume that God is the cause of the duration of things, he does not base this claim on *Ethics* I.16. (2) When he speaks of God as cause of the essences and existence of things in the later propositions of Part I, he has in mind existence in the recherché sense he later contrasts with duration ("the very nature of existence"). Existence in this recherché sense may (according to II.45 and III.7) be identified with "the actual essence" of things.

Unfortunately, this reading does leave the reference to the causing of duration in I.24 a seemingly inexplicable loose end. But such a loose end seems unavoidable in any case, for Proposition 45 of Part II does quite explicitly imply that the divine inference of I.16 is *not* an inference to the existence of things at a certain time and place.

The second problem that needs to be considered has to do with the transition in Spinoza's system from infinite attributes to finite modes. It is a problem that is often mentioned in the literature, and it does not particularly weigh against my reading of I.16 as exemplifying the concept of *scientia intuitiva*.[15] But it does have a special connection with my reading.

How is it *possible* that infinite understanding should infer adequate ideas of the essences of things from adequate ideas of the formal essences of certain attributes of God? Spinoza tells us in I.21 that "all [things] that follow from the absolute nature of any attribute of God" must be infinite and eternal, i.e., the infinite modes. In the proof he comments: "Hence that which so follows from the necessity of the nature of any attribute cannot have determinate existence or duration." The language here is highly reminiscent of that of I.16. The statement, in fact, can happily be seen as confirming my claim that I.16 does not have to do with the origin of things' duration. Unfortunately, though, I.21 goes further than this, for it seems to indicate that only *infinite* and eternal modes—or ideas of such modes—could follow or be inferred from "the necessity of the divine nature."[16]

E I.21 thus presents grave difficulties for interpreting the "*sequi*" of I.16, difficulties that arise independently of my attempt to relate that proposition to the concept of *scientia intuitiva*. To express the point bluntly, Spinoza seems both to affirm and to deny that finite things "follow from" the divine nature. But we are now in a position to see some further complications of this problem. It is not *just* a problem for the reading of E I.16 but *also* a problem for the concept of *scientia intuitiva*—even if I am wrong in regarding the two as intimately related—for we can have *scientia intuitiva* just insofar as we *can* proceed *from* adequate knowledge of the essence of God's attributes *to* adequate knowledge of the essences of things. If this is not possible in the case of finite things, then it appears that they cannot at all legitimately be included under *scientia intuitiva* any more than under the derivation of I.16 (even if these are, contrary to my suggestion, distinct). The texts do seem to me confusing and unsettled on this point—quite a fundamental one, unfortunately, for Spinoza's "necessitarianism," his epistemology, and his theory of human happiness and salvation.

In originally relating I.16 to *scientia intuitiva* I partly relied on a proposition affirming that all ideas in God are adequate. I used this to show that infinite understanding conforms to *scientia intuitiva* in that it infers to adequate ideas of the essences of things, but I also mentioned having some reservations about the point. We have just seen that there are general difficulties in understanding the possibility of inferring, or proceeding, from the divine attributes to anything finite. It is now appropriate to note that the arguments Spinoza offers for his claim about the adequacy of God's ideas (V.17; II.36) are not very satisfactory. In these arguments Spinoza in effect holds that the *truth* of all ideas as related to God entails the *adequacy* of all ideas related to God. But God's ideas are all true just because they all have corresponding ideata. And "adequacy," as Spinoza himself defines it, seems to involve the additional condition of possessing all the *intrinsic properties* of a true idea (II. Def. 4). This awkward fallacy in Spinoza's argument may be a reflection of his deep difficulty in rationalizing the transition (in God) from infinite attribute (or its idea) to finite mode (or *its* idea).

The study of I.16 brings up another problem. This again, I think, is a difficulty that could not be avoided by any reasonable reading of I.16. As I briefly suggested in a previous remark, Spinoza clearly and firmly distinguishes the status of infinite understanding from that of the attribute of thought. He argues that infinite understanding is a *mode* of the attribute of thought. This, of course, means it is not self-caused, or "conceived through itself." Yet, by I.16 it has ideas of the attributes of God. Now, surely if it is a mode, all its ideas (including our minds) must be modes. But, famously, Spinoza holds that ideas and their ideata must be causally parallel. There seems to be an outright contradiction between the latter notion and the supposition that "other-caused" ideas take as their ideata self-caused attributes. Maybe there is an obvious answer to this dilemma, but I have not been able to see it.

Thus, my interpretation of I.16 in terms of the concept of *scientia intuitiva* by no means solves the major problems of interpretation and problems of apparent inconsistency in Spinoza's system. Indeed, it leads directly into some of them.

However, I do not think it creates any *new* problems of this sort. And it does, I hope, help to illuminate the important intimacy among the metaphysical, epistemological, and ethical aspects of Spinoza's thought.

IV

In conclusion, I wish to turn to a different sort of issue. It is, I think, a common experience of students of Spinoza to feel torn between—or to alternate between—two conflicting conceptions of his philosophical stature. Some of the time, one is mainly of the opinion that his system is just too hopelessly shot through with inconsistency, fallacy, obscurity, and idiosyncrasy to *deserve* the painstaking analysis it seems to demand of the reader. At other times, one may be more impressed with the startlingly original and even powerful character of Spinoza's underlying philosophical conceptions. The treatment of creation, or the origin of dependent being, in Part I—and perhaps particularly in I.16—provides a good example of why this should be so. We have already surveyed a number of serious—indeed critical—problems with which it is bound up. On the other hand, the idea that things arise *necessarily*, as if by deductive inference, from God or nature is surely a bold and provocative one, however exactly it should be interpreted. It is a view that in some ways, at least, does provide a direct and intriguing alternative to the more traditional voluntarist conception defended (for instance) by Leibniz. Therefore, I would like finally to comment briefly on the following question: How successful is Spinozà in his attempt to establish that his form of necessitarianism is more in conformity with reason than is traditional voluntarism?[17]

The core of the voluntarist position is, I take it, the following: The world is, and is *as* it is, because God *chose* that this be so. He could have chosen differently and so things could have been different. It could even have been the case that he abstained from creating anything. Spinoza surely does make some cogent or plausible points against this view. Some are familiar, such as the argument that an all-perfect God would have no reason to decide to bring into being any thing outside himself. (This precise sort of difficulty does not arise for Spinoza, since his God does not do anything on purpose and hence *requires* no "reason for acting."[18]) Some of Spinoza's other objections to voluntarism are much more original. He argues, for instance, that, if God could have chosen differently, there must be more than one possible divine nature. But the ontological argument will apply to each. Therefore, there would have to be more than one God! (I. 33).[19] But one of the most interesting and fundamental points of contrast between Spinoza and his voluntarist opponents is found in their respective concepts of omnipotence. And, although Spinoza makes great fun of his opponents' views on this subject, I do not think he provides a coherent refutation.

Spinoza, as I read him, holds that God's omnipotence requires that everything in God's power actually comes into being. In other words: "G is omnipotent" entails "G brings about everything in its power to bring about." His opponents, he indicates, deny this proposition on the riduculous grounds that, unless there are unrealized possibles in God's understanding, his power to create would have come to an end! (And, hence, I suppose, be limited.)

> . . . [A]lthough they conceive God as actually understanding to the highest degree, they nevertheless do not believe he can bring it about that all [things] which he actually understands, exist; for they think they would in that way destroy God's power. They say that if he had created all [things] that are in his understanding, then he would be able to create nothing more, which they believe is inconsistent with God's omnipotence . . . (E I.17s).

Now, I do not know whether any of Spinoza's contemporaries actually argued like this, but certainly the voluntarist position can be stated in more plausible terms. The voluntarist may contend that omnipotence requires the power *to bring about or prevent* any possible state of affairs and therefore requires choice—the choice of which to do. In other words: "G is omnipotent" entails "it is in G's power to determine, for any possible contingent state of affairs, whether or not it obtains."[20] Presumably, Spinoza would insist that such a conception of omnipotence rests on an illegitimate anthropomorphization of God, an extrapolation from a common conception of human power to God. But he does not systematically defend or explain his own intuition. Perhaps he supposed that the demonstration of I.16 was sufficient to settle the issue in his favor. If so, I think he was wrong.[21]

Notes

1. E is used as abbreviation of *Ethica*. Translations are my own. Geb. (below) refers to *Spinoza opera*, ed. C. Gebhardt (Heidelberg: Carl Winters, 1926) (4 vols.).

2. Geb. IV, p. 46. (The peculiar punctuation follows Gebhardt's text.) Throughout the paper I translate '*intellectus*' as 'understanding'. In an earlier version of the paper, I frequently used 'the understanding'. At a Spinoza symposium, E. M. Curley objected to this phrasing on the grounds that it introduces a specious definiteness where (on Curley's view) Spinoza is only talking about *an* understanding or *some* understanding or other. (The Latin, of course, has no article at all.) Curley made this point in connection with more substantive criticisms. Although I was not persuaded by his other substantive criticisms, I have dropped the definite article throughout for the sake of consistency and to avoid the appearance of question-begging on the issue.

3. See especially H. A. Wolfson, *The Philosophy of Spinoza* (Cambridge, Mass.: Harvard University Press, 1934), pp. 142-57.

4. See especially Francis S. Haserot, "Spinoza's Definition of Attribute," in S. Paul Kashap, ed., *Studies in Spinoza* (Berkeley: University of California Press, 1972, originally published 1953); Martial Gueroult, *Spinoza: Dieu (Ethique, 1)* (Paris: Aubier-Montaigne, 1968), p. 50.

5. One exception is an unpublished manuscript by Genevieve Lloyd, *The Eternity of the Mind: A Study of Spinoza's Ethics*. Lloyd's approach is in some ways congruent to mine, though there are also substantial differences between our interpretations.

6. Cf. E II.4; 1.34.

7. Geb. II, p. 122.

8. My linking of the third kind of knowledge with E I.16 seems to entail a different understanding of 'proceeds' in the definition of the former than that sketched by Guttorm Fløistad, "Spinoza's Theory of Knowledge Applied to the *Ethics*," in Kashap, pp. 271-72. Fløistad, incidentally, provides a useful gloss on the term 'formal essence'.

9. Cf. E V.22. See also E. M. Curley, *Spinoza's Metaphysics* (Cambridge, Mass.: Harvard University Press, 1969), pp. 141-42.

10. This passage is also cited, in a similar connection, by Joel I. Friedman in "Spinoza's Denial of Free Will in Man and God," in Jon Wetlesen, ed., *Spinoza's Philosophy of Man: Papers Presented at the Scandinavian Spinoza Symposium, 1977* (Oslo: 1978), pp. 51-84.

11. It is not unreasonable to wonder whether "follow *from the necessity* of the divine nature" implies "follow *with necessity from* the divine nature"; however, both the demonstration of E I.16 and Spinoza's subsequent use of the proposition idicate that he does intend this implication.

12. I mention Leibniz in this connection because his philosophy seems to provide in some ways an excellent *example* of the *type* of view Spinoza wants to oppose (and one with which I happen to be familiar). I do not mean to suggest that Spinoza was reacting against his successor!

13. I have learned in discussion that quite a few people want to attribute to Spinoza the view that everything in God's understanding has temporal or durational existence at some time or other. This issue seems to me an obscure one textually. There is also the problem of compossibility: Is it reasonable to hold that everything in God's understanding is compossible with everything else in the sense that, given enough time, it is possible that all these things exist? Alan Donagan attributes to Spinoza a view about incompatible essences in "Spinoza's Proof of Immortality" (*Spinoza: A Collection of Critical Essays*, Marjorie Grene, ed. [Notre Dame, Ind.: University of Notre Dame Press, 1979; originally published 1973], pp. 253-55). (Like Fløistad [see n. 8], Donagan gives some help with the concept of "formal essence," which he contrasts with "actual essence." However, Donagan does not appear to take note of the fact [argued above] that Spinoza recognizes *two* senses of 'actual',)

14. '*Abstracte*' and '*abstractus*' are not words that Spinoza uses often. See, however, E I.15S (Geb. II, p. 59), and Letter XII (to L. Meyer) (Geb. IV, pp. 56-57).

15. See, for instance, Alasdair MacIntyre, "Spinoza, Benedict (Baruch)," *The Encyclopedia of Philosophy*, ed. Paul Edwards, (New York: Macmillan, 1967), Vol. 7, p. 535.

16. In "The Causality of God in Spinoza's Philosophy," *Canadian Journal of Philosophy*, Vol. II, No. 2 (December, 1972), A. J. Watt argues that the direct causality of God, as asserted in E I.16, should be understood as the causing of *essences*. On this point, of course, my own interpretation largely follows his. But Watt also seems to hold—what I would deny—that this construal of 1.16 helps avoid problems about the emergence of the finite from the infinite.

17. Of course, if my foregoing argument is right, there is not a *complete and direct* opposition between Spinoza and traditional voluntarism, at least as far as E I.16 goes (since Spinoza is not there concerned with the bringing about of temporal existence). I think there is still enough opposition to consider intelligibly the issue taken up in this section.

18. However, the problem about "how" the finite modes arise from the infinite attributes in his system may perhaps be regarded as somewhat parallel.

19. I sketch a similar line of reasoning in relation to a problem of interpreting Leibniz's views about modality in "Possible Gods," *The Review of Metaphysics*, Vol. XXXII, No. 4 (June, 1979), pp. 717-33.

20. Curiously, Spinoza himself touches on this conception at the beginning of the scholium to E 1.17.

21. I am grateful to more people than I could reasonably name here for helpful comments on earlier versions of this paper, many of which have led to changes in argument and structure. But I must particularly thank Eyjolfur Emilsson, E. M. Curley, George L. Kline, and Joel Friedman for detailed comments and criticisms.

Part III
Anthropology and Ethics

[14]

The Southern Journal of Philosophy (1991) Vol. XXIX, No. 4

SPINOZA'S DENIAL OF MIND-BODY INTERACTION AND THE EXPLANATION OF HUMAN ACTION

Charles Jarrett
Rutgers University

Introduction

The following paper is an attempt to elucidate and defend a central part of Spinoza's solution to "the mind-body problem." Section I provides a characterization of Spinoza's theory and offers a defense of it against a charge of logical incoherence. Section II endeavors to clarify the theory by considering its implications for the explanation of human action. Section III, finally, connects a long-standing interpretive controversy in Spinozism to problems faced by current attempts to develop an adequate physicalist theory of mind.

I assume in what follows that current work on the mind-body problem, and especially Davidson's "anomalous monism," can be profitably employed to help clarify Spinoza's position. I also suppose, and will attempt to make it more evident here, that Spinoza's theory can shed light on current controversies.

A more detailed account of what readers can expect is as follows.

1. In the *Ethics* Spinoza appears to have accepted an identity theory of mental and physical individuals, or items. He also seems to have maintained that there is an irreducible conceptual distinction between the mental and physical, and that there are no causal relations between them.

This gives rise to a problem. If an individual mental event (an "event-token") literally *is* a certain physical event, and this physical event has physical effects, does it not follow

---

Charles Jarrett received his Ph.D. from the University of California at Berkeley in 1974 and is currently an Associate Professor of Philosophy at Rutgers University, Camden College of Arts and Sciences. His interests include the history of modern philosophy, philosophy of mind, ethics, and philosophy of language.

that at least one individual mental event has physical effects?[1]

Davidson supposes that this is so. Spinoza denied it. They agree, however, in holding that nothing physical can be explained by citation of something mental, *qua* mental. In section I of this paper I argue that Spinoza's apparent inconsistency here is only apparent.

2. Spinoza's denial of causal interaction also appears to entail that ordinary teleological explanations of "physical acts" are unacceptable. How, if mental states have no physical effects, can citation of an agent's beliefs and desires help to explain acts that "involve" physical changes, such as walking into a store? Spinoza gave two answers to this, one of which was that motion itself can be conceived as mental. But how is this possible? An answer to this is here suggested, along with a view about the function of the conceptual division between the mental and the physical.

It is maintained, more specifically, that the *conceptual* divide revealed by the divergence between a physico-chemical description and the attribution of a goal or state with content is one that must be crossed, and in the *Ethics* is crossed (primarily in Part III), in the transition from metaphysics to ethics. The thesis, as expressed in the *Treatise on the Emendation of the Intellect*, is that nothing is good or bad except insofar as it affects the soul (*anima*).[2] So no matter how complex a body is, normative thought becomes applicable only when that body is conceived, not as a collection of chemicals, but as a being with goals or desires.

3. Spinozists are not in agreement about the ultimate source of the conceptual distinction between the mental and the physical. Everyone agrees that the (and our) concept of Thought is distinct from, and irreducible to, the concept of Extension (and vice versa), according to Spinoza. Whether Thought itself is actually distinct from Extension is in dispute, however. Must the conceptual distinction be grounded in an ontological one?

Spinoza thought not, I hold, but the interpretive problem of the attributes, and the problem of determining what Spinoza should have said, has emerged in this century as a substantive problem about how to be a good materialist.

Materialists who have attempted to develop a "token-identity theory" have been faced with the reply that an identity of mental and physical individuals can be sustained only if a dualism of mental and physical properties (universals) is granted. Because properties are conceived primarily as the correlates of predicates, this has prompted some to propose a translation of "mentalistic" talk into either "physicalistic" or "topic-neutral" vocabularies. In Spinoza's terms, the success of such a project would amount either to a reduction of the concept of Thought to that of Extension, or (almost equivalently) to the denial that the concept of thought has any distinctive descriptive content.

An alternative reply has been to suppose that there is an irreducible conceptual distinction between the mental and the physical. The most prominent current exponent of such a view, however, has conjoined it to the thesis that mental events, conceived as mental, are incapable of being connected by strict laws to anything, that is, that they are, as mental, causally inefficacious (and equally incapable of being caused). Mental events cause physical ones, on this view, but they do so only in virtue of their physical characteristics.

But if the possession of a "mental property" is neither a cause nor an effect of anything, what ultimate explanatory role could the postulation of such properties have? None, we have been invited (by some) to suppose. But then what use are they? Doesn't a conceptually non-reductive materialism then end in a denial of the existence of the mental?

Spinozism is immune to this, because the mental—like everything real—is governed by strict laws, and not causally impotent. It is immune because it is not anomalous. But if the mental has only mental causes and effects, and everything real has causes and effects, does this not entail that the mental, like the physical, forms a closed (and infinite) system?

What is a good materialist to do? I suggest myself that we must transcend a purely physicalist conception of ourselves. This means that the denial of an ontological division must be coupled with an irreducible conceptual division. (Whether this entails an objectionable dualism of properties will then depend on how we solve the problem of universals.) A good materialist will then be—broadly speaking—a Spinozist.

I. Mind-Body Identity and the Rejection of Interaction

Although Spinoza at one time accepted a dualistic account of the relationship between the human mind and body,[3] his position in the *Ethics* seems to be that there is a conceptual, not an ontological, distinction between them. He expressed this by saying, for example, that the human mind and body "are one and the same individual, conceived now under the attribute of thought, now under the attribute of extension."[4] (EIIIp2s) If we take a mind to be a thing that thinks, perceives, desires, and so on, and we take a body to be a thing that moves, has mass, and so forth, then Spinoza's thesis here amounts to the claim that that which thinks is the same individual as that which moves. Mental states are ascribed to the very same thing to which we ascribe physical states.

This characterization of Spinoza's position is importantly incomplete, however. For it is compatible with a "dual-aspect" theory, or a Strawsonian theory of persons,[5] insofar as it merely maintains that the subject to which we ascribe activities such as thinking is the very same as that to which we ascribe motion. It is clear, however, that Spinoza is committed to much more than this. He maintained not merely that the human mind and body are the same; he maintained as well that mental events and physical events are one and the same. Explicit textual support for this is found, for example, in EIII Preface, where he held that a decision (or "decree of the mind") and a "determination of the body" are one and the same thing, conceived in different ways.[6]

Thus Spinoza's theory does not merely maintain that there is a unitary subject to which we ascribe "mental" and "physical" activities; it maintains that mental events or acts are physical—and vice versa.

If we ask here whether Spinoza conceived of these decrees of the mind and determinations of the body as types or tokens, the answer seems clear. It is individual mental acts, or tokens rather than types, with which Spinoza's theory is primarily concerned. In EIIp48s, for example, Spinoza rejected the existence of "absolute faculties" of will, understanding, desire, love, and so on. "These and similar faculties are either entirely fictitious, or else are nothing besides metaphysical beings, or universals, which we commonly form from particulars. So intellect and will are related to this and that idea, or to this and that will in the same way as stonehood [*lapideitas*:being a stone] is related to this and that stone, or man to Peter and Paul."[7] Such faculties "are universal notions which are not distinguished from the singulars from which we form them."[8]

That there are individual mental acts, according to Spinoza, is clear. That there are *only* individual mental acts also seems clear: "In the mind (by the preceding proposition) there is given no absolute faculty of willing, and not willing, but only singular volitions, namely, this and that affirmation, and this and that negation."[9] But as the scholia to IIp40 make evident, in addition to the *universalia* which are "not distinguished" (*non distinguuntur*) from particulars, and which are inadequately understood, there are also certain properties of things, of which we form common notions or adequate ideas.[10] But these are not expressed by (and are not correlates of) terms such as "man," "horse," "understanding," "will," "love," or "desire."[11]

Spinoza's identity theory of mental and physical events (and of minds and bodies) thus proposes an identity of individuals, or tokens. Universals—or at least most of the ordinary specimens of them—are nothing, really, besides individuals.[12]

If minds are bodies, and individual mental events or states are physical, can we not then explain your thoughts by talking about the neurophysiology of your brain, and can we not then explain why your body moves as it does by talking about your thoughts and emotions? Spinoza's (official) answer is unequivocal. "The body cannot determine the mind to think, nor can the mind determine the body to motion or rest, nor to anything else (if there be anything else)" (EIIIp2). There are no causal relations between mental and physical events. Thus if an explanation requires citation of a cause, you will not get an explanation of your thought by citing facts about cell membrane permeability, the properties of sodium ions, and synaptic gaps. Nor will you get an explanation of the motion of your body, or of brain events, by citing facts about what you think or how you feel.[13]

The relations, or lack of them, between the concept of a mental event and the concept of a physical event are thus partially clarified by Spinoza. Conceived as physical, or under the attribute of Extension, a physical event must be (causally) explained by another physical event; conceived as a decision (*mentis decretum*), or under the attribute of thought, a mental event must be (causally) explained by another mental event (that is, by an event conceived as mental).

Spinoza's explicit argument for the denial of causal interaction relies (in EIIIp2d) essentially on the claim that each attribute of God is "conceived through itself and without another,"[14] and thus a mode of one attribute does not "involve the concept" of any other attribute. Part of what Spinoza is asserting here is that no definitional reduction of mental to physical concepts, or of physical to mental concepts, is possible. The concepts of mental and of physical events are independent of each other; they are not interdefinable, nor "reducible" to each other. But how does this help to establish that there are no causal relations between them? The answer is supplied by EIax4 (as noted in IIp6d). This axiom entails that if a mental event caused a physical one, then the knowledge or concept of the physical event would depend on and involve that of its cause. An analogous result would obtain if physical events caused mental ones. The thesis that if x causes y, then there must be a conceptual connection between them, it might be suggested, is not unlike the claim that if x causes y then there must be a *nomological* or strict law that would allow the deduction of one from the other.[15]

Spinoza's claim that a (mental) decision and a (bodily) determination are "one and the same" appears inconsistent with the thesis there are no causal relations between mental and physical phenomena. The argument for this is simple, but it is flawed.

Let "*m*" denote some (particular) mental event, and let "*p*" denote the physical event with which it is allegedly identical. Assume in addition, and in accordance with Ip28 and IIp6, that "*q*" is a physical event that causes *p*. Finally, let "Cxy" be a two-place predicate, to be read as "x causes y." Since

(1) $m=p$ and

(2) C*qp*

it follows by the principle of the indiscernibility of identicals that

(3) C*qm*.

Since Spinoza's position commits him to (1),(2) and the negation of (3), his position is logically inconsistent.[16]

This objection could be taken as a *reductio* of the "identity theory" interpretation of Spinoza. It might well be thought, that is, that since Spinoza did clearly reject causal interaction, and since this is incompatible with the thesis that mental and physical events are identical, Spinoza should not be "'saddled with" the identity thesis.

I think, in contrast, that the identity thesis is correct as an interpretation of at least part of what Spinoza said, and that what is wrong with the objection above is that causality is not correctly representable, in Spinozism or otherwise, by a predicate for which the principle of substitution *salva veritate* holds. "Causes y" is, in more recent terminology, referentially opaque.[17]

Spinoza's position here can be usefully contrasted with Davidson's "anomalous monism."[18] Davidson maintains that if x causes y, it does so no matter how x and y are described. Causality is correctly representable with the help of a referentially transparent predicate. Hence, since desires cause physical actions, on his view (and desires are mental events), some mental events cause physical events.

If there is a causal relation between x and y, however, there must according to Davidson be a law, expressible in general terms, that "connects" x and y (or "under which" x and y, described in some way, fall). There are no such laws connecting events conceived or described as mental and events conceived as physical (nor could there be); there are only laws that characterize events conceived as physical. Hence a mental event that causes—or is caused by—a physical one must be characterizable in physical terms. Any mental event that is a cause or an effect of a physical one must therefore itself be physical. (In fact Davidson's position is more general than this; since there are no strict laws relating mental events to anything, any mental event that causes or is caused by anything must be characterizable in physical terms. So all mental events are physical—if everything, or every mental event, is a cause or an effect of something.)

The physical world, so to speak, forms a closed system; mental events can "impinge" on this world just insofar as they are physical. It follows, on the face of it, that the properties in virtue of which an event is characterized as *mental*, are causally inefficacious. As Kim has recently put it in characterizing Davidson's view,

> . . . on anomalous monism, events are causes or effects only as they instantiate physical laws, and this means that an event's mental properties make no causal difference. . . .
>
> Anomalous monism, therefore, permits mental properties no causal role, not even in relation to other mental properties. What does no causal work does no explanatory work either; . . . Given all this, it's difficult to see what point there is in recognizing mentality as a feature of the world.[19]

The upshot is in one respect curiously—and perhaps surprisingly—like Spinoza's position. Nominally, there is disagreement about whether mental events cause physical events. Substantively, there is radical agreement that the explanation of a physical event must be given by reference to an event conceived as physical.

Important substantive disagreement certainly remains, however. Spinoza maintained that there are strict laws concerning events conceived as mental, whereas Davidson denies this.

Spinoza supposed that an event is explained by citing its cause. Since citation of a mental event—as a mental event—does not explain why any physical event occurred, no mental event, conceived as mental, causes any physical event (conceived as physical). The opacity of "explains" infects "causes." Davidson, in contrast, supposes that causality is relatively pure; opacity infects not this, but only "explains."[20]

It seems to me that on this issue Spinoza is correct: "causes y" is opaque, that is, the substitution of correferential expressions for "y" does not preserve truth. My reasons for this are explained elsewhere,[21] but it might still be useful to give an indication of them here. Suppose that George created a ring, at a certain time, by hammering a piece of gold into a certain shape. "George created the ring," is true, while "George created the piece of gold," is false. But the ring *is* the piece of gold. Use of a referentially transparent two-place predicate—"x created y" (or a one-place predicate, "George created y")—would portray these claims as logically inconsistent. But they are not.

An alternative formulation would employ not "creates y" but "brings it about that Phi(y)." The surface structure of "George brings it about that" is essentially that of a non-truth-functional operator. George brought it about that the ring exists at t1, but he did not bring it about that the piece of gold exists at t1; and this is so despite the fact that the ring, at t1, is the piece of gold. In fact he created the ring by bringing about a change in the shape of the piece of gold. But while he did not create the piece of gold at all, we nevertheless do not have *two* things, the ring and the piece of gold, at t1; we have one.

The thesis that there is no causal traffic between the mental and the physical is thus not inconsistent with the claim that the mental *is* the physical (or vice versa). The denial of interaction expresses the claim that events *conceived* as mental do not cause events conceived as physical, and vice versa. This entails that citation of a mental event (qua mental event) is of no use in explaining the occurrence of a physical one (qua physical). It does not entail, however, that a mental event, conceived as physical, cannot cause a physical event, conceived as physical.

II. The Denial of Interaction and the Explanation of Action

The implications of Spinoza's denial of interaction for the explanation of human action might appear to be straightforward. Desires, like *mentis decreta*, can cause physical actions only insofar as these desires can be conceived as states of or events in the body. And they can cause other desires and mental phenomena only insofar as they are conceived as mental states or events.

Here it is important to realize that while we might take a desire to be a paradigm of a mental state, Spinoza did not. He used the expression "*cupiditas*" in both a narrow and a broad sense. In the narrow sense, a desire is an appetite along with consciousness of it (EIIIp9s). In the broad sense, a desire is any endeavor, impulse, appetite, or volition (Gen. Def. Aff.1, expl). But an endeavor (*conatus*) can be "related to the mind alone, to the body alone, or to both" (IIIp9s). Put in other words, a desire (or "constitution of the essence" of a thing) can be conceived "through the attribute of Thought alone, or through the attribute of Extension alone, or lastly it may be referred at the same time to both." (Gen.Def.Aff.1,expl.). So a desire can cause a "physical action," just insofar as the desire is conceived as a bodily state. Conceived as a mental state, it can only cause things conceived as mental.

Thus the citation of hunger, or a desire to eat, can explain our eating just insofar as we take hunger to be a physiological state or event in the body. It is a state or event that is the *efficient cause* (sometimes) of eating. Consciousness of this state or event, and the event itself, insofar as it is *mental* or conceived under the attribute of Thought, could not usefully be cited to explain a physical change, on this view.

This claim, as Spinoza noted in EIIIp2s, conflicts with the ordinary thought that "the causes of buildings, pictures, and things of this sort" cannot be purely physical, "nor would the human body be able to build some temple, unless it be determined and led by the Mind."[22]

What are the implications of this concerning the acceptability of ordinary teleological explanations? If I try to explain an act such as Peter's entry into a bakery by citing his desire to get bread, have I then explained—or tried to explain—a physical change by an appeal to the Mind?

Assume that a desire is conceived as a mental state at least partly in virtue of the fact that it has *content*.[23] *If* this is correct, then the fact that a desire has content, or a certain content, plays no role, on Spinoza's account (as well as Davidson's), in the causal explanation of the action. Only physical features of the desire could be relevant to the causal explanation of physical acts.

It is clear, however, that the content of the desires and beliefs that we attribute to people plays a crucial role in our ordinary use of these concepts to explain human action. A desire to get some bread normally will, and a desire to get a pair of shoes normally will not, "make sense" of Peter's entry into the bakery. It appears to follow that ordinary teleological explanations of "physical acts" are unacceptable. For they attempt to explain physical changes by citation of a desire conceived as having *content*, i.e. as *mental.* They attempt, in short, to explain the causes of pictures and buildings and things of this sort by an appeal to the Mind.

Hence even if, as both Spinoza and Davidson maintain, desires are the efficient causes of physical actions, teleological explanation is not "reducible to" purely causal explanation. A desire to eat or a desire to get bread might well be an efficient cause of Peter's entry into the bakery. But citation of this desire makes sense of his act in virtue of its content, and it does so in almost complete ignorance of the details of the causal mechanisms of hunger. In contrast, a "purely causal" explanation of his act proceeds in terms of biochemistry and physics, with no reliance on the claim that certain physico-chemical states of the body are desires or ideas that have content.

Spinoza maintained that there are two ways in which desires can be conceived to cause "physical acts," without implying causal interaction between the mental and the physical. Either the desire and the act must both be conceived as physical, or they must both be conceived as mental. This raises the question of how a "physical act," such as entry into a bakery, is to be conceived as "mental." One answer, certainly, is that just as the human body and God's idea of it (the mind) are "one and the same thing," but conceived in different ways, so too each physical act of a body is the same as God's idea of it. A desire, conceived as mental, should not be said to cause the physical act of eating; it should instead be said to cause a representation or idea of the physical act. As he expressed it in EIIIp2s, "the order of the actions and passions of our body is simultaneous by nature with the order of the actions and passions of the mind."[24]

This, and along with it the dictum that *each* thing can be conceived under the attribute of thought, conjures up the image of a "ghostly counterpart" of each physical act, and each object. Corresponding to the causal sequence consisting of an appetite or determination of the body, followed by eating, there is also the sequence consisting of a mental decision, followed by a "representation" or idea of eating. And just as the apple itself is destroyed by being eaten, so too God's idea of the apple is destroyed by his idea of its being eaten.

Although we might imagine that there are two causal sequences here, with the same order, Spinoza's position, as we have already noted, is stronger than this. He did not merely hold that these sequences have the same order. He held that the items in these apparently diverse sequences are the same, that is, that there is just one sequence. Lest it be thought that Spinoza's remark in EIIIp2s entails that the actions and passions of the body are distinct from those of the mind, note the emendation of "are simultaneous by nature" (or "by nature exist together," as Curley puts it) in the following remark, made later in the same scholium:

> . . . both the decision of the Mind and the appetite and the determination of the Body by nature exist together—or rather are one and the same thing, which we call a decision when it is considered under, and explained through, the attribute of Thought, and which we call a determination when it is considered under the attribute of Extension and deduced from the laws of motion and rest.[25]

An alternative account of how a "physical act" or an act that "involves" motion can be conceived as mental—or at least as not "purely physical"—might be suggested, however. For acts can be described in different ways. One and the same act might be described, for example, in one way as "clenching your fist and moving your arm up and down" and in another as "hitting Paul." It might then be supposed that an act that is described or conceived simply as a physical change—a motion of your arm, say—can be explained only by events characterizable in purely physical terms; the same act, however, conceived as your hitting Paul, might be explained as a result of your hatred of Paul. The content of your desires and beliefs is then relevant to the causal explanation of certain features of your act, although not to those features that are "purely physical."

This idea, and indeed the example used, is drawn from Spinoza's remarks in EIVp59s:

> The act of beating (*verberandi*), insofar as it is considered physically (*quatenus physice consideratur*), and insofar as we attend only to the fact that the man raises his arm, closes his fist, and moves his whole arm forcefully up and down, is a virtue, which is conceived from the structure (*fabrica*) of the human body. Therefore, if a man moved by anger or hate is determined to close his fist or move his arm, that (as we have shown in Part II) happens because one and the same action can be joined to any images of things whatever. And so we can be determined to one and the same action both from those images of things which we conceive confusedly and (from those images of things?) we conceive clearly and distinctly.[26]

It is true that Spinoza is here concerned with actions that can arise both from Reason and from passions such as anger or hatred. It is also true that an *image*, which is said to determine the act in question here, is physical, and so Spinoza has not here supposed that a confused or clear *idea*, or hatred of Peter (which involves a belief) could cause a physical change. At the same time, however, it is of more than marginal interest that to illustrate how an action "to which we are determined" by a passion can also be one to which we are determined by reason, Spinoza selected an example that involves a physical change.

Thus the suggestion, at least, arises that an act such as beating can be considered physically, in which event it must be explained purely physically, and yet the same act can arise because the agent is moved by anger or hatred, on the one hand, or from reason, on the other. Put more generally, it seems to suggest, without asserting, that an act, conceived purely physically, is explicable only in physical terms, but the very same act, under a different description, is explicable in terms of the content of the agent's mental states.

In this way, citation of the content of a desire or belief can help explain acts that "involve" physical changes, without thereby supposing that mental events, conceived as mental, cause physical changes, conceived purely physically. It is true that a beating, in contrast to the mere motion of your arm, is not *mental* in any ordinary sense. Nevertheless, it is also not "purely physical" either, if this is taken to be anything definable in the vocabulary of physics.

Whether this suggestion is ultimately defensible is not a question I will attempt to resolve here. Certainly the example chosen to illustrate the thesis could be improved. For the difference between moving your arm and hitting Paul seems to be, technicalities aside, largely a physical difference. It depends on where Paul's nose is. More convincing, I expect, would be the difference between uttering certain sounds, and saying something, and the difference between harming people and punishing them. Producing those sounds *is* saying something, partly if not largely because of the intentions and beliefs of the speaker. So too, harming a person is, on occasion, a punishment, largely in virtue of the intentions and reasons of the actor.

Spinoza's great insight—or one of them—is his recognition that while there is no ontological difference between them, there is an irreducible conceptual difference between the physical and the mental. There is an irreducible difference between conceiving of people as collections of chemicals, and conceiving of them as thinking and feeling beings, and indeed between conceiving of yourself as a body and conceiving of yourself as a mind. Consciousness may be a brain process, but the concept of a conscious, living being is distinct from the concept of collection, however complex, of elementary physical particles.[27]

This difference is closely connected to the differences between "purely causal" explanations and "teleological" ones. The paradigm of the former is the citation of the motion of one billiard ball, along with a collision, to explain the motion of another. The paradigm of the latter is the citation of a desire and a belief to explain human actions and desires, and the citation, in functional explanations, of a characteristically helpful effect, in explaining the parts, processes, and features, of a (living or man-made) body. But helpful to what?

The relations between causal and teleological explanations are complex. It may well be that citation of the reasons for action, that is, specification of the agent's desires and beliefs, is the citation of the efficient causes of action. It remains true that the application of a concept of desire, as of a concept of a function, presupposes, imports, or makes possible the application of normative ideas. The concept of success or failure, and of what is good or bad *for* a being, for example, becomes applicable only when we attribute goals, desires, or functions to the thing in question. In addition, the very idea that a thing is organized, that is, that its parts can be understood functionally or can be distinguished on the basis of their function, seems to presuppose some unitary end-state—some goal (such as continued life or reproduction) to which they contribute.

Part III of Spinoza's *Ethics* provides a transition between metaphysics and ethics. It is a necessary transition because a merely physical description of objects, connected by causal relations, provides no basis on which to evaluate them. The impact of one atom or billiard ball on another is neither good nor bad (for anything, and as such), and there is no basis, in a world conceived purely physically (and without life), for the application of a concept of what should be, nor for the construction of a concept of an ideal. In Part III, however, we find that each thing endeavors to persevere in its own being, and with this attribution of a goal or desire, ethical or normative ideas become possible. This is, I think, another great insight of Spinoza, whether it is uniquely his or not.[28]

III. Materialism and the Conceptual Distinction between Mind and Body

The *ultimate* source of the conceptual distinction between the mental and the physical has been a never-ending controversy in the interpretation of Spinoza. I think myself that Spinoza operated with a theory according to which there are only three possible kinds of distinction between things—real, modal, and conceptual (i.e. *rationis*). In this he followed Descartes, who himself was indebted to Suarez. But Suarez was more subtle. He distinguished between a *distinctio rationis ratiocinatis*, and a *distinctio rationis ratiocinantis*.[29] What ontological difference, if any, underlies our different ways of conceiving a thing?

It is undisputed that there is a *conceptual* distinction between the attributes. *The* concept—and our concept—of thought is distinct, according to Spinoza, from our concept of extension, and neither concept is reducible to the other. What is in dispute is whether thought and extension themselves are in fact distinct. If they are, it is difficult to see why they would not be distinct substances or, what is almost as damaging to Spinoza, why the division of *entia extra intellectum* into substances and modes is not incomplete. If attributes exist and they are in fact not identical, not substances, and not modes, then it seems inescapable that not everything is either a substance or a mode. But "extra intellectum nihil datur praeter substantias, earumque affectiones" ("outside the intellect there is nothing besides substances and their affections"; EIp4d; G.II. 47, 28-29).

The opposing side, however, has its reasons as well—and indeed, its texts. Certainly the interpretive issue cannot be settled in this short paper, even if (what is doubtful) it can be settled satisfactorily at all. But it is not merely of passing interest, I think, that the unclarity surrounding the ultimate source, in Spinozism, of the conceptual distinction between thought and extension is matched if not surpassed in contemporary ("analytic") philosophy, by unclarity about how to be a good materialist. The central problem about this has plagued twentieth century materialism, and the identity theory, from its inception. It arose initially from the attempt by the Australian materialists to defend the identity of sensations and brain processes as a *contingent* (and "non-analytic") truth. For this seemed to require that in addition to physical properties of brain processes, there must exist *mental* properties of sensations. (For only in virtue of a distinction between mental and physical properties, it seemed, could we identify sensations in a way that is independent of our identification of brain-processes). It seemed, in short, that the identification of mental and physical items required a distinction between mental and physical *properties*. Why then is this not just a replacement of "object" dualism by "property dualism"? (And isn't this just what, on an "objectivist" interpretation, we have in Spinoza: one substance with a real plurality of attributes?)

Thus the rejection of a dualistic account of mental and physical objects, or items, seems to require a dualism of *properties*, just as, for different reasons, Spinoza's identification of thinking and extended substance, and of minds and bodies, appears to require that the attribute of Thought be (actually and ultimately) distinct from the attribute of Extension.

Although the *contingent* character of the identity theory might well be denied,[30] the problem of the relations between mental and physical *properties* seems to remain. And in reply it has proven tempting either to deny the existence of mental properties, or to reduce them to, that is, identify them with, physical properties. Since properties are taken to be correlates of predicates, materialists have often supposed either that there is no real content to our talk of the mental, or to suppose that that content must be expressible in "physical" terms—that is (ultimately), in the vocabulary of physics. In Spinozistic terms, this is equivalent to the claim that either Thought does not exist, or the concept of Thought is reducible to the concept of Extension.

The idea that all *individual* objects and events are physical, but that no reduction or identification of "mental properties"—that is, types or universals—with physical ones is possible, has been espoused by Davidson. (If we substitute "the concept of thought and the concept of extension" for "mental and physical properties" here, we get a position with which Spinoza agrees.) The problem encountered by the Australian materialists, however, seems to remain (or to re-emerge). This is clearly expressed, once again, by Kim, who argues that "nonreductive materialism is not a stable position. There are pressures of various sorts that push it either in the direction of outright eliminativism or in the direction of an explicit form of dualism." Very much in outline, these pressures arise from the supposition that if individual mental events are physical, then there must be mental properties. If these properties are causally impotent, there is little point in supposing that anything is mental. If they are causally connected to physical events, then mental properties are reducible to physical properties, or else the physical world is not a "closed system." If, on the other hand, they are not causally connected to physical events, then the mental constitutes an autonomous domain (of properties, if not individuals).

Although this ignores important parts of Kim's paper,[31] it presents in rough outline what is important for present purposes. Mental and physical *individuals* can be identified only if it is supposed either that mental *properties* are reducible to, or identified with, physical ones, or else a radical dualism of properties is accepted. Conceptually non-reductive materialism is not then viable, and so neither is Davidson's brand of monism. But for the same reason, Spinoza's substance monism would *require*, if it is to be intelligible, a dualism (or more) of attributes. The conceptual distinction between the attributes entails an ontological distinction between them. Furthermore, Spinoza's mind-body identity thesis would require a dualism of mental and physical properties.

A quick dismissal of the problem would of course be entailed by the rejection of the existence of properties, that is, of universals. If thinking of Spain is a universal, it poses no more of a problem for the materialist than moving to Spain. For neither property, on this view, would be an *ens reale*.

Although such a dismissal would be too quick, it also seems that mental properties are—in the abstract—about as objectionable as grammatical properties. "Is a grammatical sentence of French," "contains a grammatical error," "was once spoken by Napoleon" and so forth, are predicates whose meaning, I would suppose, is inexpressible without loss in the vocabulary of physics. If a property is a correlate of a predicate, the existence of mental properties threatens a physicalist ontology to the same extent that the existence of grammatical properties does. But sentences can be physical objects, such as ink marks and air vibrations, without supposing that what we say in truth about them must be reducible to the vocabulary of physics. Indeed, fences are physical objects, although "is a fence" is not definable, or so I think, in the basic vocabulary of any science of physics.

The question about causality also has an analogue here. Note first that it is not obvious what it means to say that universal properties are *causes* (or effects). How is a universal imagined to cause (or be caused by) anything? One reply would maintain that it is not the universal or property itself, but the fact that a certain thing *has* that property that is a cause (or effect). Being a thought about Spain causes nothing by itself, but the fact that a certain event has that property might cause a person to think of Barcelona—it might lead, that is, to a thought about Barcelona.

"A desire causes physical changes, but it does do so only in virtue of its physical properties," is like, "Writing a sentence causes physical changes, but it does so only in virtue of its physical properties." The existence or production of a sentence (token) on a piece of paper, for example, might cause blue ink to appear on the paper. And we might say, with Spinoza, that the citation of the fact that this is a grammatical sentence of French does as much good in explaining why it caused blue ink to appear on the paper as does the citation of the fact that you have a certain idea in an attempt to explain why your limbs are moving. So too, it is not the fact that you made marks on this paper that explains why this is a bad check, or a forgery. It is rather the fact that you had no money in the bank, or signed someone else's name to a check. "Making marks on paper" and "forging a check" are descriptions at different (conceptual) *levels*.

An act described in terms of moving muscles must be explained in terms of the biochemistry of the body. The same act described as a thrashing, or the purchase of some bread, must be explained in terms of the content of your beliefs and desires.

Although we might be muscles in motion, or chemical reactions, we are also human beings. An adequate conception of ourselves requires, as Spinoza recognized, that we transcend a conception of ourselves as purely physical, or even biological, beings.[32]

Thus if the materialist tradition has been culturally bankrupt, it is not because it refuses to admit a special type of item into the privileged category of the real.[33] It is instead because it has failed to recognize what Spinoza saw. There are irreducibly different ways of conceiving of ourselves, and the important problems of philosophy—and of life—cannot even be posed when we conceive of ourselves purely physically. In addition, more than half of "the mind-body problem" is not even addressed by resolution of the ontological problem. The central problem is conceptual, and even when it is decided or declared that there is, or is not, an ontological difference between the mental and the physical, most of the real work remains.[34]

NOTES

[1] An analogous problem also arises, for similar reasoning would appear to show that at least one physical event has a mental effect. A more general statement of the problem is also possible, since Spinoza held that everything is conceivable as mental (under the attribute of thought) and as physical (under the attribute of extension).

[2] Spinoza wrote, more specifically, "viderem omnia, a quibus, & quae timebam, nihil neque boni, neque mali in se habere, nisi quatenus ab iis animus movebatur" (" I saw that all things by which and of which I was afraid, had nothing either good or bad in themselves, except insofar as the soul was moved by them . . . " (This is my translation from Carl Gebhardt (ed.), *Spinoza Opera* (Carl Winters, 1972, reprinted from 1925), Vol. II, p. 5, lines 9-11. Gebhardt's edition will subsequently be cited as "G." followed by the volume, page, and line numbers. See Edwin Curley, *The Collected Works of Spinoza* (Princeton University Press, 1985), Vol. I, p. 7. Unless otherwise noted, Curley's translations are used in this paper (and cited as "C.I." followed by the page number).

[3] Spinoza's early dualism is discussed in my "On the Rejection of Spinozistic Dualism in the *Ethics*," *Southern Journal of Philosophy*, Vol. xx, No. 2 (1982), pp. 153-175.

[4] Translation by Edwin Curley in C.I.494. ("Mens, & Corpus una, eademque res sit, quae jam sub Cogitationis, jam sub Extensionis attributo concipitur." G.II,141, 24-26.) Cf. IIp7s, which maintains that "a mode of extension and the idea of that mode are one and the same thing, but expressed in two ways" (C.I.451; G.II.90, 8-9).

[5] Strawson maintains that the concept of a person, as a being to which both mental and physical properties are ascribed, is primitive. See P. F. Strawson, *Individuals* (London, 1959). For a discussion and criticism of a "dual aspect" interpretation of Spinoza, see Jonathan B. Bakker, "Did Spinoza Have a Double Aspect Theory?" *International Studies in Philosophy*, Vol. 14, 1982, pp. 1-16.

[6] "Mentis tam decretum, quam appetitum, & Corporis determinationem simul esse natura, vel potius unam, eandemque rem, quam, quando sub Cogitationis attributo consideratur, & per ipsum explicatur, decretum appellamus, & quando sub Extensionis attributo consideratur, & ex legibus motus, & quietis deducitur, determinationem vocamus . . ." (G.II, 144, 3-8).

[7] My translation (G.II, 129,22-27: ". . . has, & similes facultates, vel prorsus fictitias, vel nihil esse, praeter entia Metaphysica, sive universalia, quae ex particularibus formare solemus. Adeo ut intellectus, & voluntas ad hanc, & illam ideam, vel ad hanc, & illam volitionem eodem modo sese habeant, ac lapideitas ad hunc, & illum lapidem, vel ut homo ad Petrum, & Paulum." Cf. etiam EIIp49d (G.II.130,18-20), quoted in note 9 infra.

[8] C.I.484. G.II.3-4. (EIIp48s).

[9] My translation (G.II.130,18-20). "In Mente (per Prop. praeced.) nulla datur absoluta facultas volendi, & nolendi, sed tantum singulares volitiones, nempe haec, & illa affirmatio, & haec, & illa negatio."

[10] See especially IIp40s2 (G.II.2-14).

[11] Spinoza's central example of what is common to all things is expressed in EII.lemma 2. (See the reference to this in EIIp37.) Lemma 2,d maintains that all bodies agree in involving the concept of the same attribute, in moving more or less quickly, and in being able to move and to be at rest.

[12] Spinoza's nominalist tendency is forcefully expressed in his other works as well. Passages such as the following from KVII,16 are commonplace: ". . . when I consider them attentively, they seem to me to be universals, and I cannot attribute anything real to them." (C.I.122) and ". . . because man has now this, now that Volition, he forms in his soul a universal mode which he calls the Will, just as he forms the Idea of man from this and that man. And because he does not sufficiently distinguish real beings from beings of reason, it comes about that he considers the beings of reason as things that are truly in Nature . . ." (C.I. 123-4).

The attribution here of a "token" identity of the mental and physical to Spinoza diverges sharply from Bennett's interpretation. Bennett maintains in fact that an individual human mind and body—your mind and body, for example—are universals, according to Spinoza. (They are states of substance or the attributes, and states, according to Bennett, are properties, that is, universals. See Jonathan Bennett, *A Study of Spinoza's Ethics* (Hackett Publishing Co., 1984, especially pp. 92 and 94. Cf. p. 57.) I hold, in contrast, that Spinoza rightly recognized that there are individuals in categories other than that of substance. When two bodies are moving, for example, the motion of one is distinct from the motion of the other, and when two people desire or think something, the desire or thought of the one person is (numerically) distinct from that of the other. (For more on this, see my "The Concepts of Substance and Mode in Spinoza," *Philosophia*, Vol. 7, No. 1 (1977), pp. 83-105.)

[13] Here I mean by an explanation of it, an explanation of why it occurred.

[14] See EIIp6 (twice cited in IIIp2d) and the reference in IIp6d to Ip10.

[15] See Donald Davidson, *Essays on Actions and Events* (Oxford University Press, 1980), especially "Mental Events," p. 108. This essay originally appeared in Lawrence Foster and J.W. Swanson (eds.), *Experience and Theory* (University of Massachusetts Press, 1970), pp. 69-101.

[16] The apparent inconsistency in Spinoza's position is illustrated here by noting that, at first sight, if the effect of a physical event is identical with some mental event, then that mental event is an *effect* of a physical event. It could just as well be illustrated by considering the claim that if the *cause* of a physical event is identical with some mental event, then that mental event is a *cause* of a physical event. For Spinoza denies both that physical events cause mental ones and that mental events cause physical ones.

[17] The term employed is recent, but the idea is certainly not. It figures prominently in Descartes' argument for the real distinction between mind and body. (For more on this, see Charles Jarrett, "Cartesian Pluralism and the Real Distinction," *The Southern Journal of Philosophy*, Vol. xix, No. 3 (1981), pp. 347-360. Leibniz' recognition of referential opacity is noted by Benson Mates, in *The Philosophy of Leibniz* (Oxford University Press, 1986), pp. 130-132. Indeed Mates appears to be at least one of the first to have recognized that Leibniz restricted the principle that if A is the same as B, they can be substituted for each other *salva veritate*, to extensional contexts. (See Mates, *op. cit.*, p. 30.)

[18] See Donald Davidson, (*op. cit.*).

[19] Jaegwon Kim, "The Myth of Nonreductive Materialism," *Proceedings and Addresses of The American Philosophical Association* (Presidential Address delivered to the Central Division, April 1989), p. 35.

[20] See Dagfinn Follesdal, "Causation and Explanation: A Problem in

Davidson's View on Action and Mind" in Ernest LePore and Brian P. McLaughlin, eds., *Actions and Events: Perspectives on the Philosophy of Donald Davidson* (Basil Blackwell, 1984), pp. 311-323.

[21] See my "Materialism," *Philosophy Research Archives*, Vol. VIII (1982), pp. 457-497.

[22] On the face of it this denies a conception of intelligence (of planning and so on, taken as mental) as arising and developing because of the causal contribution that this makes to the physical survival of creatures. That is, it rejects an evolutionary account of the development of mental capacities, insofar as such an account requires these mental states to contribute to (cause) the physical continuation of the individual and its (successful) reproduction. It does not, however, reject such an account, insofar as these mental capacities are conceived as physical.

[23] It is a desire *that* something be the case. I suppose here that the *content* of a thing is something that can be affirmed or denied and is capable of truth or falsity. (Sentences and speech acts have content, in this sense, but are not in any straightforward sense "mental." Hence being mental cannot simply consist in "having content.") That Spinoza held that it is partly in virtue of having content that a mode is "mental" (falls under the attribute of Thought) is suggested, but not I think established, by several passages.

Spinoza held that an idea, unlike an image or picture, involves affirmation or negation. (IIp49s; G.II,132,10-11) He also held that ideas are prior to all other modes of thought, such as love and desire (II, ax3), and indeed that all affects that are passions of the soul (*animi Pathema*) are confused ideas (IIIGen Def Aff.). Actions of the mind, in contrast, arise only insofar as we have adequate ideas (IIIp58,d). It does not quite follow from these claims alone that every mental item, that is, every mode of thought *is* an idea, and hence involves content (affirmation or negation); it is strongly suggested, however.

The converse, that if something involves content that is affirmed (or denied?), then it is an idea, seems to be maintained in IIp49d. Here Spinoza attempted to show that a volition is an idea, on the grounds that a volition is an affirmation, and an affirmation pertains to the essence of an idea. (He speaks in particular of the idea of a triangle here, but the proof is intended to be general.)

[24] G.II,141, 28-29 (my translation).

[25] C.I, 497. G.II, 144, 3-8. "Mentis tam decretum, quam appetitum, & Corporis determinationem simul esse natura, vel potius unam, eandemque rem, quam, quando sub Cogitationis attributo consideratur, & per ipsum explicatur, decretum appellamus, & quando sub Extensionis attributo consideratur, & ex legibus motus, & quietis deducitur, determinationem vocamus . . ."

[26] C.I,580. G.II.255, 14-23. (Parenthetical expressions in Latin are my addition.)

[27] Compare the *Tractatus Politicus*, Ch. V, sec. 5: " I am speaking of a truly human existence, which is characterized, not by the mere circulation of blood and other vital processes common to all animals, but primarily by reason, the true virtue and life of the mind" (translated by A. G.Wernham, *Benedict de Spinoza: The Political Works* (Oxford University Press, p. 311. (G.IV.296,12-15: "vitam humanam intelligo, quae non sola sanguinis circulatione, & aliis, quae omnibus animalibus sunt communia, sed quae maxime ratione, vera Mentis virtute, & vita definitur.")

[28] That Spinoza appears to have "simply followed Hobbes" (*Leviathan*, Ch. VI) in rejecting a concept of good and evil that is independent of desire was noted by David Bidney, *The Psychology and Ethics of Spinoza* (Yale University Press, 1940), p. 107.

[29] For more on this, see my "Some Remarks on the 'Objective' and 'Subjective' Interpretations of the Attributes," *Inquiry*, 20 (1977), pp. 447-456.

[30] See Wallace I. Matson, *Sentience* (University of California Press, 1976).

[31] His discussion of recent explorations of the notion of supervenience (of the mental on the physical) has, for example, been ignored here.

[32] This should not be taken to deny that we *are* physical, and biological, beings. "Ontological transcendence" is not intended. For an illuminating account of Spinoza's transition from a biological conception of ourselves to an adequate conception, see A. Matheron, *Individu et Communaute chez Spinoza* (Les Editions de Minuit, 1969), especially pp. 243-278.

[33] See Donagan, *Spinoza* (University of Chicago Press, 1988), pp. xi-xii, but note that Donagan does not suggest that a "token identity" theory, as opposed to type-reductive materialism, is the culprit.

[34] It is not so much that we need to *justify* our "mentalistic" thought of ourselves by grounding it in a physicalist ontology, as to understand it. The "real work" to which I allude here consists largely in (i) gaining a deeper understanding of the diverse conceptual levels we employ, and of their (non-ontological) relations to each other and (ii) answering the central questions that arise from within one of these levels. These are questions, as Donagan well put it, "about why human life matters and how it should be lived." (Donagan, *op. cit.*, p. xi.)

[15]

TANQUAM NATURAE HUMANAE EXEMPLAR: SPINOZA ON HUMAN NATURE

Lee C. Rice

Introduction

RECENT interest in the role of human nature in Spinoza's moral philosophy and anthropology probably has its roots in the earlier revival of interest among social philosophers in his political theory. Human nature forms a logical "bridge" concept between the metaphysical and the socio-political accounts which Spinoza offers, and so it is natural that contemporary thinkers would look to this concept as a means of attacking a number of puzzles and seemingly contradictory affirmations in the text of Spinoza.

Perhaps the most urgent question relating to human nature in Spinoza is whether there is any such thing. Spinoza's position on universal concepts is an avowedly nominalistic one on the surface. In E2P40Schol,[1] he analyzes the transcendental predicates as products of blurred sensory images, and proceeds to apply his analysis to 'universals' in general and 'man' in particular:

> From similar considerations arise next those notions which are called 'universal,' such as 'man,' 'horse,' 'dog,' etc. So many images (such as that of a human being) are formed simultaneously in the human body that our sensory capacity is overcome, not completely, but at least to the extent of being able to visualize the small differences (such as complexion, stature, and size of each object). Since the mind is unable to imagine a determinate number of such individuals, it visualizes only what they have in common [*in quo omnes conveniunt*] insofar as the body is affected by them.[2]

This appears to leave little room for a full-blooded metaphysics of human nature, and to relegate talk about 'humanity' to somewhat the same status of vague generality as we find, for example, in avowed conceptualists such as Hume.[3] Indeed, earlier commentators such as Balz,[4] Eichberg,[5] and Rivaud[6] took seriously Spinoza's declarations in the first two parts of the *Ethics*, and tended to see later remarks about human nature as generalistic in tone as well as in epistemic status.

This interpretation was first seriously challenged by Francis Haserot,[7] and has found more recent support in Martial Gueroult,[8] Jonathan Bennett,[9] C. T. Campomanes,[10] Alan Donagan,[11] Lucia Lermond,[12] and Diane Steinberg.[13] The platonistic reading of Spinoza is most often defended by appeal to later passages in the *Ethics* where Spinoza is dealing with human 'liberation' and the several references to *humana natura* in E2 through E4. Such an approach amounts to making Spinoza's philosophical anthropology fit the structure of a posited moral philosophy, which is the strategy pursued by Haserot (487–488), Donagan (136–137), and Lermond (79–84). Its central weakness is that of reversing the very order of knowledge which, according to Spinoza, is integral to understanding his philosophy. A principal theme of the *Ethics* is that human well-being is integrally dependent upon an adequate knowledge of nature itself (i.e., metaphysics precedes ethics).

Gueroult and Bennett take a different tack. The former, relying heavily on Spinoza's earlier work, *Cogitata Metaphysica,* argues for a distinction between *entia imaginationis* (blurred general images) and *entia rationis* (adequate general concepts).[14] Bennett interprets E2P40Schol as an attack on some, but not all, general terms; and goes on to reaffirm the appeal to Haserot's notion of 'rational universality.'[15] These efforts have not been without their detractors. E. Giancotti Boscherini[16] has cast serious, and in my view decisive, doubt on the

[1]I would like to express my gratitude to Fr. Roland Teske, S.J., and to Dr. Michael Wreen, both of Marquette's Philosophy Department, who read an early draft of this study, for their sympathetic criticisms and suggestions. Translations from the text of Spinoza, where given, are my own. I prefer the edition of J. Van Vloten and J. P. N. Land [*Benedicti de Spinoza opera quotquot reperta sunt,* 3rd edition. 4 vols. The Hague: M. Nijhoff, 1914] to that of Carl Gebhardt [*Opera, im Auftrag der Heidelberger Akademie der Wissenschaften.* 4 vols. Heidelberg: Carl Winters Verlag, 1925]. A new and critical edition of the *Opera Postuma* is under preparation in the Netherlands, and will probably resemble closely the Van Vloten and Land edition. References to the *Ethica* are internal. E2P13Cor is the corollary to Prop. 13 of Part 2. Other abbreviations are Dem(-onstration), Schol(-ium), App(-endix), and Def(-inition).

[2]E2P40Schol. I take *imaginare* as sensory perception in general, for which see Michèle Bertrand, *Spinoza et l'imaginaire* (Paris: Presses Universitaires de France, 1983), 37–43; and G. H. R. Parkinson, *Spinoza's Theory of Knowledge* (Oxford: Clarendon Press, 1954), 139–140.

[3]See Douglas Den Uyl and Lee Rice, "Spinoza and Hume on Individuals," *Reason Papers* 15 (1990), 91–117.

[4]A. G. A. Balz, *Idea and Essence in the Philosophy of Hobbes and Spinoza* (New York: Columbia University Press, 1918).

[5]Waldemar Eichberg, *Untersuchungen über die Beziehungen der Erkenntnislehre Spinozas zur Scholastik mit besonderer Beruksichtigung der Schule Okkams* (Leipzig: Robert Noske, 1910), esp. 15 and 30–35.

[6]Albert Rivaud, *Les notions d'essence et d'existence dans la philosophie de Spinoza* (Paris: Alcan, 1906).

[7]"Spinoza and the Status of Universals," *Philosophical Review* 59 (1950), 469–492.

[8]*Spinoza: l'Ame (Ethique II)* (Paris: Aubier, 1974), especially Appendix 3. Gueroult distinguishes no fewer than ten senses of 'essence' and 'nature' in Spinoza, breaking them down into general and particular essences or natures.

[9]See *A Study of Spinoza's Ethics* (Indianapolis: Hackett, 1984), 39; and "Spinoza's Mind-Body Identity Thesis," *Journal of Philosophy* 78 (1981), 573–584. See also the critical reviews of the first study by L. Rice in *Bulletin de l'Association des Amis de Spinoza* 16 (1986), 8–11, and in *The Modern Schoolman* 63 (1986), 218–220; and L. Rice, "Spinoza, Bennett, and Teleology," *Southern Journal of Philosophy* 23 (1985), 241–254.

[10]See *Una antropologia del conocimiento: Estudio sobre Spinoza* (Madrid: UPCM, 1981), esp. 111–122.

[11]*Spinoza* (Chicago: University of Chicago Press, 1988), 51–52. See also his earlier study, "Essence and the Distinction of Attributes in Spinoza's Metaphysics," in M. Grene, Ed., *Spinoza* (Garden City: Doubleday, 1973), 164–181.

validity of Gueroult's analysis of Spinoza's earlier works; while Errol Harris[17] has argued convincingly against any notion of 'general' essence in Spinoza, thereby setting aside both Gueroult's and Donagan's platonistic readings.[18] Finally, Daniel Flage has suggested that the analysis of general essences (upon which appear to be based the interpretations of Gueroult, Donagan, and Bennett) rests on a confusion between metaphysical (or constitutive) and epistemic (or expressive) notions of essence in Spinoza, but argues that both of these uses are nominalistic in structure.[19] Of all the recent attempts to assert the existence of a unified human nature in Spinoza, only Steinberg accedes to Spinoza's nominalism, thus separating her from Haserot, Bennett, Gueroult, and Donagan, as well as from the crucial problems which their interpretations face. Her ingenious reading of Spinoza makes essences individual, but posits human nature as a higher-order essence. In what follows, I will suggest that this interpretation runs up against difficulties which, albeit not identical to those faced by Haserot and friends, are equally unresolvable. I will conclude with a putatively nominalistic reinterpretation of those passages upon which her interpretation is claimed to rest.

It might be claimed that my procedure, like Steinberg's and Haserot's, is wrong-headed, and that the question of human nature should be raised within the larger context of Spinoza's purported nominalism. After all, if there are no general natures at all, then there is no general human nature. So, if one could show that Spinoza was a consistent nominalist, the problem of human nature would evaporate. The problem with that approach is twofold. First, the evidence offered for Spinoza's platonism is always based on his remarks about 'human nature': there is no other textual evidence on which the thesis could be supported. Secondly, approaching the problem from the "top down" (via his general theory of essences and attributes) takes us into the swampland of Spinoza's theory of attributes and predication. I shall argue that what Spinoza has to say about human nature is, on a nominalistic reading, tolerably clear; and there is no reason to explicate *obscurum per obscurius* by subsuming the issue under the larger and more problematic question of the attributes. Such a top down approach would not be decisive in any case; for, even if Spinoza were not, on such a reading a nominalist, the claim that there exist some general or universal natures or essences would hardly entail that human nature were one of them.

Human Nature as Individual

Steinberg's interpretation of Spinoza's remarks about human nature is motivated by a desire to avoid the egoistic (or what she perspicuously calls the 'instrumentalist') interpretation of his ethical theory, whereby my relationships with other human beings are based upon their perceived utility to me. She does not make it clear why she objects to this interpretation. One might, of course, take a kantian line and argue that such a view of human relationship constitutes a primary example of the (immoral mis-) use of persons as means.[20] Alternatively, one might follow Bennett's lead[21] in claiming that the move from individual conational activity to genuine interpersonal relationships is not valid within the context of Spinoza's method in any case. I suspect that it is the latter direction which Steinberg intends, since she bases her analysis on E4P29-E2P31. This cluster of propositions constitutes the basis for Spinoza's claim in E4P35 that, "*Quatenus homines ex ductu rationis vivunt, eatenus natura semper necessario conveniunt,*" whose second corollary is a version of Adam Smith's invisible hand.[22] These passages also form the focal point for Bennett's critique.

The notions of harm and utility are explicated in E4P30, whose demonstration connects them with "what is common to us":

> We call something evil which is a cause of pain [*tristitia*] (by E4P8), i.e. (by E3P11Schol), when it decreases or reduces our power of acting. If, then, an object were evil for us because of something in common with us, it could decrease or reduce that which it holds in

[12]*The Form of Man: Human Essence in Spinoza's Ethics* (Leiden: E. J. Brill, 1988), 55–56.

[13]"Spinoza's Ethical Doctrine and the Unity of Human Nature," *Journal of the History of Philosophy* 22 (1984), 303–324. The notion of an essential property of an individual is developed further in her "Necessity and Essence in Spinoza," *The Modern Schoolman* 64 (1987), 187–195.

[14]Martial Gueroult, *Spinoza: Dieu (Ethique I)* (Paris: Aubier, 1968), 413–425.

[15]Bennett, *Study*, 40.

[16]"Man as a Part of Nature," in J. Wetlesen, Ed., *Spinoza's Philosophy of Man* (Oslo: Universitetsforlaget, 1978), 85–96. See esp. 86–88.

[17]"The Essence of Man and the Subject of Consciousness," in J. Wetlesen, Ed., *Spinoza's Philosophy of Man* (Oslo: Universitetsforlaget, 1978), 119–135 (esp. 125–127). See also his earlier "The Mind Body Relation," in James B. Wilbur, Ed., *Spinoza's Metaphysics* (Assen: Van Gorcum, 1976), 13–28.

[18]I am using 'platonism' with lower case here in deference to the distinction between Platonism and platonism which is made in modern semantics. No connexion, historical or otherwise, to Plato's philosophy is asserted or intended by the supporters of a platonic reading which I have mentioned. In contemporary jargon, platonism (lower-case-p) is a position which allows quantification over sets or properties. The denial of platonism admits quantification over only individuals. See Nelson Goodman, *The Structure of Appearance* (Cambridge: Harvard University Press, 1951); and Nelson Goodman & W. V. O. Quine, "Steps Toward a Constructive Nominalism," in *Problems and Projects* (Indianapolis: Bobbs-Merrill, 1972), 173–198.

[19]"The Essences of Spinoza's God," *History of Philosophy Quarterly* (1989), 147–160.

[20]This criticism is raised by A. E. Taylor, "Some Incoherencies in Spinozism (II)," in S. Paul Kashap, Ed., *Studies in Spinoza* (Berkeley: University of California Press, 1972, 289–309 [reprinted from *Mind* 46 (1937), 137–158]. See esp. 301–304.

[21]Bennett, *Study*, 299–306.

[22]E4P35cor2: "Cum maxime unusquisque homo suum sibi utile quaerit, tum maxime homines sunt sibi invicem utiles." E4P37Schol1 and E4P37Schol2 speak of the construction of civil society, and are in fact introductions to Spinoza's political works. See Etienne Balibar, *Spinoza et la politique*, 2nd ed. (Paris: Presses Universitaires de France, 1990), 91–105.

> common with us, which (by E3P4) is absurd. Nothing, therefore, can be evil for us because of what it has in common with us; but, on the contrary, insofar as it is evil . . . it is contrary to us.

The problem, as Steinberg notes, lies with the notion of 'possession in common.' She argues that the demonstration is invalid if we construe this phrase in any of the following three manners: nominalistically (a single predicate, which applies to two objects only conventionally), conceptualistically (a common idea), or realistically (as an aristotelian common nature). We must, she goes on to argue, construe such common possession as one of *absolute identity* (309). This will in turn imply (315) that all human beings possess human nature in common in the sense that: (1) human nature is "involved" in the nature of each human, and (2) two human beings do not constitute two instances of human nature.

I will argue in closing that Steinberg's view of what constitutes a nominalistic interpretation of 'possession in common' is insufficiently general or conventionalist (it provides sufficient conditions which are not necessary). What I want to suggest at this point, however, is that, regardless of how we interpret 'possession in common,' the interpretation of human nature which she claims to follow from it cannot be attributed to Spinoza in any case. That interpretation is aptly summarized by Steinberg (319) as follows:

> What I want to suggest now is that Spinoza held mankind or humanity as a whole to be a complex individual whose parts are individual human beings. If this is so, then it must be the case that the parts — individual human beings — are partially governed in their interactions by certain laws which constitute the nature or essence of the individual which is mankind . . . just as hearts would not be hearts if they did not exist as parts of bodies, human beings would not be human beings if they were not each a part of mankind.

The notion of complex individuals upon which such an interpretation rests is to be found in the discussion of extensional individuation following E2P13, but Spinoza also provides a useful example of what he has in mind in Letter 32. Writing to Oldenburg, he takes the example of a worm (we would say 'micro-organism' today) living in the blood, and endowed with sufficient sight and reason to understand the particulate interactions within the bloodstream. Such a worm would live in the blood much as human beings inhabit the universe, regarding each particle as a whole rather than as a part *(et unamquamque sanguinis particulam ut totum, non vero ut partem consideraret),*[23] and thus unable to see in what manner the laws governing interactions within the blood are governed by (i.e., are consequences of) still more general laws relating to the nature of the blood. In E2P13Lemma7Schol Spinoza goes on to argue that every individual at any level of interaction is always part of a yet more complex individual at the next order up; and that this order can be extended *ad infinitum* to embrace the whole of nature, *"cujus partes, hoc est omnia corpora, infinitis modis variant, absque ulla totius individui mutatione."*

Without, accordingly, having to worry about the interpretation of E4P-30Dem, we can assert that, for Spinoza, human beings are real parts of a larger and more complex individual, of the laws of whose nature the laws governing our behavior are consequences. So far, so good. Steinberg is clearly correct here in seeing no departure from Spinoza's avowed nominalism. As Goodman and other logicians are constantly reminding us, the nominalist may quantify only over individuals, but is not limited in the choice of what constitutes an individual. The big question, of course, is that of the nature of the 'super-individual' of which we are parts (and which is in turn part of still higher orders of individuals, up to god/nature itself).

[23]Ep. 32, Van Vloten and Land, III, 120. See also A. Wolf, *The Correspondence of Spinoza* (1928; reprinted New York: Russell & Russell, 1966). A vitalistic interpretation of Spinoza's account of individuation is given by Hans Jonas in "Spinoza and the Theory of Organism," in M. Grene, Ed., *Spinoza* (New York: Doubleday, 1973), 259–278; and by William Sacksteder, "Spinoza on Part and Whole: The Worm's Eye View," *Southwestern Journal of Philosophy* 8 (1977), 139–159. This interpretation is criticized by L. Rice, "Spinoza on Individuation," in M. Mandelbaum and E. Freeman, Eds. *Spinoza: Essays in Interpretation* (LaSalle: Open Court, 1975), 195–214 [reprinted from *The Monist* 55 (1971), 640–659].

[24]*Individu et communauté chez Spinoza* (Paris: Editions de Minuit, 1969), esp. 340–350. Nothing which I write in what follows should be taken as deprecatory of this monumental work, which is single-handedly responsible for the revival of interest in Spinoza's social psychology among francophone philosophers. The absence of an English translation is a great loss.

[25]See Matheron, *Individu*, 273: "La définition spinoziste de l'individualité, en effet, comporte deux terms: d'une part, le nombre et la nature des éléments composants; d'autre part, la loi selon laquelle ils se communiquent mutuellement leurs mouvements."

[26]"Man and Nature in Spinoza," in J. Wetlesen, Ed., *Spinoza's Philosophy of Man* (Oslo: Universitetsvorlaget, 1978), 19–26. See esp. 24–25.

[27]"Individual and Community in Spinoza's Social Psychology," in E. Curley and P.-F. Moreau, Eds., *Spinoza: Issues and Directions* (Leiden: Brill, 1990), 271–285. This article offers a number of additional examples of authors holding a literal interpretation of civil society as an individual, and also a discussion of the thesis of methodological individualism; for which also see May Brodbeck, "Methodological Individualism," in W. Dray, Ed., *Philosophical Analysis and History* (New York: Harper & Row, 1966), 319–332.

[28]See *Power, State, and Freedom* (Assen: Van Gorcum, 1983), 70–73.

[29]"Spinoza, Human Nature, and History," *The Monist* 55 (1971), 602–616.

[30]A good example is in the *Tractatus Politicus*, Ch. 6; where he speaks of a multitude of people coming together into a society, *"sed ex communi aliquo affectu naturaliter convenire et una veluti mente duci velle."* The *"veluti"* here clearly suggests that the passage should not be taken literally.

[31]A good discussion of what I am calling a top down explanatory model is provided by Rivaud, 165–168. "L'ordre suivi par Spinoza est exactement inverse de l'ordre suivi par les philosophes du mécanisme, qui explique par les éléments simples les choses complexes. l'être est d'autant plus parfait qu'il se suffit mieux á lui-même, et en ce sens l'indépendance appartient à l'individu total seul" (165).

The popular candidate has been, not [Steinberg's or some other version of] human nature, but rather civil societies (plural); since Spinoza certainly does describe our membership in such societies in terms of a part-whole relationship. Alexandre Matheron's remains the most masterful exegesis of this interpretation.[24] Individuals are components of civil society in much the same way as organisms are parts of larger organisms, and with the consequence that physical and civil law experience a *rapprochement.*[25] Edwin Curley appears to adopt a variant of this interpretation of civil society also, but he is unclear on the issue of whether this notion of individuation is to be taken literally or metaphorically.[26] A literal reading makes Spinoza a precursor of Hegel or even Marx, whereas a metaphorical reading places him squarely in the tradition of political liberalism and the epistemological thesis of methodological individual. This metaphorical interpretation is supported by L. Rice,[27] Douglas Den Uyl,[28] and Robert McShea.[29] In the few passages in the political writings which could be produced to support the claim that civil society is an individual whose parts are its citizens, Spinoza usually appears to be talking loosely, and frequently so indicates.[30] And finally, the nomological relation of individuals to civil society in Spinoza's political theory appears to be a 'bottom up' relationship, whereby the particular characteristics of a society or culture are explained by appeal to the appetites and characteristics of its citizenry. But Spinoza's example of the worm in the bloodstream is clearly intended to be a 'top down' model, whereby individual behavior would be explained by appeal to higher-order social laws.[31] Yet nowhere in Spinoza's discussions of human behavior do we find an explanation which even faintly resembles such a model. That human beings are social animals arises, in the *Ethics,* from their individual weakness combined with conational drive to survive. And it is out of this individualistic nomological fabric that the garment of civil society is woven, both in E4P37 and in the two political tractates.

Some of the problems effecting the civil-society reading of higher-order individuals carry over to Steinberg's human-nature reading, whereas others do not. The 'top down' objection is one which, I believe, also applies to her interpretation. Conceive, if you will, a super-individual whose members are all and only members of the species *homo sapiens.* If we are religiously goodman-like nominalists, there is certainly no objection to such a bundling. But Spinoza, unlike Goodman, wants his higher-order individuals to provide a framework of laws which explain the behavior of their component individuals. Pending any examples of what such explanations would even look like within a spinozistic system, the notion of such an individual need not be accorded any interpretive credibility.

There is another problem, which Steinberg's interpretation does not share with the civil-society readings; and that takes us back to E2P13Lemma7Schol:

> We thus see how a complex individual can be affected in many ways and yet preserve its nature. Up to this point, we have conceived an individual composed solely of bodies distinguished one from another by energy [*motu et quiete*] and velocity [*celeritate et tarditate*], i.e., an individual composed of the simplest bodies. If we now conceive another individual thing composed of several individuals *of different natures,* we shall find that this thing can be affected in many other ways while still preserving its nature. [emphasis mine]

As this passage makes clear (and this is where Spinoza's physicalistic nominalism parts company with Goodman's logicistic nominalism), higher-order individuals are constituted by collections of lower-order individuals *ex pluribus diversae naturae,* not of lower-order individuals of the same nature. This is not a problem for the civil-society reading, or at least not for Matheron's variant thereof; since for him there is no unifying human nature, and each individual which is a part of civil society has truly a different nature from every other. Steinberg's reading, however, would have us say that each human being has a nature different from every other, but that each of us is a part of a human nature which is an individual with a "unifying nature." It strikes me that, pending further explanation of the role which such a super-individual human nature is to play, and why we need to hypothesize it in Spinoza's world, such a claim is open to a standard form of Aristotle's "third-man" argument.

Return momentarily to the micro-organism in the bloodstream. The individual of which it is a part is not a super-individual of such micro-organisms, but rather (in Spinoza's own explanation) the bloodstream itself, from whose nature the behavior of that micro-organism is (or at least, for Spinoza, is supposed to be) a deductive consequence. A candidate for such status as a super-individual relative to us is neither Steinberg's human nature viewed as individual nor Matheron's organic civil society, but rather (and in contemporary jargon) an ecology or ecosystem. Such systems are partially closed, and meet the spinozistic requirement of *diversae naturae* by the heterogeneity of subindividuals which make them up. I am suggesting that the notion of a higher-order individual (relative to us) may be a place-holder[32] for a physicalistic model which was not available in Spinoza's time.

I do not claim decisiveness for this positive interpretation of higher-order individuals, but only that it is at least as probable as Steinberg's super-

[32]See L. Rice, "Individual and Community," 280–281, for a more detailed, but hardly adequate, exposition of this theme. The notion of a 'placeholder' is taken from Bennett (*Study*, 139–143).

[33]Steinberg, "Spinoza's Ethical Doctrine and the Unity of Human Nature," 320.

[34]See Rivaud, 128–130, for a discussion of this principle and the role which it plays in Spinoza's account of the causality of god/nature generally.

individual, and appears to fit Spinoza's account of scientific explanation more closely. Offering it, however, commits me to return to E4P30 and its progeny, and to attempt to make sense of that cluster of claims differently. It is to that project which I now turn.

Id Quod Res Nobiscum Commune Habet

Having now argued against what Steinberg herself fears might be construed as a "giant ant colony" interpretation[33] of human nature, I must now offer an at least plausible reconstruction of E4P30 which makes sense of 'possession in common.' My reconstruction is as follows:

'x has something in common with y' = def 'x is similar to y'

Note that this is a definition in use, and does not quantify over the thing in 'something' — i.e., two things may have something in common without there being any third *thing* which they have in common (although there would be true statements which apply to both). In my definiens similarity is reflexive, symmetric, but not transitive. My suspicion is that Spinoza did not give much thought to the notion of similarity, and that some of his remarks about 'brutes' are due to mistaken assumptions about some sort of tidy underlying transitivity, but more of that later.

While similarity is not transitive, there may well be a set of propositions which is such that each of its members applies to individual human beings, and that there exists no non-human being which is such that each sentence in the set applies to it; although, for any proper subset of this set, there will be non-human beings which are such that each element of this subset is true for them. The set of propositions which would constitute necessary and sufficient conditions for human nature would also not define any super-individual: indeed, because of the *diversae naturae* requirement for super-individuals, any super-individual some of whose parts were human beings would, on Spinoza's count, *not* satisfy some of the conditions in the set.

Under this interpretation, saying that two or more things have something in common is to say no more than that one or more propositions apply to each of them, and (because of Spinoza's insistence on the nomological-deductive model) that these propositions describe certain lawlike conditions. Two human beings are human, in this sense, because of certain relevant similarities in their essences. But the essences are individual, and there is no higher-order essence (i.e., essence of a super-individual) which confers this state of being human upon them.

I now want to suggest that one apparent counterexample to this similarity explanation is in fact a confirmation. That is E1P17, where Spinoza is talking about the question (which he answers negatively) whether intellect and will pertain to the essence of god/nature. He uses the 'self-evident' (at least to him) axiom that any x which is the cause of both the essence and existence of some y must differ from that y both in essence and in existence.[34] He gives an example:

> Ex. gr. homo est causa existentiae, non vero essentiae alterius hominis; est enim haec aeterna veritas: et ideo secundum essentiam prorsus convenire possunt; in existendo autem differe debent; et propterea, si unius existentia pereat, non ideo alterius peribit; sed si unius essentia destrui posset, et fieri falsa, destrueretur etiam alterius essentia. (E1P17Schol)

A platonistic reading of this passage[35] would have it mean that there is a single human nature in which all (and only) humans agree *(convenire)*. But Spinoza says only "*convenire possunt*" — not that two human beings do agree, but that they *can*. Matson[36] suggests that Spinoza has identical twins in mind here, and his interpretation would certainly explain the "*possunt.*" It would also, and incidentally, provide a partial explanation of my similarity set of propositions above, which would be a place holder for a set of statements about the human genetic makeup. While we can certainly not foist such a larger interpretation onto the passage, it would certainly accord well with Spinoza's physiological explanations of human behavior in E3 and E4.[37] Further confirmation of such a similarity reading is also found in Spinoza's letter to Schuller of 29 July 1675 (Letter 54). Schuller had addressed two questions to Spinoza, of which the second is relevant to us. Spinoza writes:

> I pass on to the second, which is whether something can be produced by another which is different [*discrepat*] from it in both essence and existence, since things which are so different from one another appear to have nothing in common. But since all individual things, except those which are produced by things similar to them [*quae a suis similibus producuntur*], differ from their causes both in existence and in essence, I can see no reason for raising any question on this point.[38]

Here we have sameness of essence joined with, and indeed explained by, the

[35]See, for example, Lermont (55) and Haserot (487–488).

[36]Wallace Matson, "Body Essence and Mind Eternity in Spinoza," in *Spinoza: Issues and Directions*, ed. E. Curley and P.-F. Moreau (Leiden: E. J. Brill, 1990), 85–87.

[37]See Bennett, *Study*, 81–84; and Boscherini, 86. The former refers to this tendency in terms of "the primacy of the physical," and the latter simply (and probably more correctly) as materialism.

[38]Letter 53, Van Vloten and Land III, 206. Spinoza here uses an example from intermodal causality to exemplify the causal relation between attributes and their modes. See Rivaud, 127–130, for a discussion of the problems in this procedure.

[39]He also adds that horses differ from humans in their affects or emotions as much as equine nature differs from human nature; but then, almost anticipating a platonic misinterpretation, adds that, *for the same reason*, philosophers differ from drunkards (wrongly assuming that the set intersection is null).

notion of similarity. There is nothing whatever in this passage to suggest that strict identity, or indeed any relation stronger than similarity as I characterized it above, is in Spinoza's mind.

The other leading passage, cited by both Haserot and Steinberg, where Spinoza seems to deal with a stronger (i.e., than that of similarity) concept of human nature is in E4Pref, where he mentions the "*exemplar naturae humanae.*" Here the question is that of applying evaluative terms in an objective (non-relative) context. Spinoza explains:

> . . . these terms ought to be retained. For, since we want to form the idea of a human being [*ideam hominis*] which we may look to as a model of human nature [*tanquam naturae humanae exemplar*], we will find it useful to retain these terms in the sense I have explained.

There is nothing here to suggest that human nature, for normative purposes, is anything but a construct: what it references is, not some higher-order essence or nature, but rather an individual self-image. This is not the place to raise the question how such an individual construct could forge the necessary connexion between descriptive and evaluative concepts (or what Spinoza would have to say about charges of being on the verge of committing the 'naturalistic fallacy'), but only to insist that there is nothing in it which violates a prima facie nominalistic view of human nature.

My reconstruction, however, is not without its own problems; and it is to these which I will briefly turn in closing.

Closing Notes: Animals and Extra-Terrestrials

If I am right in saying that Spinoza neither has nor wants a univocal concept of human nature, and that he can make do quite handily with a working set of generalizations about what distinguishes human beings in general from things which are not human, it must still be conceded that he comes up short, at least in the *Ethics*, in even this more modest effort. E2Ax1 says that "*homo cogitat*," and then Spinoza proceeds (in E3P57Schol) to remind us that animals (or at least some of them) think and feel also.[39] The propositions which make up E3 and E4 concerning human behavior, and even those concerning the eternity of the human mind in E5, do not, at least on their face, offer any claims which are strongly distinctive for human beings. Where do we find Spinoza's anthropology?

In a seminal article on this question, Alexandre Matheron[40] argues the twofold thesis that (1) no such anthropology is to be found in the *Ethics*, and that (2) the lack of such an anthropology is no real weakness in Spinoza's procedure.

His evidence for the first claim is, as I see it, decisive. Not a single claim in the *Ethics* relating to human nature (or, more often, to the human *body*) provides anything more than a necessary condition for being human; and all of them taken together are not jointly sufficient either. Spinoza is, in Matheron's view, unable to distinguish human beings from rational non-humans either below (higher animals) or above (extra-terrestrials).

For what Matheron (182) calls an *"éthique de la similitude,"* such an anthropology is not required. The root of morality lies in the desire of each individual to maximize his/her own power, which is precisely the thesis of instrumentalism which Steinberg opposes.[41] To so maximize my own power and control, I select beings as co-operants whose essences are *similar* to mine. Such selection is based on a similarity of needs and a similarity of operation. This is all that is required for social communities. As E4P29 tells us, "No individual whose nature is quite different from ours can either assist or check our power to act; and nothing whatsoever can be good or evil for us unless it has something in common with us." Once again we see the notions of similarity and difference explained (or rephrased) by the notion of 'having something in common.'

But, if this is so, why is Spinoza so unsympathetic to animals? E4P37 is typical of his attitude. "I do not deny that animals feel, but I do deny that we are for that reason prevented from attending to our advantage and from making use of them as pleases us; since their natures do not agree with ours [*nobiscum naturae non conveniunt*], and their affects are different in nature from human affects." Matheron (182–183) concedes that this is a problem, especially since at least some members of the species *homo sapiens* may certainly display fewer human features (e.g., those with genetic defects, etc.) than some higher animals. I will have to follow Matheron's lead, and construe Spinoza's several remarks about animals as representing a "blind spot" rather than a consequence of anything he says in the *Ethics*. These remarks are few, and usually offhand: it would be mistaken to attempt to wrest a theory of 'animal nature' from them, especially in light of the fact that Spinoza has so little to say about any distinctively human nature.

A larger problem also looms. The platonist may claim that my (and, I believe, Spinoza's) account of 'possession in common' based upon similarity in fact presupposes some underlying nature as an objective basis for such a commonality. Recall that, while acceding to Spinoza's nominalism generally, Steinberg argues that a nominalistic reading of E4P30 would invalidate the demonstration thereof. Her argument in fact takes nominalism as conventionalism; so that, without a real objective underlying and identical nature, the

[40]"L'Anthropologie spinoziste?," in *Anthropologie et politique au xviie siècle* (Paris: Vrin, 1986), 17–27.

[41]See Matheron, 21. "Spinoza, en toute rigueur et théoriquement, ne sait pas ce que c'est que l'homme, et il s'en passe très bien: il n'a pas besoin de le savoir pour édifier son système."

similarity predicated of all humans would rest upon an arbitrary convention. Her claim is certainly not true for modern versions of nominalism such as those of Goodman: one can be a nominalist and still argue that claims about similarity are not merely verbal. Such has been my underlying assumption.

The unanswered problem is that of the ontological baggage which such an epistemic claim about the non-arbitrariness of similarity claims may bring with it. This is, of course, the problem of nominalism in the fullest sense: ontological and logical. I argued at the beginning of this paper that the question of human nature in Spinoza is best answered by circumventing this issue in favor of looking at more localized problems. If my analysis is correct, it will have shown that Spinoza at least thought that one need not eschew the objectivity of similarity as a cost of upholding nominalism; but, of course, it will not have shown that he was correct in so thinking. This raises the general question of the status of nature and essence in Spinoza. My analysis in this essay suggests that deferral does not yield cirumvention as a byproduct, and that the question of essence and nature from a perspective wider than that of human nature is one well worth raising.

Tanquam Naturae Humanae Exemplar: Spinoza on Human Nature
Lee C. Rice

[16]

Heidi M. Ravven

Spinoza's Individualism Reconsidered
Some Lessons from the *Short Treatise on God, Man, and His Well-Being*

Overview

Spinoza is generally considered to be a psychological egoist whose starting point, like that of Hobbes, is the atomic individual pursuing his or her own interests in competition with other such individuals. Stanley Rosen, for example, offers a quintessential statement of the standard atomistic interpretation of Spinoza's account of the human person:

> Spinoza dissolves this order into its individual atoms in such a way that natural man is strictly presocial. For both Hobbes and Spinoza, the state of nature is consequently characterized by a primacy of the individual, that is, the multiplicity of individuals, each striving to exist in so far as its power permits, constitutes the initial datum. Society arises when man, the intelligent individual, recognizes the advantages of union in compromise; he recognizes that society, or the man-made instrument for the satisfaction of desires, is in general the efficient augmentation of his individual power.[1]

Ethics on this view consists in the correction of the egoist pursuit from aiming at mistaken and irrational self-interests to truly advantageous and rational ones. This kind of reading suggests that Spinoza's philosophy begins with and also has as its goal a psychological and ethical egoism, albeit one that is corrected toward truly, and away from merely seemingly, prudential interests. William Frankena characterizes Spinoza's ethical theory along these lines as a naturalism which, in the end, dispenses with all norms of moral judgment in favor of strict prudence.[2] And Jonathan Bennett also

[1] "Spinoza," in *The History of Political Philosophy*, ed. Leo Strauss and Joseph Cropsey, 2nd ed. (Chicago: Rand McNally, 1972), 433–434.

[2] William Frankena, for example, takes this position in "Spinoza's 'New Morality': Notes on Book IV," in *Spinoza: Essays in Interpretation*, ed. Eugene Freeman and Maurice Mandelbaum (La Salle, Ill.: Open Court, 1975), 85–100.

conceives of Spinoza's ethical doctrine along prudential lines when he writes that for Spinoza "the only possible value system is egoism, the only rational value system is informed egoism." He continues: "I see [Spinoza] as presenting a morality which enjoins us to be informed, to use our heads, to pursue what is really in our interests instead of going by momentary feelings or careless calculations."[3] Lee Rice, who ascribes to Spinoza an 'ecoethic', nevertheless belongs in this camp for he regards "Spinoza's political philosophy as radically individualistic and more exactly as a precursor of contemporary libertarianism."[4] For any "communal unities" that Spinoza refers to are in his estimation only "the products of perceived utilities and based upon a limited principle of self-defense."[5]

Another reading suggests instead that although Spinoza begins with Hobbesian psychological atomic egoism, his final ethical standpoint is the transcendence of such egoism (resulting in intellectual love) through the overcoming of the boundaries of individuality.[6] Jonathan Bennett, for example, cites as Stuart Hampshire's position with which he agrees, that Spinoza's egoism is the impetus toward self-preservation of an individual who has burst its bounds, so to speak. Bennett writes:

> But perhaps [Spinoza] meant something like the thought I have taken from Hampshire, namely, that our decisions about which bits of the world count as single individuals depend in part not on common effects but on common goals. Spinoza virtually says as much ... [Spinoza] should admit that his doctrine of individual egoism is true only because of how he defines 'individual', and that it may well be false when understood in terms of our normal concept of a single individual, person, or the like.[7]

[3] Jonathan Bennett, *A Study of Spinoza's "Ethics"* (Indanapolis: Hackett, 1984), 298.

[4] Lee C. Rice, "Individual and Community in Spinoza's Social Psychology," in *Spinoza: Issues and Directions*, ed. E. Curley and P.-F. Moreau (Leiden: E. J. Brill, 1990), 274.

[5] Ibid. Rice also distinguishes here between egoism and the radical individualism he attributes to Spinoza.

[6] Kenneth Blackwell in *The Spinozistic Ethics of Bertrand Russell* (London: George Allen & Unwin, 1985) attributes the second position I raise to Bertrand Russell, for example.

[7] Bennett suggests (306) that there is a "deep incoherence," "a split" in Spinoza's psychological egoism, on the one hand, and his doctrine of the transpersonal nature of the individual as bursting the boundaries of the biological individual proper.

This kind of reading relies on Spinoza's account of how complex individuals are composed of simpler individuals and even higher orders of more complex individuals[8] are composed of the less complex individuals until, as Spinoza writes, "if we thus continue to infinity, we shall readily conceive the whole of Nature as one individual whose parts — that is, all the constituent bodies — vary in infinite ways without any change in the individual as a whole" (E2p13s2).[9] For the maintenance of a certain "mutual relation of motion and rest" (E2p13L5), an equilibrium, renders them an individual. Spinoza himself hints that this metaphysical position has some bearing upon social relations:

> If two individuals of entirely the same nature are joined to one another, they compose an individual twice as powerful as either one. To a man, then, there is nothing more useful than man. A man, I say, can wish for nothing more helpful to his staying in existence than that all should so agree in all things that the minds and bodies of everyone would compose, as it were, one mind and one body; that all together should seek for themselves the common advantage of all. (E4p18s)

Bennett suggests that with this doctrine "Spinoza is trying to give metaphysical support for a certain moral vision" (306). This is a point that has been made by others but, oddly enough, rarely connected to Spinoza's ethical doctrine.[10]

[8] Lee C. Rice, "Individual and Community," analyzes exactly the same passages of the *Ethics* and comes to strikingly opposite conclusions; cf. his "Spinoza on Individuation," in *Spinoza: Essays in Interpretation* (note 2 above), 195–214.

[9] All references to Spinoza's works will be to the J. Van Vloten and J. P. N. Land third edition of *Benedict de Spinoza: Opera* (The Hague: Martinus Nijhoff, 1913). Translations of the Ethics will be from *Baruch Spinoza: The Ethics and Selected Letters*, trans. Samuel Shirley and ed. Seymour Feldman (Indianapolis: Hackett, 1982) unless otherwise noted.

Spinoza's doctrine of egoism as the impetus toward self-preservation of an individual who has burst its bounds because "the whole of Nature [is] one individual whose parts ... vary in infinite ways without any change in the individual as a whole" may owe a debt to Gersonides. Gersonides cites Averroes as one who, unlike himself, held the possibility of the unitary and complete knowledge of all things. Averroes, he goes on, also linked that consummate knowledge to social and ethical perfection insofar as he maintained that the acquisition of that knowledge depended upon the prior existence of "a perfect society where people cooperate with each other" (*Wars of the Lord* I, 12 [Riva di Trento ed., fol. 15, col. b; Feldman trans., 218]).

[10] English language treatments of Spinoza's ethical theory have focussed mainly on arguing pro and con regarding the question raised by G. E. Moore in 1903, Is Spinoza an ethical naturalist? The attempt within Anglo-American philosophy to characterize

Stuart Hampshire, for example, although he holds that Spinoza maintains the transcendence of ego boundaries, nevertheless, does not connect that doctrine to Spinoza's ethical theory. Hampshire, in his extended treatment of Spinoza's moral philosophy[11] in which he systematically contrasts it with Aristotle's, interprets Spinoza's position in entirely different terms, namely, as the moral conversion of the commonly understood individual person and his or her liberation from illusion. Hampshire suggests that "the language of salvation and beatitude that [Spinoza] uses enforces the analogy [i.e., with "religious conversion"], a shaking off of the burden of illusion and anxiety; there is an echo of Lucretius" (64). Thus Hampshire does not treat Spinoza's ethics in terms of the extension of the scope of the individual even though he may note it in another context. Bennett, in contrast, does seem to believe that it is individualist egoism pro or con that is at stake in assessing Spinoza's ethical theory. Bennett opts for the prudential egoist model as predominant. He holds that Spinoza argues for both the first version, prudential egoism, and also for

---

and situate Spinoza's ethical theory begins with G. E. Moore's passing references in *Principia Ethica* (1903), e.g., p. 113, to Spinoza's *Ethics* as a paradigmatic example of what Moore characterizes as committing the Naturalistic Fallacy in ethical theorizing. Moore maintains that "[t]o hold that from any proposition asserting, 'Reality is of this nature' we can infer, or obtain confirmation for, any proposition asserting 'This is good in itself' is to commit the naturalistic fallacy" (114). Moore's characterization of Spinoza was developed in a full length treatment by C. D. Broad in his *Five Types of Ethical Theory* (1930). Broad set off a debate about Spinoza's (meta-)ethics which is still continuing among some Anglo-American philosophers ninety years after Moore published his book. Whether or not, and if so in what sense, Spinoza ought to be considered an ethical naturalist has been a major concern of the majority of English language scholars who have addressed the issue of the nature and success of Spinoza's ethical theory. Those who have addressed this issue include C. D. Broad, David Bidney, Edwin Curley (1973), William Frankena, Robert McShea, and Herman de Dijn.

Another type of treatment of Spinoza's moral philosophy compares it to an account of religious conversion. Stuart Hampshire's *Two Theories of Morality* is the classic example and the most recent work falling in this category is Pierre-François Moreau, *Spinoza: L'Expérience et l'Éternité* (Paris: PUF, 1994), chap. 1. A close parallel to Hampshire's characterization of Spinoza's moral philosophy as conversionary in structure — and that conversion as shaped by its central notion of a transition to freedom — comes, perhaps, from an unlikely quarter: Jon Wetlesen's interpretation of Spinoza's ethics as akin in some important respects to Buddhism.

[11] *Two Theories of Morality* (Oxford University Press, 1977).

this second version, i.e., universal harmony and collaboration emergent from the ultimate overcoming of the bounds of atomic individuality. He maintains, however, that Spinoza's prudential egoism, on the one hand, and the expansive boundaries of individuality, on the other, are irreconcilable and represent a "deep incoherence" in Spinoza's thought (306).

Andrew Collier, in a recent paper on Spinoza's ethical theory, "The Materiality of Morals: Mind, Body and Interests in Spinoza's *Ethics*"[12] has argued for a third alternative, namely, that the bounds of the individual person whose survival and flourishing are the object of his ethics are *never* atomic. The individual is never the Hobbesian atomic individual but always is *in practice* wider both in body and in mind than the bounds of a person's skin, so to speak. This is because the individual in the Spinozist sense includes and encompasses its bodily and mental interactions with the world. He holds that "we must consider the body as extendible, in the sense that the more the body in the narrow sense interacts with the world about it, the more the world is to be counted as part of the person's 'inorganic body'" (76). Collier in his paper ascribes to Spinoza the metaphysical doctrine of the transpersonal nature of the individual and argues generally that that doctrine is the cornerstone of Spinoza's moral philosophy. Collier begins by raising the question:[13] "For Spinoza, what body is identical with the human mind?" He answers that the body that the mind is identical with is not our narrow body proper but what he calls the "inorganic body," by which he means "inorganic structures centered on our bodies in the narrow sense but extending indefinitely beyond them" (77). He frames his argument as an interpretation of E2p16d,c.[14]

[12] In *Studia Spinozana* 7 (1991), devoted to "The Ethics in the *Ethics*," 69–93.

[13] Introduced by Douglas Odegard in "The Body Identical with the Human Mind: A Problem in Spinoza's Philosophy," in *Spinoza: Essays in Interpretation* (note 2 above).

[14] "P 16: The idea of any mode wherein the human body is affected by external bodies must involve the nature of the human body together with the nature of the external body.

D: All the modes wherein a body is affected follow from the nature of the body affected together with the nature of the affecting body (Ax.1 after Cor. Lemma 3). Therefore the idea of these modes will necessarily involve the nature of both bodies (Ax.4,I). So the idea of any mode wherein the human body is affected by an external body involves the nature of the human body and the external body.

Cor. 1: Hence it follows that the human mind perceives the nature of very many bodies along with the nature of its own body.

Collier bases his argument on the further claim that one ought to understand Spinoza's concept of the conatus not as implying panpsychism but instead as a doctrine of inertia in simple bodies and of the mutual relations of motion and rest in complex bodies, and thus his position of an individual conatus that is extendible beyond a given organic body proper is tenable (71–72). Collier cites as further textual evidence for his position Epistle 66 in which Spinoza writes: "The essence of the mind consists in this alone that it is the idea of an actually existing body; and accordingly the mind's power to understand extends to those things only which this idea of the body contains in itself or which follow from it."[15]

From these texts and from the interpretation he mounts, Collier argues that we can draw important conclusions about Spinoza's moral theory: it relies on a "conception of interests transcending 'ego boundaries'" (83). And thus it "removes the conatus theory from Hobbesian egoism" very far indeed (82). "On this account," Collier writes, "the interest that a person has in the wellbeing, not only of other people, but even of inanimate parts of their world, is no more problematic than their interest in the wellbeing of their own physiological parts" (82). Collier concludes by explaining Spinoza's anti-moralism in terms of the materiality of morals[16] inherent in this position: "We will not make people more moral by telling them to be moral: we will not even make ourselves more moral by trying to be more moral. Better moral values (under the attribute of thought) cannot exist without a

Cor. 2: Secondly, the ideas that we have of external bodies indicate the constitution of our own body more than the nature of external bodies."

Collier maintains: "By Spinoza's definitions, this constitutes the environment as part of the organism. ... An organism's eco-system constitutes its essence" (76). And: "The human body, in the narrow sense, is certainly a complex individual of a higher order of complexity. But (a) we must ask ourselves the question: on these definitions, is not the same true of that person's world, their 'Umwelt'—'world' in an almost Heideggerian sense: an individual of a higher order of complexity, since it includes the body narrowly conceived, and other bodies as well? And if so, surely it is this body, not the body in the narrower sense that is trans-attributively identical with the person's mind" (75).

[15] Quoted by Collier (74) as quoted in Hans Jonas, "Spinoza and the Theory of Organism," in *Spinoza: A Collection of Critical Essays*, ed. Marjorie Grene (Garden City, NY: Anchor Books, 1973), 276.

[16] Cf. my article, "Spinoza's Materialist Ethics: The Education of Desire," *International Studies in Philosophy* 22/3 (1990): 59–78.

better set of causal relations between people, nature, ... (under the attribute of extension)" (85).

I shall argue in this paper that the extendible character of body and mind made actual in one way or another in every psychophysical interaction is more explicitly stated and unambiguously maintained by Spinoza in the *Short Treatise* than in the *Ethics,*[17] a text that Spinoza wrote before he introduced his conatus doctrine in the *Ethics*. The *Short Treatise* provides evidence that Spinoza initially held an unambiguous position on the expandable boundaries of mind and body which he retained but also later modified and tried to reconcile with a stronger version of individual interests. It alerts us that the position that Spinoza is an individualist for whom prudential reasoning is the only means to sociality is not the whole story. Whatever tendencies there are in that direction evident in both the *Ethics* and in the political works[18] are balanced by the alternative tendency, as Bennett has shown, which tendency the *Short Treatise* emphasizes. I also call attention in this paper to Spinoza's identification and contrasting of two kinds of mind and body expansion. For I show that Spinoza in the *Short Treatise* points to a contrast between a local and a global extension of personal boundaries. Rather than conceiving of the extension of mind and body as a smooth process beginning from a narrow core outward, so to speak, and toward higher and higher (and wider and wider) orders of individuals,[19] Spinoza invariably contrasts an adequate version of expansion with an inadequate one.[20] The expansion to only the immediate environment characterizes the inadequate version. This distinction would seem to be illuminating of the contrast that Spinoza later draws between the perfected

[17] Bernard Rousset notes this late introduction of the conatus doctrine as well in his essay, "Le poids de l'éthique dans l'Ethique," *Studia Spinozana* 7 (1991): 48.

[18] When Spinoza maintains that "[i]n the state of nature men are enemies" (*Tractatus Politicus* 2 §14, Van Vloten and Land, 2: 18 / *Benedict Spinoza: The Political Works*, ed. and trans. A. G. Wernham (Oxford: Clarendon Press, 1958), he thereby suggests that this is true only at some theoretically imaginary presocial starting point.

[19] Alexandre Matheron in *Individu et Communauté chez Spinoza* (Paris: Les Editions de Minuit, 1969), argues that the state is a true Spinozist individual in the hierarchy of individuals culminating in the "face of the universe." Lee Rice argues against this position of Matheron in his essay, "Individual and Community" (note 4 above), maintaining instead that the kind of unity that the state embodies is not at all a genuine individual by Spinoza's standards.

[20] This is a point that both Collier and Matheron miss.

globalization of sociality emergent from his ethics and the limited one he attributes to political life. Even though the immediate environment that the mind and body in the *Short Treatise* expand to encompass is natural and not social or historical, when Spinoza later in his two political works takes up the extension of ego boundaries as a characterization of the person in the political community, the local extension of boundaries elaborated in the *Short Treatise* would seem to explain such a move and also account for its inadequacy.

In the *Tractatus Theologico-Politicus*[21] and the *Tractatus Politicus,*[22] Spinoza uses language that suggests that the boundaries of the individual body and mind are extended to encompass the body politic and the group mind, concepts which include law and constitution, on the one hand, and, on the other, group emotions, collective desires, and also communal moral values (e.g., TP 3 §§5–7). "Where men have common laws, [they] are all led as if by one mind" (TP 2 §16). In the *Tractatus Theologico-Politicus* and the *Tractatus Politicus*, Spinoza characterizes the normal civil and political condition as one of irrational group cohesion and collective obedience in which the individual is merged into the group. In the *Tractatus Theologico-Politicus* (3), he attributes to imaginative religious control the irrational group cohesion and group emotions of the multitude—for collective religious obedience is the means to "concentrate the strength of all ... members [of society] into one body, as it were, a social body" (2:24/90). And, in the *Tractatus Politicus* (4 §1), Spinoza attributes to political management the capacity to render common the irrational passions of the multitude, to create thereby one group mind—for "the supreme powers ... are the mind of the state" (TP 4 §1). The

[21] "This, then, was the object of ceremonial observance, that men should never act of their own volition but always at another's behest, and that in their actions and inward thoughts they should at all times acknowledge they were not their own masters but completely subordinate to another. With regard to Christian ceremonies, ... they were instituted only as external symbols of a universal Church, not as conducing to blessedness or as containing an intrinsic holiness. ... [T]heir only purpose was the unification of a particular society" (*Tractatus Theologico-Politicus* (TTP), chap. 5, Van Vloten and Land, 2: 151/ Shirley trans., *Baruch Spinoza: Tractatus Theologico-Politicus* (Leiden: E. J. Brill, 1989), 119).

[22] And Spinoza writes even more baldly in the TP 6 §1: "Since men, as I said, are led more by passion than by reason, their natural motive for uniting and being guided as if by one mind is not reason but some common passion; common hope, or common fear, or a common desire to avenge some common injury."

doctrine of the *Short Treatise* can also suggest to us, against Collier and others, why the extension of the individual body and mind in the political individual, the national community, is only a half-way measure and not an adequate expression of ethics, according to Spinoza. The doctrine of the *Short Treatise* would seem to imply that when Spinoza speaks of the group mind and body politic[23] in his political works he is neither speaking merely metaphorically of an underlying individualist prudentialism, on the one hand, as Lee Rice and Robert McShea suggest,[24] nor, on the other, is he going so far as to claim that the political state is a natural individual in the sense that Nature as a whole is, albeit a subordinate one as Alexandre Matheron claims.[25] Instead, the mind and body politic could be said to have the same status that Spinoza gives to the expansion of the body-mind in and through its local interactions in the *Short Treatise*, which is to say the expansion is real but inadequate both in scope and in form.

I shall focus particularly on a text forming part of the preface to part II. Here, Spinoza offers an overview in fifteen points of his account of the human person, body and soul. In the initial section of the preface, Spinoza argued obliquely against Aristotelian teleology and in the last chapter of book I he rejected as untenable any standard ethical valuation based on an account of a natural human species essence. Now he turns to a 15-point enumeration of all that we can now know about the human person based on the relevant metaphysical claims of book I.

23 "Where men have laws in common, [they] are all led as if by one mind" (TP 2 §16). Also, "Since men, as I said, are led more by passion than by reason, their natural motive for uniting and being guided as if by one mind is not reason but some common passion; common hope, or common fear, or a common desire to avenge some common injury" (TP 6 §1).

24 As Lee C. Rice maintains in his essay, "Individual and Community": "What I call the metaphorical interpretation insists that the state has no ontological status as an individual, but functions only as an historical aggregate in the loosest sense of that term. The interpretation is aptly summarized by McShea: 'Spinoza holds the individualist view of man on several counts besides that of his involvement in the thought of his time. The metaphysics is nominalist, at least in intent, so that "society" or "mankind," for instance, cannot for him be anything but collective nouns that point to numerous concrete individual relations'" (274). Rice quotes here from Robert McShea's article, "Spinoza on Power," *Inquiry* 1/12 (Spring 1969): 133–143.

25 In *Individu et Communauté chez Spinoza* (note 19 above).

1. Our soul is either a substance or a mode; not a substance, for ... there can be no limited substance in Nature. Therefore a mode.
2. Being a mode, then, it must be a mode either of substantial extension or substantial thought; not of extension because etc.; therefore, of thought;
3. Because substantial thought cannot be limited, it is infinitely perfect in its kind, and an attribute of God.
4. A perfect thought must have a knowledge, idea, mode of thinking, of each and every thing that exists, both of substances and of modes, without exception.
5. We say "that exists" because we are not speaking here of a knowledge, Idea, etc., which knows the whole of Nature, the connection of all beings according to their essences, without knowing their particular existence, but only of the knowledge, Idea, etc., of particular things which continually come into existence.
6. This knowledge, Idea, etc., of each particular thing which comes to exist, is, we say, the soul of this particular thing.
7. Each and every particular thing that comes to exist becomes such through motion and rest. The same is true of all modes in the substantial extension we call body.
8. The differences between [one body and another] arise only from the different proportions of motion and rest, by which this one is so, and not so, is this and not that.
9. From this proportion of motion and rest, then, there comes to exist also this body of ours, of which (no less than of all other things) there must exist a knowledge, Idea, etc., in the thinking thing. This Idea, knowledge, then, is also our soul.
10. But our body had a different proportion of motion and rest when we were unborn children, and later when we are dead, it will have still another. Nevertheless, there was before our birth, and will be after our death, an Idea, knowledge, etc., of our body in the thinking thing, as there is now. But it was not, and will not be at all the same, because now it has a different proportion of motion and rest.
11. To produce in substantial thought an Idea, knowledge, mode of thinking, such as [this soul of] ours now is, not just any body whatever is required (for then it would be known differently than it is), but one which has this proportion of motion and rest and no other. For as the body is, so is the soul, Idea, knowledge, etc.
12. So if such a body has and preserves its proportion—say of 1 to 3—the soul and the body will be like ours now are; they will, of course, be constantly subject to change, but not to such a great change that it goes beyond the limits of from 1 to 3; and as much as it changes, so also the soul changes each time.
13. And this change, which arises in us from the fact that other bodies act on ours, cannot occur without the soul's becoming aware of it, since it, too, changes constantly. And this change [i.e., in the soul] is really what we call sensation.
14. But if other bodies act on ours with such force that the proportion of motion [to rest] cannot remain 1 to 3, that is death, and a destruction of the soul, insofar as it is only an Idea, knowledge, etc. of a body having this proportion of motion and rest.

15. However, because it is a mode in the thinking substance, it has been able to know and love this [substance] also, as well as that of extension; and uniting itself with these substances (which always remain the same), it has been able to make itself eternal.[26]

Local, Interactive Body and Mind

Spinoza begins the enumeration (1) with his new definition of the human soul—the soul is a mode of substantial thinking, a thinking mode.[27] The soul of each particular thing, moreover, is (6) "this knowledge, Idea, etc., of each particular thing which comes to exist." This means that the soul, first of all, is the awareness of the body: it monitors the body's changes (12), itself changing correspondingly when external bodies "act on ours." This is sensation, Spinoza tells us (13). The knowledge or idea of this particular

[26] All English translations are taken from the *Short Treatise on God, Man, and His Well-Being,* in *The Collected Works of Spinoza,* ed. and trans. Edwin Curley (Princeton University Press, 1985) (henceforth, abbreviated, KV from the Dutch title of the work: *Korte verhandeling van God, de mensch en deszelfs welstand*). The enumeration forms an extended note to the text of II, Preface (Van Vloten and Land, 4: 36–37). Curley includes the enumerated statements after [§5], 95–96.

The reliability of the *Short Treatise* and its precise origins and conditions of composition are a matter of considerable and continuing dispute. On this matter and the current state of scholarship on the treatise see Curley, Editorial Preface, 46–53. See also Joseph Ben-Shlomo's (Hebrew) Introduction in *Baruch Spinoza: Short Treatise on God, Man, and His Well-Being,* ed. Joseph Ben-Shlomo, trans. into Hebrew Rachel Hollander-Steingart (Jerusalem: Magnes Press, the Hebrew University, 1978).

[27] In this Spinoza parallels, but also asserts his difference from, Descartes's philosophical starting point, the doctrine of the soul as a thinking substance. Spinoza here precisely challenges Descartes's claim of the relative independence of the human mind and body, their capacity to exist and also to be conceived independently to some degree. The position of Descartes that Spinoza takes issue with is presented succinctly as follows by Edwin Curley: "What Descartes implies, in calling these things [mind and body] substances, is that they all have some measure of independent existence. No doubt my body depends causally on a great many things for its continued existence. It would not remain in existence for long if it were not sustained by food and drink and air; and Descartes believes that it would not remain in existence for one moment if it were not sustained by God, who must continuously recreate it from one moment to the next. But it does not depend logically on any of these things" (Edwin Curley, *Behind the Geometric Method: A Reading of Spinoza's "Ethics"* ([Princeton University Press, 1988], 7).

body — i.e., the awareness of the ongoing bodily changes — is the human soul (6, 12, 13). (In the *Ethics,* Spinoza will add to this account reflexivity — i.e., that the soul is awareness and its (self-)knowledge of its awareness.)[28] But Spinoza implies here in the enumeration that there are two poles of cognition: an initial bodily awareness and (in 15) an ultimate knowledge that expresses the extensional explanatory order. The human soul or mind is first of all a case of (ongoing) cognitive activity, reflective and expressive of its own body but only as that body encounters others which affect it and change it — i.e., impinge upon it, on the one hand, but also contribute to its ongoing survival, on the other. It is not, as it is for Descartes, an immaterial substance, a self, that thinks. Spinoza is, first of all, defining the person as identical to her body's constitution by its interactions with the world plus her moment-by-moment awareness of that interactive process. Ultimately, however, the person is identical to her body as it is caused by the extensional order and known as such in her expanded and corrected causal understanding of it. The person is a cognitive body in interactive process — a ratio, a self-maintaining equilibrium — and not a static entity, a 'self,' that *has* a body and *has* thinking. In this, Spinoza is both displaying a similarity with the Cartesian position and also establishing where he differs. For we can infer that Spinoza wishes to suggest — and this is borne out subsequently in the text — that (what we would normally call) the subject, that which is the object of ethical concern and transformation, is identical to its bodily and cognitive interactive processes and not an entity conceptually distinguishable from them, which contains that knowledge and does the thinking.[29] That is to say, a person

[28] See, E2p20 and p21, s. At the very end of the manuscript, "Of the Human Soul," appended to the KV (Van Vloten and Land, 4: 90f/Curley [17] 156) Spinoza introduces a rudimentary notion of reflexivity. Spinoza tries out many of the innovations that he will introduce in the *Ethics* in the material Curley includes as two Appendixes and at the very end of the last chapter (from Curley [7] 147ff.).

[29] This position has Stoic precedents. A. C. Lloyd, for example, in "Emotion and Decision in Stoic Psychology," in *The Stoics,* ed. J. M. Rist (Berkeley: University of California Press, 1978), argues that "according to Zeno, Chrysippus, and so far as we know all Stoics before Posidonius, the soul had no nonrational part; *the self, let alone the soul, was identical with the capacity to think, reason and use language;* other activities or faculties were really special conditions or dispositions of this rational capacity, the ruling principle or hegemonikon" (235). The Neoplatonic tradition maintained the identity of knower, known, and the activity of knowing in the act of knowledge — a position derived from Aristotle — which also suggests the absorption

includes her bodily exchanges with the world and her awareness of them.[30] Spinoza has said this in (6)[31] and he elaborates on it in (9): "From this proportion of motion and rest, then there comes to exist also this body of ours, of which (no less than of all other things) there must exist a knowledge, Idea, etc., in the thinking thing. This Idea, knowledge, then, is also our soul."

Human cognition at either pole, namely, as a particular ongoing awareness of bodily interactions or as an extensional causal account of this particular ratio, lacks independent, originative force. *The mind always registers and is determined by some reality or other—either immediate and local or theoretical and infinite—provided by extension, modal or substantial, respectively.* Spinoza has already argued early in the *Short Treatise* that, "a finite intellect can understand nothing through itself unless it is determined by something external" (KV I [5] 3). Spinoza is clearly arguing against Descartes, who characterized his own doctrine of the cogito (e.g., in

---

of the self into its cognitive activity. Such absorption, in fact, was the basis for the Neoplatonic claim of the (mystical) unity of the human person with the divine thinking. If there were a self that was doing the thinking and thereby constituted a remainder, such unification would be impossible. A. C. Lloyd notes this point in Plotinus (as well as above in the Stoics) in his *Anatomy of Neoplatonism* (Oxford University Press, 1990): "Perfection coincides, and is even identical with consciousness. For except at the limits of thought consciousness is what Plotinus calls thinking; and every real thing is a thought. Secondly, a true thought is not just an object of thought but the thinking of it; and thirdly, in order to be a complete or perfect thinking, it must be a self-thinking. This is equivalent to a thought without a thinking subject ... ; and if this is a strange idea, it has been made less so by the history of philosophy, where it is often attributed to eighteenth-century idealists. It is what, in chapters 6 and 7 of the tract [*On Nature, Theoria, and the One*] Plotinus refers to, and probably means by, theoria, which I have called consciousness" (182–183).

[30] Gilles Deleuze makes this same point in reference to both Spinoza and Leibniz and quotes Leibniz as evidence (in his *Expressionism in Philosophy: Spinoza,* trans. Martin Joughin [New York: Zone Books, 1990]): "We are ourselves ideas, by virtue of our expressive capacity: We can therefore define our essence or idea as that which includes everything which we express. And since it expresses our union with God himself, it has no limits and nothing is beyond it" (316). The quotation is from Leibniz, *Discourse on Metaphysics,* §16 [314].

[31] We should note here that this is not an account of *human* nature and of the *human* soul but of *all things* as ensouled. In the *Ethics* Spinoza will maintain that all bodies have ideas but not all ideas are able to register awareness of their bodies due to lack of bodily complexity; see, e.g., E2p13s2 and p14.

distinguishing it from Augustine's proto-doctrine) as "show[ing] that this I, which thinks, is an immaterial substance which has nothing corporeal about it."[32] There is no independent (mental) will for Spinoza that performs and controls the thinking process and its content as there is for Descartes. Rather, thinking necessarily registers passively its own body knowledge of one kind or another. Individual thinking is not characterized by volition as it is for Descartes, but instead (in the Short Treatise but not, of course, in the Ethics) it is passive to both local and global determination. The idea of the body thus consists in propositional statements that make claims about how a given encounter between self and world enhances or diminishes the economy of the ratio between self and environment that is the body.

Spinoza insists that the soul's character as a process of bodily awareness is also consistent with its individuality and internal coherence. For, Spinoza emphasizes here, as we have seen him do elsewhere, that individuality is not lost in substance, in his claim of the convertibility of definition, that individuality must figure in definition and not only generality: to wit, "that what is said also can neither be nor be understood without the thing" (KV II Preface [5]). Thus, this idea (our soul) is in substantial thought, namely, as the idea of this particular body as it changes (10), but still always maintaining some self-identity or other. And, since my soul is a particular ongoing awareness of my particular body's interactions and changes, Spinoza also regards the soul as unified, i.e., it is one cognitive activity and not divisible into parts or discrete functions—although we later find in the text that it can have what we might identify as phases or stages. For Spinoza, as for Descartes[33] and the Stoics,[34] the soul's thinking is broadly conceived,

[32] In a letter of November 14, 1640 to Andreas Colvius quoted by Bernard Williams in his article, "Descartes," in the *Encyclopedia of Philosophy,* ed. Paul Edwards (New York: Macmillan, 1967), 2: 348.

[33] Our soul is, according to Descartes, in fact precisely and exclusively a mind for "there is nothing in us which we must attribute to our soul except our thoughts " (*Passions of the Soul,* in *The Philosophical Writings of Descartes,* trans. John Cottingham, Robert Stoothoff, and Dugald Murdoch [Cambridge University Press, 1985], I, §17, 335. All quotations will be taken from this translation (by Robert Stoothoff) and it will henceforth be cited as, *Passions.*

[34] Michael Frede, for example, in "The Stoic Doctrine of the Affections of the Soul" (in *The Norms of Nature: Studies in Hellenistic Ethics,* ed. Malcolm Schofield and

including all of what we generally regard as psychic functions within its own psychic unity. From Spinoza's definition of the soul as a mode specifically of thinking, and from his implied position of the unity of the soul—a position he makes explicit a bit later[35]—we can conclude that there is, according to him, nothing that goes on in one's soul or mind that is not at bottom some sort of thinking. This position is unlike either the Platonic or the Aristotelian account of the soul but like the Stoic (and similar to Descartes's who also follows the Stoics here).[36] And like the Stoic unified soul that was identical to its reasoning, or the Neoplatonic soul that had various levels of activity but all approximated its highest form of knowing, for Spinoza, too, it is the soul's cognitive activity that clearly characterizes it as a whole. And this cognitive activity that is the soul grasps the process of maintaining ongoing equilibrium—i.e., the relation that one maintains—between oneself and the environment that is the coherence and self-identity of the body.

Since the soul is an individual cognitive process grasping and expressing the ongoing changes in the body in response to external contributions to it which it incorporates and impingements upon it to which it reacts, it thinks not only itself (its body) but also, necessarily, the world via itself. That is, one becomes aware of the world precisely when and insofar as it affects oneself. Thus, the mind registers an ongoing relation, precisely, the permeable boundaries, between body and immediate environment. It registers the boundary conditions and it expresses the relation from the perspective of the self rather than of the other. And the mind is that relation since the mind is identical with its thinking and not a thing that thinks, a container of thoughts. As a result, the soul is not in atomic isolation—a self for which all otherness is external to it—as is the case for Descartes's substantive self, for example. Spinoza clearly rejects not only Descartes's attribution of substaniality to the mind but also his doctrine of the body's externality to the self-mind that follows therefrom. He instead maintains that both body and mind are internal to oneself. In fact, it is always the interactive body—either locally

Gisela Striker [Cambridge University Press, 1986], 102) writes of the Stoic soul: "What distinguishes the Stoics from the rest of the [ancient] tradition is their assumption that all motions of the soul are motions of reason and, ultimately, judgments."

[35] In KV II, 16–17, Spinoza denies that the soul has separate independent faculties.

[36] Descartes writes in *Passions* (art. 47; AT 11: 364): 346: "For there is only a single soul in us, and this soul has within itself no diversity of parts; one and the same soul is sensitive, rational, and all its appetites are volitional."

experienced or globally understood—that is the 'self' to which the mind provides access. As long as the mind is united to its particular body-awareness, it can never be dissolved into a nature in which it has no particular location and locus of awareness. Nonetheless, the body that figures in the mind's awareness is never clearly, nor once and for all, bounded although it, at the same time, maintains its identity and coherence. The scope of the mind, thus—and it would seem, from its birth necessarily—shifts with the shifting of the boundaries of the body, always grasping—and thereby coming to encompass and also be in part constituted by—even at its very narrowest, some external (albeit local) relation. The inadequacy of this kind of knowledge is not that its grasp errs. Its inadequacy is that the reality that figures in mere awareness is both local, and thus partial, and also non-causal. The problem is material and not only cognitive: Spinoza in the *Short Treatise* insists that the mind must shift to a different object, namely, to the extensional and mental causal order, for true understanding and scientific explanation to be possible.

Genetic Explanation

In his rule of definition Spinoza shows how his metaphysics is to govern the actual cases of the explanation of particulars.

> The rule is this: That belongs to the nature of a thing [the causes] without which the thing can neither exist nor be understood [and this, we have been told, is God]: but this is not sufficient; it must be in such a way that the proposition is always convertible, viz. that what is said also can neither be nor be understood without the thing. (KV II Preface [5])

Spinoza's rule of definition assimilates definition to genetic explanation, i.e., to causal accounts and complex theoretical explanations in terms of the causal principles that define substance—namely, thought and extension. Edwin Curley writes that for Spinoza the definitions of things "should give the proximate cause of the thing and that it should suffice for deducing all the properties of the thing."[37] That is to say, a thing embraces in its definition the causes that engender its coming to be what it is essentially and its having the properties it does.[38] *That genetic explanation serves Spinoza as definition*

[37] Edwin Curley, "Experience in Spinoza's Theory of Knowledge," in *Spinoza: A Collection of Critical Essays* (see note 15 above), 43.

[38] Curley makes the point that Spinoza's account of definition is genetic—i.e., it demands a causal account—and involves systematic theoretical explanation:

implies that the boundaries between a thing and both its (extensional) causal sources and its scientific explanation are broken down. A person, for example, in knowing the causal sources of her own particular mind and body, mentally incorporates them, thus becoming them mentally or consciously—she already had to have been them (i.e., had them immanent) extensionally. They become immanent and operative within (and as) the mind. That is what happens when she rediscovers the individuality of her body (and thereby her mind) within the substantial order.

We can understand this better if we turn to what Gilles Deleuze contributes to the discussion of Spinoza's use of genetic explanation as definition. Deleuze points out that Spinoza's "conception of knowledge," that is, his use of genetic explanation, is in itself Aristotelian even though he rejects Aristotelian definition by species essences.

> [An idea] is only true knowledge to the extent that it comprehends the thing through its proximate cause ... This conception of knowledge is thoroughly Aristotelian. Spinoza does not merely mean that the effects known depend on causes. He means in Aristotelian manner that knowledge of a thing itself depends on a knowledge of its cause.[39]

Deleuze explicates the Spinozist position with reference to his reading of Spinoza's philosophy in terms of 'expression':[40] A true idea "must express

---

"Where do we look for a definition or statement of the nature of vision? Well, first of all, not in the dictionary. The kind of definition Spinoza is after is not going to be a statement about the correct use of language. Probably we look for it in a work like Descartes's Dioptrique. And what we find given there as an account of the nature of vision is not a simple statement which comes at the beginning of the investigation, but a complex theoretical explanation of the process by which we see things. [...] We can see this if we look at the requirements he lays down in the Treatise for a good definition of a created thing. The principal ones are that it should give the proximate cause of the thing and that it should suffice for deducing all the properties of the thing.[...] [Spinoza's] causal definition of the circle [namely, as that figure which is described by any line of which one end is fixed and the other moving] gives you its essence, not just a property" (ibid., 42–43). And Curley (43) refers to Spinoza's defintions as genetic definitions.

[39] Gilles Deleuze, *Expressionism* (see note 30 above), 133.

[40] Edwin Curley (*Behind the Geometrical Method,* 38) parses Spinoza's claim that particular things are modes by which God's attributes are expressed in a certain and determinate way (E1p25c) as articulating the relation between natural laws and what follows from those laws. Deleuze also shows how 'expression' works within thought as well as extension, thus capturing its role in language, too.

this cause itself, must, that is, 'involve' a knowledge of the cause" (133). Deleuze:

> It is not a matter, as in the Cartesian Method, of knowing a cause from its effect. Rather is it a matter of understanding the knowledge we have of the effect through a knowledge, itself more perfect, of its cause.
>
> [...] We do not proceed from properties of the effect to certain properties of the cause, which would be only, as it were, necessary conditions in relation to this effect. Starting from the effect we determine the cause, ... as the sufficient reason of all the properties we conceive the effect to possess. It is in this sense that we know through the cause, or that the cause is better known than its effect. Cartesian Method is regressive and analytic. Spinoza's Method is reflexive and synthetic: reflexive because it involves knowledge of an effect through knowledge of its cause; synthetic because it generates all the properties of the effect from the cause known as sufficient reason. We have an adequate idea to the extent that, from a thing, some of whose properties we conceive clearly, we give a genetic definition, from which follow all of its known properties (and still others that we do not know).
>
> [...] To the extent that a thing's definition expresses its efficient cause or the genesis of what it defines, the thing's idea itself expresses its own cause, and we have rendered the idea adequate. (134–135)

That is to say, in the definition or the nature of a thing, the cause is prior to (more essential to and better known than) the effects and the knowledge of the cause is also prior to (more essential to and better known than) the knowledge of its properties, which are its effects.

That Spinoza's method is genetic, Deleuze continues, "leads us as quickly as possible to the idea of God" since an idea is adequate "insofar as it expresses its own cause, and insofar as it expresses God's essence as determining that cause" (139). The convertible character that Spinoza emphatically attributes to definition, namely, that the causal explanation cannot be understood without the thing, entails that the person has her own particular individuality in herself only modally but her very particularity is in the causal order substantially, that is, essentially. Thus, a given thing's individuality is manifested and grasped better and more completely in its (substantial) causes and causal explanation than in its local modal existence. Spinoza has already made it clear in the narrative section of the preface that particulars have their defining characteristics in the causal order as a whole. Let's recall what he says there in denial of a human species nature: "And we deny what they make a fundamental principle: that that belongs to the nature of a thing without which the thing can neither exist nor be understood" (KV II Preface [5]). Spinoza's position is that the modal nature of particulars places their individual ideas in a causal order that defines them and with which each can come to be identified through causal self-knowledge—i.e., through the knowledge of its body which is its mind. And we also know what the nature and scope of causal knowledge is since we recall (4) that "a perfect thought must have a knowledge, idea, mode of thinking of each and every thing that exists, both of substances and of modes, without exception."

The causes that define and engender a particular body and a particular mind already exist if the individual itself does. What the human person can do is to reproduce in her own thinking the causal extensional order as a whole as it gives rise to (produces) one's particular body as a mode of extension instead of reproducing in idea the body's ongoing experience of its own responses to the immediate surroundings. That change in thinking happens through knowing true definitions, genetic explanations, and ignoring sensations. When that happens, the content and order of the individual's mind is the same as the substantial order of thought. It is a perfect thought. Moreover, that cognitive activity renders the human mind "perfect" not only in terms of its content but also insofar as it is now globally determined thought, instead of locally determined. In knowing, it does not overcome its external determination per se but only its local external determination. Spinoza in the *Short Treatise*, in contrast with the *Ethics,* consistently and repeatedly characterizes the mind as standing in a passive relation to the objects which provide its mental content. He here interprets his principle of the intelligibility of all things only through the prior knowledge of substance to entail the external causal determination of the mind as well as of the body. Spinoza puts the matter explicitly in the latter part of book II:

> The intellect is wholly passive, i.e., a perception in the soul of the essence and existence of things ... it is never we who affirm or deny something of the thing, it is the thing itself that affirms or denies something of itself in us. (KV II 16 [5])

The passivity of the mind includes both that its mental content is externally determined—i.e., neither internally generated nor freely willed—and also, that it is, at least in part, of some reality initially presented as external to it.

Thus, knowledge is a sort of coming to consciousness of the causal order already obtaining, for the rendering of the global causal order within a person's thinking is, at the same time, a real manifestation of substantial thought (insofar as it grasps substantial extension) determining one's mind. While the human mind remains passive, because it can discover its own causes and individual definition within the substantial order, it can become identical with it. Through reproducing as its own self-explanation (i.e, the true knowledge of its body) that ideational order as a whole, it unites with it. "Because it is a mode in the thinking substance, it has been able to know ... this [substance] as well as that of extension; uniting itself with these substances," Spinoza tells us in the last of the 15 points. The human mind in knowing itself in (and through) its substantial causes, gives an account of its particular body as that body is a mode of extension. The particular body is thus rendered intelligible within its mind's thinking. In knowing, the mind knows the causal explanation of the body as a mode of extension and therefore the person forms a union with its own bodily causes insofar as they are manifestations of extension and grasped in thought. Thus the human mind (and person) according to Spinoza overcomes its modal fragmentation and localism while it, at the same time, thereby reembraces its modal passivity and cognitive dependency but now from a global perspective. It conceives itself — this individual body and mind — as the product of the (substantial) order. In the *Short Treatise* Spinoza puts it thus:

> If we were left to ourselves, and so do not depend on God, there would be very little, or nothing, that we could accomplish. ... But that would be quite the contrary to what we now see, viz. that we depend on what is most perfect in such a way that we are a part of the whole, i.e., of him, and so to speak contribute our share to the accomplishment of as many well ordered and perfect works as are dependent on him. ... We ascribe everything we do to God, who is the first and only cause of all that we accomplish. (KV II 18 [2–3])

Later on, Spinoza will express the ongoing and reembraced passivity of the human mode toward the causal substantial order in quite dramatic theological and ethical terms when he writes that "we are truly God's servants — indeed his slaves — and that our greatest perfection is to be such necessarily" (KV II 18 [2]). "True freedom," Spinoza writes in the last chapter of the *Short Treatise,* "is to be and remain bound by the lovely chains of the love of God" (26 [5]). Because it is the nature of substance to incorporate a person's particular causes, when she has (causal self-) knowledge, she thinks her own definition as this expression of thought and extension and that is to think her own idea (of her own body) as it is in substantial thought. The mind in knowing unites itself with its own causes — by making them explicit to itself and as defining itself. It thus perfects itself as uniquely itself — rather than in terms of a species explanation and standard—but in doing so also overcomes its immediate ongoing self- (and local) awareness. At the same time, however, the mind also embraces as itself the entire causal order of which it is part and product, since they are contributory to its own genesis. Hence, it is precisely in thus reconceiving her own individuality that a person transforms her initial perspectival localism and isolation into a global causal explanation in which the individual is not lost but fully manifested for the first time. We recall again that according to (4), "A perfect thought, etc."

But, what about my narrow self, the body proper—not the known causal order of extension as it engenders my individual body in an ongoing manner but the body whose boundaries are coextensive with, and whose awareness and desires remain rooted within, my narrow flesh? Are my body's experiences and desires so easily assimilated to their causal explanation within the substantial extension and thought? Can the global causal explanation of one's unique individuality exhaustively grasp the individuality that is my body?

Global Body and Mind

For Spinoza, the mind operates as a process that includes within itself and as its expanded self the connection between self and world both in rudimentary cognition (bodily awareness) and also in perfected knowledge. He proposes (as we have just seen) that thinking (a) qua awareness spans the external object—as it impinges on the body—and oneself and also that (b) qua perfected knowledge, thinking becomes united or identified (through an intellectual love) with the causal order now seen to be continuous with, and explanatory of, oneself (15). In neither case is thinking ever regarded by Spinoza as grasping the interior condition of the untouched, isolated soul. Nor does he regard it as ever providing a grasp of an external reality completely unrelated to oneself (9). Soul and body have and manifest boundaries that are constantly shifting in relation to what at first appears to be beyond them. In both cases, the mind grasps the individuality, the ratio[41]

[41] Edwin Curley (*Behind the Geometrical Method,* 48) attributes to Spinoza two types of global causal order, namely, an infinite order of particular causes and a finite order of infinite causes or natural laws, which together provide a complete causal account: "To understand finite particulars—the motion of this particular body, at this particular speed, in this particular direction—we must appeal not only to the other finite particulars which conditioned them, but also to the laws of nature. The

that its body maintains (11)—I have argued elsewhere that this 'ratio' ought to be understood in terms of systems theory as the maintenance of an equilibrium and systemic coherence amidst constant inputs and outputs[42]—first as it manifests itself in ongoing sensation, simple experience, and then theoretically and exhaustively. According to the first form of cognition, awareness spans and reflects the ongoing interaction of self-identity and immediate environment that a given body maintains and experiences. In the second kind of thinking, Spinoza emphasizes that the mind becomes identical with its infinite extensional causes as grasped in substantial thought, thus overcoming its separateness from all things beyond it. Yet the price the mind pays for its expression of bodily infinity—in the *Short Treatise* but not in the *Ethics*—is that of losing its special connectedness to its own sensation, its awareness of its own body's interactions with the immediate, local environment.

15. However, because it is a mode in the thinking substance, it has been able to know and love this [substance] also, as well as that of extension; and uniting itself with these substances (which always remain the same), it has been able to make itself eternal.

This transition is not a mere transitory occurrence for an eternal union is its result.[43] Furthermore, it is not only that one's *knowledge* of reality has changed but the very *reality* of one's mind has changed. It has now extended itself to be the mind of the world body, so to speak. Moreover, the better it thinks, the more substantial it becomes—that's immanence—and not only the more of the substantial order it grasps or mirrors.[44] Thus does knowledge

---

finite modes follow from, are deducible from, a finite series of infinite causes, the laws of nature taken in conjunction with an infinite series of finite causes, the other prior finite modes. That is, the finite modes follow from God's attributes (via the infinite modes), but do not follow from God's nature absolutely, only conditionally."

[42] See my article, "Notes on Spinoza's Critique of Aristotle's Ethics: From Teleology to Process Theory," *Philosophy and Theology* 4/1 (Fall 1989): 3–32.

[43] We recall that Wolfson points out that the characteristic of love to produce a union with God is mentioned here but not in the *Ethics* (Harry A.Wolfson, *The Philosophy of Spinoza,* two volumes in one [1934; reprint, New York: Meridian Books, 1958], 2: 281).

[44] Deleuze makes this point the center of his treatment of Spinoza in *Expressionism* (note 30 above). He writes (14–15), for example, that "the range of the notion of expression is not merely ontological; its implications are also epistemological. ...

effect a union of mind and the extensional order and not merely an adequate reflection.

Spinoza had already introduced a rudimentary account of individual and universal striving in book I of the *Short Treatise.* In (I 5 [1]), he had written that there is a "striving we find both in the whole of Nature and in particular things, tending to maintain and preserve their being," not only "to preserve [each] in its state" but to "bring it to a better one."

> We posit a universal and a particular Providence. The universal is that through which each thing is produced and maintained insofar as it is a part of the whole of Nature. The particular Providence is that striving which each particular thing has for the preservation of its being insofar as it is considered not as a part of Nature, but as a whole. (KV I 5 [9])

The person thus has an inner force maintaining itself as an empirical, proximate, and local whole as well as a presence in the universal. When Spinoza refers here in the enumeration (12 & 14) to the body's self-maintenance as an ongoing unity and identity as an equilibrium of motion (interaction and change) and rest (internal coherence, self-identity),[45] he is therefore recalling his earlier account of individual striving. In the *Short*

Spinoza's conception of the adequacy of ideas seems always to involve this expressive character. From the *Short Treatise* onward he was seeking a conception of knowledge that would account for it, not as some operation on an object that remained outside it, but as a reflection, an expression, of an object in the mind. The word 'express' has various synonyms. The Dutch text of the *Short Treatise* does employ *uytdrukken* and *uytbeelden* (to express), but shows a preference for *vertoonen* (at once to manifest and to demonstrate): a thinking being expresses itself in an infinity of ideas corresponding to an infinity of objects; but the idea of the body directly manifests God; and attributes manifest themselves in themselves *(Short Treatise* II.xx.4 *(uytdedrukt);* I. Dialogue II.12 *(vertoonen);* I.vii.10 *(vertoond).*" "The concept of expression," Deleuze further suggests (322), "is in short a specifically philosophical concept of immanence, which insinuates itself among the transcendent concepts of emanative or creationist theology. ... [There is] the immanence of its expression in what expresses itself, and of what is expressed in its expression."

[45] For an excellent discussion of both Spinoza's notion, and also the history of the notion, of physical identity as a 'ratio,' a homeostasis, see Jacob Adler's (unpublished) essay, "Strange Bedfellows: Spinoza and Alexander of Aphrodisias on the Immortality of the Soul." Adler argues that Spinoza's account of the soul is as corporeal attunement set to a specific ratio of motion and rest, which account is ultimately attributable to Alexander.

Treatise, this "striving" does not include any affective dimension — i.e., any bodily based desire — that effects and maintains the equilibrium or even registers the experience in awareness as desire. This is unlike the mature account of the conatus in the *Ethics* where the conatus is identified as "desire"— a central and general desire to maintain self-identity and to pursue self-furthering of which all other particular desires and emotions are specifications — and that desire is said to to be each person's essence (e.g., E3 def. of emotions 1). Rather than a bodily based and bodily maintaining desire, affectivity in the *Short Treatise* enters in only within the process of cognition and not within that of maintaining and furthering the body's individuality. Spinoza introduces affectivity at the end of the enumeration, in (15), with his account of the cognitive union with the substantial, the knowledge that is love. In the *Short Treatise,* affectivity is, alternatively, a consequence or a concomitant of cognition, but never an expression of the body and its purposes. Later, in the *Ethics,* Spinoza will no longer speak of 'love' in the way he does in the *Short Treatise* but instead of 'desire' as more basic than, and the source of, love and all the other emotions. There, Spinoza maintains a threefold distinction: desire, love, and, ultimately, intellectual love. Unlike the *Short Treatise*'s unitive love, which is exclusively a concomitant of cognition merging the knower into her mental-affective object, 'desire' is, instead, the essential nature and expression of the conatus, spanning body and mind. Desire is the impulse toward self-development, expressive of the striving for self-preservation and self-furthering. And thus rather than exchanging one object of love (the body) for another (the extensional order) as Spinoza will propose here in the *Short Treatise,* the desire that first and foremost preserves its own body and awareness will itself develop beyond that narrow scope in its very pursuit of its (seemingly narrow) aim. The conatus will be shown to be inherently self-evolving from its early focus upon, and concern for, the most narrowly enclosed self, to a refocussing upon, and concern for, a self inclusive of all its causal relations.

In the *Short Treatise,* Spinoza succumbs to a modified dualism. Cognitive love can have either of two objects to know and love and to unite with. Each person, therefore, stands before a dilemma that is at once cognitive and moral: is it to be the body (as mode) or, instead, substantial extension and thought, with which my mind is to unite? Spinoza tells the reader near the end of the book that "the soul can be united either with the body of which it is the Idea or with God, without whom it can neither exist nor be understood"

(KV II 23 [1]). Spinoza proposes repeatedly throughout the *Short Treatise* that the person has the choice[46] of uniting herself with the body or with God through focussing her mind on one or the other. The posing of such a choice, namely, between psychic union with the body or with God, would not be intelligible if Spinoza were not suggesting that cognition unites the person with the object of her mental content — a theoretical option in regard to the body that Spinoza suggests here in the *Short Treatise* but abandons in the *Ethics,* where the body and mind are characterized by identity rather than as two-in-union.

And because the body is the very first thing our soul becomes aware of — for as we have said, there can be nothing in Nature whose Idea does not exist in the thinking thing, the idea which is the soul of that thing — that thing must, then, necessarily be the first cause of the idea.

But this Idea cannot find any rest in the knowledge of the body, without passing over into knowledge of that without which neither the body nor the Idea itself can either exist or be understood. Hence, as soon as it knows that being, it will be united with it by love.

[6] To grasp this union better and infer what it must be, we must consider the effect [of the union] with the body. In this we see how, by knowledge of and passions toward corporeal things, there come to arise in us all those effects which we are constantly aware of in our body, through the motion of the ['animal,' i.e., bodily] spirits; and so (if once our knowledge and love come to fall on that without which we can neither exist or be understood, and which is not at all corporeal), the effects arising from this union will, and must, be incomparably greater and more magnificent. For these [effects] must necessarily be commensurate with the thing with which it is united. (KV II 22 [5–6])

The dichotomy that Spinoza presents here is between the union of the mind with its empirical bodily awareness (sensation) or with the body (and soul) substantially conceived.[47] We have seen that Spinoza stresses repeatedly that

[46] Whatever that might mean for Spinoza since there is, in fact, no choice according to him. I shall defer a discussion of this matter for the present.

[47] Wolfson (see note 43 above) points out that the account of knowledge as producing a direct merger with its mental objects is prominent in the *Short Treatise,* whereas it is muted, never explicitly articulated, in the *Ethics.* Wolfson further points out that neither the characteristic of love to produce a union with God, nor the categorization of love according to its objects, appears in the *Ethics.* These are characteristic doctrines of the *Short Treatise,* but not of the *Ethics.* Wolfson holds, however, that "even in the *Ethics* there is an indirect allusion to both these views" (Wolfson, 2: 281).

it is the (essence of the) individual body, this particular proportion of motion and rest, and not only the soul—the awareness of this equilibrium between self and environment—that is taken up into the substantial order in and through knowledge. The dualism is, thus, between the empirical body and the substantial orders, thought and extension, not between the body and any mental order. It is the empirical and local—i.e., not just the body but the empirical mind, too (namely, awareness, sensation) insofar as it is rooted in and expressive of the local body—that is in danger of being lost in the substantial and universal. They are subsumed to the extent that they are taken up not by an exclusively mental realm but rather by and in substantial thought and extension. The soul, "uniting itself with these substances ... has been able to make itself eternal," Spinoza concludes the enumeration (15). Thus, in unitive cognition, the local body and any cognitive states expressive thereof are retained only in theory but not as an ongoing source of awareness and resistance to the knowledge that submerges the empirical particular in the explanatory order.[48] Both of Spinoza's types of cognition save the individual from loss in Aristotelian species natures (species essences). The cognition that is union does not, however, retain any rootedness of the individual in the cognitive states that are inextricably bound up with and emergent from the local body's day-to-day experiences.

According to Spinoza's account of unitive cognition, it is the external object with which the mind forms a union and identification when it appears as one's mental content. In this case, cognition seems to lose itself—its local

[48] Spinoza's fully worked out position in the *Ethics*, according to which (1) all knowledge is of the body together with the external world, and (2) all knowledge is also via the body, and (3) the emotions are the ongoing registering and expression of that relation, anticipates Antonio Damasio's conclusions from his own and other recent neurobiological research summarized for a popular audience in his book, *Descartes' Error: Emotion, Reason, and the Human Brain* (New York: Grosset/Putnam, 1994). Damasio maintains that "feelings are just as cognitive as any other perceptual image" (159). At the same time, "feelings let us mind the body" and "are first and foremost about the body," offering us "the cognition of our visceral and musculoskeletal state as it becomes affected." He continues here: "Feelings offer us a glimpse of what goes on in our flesh, as a momentary image of that flesh is juxtaposed to the images of other objects and situations. By dint of juxtaposition, body images give to other images a quality of goodness or badness, of pleasure or pain." Damasio concludes that "it is as if we are possessed by a passion for reason, a drive that originates in the brain core, permeates other levels of the nervous system, and emerges as either feelings or unconscious biases to guide decision making" (245).

empirical awareness — in the embrace of the explanatory object as its own true content. Substance is as ambiguously external as is the immediate environment. In the first account, it is one's own local bodily interactions that provide the content of awareness, and in the second, it is one's own extensional individuality that one rediscovers within substantial thought. Thus, thinking manifests either (1) the spanning and inclusion within itself of the relation to immediate otherness, namely, to the external environment impinging upon and changing the body, or, instead, (2) the union of the mind with an object which is also itself — namely, its own causes, the extensional causes that constitute and explain one's body and their expression in thought. According to this second way cognition includes the connection to what is beyond it within itself, the soul comes to be in seamless unity and identity with the infinite thought and extension rather than manifesting (as itself) the relation between self and world. The body proper and the mind's ongoing awareness of it in sensation, while never atomic, are nevertheless at best only limitedly expansive and therefore to be forsaken. While both body and mind manifest permeable boundaries, expanding always to include aspects of what is outside them, Spinoza considers bodily awareness as still too confining, too limitedly local, and also too irrational to express the body's true manifestation and explanation in the causal order.

Concluding Remarks

If we return briefly to the passages from the *Tractatus Theologico-Politicus* and the *Tractatus Politicus* referred to in the beginning, we now have some rudimentary sense of the status of the 'group mind' and 'body politic.' These social entities are neither strictly metaphorical and thereby allegedly mask Spinoza's true atomic individualism, nor are they "natural individuals" and as such constitutive and descriptive of the way things really are according to the order of nature and explanation. They instead convey the real (yet limited) character of the local extension of the body and mind as encompassing their immediate environment — a phenomenon that also sheds some light on the concept of the 'common order of Nature' in the *Ethics*. This limited extension is true — it is factually existent — but not ultimately adequate either materially or cognitively for it neither makes manifest nor grasps the underlying causal system and explanation. We can now approach Spinoza's political thought armed with the understanding that the immediate local world is the first actual extension of one's own body and one's own mind. The group mind and the body politic are not mere metaphors masking an individualist reality even though they are surely inadequate both materially and conceptually. They signal both the ongoing need for an imaginative political discourse and way of life suited to the less than complete and fully rational character of that extended body, the political community, and also, the ultimate need for philosophy and ethics.

Hamilton College
Clinton, NY

[17]

British Journal for the History of Philosophy 8(1) 2000: 1–19

BEYOND SUBJECTIVITY: SPINOZA'S COGNITIVISM OF THE EMOTIONS

Gideon Segal

ABSTRACT

In what follows I try to show that Spinoza modelled his project of rational psychology, in some of its major respects, upon Descartes's metaphysics of matter. I argue further that, like Descartes, who paid for the rationalization of the science of matter the price of having to leave out of his description non-quantifiable qualities, so Spinoza left out of his psychology the non-rationalizable aspects of emotions, i.e. whatever in them could not be subsumed under common notions. He therefore was left with the cognitive aspects of emotions, keeping outside of his report the inner feeling which accompanies them. Spinoza's psychology, I claim, disregards any non-cognitive aspect of emotions.[1]

For most of us, emotions seem to be of a distinct nature among mental states and events. In particular, being affected by an emotion is, phenomenally speaking, a different sort of experience than having, or entertaining, a thought. Indeed, an emotion can – though perhaps it must not – involve cognition. Thus, when I love someone, I can state her look, demeanor or moral virtues as causes of this feeling. But even so, love has more to it than the cognition that provides it with an object, a reason or a causal basis. Love, and similarly other emotions, seem to have an inherent hard-core mental content peculiar to each, a content which is purely experiential, one that cannot be transcribed in words or any other means of expression that usually conveys cognitive content. Not that people cannot express their feelings – they can and they do so successfully. But there is an ineliminable gap between the content expressed and what is originally experienced. This gap is manifested in the fact that the person to whom an emotion is expressed

[1] The term 'cognitivism' with regard to Spinoza's theory of the affects was already used by Amihud Gilead. See his 'Human affects as properties of cognitions in Spinoza's philosophical psychotherapy', in Y. Yovel (ed.) *Desire and Affect: Spinoza as Psychologist (Spinoza by 2000 – The Jerusalem Conferences, vol. III)* forthcoming.

British Journal for the History of Philosophy ISSN 0960-8788

does not know what the thing expressed is, unless she or he had at least once experienced a similar feeling – which fact shows that, while conveying the cognition involved in the emotion, the expression also indicates to the listener an inexpressible experiential content bound with this cognition. The listener is then supposed to regain acquaintance with the emotion out of his own inner resources – for its hard-core experiential content cannot be imparted from outside. Thus, when I say: 'I lost my hat yesterday', you gather a piece of information you did not have before I expressed it. But when I say: 'I am very frustrated having lost my hat yesterday', a feeling is successfully indicated to you *if* you know its kind from your own inner experience in the past. A propositional or any other cognitive content cannot replace the unique feeling one has when affected by frustration, love, envy or disappointment. Hence by expressing such content one can at best draw a recollection in the listener of his own past love, envy etc. That attempts to express an emotion often involve metaphorical usage, is a further indication of the inaptness of a report about the cognitive content involved in the emotion to conveying what the emotion truly is. Had there been to emotion nothing beyond the cognition it involves, such usages would probably have been less frequent.

Now, we should distinguish two major assertions comprised in this depiction of the affective realm. One is that, having an emotion essentially involves an inner irreducible feeling that neither words, nor any other means of communicating an experiential content, can express. The other is that this hard-core content is peculiar to emotions, and that it is either lacking in other types of mental occurrences such as cognitions, or that, if other mental occurrences do similarly involve an experience not reducible to cognizable content, it is different in kind from the content involved in emotions.

Coming now to Spinoza, my main point in the present paper is the following: there is no textual evidence allowing us to suppose that Spinoza would have denied that emotions – a term I will use henceforth to denote the mental aspects of Spinozistic affects, i.e. the affects considered under the attribute of thought[2] – do involve an inner feeling, irreducible to cognitive

[2] I follow Curley in using 'affect' for Spinoza's *affectus*. In his explanation for this choice Curley criticizes Elwes' and Shirley's choice of 'emotion' for this purpose, claiming that: (a) it 'suggests a passive state, whereas an *affectus* may well be active'; (b) it turns exclusively upon the psychological aspect of *affectus*, while in Spinoza the latter term refers to bodily as well as mental aspects; and (c) it seems too narrow for it seems unnatural to call desire an emotion (E. Curley (tr. & ed.) *The Collected Works of Spinoza* (Princeton, Princeton University Press, 1988) p. 625). However, the etymology of 'emotion' (e-motion) sustains a judgement contrary to Curley's point (a), and since I leave desire out of my main argument, point (c) is irrelevant to my purposes; finally, point (b) is my specific reason for using 'emotion' in my main argument, which deals with the mental ('psychological') side of the affects. The legitimacy of a separate treatment of the affects as modifications of the mind,

content. Furthermore, since the presence of such experiential content is a matter of everyday mental life, immediately accessible to any normal person, we must suppose that its denial never occurred to him as a plausible claim (if he ever gave it any thought at all). However, according to Spinoza, emotions, i.e. affects considered as modifications of our mind, are nothing but cognitions of bodily occurrences of a certain kind. Being affected by joy, love, envy, etc. is in fact having, in each case, an idea with a particular content; therefore, our inner experience of an affect or emotion is in fact our inner experience while having a certain cognition that constitutes this emotion. What we feel when we are joyous, envious etc. is an awareness of a cognitive content specific to each affect.

Thus Spinoza would reject the second of the two assertions mentioned above, i.e. he would deny the peculiarity of the inner experience of emotions. Our experience of being aware to the cognitive content that constitutes an emotion is specific to each occurrence of an emotion, but it is not of a special kind distinguished from our experience of awareness to any other cognitive content, which as such is necessarily accompanied by the inner experience of awareness to it.

Before pursuing my main argument, one preliminary qualification should be met with regard to the nature of the affects. Taking the emotions to be nothing but cognitions might seem to disregard their dynamic aspect, an aspect so central to Spinoza's metaphysical account of the modal realm. However, it should be born in mind that cognition, rather than a *state* of the mind, is itself imbued with dynamism. In fact, Spinoza emphatically claims this when, defending his definition of idea as 'a *concept* of the Mind that the Mind forms because it is a thinking thing' (2def3,exp), he explains that he preferred the term 'concept' (*conceptus*) to 'perception' (*perceptio*) specifically because the former 'seems to express an action of the Mind' whereas the latter does not. It thus turns out that to have an idea is to conceive of a

i.e. as modes of substance under the attribute of thought, despite their having a complementary ontological status as modifications of the body, or modes of substance under the attribute of extension (and similar status in each of the infinite attributes of the Substance), rests on Spinoza's own methodological principle expressed succinctly in the *Ethics*, Part II, Scholium to Proposition 7:

> so long as things are considered as modes of thinking, we must explain the order of the whole of nature . . . through the attribute of Thought alone. And in so far as they are considered as modes of Extension, the order of the whole of nature must be explained through the attribute of Extension alone. I understand the same concerning the other attributes.

(Spinoza's writings are quoted from Curley's translation, using the following abbreviations: A number opening a citation denotes the part of the *Ethics* cited; def = Definition; ax = Axiom; p = Proposition; d = Demonstration; exp = explanation; c = Corollary; s = Scholium; AD = Affect Definition [in *Ethics* III]; pref = Preface; TIE = *Treatise on the Emendation of the Intellect*.)

cognitive content, or simply to cognize, and this should be taken to signify an activity rather than the mere having or inclusion of a content as part of some static mental inventory. Consequently, by claiming that emotions are cognitions, I grant them the active nature the latter have *qua* Spinozistic concepts; later on, when I treat Spinoza's definition of the affects, the specific content of their dynamic nature, in both its mental and bodily aspects, will come to the surface.[3]

I start my argument by describing what I believe is a central motivation in Spinoza's theory of affects, namely, his adoption of Cartesian methodological principles, which originally operated in Descartes's metaphysics of matter. I will then try to show that it is by way of adapting these principles to his theory of mind that Spinoza reached his cognitivistic psychological theory.

SPINOZA AND CARTESIAN SCIENCE OF MATTER

Spinoza's theory of the affects follows the often-quoted programmatic statement in the Preface to *Ethics* III. He starts by denouncing those who treat the affects morally rather than in a disinterested fashion paradigmatically typical to scientific research; he then puts forward his basic methodological assumption, that everything that happens in nature is subject to the same type of laws; and since, as he says, man is part of nature (rather than a separate kingdom within it), the affects should be studied as any other natural phenomenon, with the aim of discovering the laws that govern them. A partial quotation will serve our purpose here:

> The Affects . . . of hate, anger, envy, etc. considered in themselves, follow from the same necessity and force of nature as the other singular things. And therefore they acknowledge certain causes, through which they are understood, and have certain properties, as worthy of our knowledge as the properties of any other thing, by the mere contemplation of which we are pleased. Therefore, I shall treat the nature and powers of the Affects, and the power of the Mind over them, by the same Method by which, in the preceding parts, I treated God and the Mind, and shall consider human actions and appetites just as if it were a Question of lines, planes, and bodies.

Readers of these renowned sentences tend to emphasize two things. The one is Spinoza's stress on the affects' having 'certain causes', and hence his intention at laying out causal laws governing the affective realm. The other is his announcement that he will deal with the affects as one does with lines,

[3] I thank the reader of the *British Journal for the History of Philosophy* for prompting this clarification, as well as the one on p. 2 and note 2 above concerning the terminology I use with regard to the affects.

planes and bodies, i.e. by the Geometrical method.[4] As has already been observed by Bennett,[5] between these two points the centre of weight is on the first one.[6] Rather than the Geometrical apparatus of demonstration, the fitness of the affects for scientific study is what Spinoza ascribes importance to. After all, the grave weaknesses and fallacies he detects in previous studies of emotions could not be set aright by adopting a new formal tool of demonstration. Spinoza does indeed present his system of affects in 'geometrical' order, and one should not belittle its importance for him. But he was not going to revolutionize the traditional study of affects by casting old conceptions into the mold of definitions, axioms and demonstrated propositions.

Yet, when speaking of lines, planes and bodies, does Spinoza necessarily have in mind his 'geometrical' method of exposition and demonstration? A careful reading of the paragraph quoted above may reveal a different, or at least an additional point, which is perhaps more important than those mentioned above. The key is, I suggest, in the somewhat obscure remark concerning 'certain properties, as worthy of our knowledge as the properties of any other thing'. Its meaning is elucidated by the reference, in the same paragraph, to lines, planes, and bodies as typical objects of the kind of study to be employed according to Spinoza to the affects. Put in a nutshell, my suggestion is that when Spinoza spoke about studying the properties of the affects in the way one studied lines, planes and bodies, he may have had in mind Descartes's treatment of the physical realm. We know the tremendous influence of Descartes's metaphysics of matter and spirit upon Spinoza, evidenced most strikingly by his endorsement, although in a modified version, of the Cartesian identification of thought and extension as basic ontological kinds, making them the two attributes of God-Substance that are within the reach of human knowledge. Yet Cartesian metaphysical conceptions could not enter Spinoza's reflective thinking without the methodological considerations that led to Descartes's conceptual achievements having their impact on him as well. Thus it is quite plausible that, when Spinoza refers in the Preface to *Ethics* III to the method by which he has studied God and

[4] H. A. Wolfson, *The Philosophy of Spinoza* (New York, Meridian Books, 1958 (reprint of 1934 edition)), Vol. II, 181; J. Bennett, *A Study of Spinoza's Ethics* (Cambridge, Cambridge University Press, 1984), p. 20. Still others focus on the Preface mainly as stating Spinoza's general attitude in the study of the passions, proclaiming it to be dispassionate (see Paul Weiss, 'Some pivotal issues in Spinoza', in R. Kennington (ed.) *The Philosophy of Baruch Spinoza* (Washington, The Catholic University of America Press, 1980), p. 4; Y. Yovel, *Spinoza and other Heretics – The Adventures of Immanence* (Princeton, Princeton University Press, 1989), p. 142 and non-moralizing (Yovel, ibid.).

[5] Op. cit. 20.

[6] However, the apparent emphasis on causal laws governing the affects may be interpreted in more than one way. In my 'Ideas, Affects and Causality', *North American Spinoza Society Monograph*, No. 6 (1997), 3–21, I claim that the interpretation of the relations between the affects as causal *according to the paradigm of physical sciences* is mistaken.

the human mind, what he thinks of is not only the geometrical method of exposition and demonstration but the more significant analytical tool that Descartes employed with regard to the body and the soul. To explain this reading of 3pref, some elaboration on Descartes is inescapable.

One of the peaks of Descartes's treatment of the concepts of mind and body is his few paragraphs about the piece of wax in *Meditations* II. Reflecting on Descartes's motives in it, one may recall the biographical account at the outset of the *Meditations*, where he proclaims the main motive of his immersing himself in contemplation, namely, 'to establish [scientific knowledge] that was stable and likely to last'.[7] Firm scientific knowledge is what Descartes aspires to, and his philosophical project aims at laying out evident principles that would provide him with a sound basis for its attainment.[8] The wax analysis must have seemed to Descartes as a great step ahead on the way to achieve this goal. In it (and again in *The Principles of Philosophy*, II, 4 ff) Descartes searches for the essential properties of bodies.[9] His analysis results in an insight to the nature of physical reality as purely spacial. The physical realm consists of entities of which all the inherent qualities stem from their spacial attributes (with the addition of the temporal dimension requisite for the description of change). If indeed this is an adequate analysis of bodies, Descartes's achievement is a sumptuous one; for spacial properties are clearly and distinctly conceived, and, moreover, they are amenable to mathematical treatment. Further yet, since Descartes discovered a promising way of reducing geometry to algebraic equations, by what has later been entitled the Cartesian Coordinate System, the combination of this discovery with the reductive result of the wax analysis renders the realm of bodies amenable to rigorous quantitative treatment, no doubt a promising advancement towards Descartes's goals.[10]

Spinoza's ultimate goals lie far beyond a firm grounding of scientific knowledge, but the latter is required for their achievement. First, one needs to develop mechanics and medicine as basic *pre*conditions to a moral endeavor (TEI 15, 16). But more importantly, metaphysics, a theory of mind and psychology are required as intrinsic constituents of the very process of moral

[7] Descartes, *Meditation I*, *The Philosophical Writings of Descartes*, translated by J. Cottingham, R. Stoothoff and D. Murdoch (Cambridge, Cambridge University Press, 1984), Vol. II, p. 12. (henceforth *Writings II*).

[8] Descartes's other goal is to rebut the newly emergent skepticism of the seventeenth century; cf. Marcelo Dascal, Preface to the Hebrew translation of Descartes's, *Principles of Philosophy*, I (Tel Aviv, University Publishing Project, 1979), p. 7.

[9] The piece of wax, of which Descartes speaks in *Meditations* II, only serves as an example, which is good as any other, of a physical body (I follow here Margaret D. Wilson, *Descartes* (London, Routledge & Kegan Paul, 1978), Chap. II, sec. 4, and Anthony Kenny, *Descartes – A Study of His Philosophy* (New York, Random House, 1968), Chap. 9).

[10] My description of Descartes's aims and conclusions is deliberately partial and sketchy. A more comprehensive treatment is out of the scope of the present discussion.

improvement, as the structure and content of the *Ethics* make crystal clear. Central among these undertakings is a systematic study of the affective realm. Therefore, Spinoza pursues an analysis of this realm in terms that would render it amenable as much as possible to rigorous scientific treatment. And although the properties of the affects cannot as a rule be quantified, and hence cannot be treated mathematically in the way bodies as *rei extensae* can, Spinoza is going to show that once their true essence is defined, they do lend themselves to rational procedures[11] and as such they are amenable to reasoning in geometric fashion.

These Spinozistic interests kept in mind, one may quite plausibly conjecture that when he proclaims his study of affects to follow the study of *lines, planes and bodies*, Spinoza may be revealing his intention to model his psychological research, in terms of the methodological considerations to be applied to it, upon Descartes's study of bodies as spatial entities. Thus, Spinoza's crucial point here is the method to be chosen for the analysis of psychological concepts, rather than the method by which their nature and governing laws is to be presented and demonstrated. In fact, the principal importance of the applicability of the geometrical method of representation and demonstration to the affects is its being a celebrated outcome and a manifestation of the scientific nature of their study.

One must however keep in mind that Descartes's grand philosophical move has its inescapable conceptual price. For, aiming at a theory of the material universe in which every assertion would be evident, Descartes would allow no terms into his metaphysics of matter, but ones that refer to items that can be clearly and distinctly known. Moreover, Descartes's litmus test for such scientifically approved terms was their quantifiability. Whatever cannot be quantified and measured can only be a vague quality, of which we cannot form a clear and distinct idea, and, by Cartesian terms, none but clear and distinct ideas can count as knowledge. Consequently, any term denoting properties that cannot be grasped quantitatively is eliminated from his concept of body. Thus, qualities like hardness, redness or bitterness, which Descartes considers not to be expressible in quantitative terms, are ruled out from this bottom-line description of the material Universe.[12] The resulting concept of body (a piece of wax, say) is that of an extending entity stripped of all its sensible qualities, such as colour, taste, emitted sound, etc. The only real changes a body can endure are those of form and place, and whatever real properties to be found in it, e.g. heaviness, attraction, should be reducible in the final analysis to combinations of these two properties in bodies or their parts.[13] Any property that neither

[11] See 3pref, where he contrasts his position with one holding the affects to be 'things contrary to reason'.

[12] Descartes, *Principles of Philosophy*, II, 64. Cf. J. Cottingham, *Descartes* (Oxford, Basil Blackwell, 1986) p. 83, 89.

[13] cf. Descartes's replies to the 6th set of objections, in *Writings II* (see above note 7), p. 297.

falls under extension and its modifications, nor is reducible to them, is banished to the domain of 'secondary qualities', effected *in the perceiving subject* by the two primary ones stemming from changes of form and place.[14] This is the price paid for the rationalization of the physical realm.

THE RATIONALIZATION OF PSYCHOLOGY

It should be emphasized that, when speaking of quantifiability, what Descartes is up for is amenability to rational treatment. As Cottingham notes, '[w]hat appeals to Descartes about quantifiable properties such as size and shape, is their lack of vagueness, their clarity and distinctness or "openness" to the attentive mind'.[15] Yet Descartes correlates these characteristics with amenability to 'mathematical' demonstration, meaning thereby the rigorous method of demonstration used by Euclid. Thus, his emphatic statement of the reductive principle of his metaphysics of matter, in *Principles of Philosophy* II, 64, ends by stressing that such properties can be dealt with in mathematical demonstrations. My main point here is that a similar consideration guides Spinoza's method for the study of the affective realm. Undertaking, as announced in 3pref, to treat the affects as Euclid does with geometrical entities, what he is up for is the amenability of the subject-matter of his study to rational procedures. Moreover, in taking Descartes's study as his paradigm, Spinoza endorses its essential characteristics, the most important of which was that the object of the new science should be comprehended clearly and distinctly.

How could this austere epistemological standard be applied to the study of entities, or rather occurrences, such as love, envy or ambition? It seems that the very essence of emotions does not allow the exclusion of vague qualitative aspects – vague rather than clear and distinct. For what would be left of an emotion once you remove from it the *feeling* it involves, which is by nature a vague qualitative experience?[16]

To be sure, integrating Cartesian methodological principles into the study

[14] Cf. Kenny, *Descartes*, pp. 212–13, and his quotation from the 3rd Meditation (see *Writings II*, 30). Note that the primary/secondary qualities terminology is Locke's, although the conceptual distinction is already anticipated by Descartes.

Burtt contends that 'Descartes's real criterion (in his distinction between secondary and primary qualities) is . . . the possibility of mathematical handling' (E. A. Burtt, *The Metaphysical Foundation of Modern Physical Science* (London, Routledge & Kegan Paul, 1967) p. 117). The implied suspicion with regard to the truthfulness of Descartes's piece-of-wax analysis should be dealt with elsewhere.

[15] Cottingham, *Descartes*, p. 89.

[16] See in this regard Descartes's definition of the passions as 'the perceptions, feelings, or emotions of the soul which we relate specially to it . . ', *The Passions of the Soul*, I, 27, *The Philosophical Works of Descartes*, in E. S. Haldane and G. R. T. Ross (tr. & ed.) (Cambridge, Cambridge University Press, 1931), vol. I, p. 344.

of the affects could only succeed provided some necessary adjustments be made. Thus, while the outcome of the analysis of bodies led to a purely mathematical concept of them, the analysis of the affects, no less rigorous in its outcome, could not be expected to reach a similar result. Quantification could only have a secondary role, e.g. with regard to the temporal dimension of emotions, as in the 4p10: 'We are affected more intensely toward a future thing which we imagine will quickly be present, than if we imagined the time when it will exist to be further from the present.' Nevertheless, Spinoza's treatment of the affects clearly manifests his effort at logical rigor, e.g. when he distinguishes between affective states on the basis of the medieval logical modalities of necessity, possibility and contingency (see 3p49, 4p11,12,13, 5p5).

Exact quantification and measurement set aside, what concept of emotions can we endorse which would enable our ideas of each of them to become clear and distinct? Spinoza's answer to this was, first and foremost, an elimination of his own, which is, as we shall see, analogous in certain respects to the Cartesian elimination of non-quantifiable qualitative terms from the essence of bodies. Throughout his dealings with the affects in definitions, propositions and proofs, he will refrain from referring to any of their aspects which cannot be analysed so as to render them amenable to rational thinking. Before we define the specific nature of this elimination, we must touch upon two additional points which are central to Spinozistic epistemology, and are therefore essential to Spinoza's treatment of the affective realm. I shall speak, in turn, about common notions and the mind's knowledge of itself.

COMMON NOTIONS

In 2p29s Spinoza describes very succinctly the method for acquiring systematic rational knowledge of things. The mind has adequate knowledge '*of itself*, of its own Body, and of external bodies . . .' inasmuch as, *qua* the subject of knowledge, it is 'determined internally, from the fact that it regards a number of things at once, to understand their agreements, differences, and oppositions. For . . . then it regards things clearly and distinctly' (emphasis mine). This description strongly suggests as its outcome the stock of what Spinoza later on in *Ethics* II terms 'Common Notions'.[17] The understanding of the 'agreements, differences, and oppositions' of things provides

[17] Note that Spinoza ends the Scholium of 2p29 saying that what it claims is going to be shown 'below', and that what follows the scholium is a discussion of the epistemological conditions of inadequate and of adequate cognition. Now, the latter is, at least on the level of *ratio*, i.e. knowledge of the second kind, nothing but the conceiving of Common Notions and their derivatives. Hence the allusion in 2p29s to a further discussion of its subject undoubtedly refers to the treatment of common notions.

us with knowledge of their common features, those that can only be conceived adequately.

By making Common Notions the basis of all rational thinking (cf. 2p40s2),[18] Spinoza introduces a universal requisite for scientific knowledge, including that of the affective realm.[19] To be rationally conceived, all objects must be comprehended through those of their features that are, in Spinoza's words, 'common to all things and are equally in the part and in the whole' (2p38). We should therefore turn to Spinoza's principle of common notions. Later on we will see that it is by applying this principle to the study of the affects, thereby rationalizing psychology, that Spinoza reached his cognitive theory of emotions.

Common notions, as they are defined and discussed in *Ethics* II, Propositions 37–40, are 'the foundations of our reasoning' (2p40s1,s2). They provide, so to speak, the cognitive material for *ratio*, or the second kind of knowledge. Descartes, too, used this term, apparently identifying 'common notions' with clear and distinct ideas, but the term had a long history, starting with Aristotle. Of special interest for our purpose here is the fact that in all previous writers, the adjective 'common' in 'common notions' refers to their being shared by all men, i.e. they are termed 'common' because they are affirmed by whoever attends to their content.[20] Unlike all these writers, Spinoza takes the commonness of common notions to have two meanings rather than one: first, common notions are common in that what they denote – according to 2p40s2 it is 'properties of things' – is 'common to all *things* and is equally in the part and in the whole' (2p37; italics mine). Thus, e.g. the concepts of motion and rest and their laws are common notions in that all bodies and their parts are extending modes of substance, and as such all their states and episodes are defined by motion and rest, whose laws govern them as wholes, as well as each of their parts.[21] Second, Common Notions are common to all *men*, i.e. they are shared by all human minds, their content being identical in each of them (2p38c). It should be noted that an item of knowledge being necessarily common to all men (i.e. its content being shared by all of them), is a result of its object – e.g. a property, a lawful change etc. – being common to all things and equally in the part and in the whole of each (2p38c). For, according to Spinoza, what is common to all things and their parts cannot but be adequately known (2p38), and it is owing to a thing's being perceived adequately by whoever attends to it that

[18] The same role assigned to common notions by Descartes, *Principles of Philosophy*, II/64.

[19] Among other requisites are the 'genetic' definition of entities (TIE 94–6), the subsumption under causal laws (1ax4, 3pref) and the exclusion of teleology from all causal explanations (*Ethics* I, Appendix).

[20] See quotation from Simplicius in note 22 below.

[21] See the section on physics starting with Axiom 1 after 2p13. An example from a different field is Euclid's calculus of a fourth proportional, which is based on 'the common property of proportionals' (2p40s2).

its notion is *necessarily* (rather than by coincidence, cultural circumstances etc.) common to all men (2p38, 2p38d).[22] Thus 'common notions' denote the invariant, all-prevailing ontological features of reality which, as such, are adequately, hence similarly, conceived by any attending mind.

Universals, such as a horse or a storm also appear to be common to all perceivers, yet they in fact vary according to each perceiver's singular cognitive biography, and each applies a commonly used term to his/her unique image of a horse or a storm (2p40s1). The implementation of the analytic procedure described in 2p29s, enables one to sort out adequate common notions, which provide the basis for the second kind of knowledge (2p40s2), ignoring ill-founded 'common' universals, which stem from knowledge of the first kind (ibid.).

By laying out his Common Notions theory, Spinoza introduces into his system a general methodological principle, one that applies to all fields of rational knowledge. According to this principle, adequate cognition of things (at least at the second kind of knowledge[23]) depends on their being conceived through their common elements. The methodological note in the scholium of 2p29, which anticipates the common notions section (2p37–40), contrasts the pattern of thought that leads to common notions with the passive affection of the mind by random experience: common notions are reached when the mind, rather than accumulating series of perceptions as they haphazardly affect the body, actively employs its cognitive faculties, and 'regards a number of things at once, to understand their agreements, differences, and oppositions'. In this way rational, systematic thinking replaces random perception and knowledge based on idiosyncratic, uncritically-formed universals. The question now arises, how can this universal principle of common notions, a requisite of rational thinking in all fields of knowledge, apply to the cognition of the special realm of the affects? And since this principle determines the nature of cognition admissible to any science, what constraints does it impose on the contents open to adequate knowledge of the affective realm?

[22] We already find this idea schematically expressed in Simplicius:

> These notions are common because they belong to all men with respect to the things of nature; *for [having these notions] men agree with their object* and these notions put in men no difference, like the universal judgements: the good is useful, the useful is good, etc.

Simplicius, *Epicteti Enchiridion*, quoted in M. Gueroult, *Spinoza, II, L'Ame (Ethique II)* (Olms, Hildesheim, 1974), p. 582 (English translation and emphasis mine).

[23] I here deliberately sidestep the issue of the nature of Intuitive Knowledge, and its treatment of emotions. That intuitive knowledge must be mediated by *Ratio* (the 2nd kind of knowledge) and hence that it similarly depends on common notions, is to be argued for in another occasion.

12 GIDEON SEGAL

KNOWLEDGE OF ONE'S OWN MIND

Before answering these questions, an additional consideration regarding the cognition of affects should be put to light. Apparently one can gain an insight into the nature of affects simply by a direct acquaintance with one's own mind. To everyday thinking it might seem that one's own emotions can best be approached directly through introspection. Can't I grasp the true nature of my own feelings by simply perceiving their occurrence in my mind? Do I not know, from my intimate acquaintance with my own mind, what it means to love, to envy or to feel disappointed? In fact, a positive answer to this question is typically assumed by Descartes, who maintains that 'since every one has experience of the passions within himself, there is no necessity to borrow one's observations from elsewhere in order to discover their nature'.[24] And whereas mental representations of our body and its parts may be deceptive, 'we cannot be so deceived regarding the passions, inasmuch as they are so close to, and so entirely within our soul, that it is impossible for it to feel them without their being such as it feels them to be'.[25] To be sure, these suggestions are rejected by Spinoza. His theory with regard to knowledge of one's own mind amounts to a flat denial of Descartes's claim that 'there is so much . . . in the mind itself which can serve to make my knowledge of it more distinct, that it scarcely seems worth going through the contribution made by considering bodily things'.[26] A full explanation of Spinoza's approach would take a lengthy interpretative work. Here I shall limit myself to what seem to me to be the main pertinent points.

The human mind is a complex idea of the human body, which itself is a complex mode of extension (2p15,d), and as such it *perceives* each of the body's modifications (2p12). Whenever the body is modified, either through an external cause or by its own processes, the mind is modified by an idea of that bodily modification. Now, whereas this mental modification is a *perception* of the bodily modification, it is not eo ipso *knowledge* of it. When e.g. someone is pinpricked, the change that occurs in his bodily tissues is correlated, according to Spinoza's thought-extension parallelism, by an idea of this change (e.g. he feels a sting).[27] However, he does not know, from this perception alone, that he was pricked by a pin, or even that he was pricked at all. In general, the mind's modifications, being mere perceptions of bodily modifications, do not constitute knowledge of these modifications (2p27). Rather, knowing one's body and each of its modifications requires the discernment of features that this bodily modification shares with other such occurrences. Thus, someone being pinpricked knows his body as something that is being pricked by a pin only when his mind is modified by the idea of

[24] *The Passions of the Soul*, I, 1, *The Philosophical Works of Descartes*, 331.
[25] *The Passions of the Soul*, I, 24, op. cit., 343.
[26] *Meditations* II, *Writings II*, 22.
[27] 2p7, 2p7s.

a pin pricking him, and this idea is more complex than the mere sensation of a prick, for it contains at least a partial scheme of its causal circumstances.[28] This scheme can be formed if, when the prick occurs, the person has additional sources of information, such as a visual image of the penetrating pin, with which he can combine the tactile perception to construct a clear idea of the event or else a recollection of a similar occasion in the past, which he can now associate with the actual stinging feeling. Note that more than the mere Humean conception of causal knowledge is assumed here; for interpreting the visual data as a pin penetrating one's skin involves, in addition to a simple conjunction of 'impressions', at least their sorting and reconstruction to distinguishable interrelating objects, such as the pin and the penetrated skin, as well as an idea of parts of one's body as being parts of *his own* body. This kind of complex achievement is involved even in common daily cognitions on the level of *ratio*, to which Spinoza refers in 2p29s, saying that to adequately know things we must 'regard a number of (them) at once, to understand their agreements, differences, and oppositions'.

These requirements apply also to the mind, and specifically to the affective realm. For the mind is a complex idea of the body, and for each of its ingredient ideas to be adequately known, it must be an adequate idea of its bodily correlate (2p29d). Conversely, inasmuch as each mental modification is an inadequate perception of the body's nature and condition, it is also an insufficient basis for knowing the mind itself. This indeed is the case when the ideas of the bodily occurrences take them as single items, isolated from their causal context (2p29).[29] How do I know, e.g. that I feel envy toward someone who won a coveted prize? The immediate, intimate awareness of an inner feeling is nothing more than an indication of some inconvenience in me. That this is *envy* is known to me only through my knowledge of a whole situation involving me, the winner and the prize. Generally speaking, adequate knowledge of the affects is impossible on the basis of the immediate, intimate experience we have of them. Each affect, as it is internally experienced, being a discrete inner perception isolated from a causal rationale, cannot be known adequately. To be adequately

[28] For the idea of a *partial* scheme operating on the level of *ratio* see Y. Yovel, 'The second kind of knowledge and the removal of error', in Y. Yovel (ed.) *Spinoza on Knowledge and the Human Mind (Spinoza by 2000 – The Jerusalem Conferences, vol. II)* (Leiden, Brill, 1994), pp. 101–3. For the necessary involvement of the idea of a cause in any adequate cognition see TIE 91–2 and *Ethics* I, axiom 4.

[29] Note that Spinoza denies the Cartesian idea of clarity and distinctness of single isolated cognitive items as warrantee of their truth. Adequacy is for Spinoza always context-dependent, and isolation of a cognitive content is the cause of error, even though the isolated cognition cannot, by itself, be regarded as an error. Thus, e.g. a teaspoon seen broken in a glass of water is the cause of error when cut off from the idea of the optical constraints that compel us to *see* a sound teaspoon *as if* broken. The illusion loses its deceptive status once its causal circumstances are put to light (4p1,d,s).

known, each occurrence of envy, joy, pity etc. must be correlated with an intelligible causal picture of its circumstances.[30] This does not mean that a systematic and explicit formulation of a causal scheme is required, and far less its subsumption under causal laws of a scientific type. One may be aware of feeling joy or pity without having fully adequate cognition of the nature of one's emotion. But even partially adequate cognition of these affects depends on some knowledge of the causal scheme responsible for their generation, based on an (inexplicit and usually non-conceptualized) cognition of the regularity of phenomena, 'their agreements, differences, and oppositions' (2p29s).

To adequately know an affect, one must gain adequate knowledge of the bodily affection which this affect is the idea of, which means its comprehension within its causal circumstances. This requirement takes us back to the more basic one of common notions.

COMMON NOTIONS IN THE AFFECTIVE REALM

The raw material of knowledge is a flood of sense data affecting the mind correlatively to the chance affection of the body by external bodies.[31] This 'random' or 'vague' experience (*'experientia vaga'*, TIE 19) of unsystematic series of modifications can turn intelligible through the active operation of the intellect (TIE 19; 2p29s). By the latter's analytic activity we come to conceive our own body and the external bodies affecting it through their common features, and to link them into causal chains. Thus, as we saw, a sum of experienced sensations can be conceived as indicating an event of a pinprick once the singular sensations involved are put together in a unified spatio-temporal context of distinct interrelating extended objects. And for the single sensations to form such complex ordered pictures, their common features as extended modes must be extricated from the sum of daily experiences we accumulate more or less haphazardly. Generally speaking, to adequately know the body, the mind, and their modifications, a comparative method must be employed, which aims at the common notions of each area, be it the realm of physical bodies, arithmetics, the affects or any other.[32]

[30] Therefore, inasmuch as a person feeling joy adequately knows what he feels, he is also affected by love, defined by Spinoza as 'Joy, accompanied by the idea of an external cause' (AD6; by AD7 a similar inference obtains with regard to sadness and hate).

[31] Spinoza terms this passive serial experience 'the common order of nature' (2p29c), 'common' meaning here general, hence raw, uncultivated.

[32] Both Wolfson and somewhat implicitly Deleuze contend that common notions are limited to modes of extension, i.e. to physics; Wolfson, *The Philosophy of Spinoza* (see above note 4), II, 125f); Gilles Deleuze, *Spinoza – Philosophie pratique* (Paris, Les Edition de Minuit, 1981), 154ff. For an example that apparently attests to the applicability of the common notions paradigm to mathematics see above note 21.

Since adequately conceiving the nature of any object of knowledge requires that it be approached through its 'common' elements – those that it shares with all others (under the same attribute), and those that would be equally conceived by all men, again we are presented with the question, how can this universal principle apply to the cognition of the special realm of the affects? Whatever the answer to this challenge, it must meet the two requirements implied in the principle of *common* notions, namely: (a) that the content of the notions by which one accounts for affective phenomena be *common* to these phenomena; and (b) that these notions be shared alike by whoever attends to them. *It is at this point that the Cartesian methodological assumptions, originally employed in Descartes's analysis of matter, become relevant to Spinoza's analysis of the affects.* For, to comply with these requirements, Spinoza, implementing a Cartesian method, expels from his concept of emotion whatever is not amenable to rational treatment, and retains only those aspects of the emotions that lend themselves to cognition through common notions.

Under this rigorous criterion, the analysis of emotions must eliminate the immediate, intimate experiences we have of them. For each occurrence of envy, joy, pity etc. in so far as it is regarded as a feeling internally experienced, or rather *felt*, is unique and incomparable to any other. As such it cannot be measured, graded or logically analysed. Thus our experience of the emotions is doomed to remain a bunch of discrete perceptions with no rhyme or reason, a flood of mental data which cannot interconnect to an intelligible coherent picture. This then is the upshot of our discussion thus far: our discrete, singular inner feelings of joy and sadness, love and hate, and all the other affects, cannot be handled by the paradigm of rational thinking.

SPINOZISTIC FEELINGS AND CARTESIAN QUALITIES

Just as Descartes ejected secondary qualities such as colours from the realm of bodies clearly and distinctly conceived, because they did not lend themselves to rational thinking, so Spinoza ruled out from his psychology the feelings that accompany the affects. You cannot 'regard' a number of inner experiences of joy at once, 'to understand their agreements, differences, and oppositions', for they are mere perceptive emotional 'colours', which by their nature cannot be logically handled, and far less treated in any way that leads to evidence.

Nevertheless, the dismissal of the feeling-aspect from scientific psychology should not be interpreted as amounting to a denial of the very presence of this aspect in joy, envy, love, etc. – neither did Descartes expel hardness and whiteness from our experience of a piece of wax. Spinozistic emotions and Cartesian sensations alike are ineliminable aspects of experience, for obviously, we cannot evade sweet taste when we taste a sweet thing, or avoid *feeling* angry when we *are* angry;[33] both experienced qualities should, however, be eliminated from our ultimate ontological report of the entities and events that constitute their epiphenomenal appearance.

Note that, in comparison with colours and sounds, the feeling aspect of emotions has an additional import, since experiencing it is the way one intimately meets oneself from within, and some would say that nothing in our emotional life is more important. Nevertheless, what an emotion feels like is doomed to remain forever in the undiscernible darkness of subjectivity, since it cannot count as candidate for a rational discourse concerning mental phenomena, which, like any rational discourse, must have common notions as 'the foundations of our reasoning' (2p40s1).

Still, one may easily sympathize with someone who, like Deleuze, takes the Spinozistic affects to be 'feelings'.[34] For admittedly, such interpretation well accords with our modern picture of the human soul, which, owing to the indelible impact of nineteenth century Romanticism, gives much significance to the subjective inner feeling of an emotion. Moreover, the reduction of emotions to sensations and feelings goes further back to Hume, who decisively denies that emotions can be discursively explained, since each is a unique feeling not derived from or otherwise intrinsically related to any other.[35] However, this line of interpretation is by definition anachronistic. Spinoza's thinking was miles away from Hume's empiricist theory of mind and he cannot be thought to have anticipated the romantics. *Pace* Deleuze, his conception of the nature of the affects was, and, given his Cartesian methodological assumptions, must have been, purely 'intellectualist'.[36] In that respect he is diametrically opposed to Romanticism, notwithstanding the large extent to which romantic writers draw upon his other conceptions, about nature as a whole, God's immanence and man's place within it.

THE COGNITIVISM OF THE EMOTIONS

Summing up the above, Spinoza's cognitive theory of emotions results, among other things, from constraints set upon it by the principle of common notions. In it Spinoza describes the mental aspect of the affects solely on the basis of their corresponding cognitive content, omitting from his analysis the non-rationalizable feeling aspect. To show that this approach permeates and

[33] Yet, under certain physiological conditioning, feeling sweetness while tasting a thing that regularly tastes sweet *is* avoidable. There can be found analogous examples in the affective realm, but their discussion must be left to another occasion.

[34] Deleuze, *Spinoza – Philosophie pratique*, pp. 68–71.

[35] This Humean position is concisely summarized and criticized in Bennett, *Study*, p. 272, citing Hume, *A Treatise of Human Nature*, in Selby-Bigge (ed.) Oxford, pp. 286–7, 329, 415. Regarding Bennett's own interpretation of Spinoza's affect theory see below n. 41.

[36] Cf. Deleuze, op. cit., p. 70.

dominates all of Spinoza's psychological theory would require much more space than can be afforded here. Therefore, rather than attempting an extensive description, I shall limit myself to analysing an example, hoping thereby to outline the very basic principles of Spinoza's cognitivism.

Suppose someone is sad because he just lost ten years' savings. The losing of the money is an occurrence in space and time, which affects the person's body in certain ways. There are things he would do with the money and which he cannot do now that the money is gone, various types of behaviour change due to his new financial situation, and all is accompanied by physiological reaction (cardiac, cerebral and the like). According to the General Definition of Affects (GenDA) in *Ethics* III, the sadness that a person feels is, in its mental aspect, an affirmation of bodily events, inasmuch as they constitute a decrease in his body's perfection or power.[37] If, when asked why he is sad, someone answers that this is because he lost his savings, this is a partial expression of his cognizing a regress in his bodily powers which losing the money involves.[38] Note that whatever is involved in this description of the nature of sadness is amenable to the method of common notions: the decrease of the body's power is a physical (and perhaps biological) process, comprehended within its causal circumstances, thereby making the occurrence of the affect an event governed by general laws. As to the emotion (the mental affect) of sadness – this is none but a complex cognitive correlate of what happens in the body, and it is as such that the emotion is accessible through generalizable features, common to all occurrences of sadness, hence also commonly known to all minds. And since all the psychology of emotions turns upon cognition of what happens, according to general laws, to the body's power or perfection, all of it is amenable to rational cognition.[39]

[37] 'An Affect that is called a Passion of the mind is a confused idea, by which the Mind affirms of its Body, or of some part of it, a greater or lesser force of existing than before'. The rest of the *definiens* treats Desire, as Spinoza himself makes clear in the last paragraph of the Explanation of the definition (= the last paragraph of *Ethics* III). I treat briefly the cognitive nature of Desire in p. 18f of my paper cited above, note 6.

With respect to the apparent limitation of GenDA to passions (passive affects) only, I concur with Parkinson that by omitting the word 'confused' from the definiens we get a really *general* definition of the affects, one which applies, consistently with Spinoza's system of affects, to 'actions' (active affects) as well (G. H. R. Parkinson, 'Language and Knowledge in Spinoza', *Spinoza – A Collection of Critical Essays*, in M. Grene (ed.) (Garden City, Double-day Anchor, 1973), p. 78 and note 21, p. 84. See also 5p3d).

[38] In this case also hate arises, since the affirmation of change of perfection is accompanied by an idea of its cause (see above, note 30). Here, however, the object of hate is an event rather than a thing.

[39] See in this regard 2p38d, where knowledge of the mind turns upon common notions of its body. Spinoza concludes this demonstration with the following generalization: 'The Mind therefore . . . necessarily perceives *A* adequately (*A* being "something that is common to all bodies"), *and does so both in so far as it perceives itself* and in so far as it perceives its own or any external body' (emphasis mine). This seems to confirm that the common notions principle applies not only to bodies but to minds as well.

To know what an occurrence of sadness is for a sad person, one must know the circumstances of that sadness, or simply *why* that person is sad, or what he is sad about. These being stated, they can be analysed to show that what they constitute *qua* sadness is a deterioration of physical capabilities in the specific stated circumstances, causally analysed.

A query arises here, since feelings such as joy in view of an imminent meeting, or envy of a successful colleague, do not seem to us in any way connected with affirming changes in the perfection and power of our body. From this point of view, the Spinozistic system of affects, admittedly an unsurpassed architectural masterpiece, may seem far removed from our everyday experience of emotions, a curiosity rather than an adequate report of the psychology of real human beings. This would be, however, a rash judgement. For the 'affirmation' that constitutes the emotion does not necessarily involve a thought about the content that is affirmed *qua* a decrease in the body's power or perfection. A sad person does not as a rule think of the deterioration of his body that constitutes the body-aspect of his affect of sadness.[40] Rather, his emotion is a mental correlate of a bodily change which involves a deterioration (i.e. a decrease in power or perfection). His focus of awareness is naturally set on the specific circumstances over which he is sad, rather than on their constituting a causal process that involves a decrease in his body's power or perfection.[41]

CONCLUSION

Following his claim that, rather than a 'dominion within a dominion', man inheres in nature and is subject to its laws (3pref), Spinoza endorsed in his inquiry of emotions the same canon of rationality which Descartes had employed in treating the material world. This meant that, to know what sadness and joy, love and hate, and all other human emotions are, one had to pursue their common features and search for properties and governing

[40] An exception to this would be the case of someone who is sad *about* the deterioration of his body's power or perfection, e.g. someone sad over being ill.

[41] Bennett seems to recognize that, defining Joy, Sadness and their kinds as 'affirmations' of changes in the body's perfection, Spinoza's deliberately maintains his cognitive terminology (this, he implies, is part of Spinoza's 'intellectualist' description of the mind). But, as he somewhat obscurely puts it, 'we ought to see Spinoza as making a place [in his talk of emotions] for bodily sensations' as well (op. cit., p. 267). But if, as Bennett maintains, the affirmation of the change in the body's perfection is none but a 'marked bodily sensation', which he contrasts with cognition (see his list on p. 267), how can it ever become clear and distinct, which by 5p3 it must if the affect should ever turn from Passion to Action (active affect)?

Bennett further maintains that Joy and Sadness involve feeling, and this in conformity with his terming these two basic affects – from which, together with Desire, all the others derive – Pleasure and Unpleasure correspondingly (ibid. p. 270). As I argued above, Spinoza could not have seen feelings but as epiphenomena, and in this regard Bennett's terms are misleading, despite his justifications on pp. 253–4.

rules cognizable to any attentive mind. The application of the common notions doctrine, that provided the methodological framework for the achievement of rational knowledge in all fields, to the study of the human mind, resulted in Spinoza's cognitivist theory of the emotions. Interpreting all emotions ultimately as cognitions of a certain kind, Spinoza purified them of any ingredient that resisted the canon of rationality. Thus, in addition to denying the ontologically inconceivable cross-attributive causal relations – for in no way would such causal relation lend itself to common notions – Spinoza's theory of emotions ignores the subjective feelings bound with emotions, although the epiphenomenal presence of this ingredient in each emotion is not denied. Not being amenable to the paradigm of common notions, the inner, subjective awareness to emotions had to remain outside the scope of rational psychology, and therefore must have seemed to Spinoza valueless from a philosophical point of view.

Holon Academic Institute of Technology

[18]

THE MATERIALITY OF MORALS: Mind, body and interests in Spinoza's "*Ethics*"

Andrew Collier

I want to argue for a particular reading of Spinoza's conception of a person's body, from which follows an interpretation of his moral philosophy. The great paradox of his moral philosophy is that he bases everything on seeking ones own advantage, yet arrives at a morality almost as far removed as can be imagined from "materialism" in the popular, non-philosophical sense of the word. My reading of Spinoza removes some of this paradox, though at the expense of attributing to him a conception of the body that some may find even more paradoxical, though I would defend it. (I speak of attributing views to Spinoza, but I am claiming that these views follow from or make sense of what he says, rather than that he consciously held them. I should perhaps also add that, while I am defending several Spinozist positions, I am doing so in different terms from those I would use were I speaking in my own voice, rather than discussing Spinoza.) The conclusions of my argument are used in my paper "The Inorganic Body and the Ambiguity of Freedom" (Collier 1991), though the grounds for them are only very sketchily indicated in that place.

1 Mind

As a preliminary to the discussion of the body, I need to take sides in a dispute about the status of the attribute of thought, and hence of the

ideas that belong to it (a dispute in which Albert Balz, Edwin Curley and Jonathan Bennett have taken their various stands — see Balz 1967, Curley 1969, Bennett 1984). Spinoza's parallelism of the attributes of thought and extension serves (among others) two functions, which pull us in opposite directions as to how to interpret it. On the one hand, it is the foundation of his theory of truth — an identity-theory, as it has been called. On the other, it founds an account of the relation of the human mind to the human body. One interpretation starts from the latter, and assumes that he means much what most other people mean by "mind" and "body", and that "thought" pertains to mind, and so is **psychological**. The paradox generated by this view is that, since he says that everything exists under both attributes, he appears to be committed to a sort of animism, ascribing minds to rocks and trees and water. The other interpretation assumes that he means what most other people mean by an idea and its object. This avoids the appearance of animism, but has the paradoxical consequence that a person's mind is related to their body as a concept to its object. On this view, ideas are **logical** entities, not, as on the other, psychological ones (or not primarily so; insofar as they double as the subject-matter of psychology, it is psychology that is being reduced to logic, rather than **vice versa** — much as Wittgenstein says in his *Tractatus Logico-Philosophicus* (5.542): "'**A** has the thought **p**' and '**A** says **p**' are of the form ' "**p**" says **p**' ". "Thought" becomes a sort of Fregean third realm, conceptual space, neither subjective nor material.

I am opting for this logicist interpretation, though I admit that on **either** interpretation there are going to be difficulties, possibly insuperable ones, about the relation between an idea and the having of an idea. Very sketchily, my reasons for accepting it are, firstly, that it explains some things in Spinoza's philosophy that the psychologistic interpretation does not. For instance, his account of truth and error: thought is taken to be related to extension like an infinite, perfectly accurate map to an infinite terrain; every element of the map (i.e. every idea) is true, but one may misalign the map to the mapped, and thus be in error. I would also suggest that if any sense is to be made of Spinoza's idea of the eternity of mind, it has to be on a logicist interpretation. But secondly — a stronger ground — Spinoza explicitly accepts the paradoxical implication of the logicist reading: "the object of the idea that constitutes the human Mind is the Body" (*E* 2P13D2 / *C* p. 457). This statement is to be the starting point of my main argument in what follows. There is no equivalent statement of Spinoza's accepting the paradoxical implication of the psychologistic reading, i.e. animism. Someone may cite the scholium to the same proposition:

> "For the things we have shown so far are completely general and do not pertain more to man than to other Individuals, all of which, though in different degrees, are nevertheless animate" (*E* 2P13D2 / *C* p. 458).

However Spinoza immediately makes it clear that they are animate only in the sense of having an idea corresponding to them, not in a psychologistic sense:

> "For of each thing there is necessarily an idea in God, of which God is the cause in the same way as he is of the idea of the human Body. And so, whatever we have said of the idea of the human Body must also be said of the idea of any thing" (*E* 2P13D2 / *C* p. 458).

Despite this, even some of the most rigorous commentators have found the temptation irresistible, to unite this view to the idea that everything has a **conatus**, a tendency to persist in its being. These two doctrines, the "animate" nature of things as being the objects of ideas (to which they are trans-attributively identical), and the universality of **conatus**, are in fact as near unrelated as any two doctrines can be in a system like Spinoza's. After all, he has referred to "the things we have shown so far" — which do not include either the notion of **conatus** or the account of degrees of complexity of individuals, which follows two pages later, and explains the **difference** between people and other things.

The idea that every thing has got a **conatus** is based on quite different grounds from the idea that every thing is "animate". The latter is part of the doctrine of the two attributes; the former I take to be one of those instructive tautologies, like "the fittest survive" — to which it is closely related: those things exist which have a tendency to persist in their being; if they had no such tendency, they would not exist. Hence, in simple bodies, **conatus** is simply inertia; in complex bodies, it is the set of mutual relations of motion and rest between the parts, without which the complex would fall apart.

The only problem for my reading of the passage about universal animateness is the reference in it to "different degrees". Surely, the mere appearance of an entity under the attribute of thought is an all-or-nothing thing? But perhaps all Spinoza means is: the more complex the body, the more complex the idea that corresponds to it. Likewise when Spinoza says:

> "Both the man and the horse, for example, are swayed by the lust to propagate, but the horse is swayed by an equine lust and the man by a human one. The lusts and appetites of insects, fishes and birds must vary in the same way; and so, although each individual lives contented with its own nature and delights in it, nevertheless the life with which it is contented and its joy are nothing else but the 'idea' or soul of the individual [body] in question, and so the joy of one differs in character from the joy of the other as much as the essence of the one differs from the essence of the other" (*E* 3P57S, quoted Jonas 1973, pp. 276–7).

Jonas comments that this establishes

> "the principle of an infinite gradation of 'animateness', co-extensive with the gradation of physical composition, for which the entirely simple is merely a limiting case: even this would not be devoid of a minimum of inwardness, since to its distinctness, such as it is, there must correspond the idea 'of' it in God — and this is its 'thought' or 'soul'" (Jonas 1973, p. 277).

But I think this reading is mistaken: there is no reference to "inwardness" here. The gradation is, on the one hand (under the attribute of extension) a matter of complexity of organisation; and on the other (under the attribute of thought) of the idea of that complexity. But the idea of a complex individual is no more of an idea than the idea of a simple individual. The relation of idea to body is the same in both cases: the idea is the idea **of** the body (objective genitive — i.e. the idea that refers to the body, not the body's idea. Jonas claims that in Spinoza the objective genitive "turns into" the subjective genitive, but while there is a real problem here, it is not that Spinoza uses the genitive equivocally, but that he does not explain how "the idea of the body" [objective genitive] can do all the work he wants it to).

I conclude that Spinoza held no paradoxical views about the "psychic" nature of things. He was not an animist or a hylozoist. The idea that everything that exists under the attribute of extension also exists under the attribute of thought may be mistaken; but if so the mistake is about the reality of a "third realm", not about the nature of rocks and trees and water. But he **did** hold paradoxical views about the relation of the human mind to the human body — namely, that the mind is the idea of the body. To this I now turn.

2 Body

In Odegard 1975, p. 61, Douglas Odegard asks: "For Spinoza, what body is identical with the human mind?". He rightly sees this as a neglected question: for the most part, people have assumed that it is that which a doctor or a lawyer would call my body: that patch of the attribute of extension that is bounded by my skin. (I shall sometimes use the term "body-actual" for the body in this sense;[1] alternatively "body in the narrow sense".) In the end, Odegard endorses this view, after having considered only one alternative: that the relevant body is my brain. So

[1] I have borrowed this term from Michael Westlake's novel *The Utopian* (Westlake 1989), and have used it in Collier 1991, where I contrast it with the "body-cosmic", i.e. the inorganic body in Marx's sense. "Body-cosmic" is the more accurate phrase, since, in one sense, parts of the body-cosmic are organic, and in another sense, it all is. But it is difficult to exclude the mystical connotations of "cosmic" — which is intended simply as an adjective formed from "world", since "worldly" has another meaning.

long as we are talking about the body-mind relation simply as *identity*, this is the obvious candidate; twentieth century identity theories virtually all take the brain as the body to which the mind is identical. But Spinoza's identity theory is of a special type: the defining characteristic of an idea is that it is the idea **of** the body with which it is trans-attributively identified. This is a relation, in modern (and medieval) terms, of **intentionality**. And on the face of it, my intentional objects are, for the most part, not parts of my body in the ordinary sense. I have no idea or no clear idea of most of my internal organs, while I have many ideas of the world about me. Spinoza's account of the body-mind relation seems to imply not only that all knowledge is perception, but also that all perception is **proprioception**, the perception of my own body. How then can we account for outer perception?

One — not very promising — way might be to say that Spinoza, like many of his contemporaries (Locke, for instance), thought that we have direct knowledge only of our own states (i.e., in this case, our physical states), but since these states are caused by our interaction with the outer world, we can have indirect knowledge of its outer causes.

What are we to say about this? While it may readily be granted that a **physical** description of perception will be a description of some such causal interaction, we will require something different of the description of perception under the attribute of thought. To the changes wrought on our sense organs by the outer object, there must correspond changes in our ideas, by which they are related to the ideas of the outer objects. That perceptual relations are causal does not, for Spinoza, mean that they are external relations.

> "The essence of the mind consists in this alone that it is the idea of an actually existing body; and accordingly the mind's power to understand extends to those things only which this idea of the body contains in itself or which follow from it" (*Ep* 66, quoted Jonas p. 276n.)

When I see an old oak tree, it is not that I proprioceive the physical image of the old oak tree on my retina; it is that, corresponding to the casting of that image (and whatever other causal processes are involved in perception under the attribute of extension), the complex idea which is my mind comes to include the idea of the old oak tree. But is that not to say that the old oak tree has become part of my body, since my body is the object of that complex idea that is my mind? How can a causal account of an act of perception under the attribute of extension be equivalent to any such statement? To answer this we must leave the problem of perception for a while, and attend to Spinoza's account of complex bodies.

That account can be summarised as follows:

1. Simple bodies can interact causally, can have mutual relations of motion and rest; insofar as there is some continuity of equilibrium in this interaction, they constitute a complex individual.

2. A complex individual may retain its identity through replacement of the parts it is composed of.

3. A complex individual may be part of other, higher order complex individuals. Hence there is a hierarchy of complex individuals in nature, each composed of simpler, and component of more complex, individuals. (We may fill in part of the sequence as ... atoms, molecules, cells, multicell organisms, social institutions ...) The greater the order of complexity, the greater the power to interact with the rest of nature.

4. Each individual, of whatever complexity, has a **conatus** which "is its essence", i.e. just is the set of mutual relations of motion and rest by virtue of which that individual persists rather than disintegrates. The **conatus** of individuals of higher orders of complexity, however, promotes such persistence by actively tending towards greater interaction with nature.

The human body, in the narrow sense, is certainly a complex individual of a high order of complexity. But (a) we must ask ourselves the question: on these definitions, is not the same true of that person's **world**, their **"Umwelt"** — "world" in an almost Heideggerian sense; an individual of a higher order of complexity, since it includes the body narrowly conceived, and other bodies as well? And if so, surely it is this body, not the body in the narrower sense, that is trans-attributively identical with the person's mind. And (b) we must consider the body as extendible, in the sense that the more the body in the narrow sense interacts with the world about it, the more that world is to be counted as part of the person's "inorganic body".

For on this account of organisms, I think there may sometimes be doubt as to what should be included in any given organism, and the correct answer may turn out to be that membership of an organism is a matter of degree. For every organism smaller than "the whole body of the universe", interacts causally with things outside it, and depends for its continuance in being on such interaction. Let us look at several instances:

1. Two symbiotic organisms which can only exist in tandem, as in a lichen. Are they two organisms in Spinoza's terms, or one? We should not pose this question as an either/or. They are one organism composed of two organisms.

2. Now consider an organism which can only exist in a very specific kind of environment. It interacts with that environment, and exists only so long as a certain equilibrium in that interaction is maintained. By Spinoza's definitions, this constitutes the environment as **part** of the organism. However an organism is not equally dependent on interaction with all the different aspects of its environment with which it does interact. Some are more essential to it than others. I am not using "essential" as a mere synonym for "necessary": I mean "pertaining to essence". An organism's eco-system constitutes its essence.

3. What of more adaptable organisms, footloose with respect to any given environment? Once it is admitted that membership of an organism is a matter of degree, I see no reason why not to say that their environment is part of their existence too, only they so to speak spread themselves more thinly.

4. Finally, what of us? Our greater adaptability depends on our capacity to transform the environment; our defining ecological niche is a

niche for workers. But that is not of course to say that we are in any absolute sense footloose. Rather, we are "more perfect" than other organisms in that we are able to affect and be affected by more of nature in more ways than others. In a sense, we spread ourselves more thinly over nature, but in a sense nature is more part of us than it is of other organisms. It is, as Marx puts it, our inorganic body (Marx 1975, p. 328).

Now it may be said that this is a striking metaphor, and as such belongs to poetry rather than philosophy. But I think that if we come to regard ourselves as incarnate not merely in the bodies enclosed by our skins but in our worlds — in organic structures centred on our bodies in the narrow sense but extending indefinitely beyond them — we can make sense of many phenomena of human life that will be obscure on any other view. First, though, let us see what sense it makes of Spinoza. Consider the following passage from Part II of the *Ethics*

> P16: "The idea of any mode in which the human Body is affected by external bodies must involve the nature of the human Body and at the same time the nature of the external body."
>
> D: "For all the modes in which a body is affected follow from the nature of the affected body, and at the same time from the nature of the affecting body ... So the idea of them ... will necessarily involve the nature of each body."
>
> C1: "From this it follows, first, that the human mind perceives the nature of a great many bodies together with the nature of its own body."
>
> C2: "It follows, second, that the ideas which we have of external bodies indicate the condition of our own body more than the nature of the external bodies."
>
> (*E* 2P16D+C / *C* p.463)

Now this proposition, demonstration and first corollary indicate two things: (1) that Spinoza had not actually drawn the conclusion that the body corresponding/identical to the human mind is of greater extent than the body in the narrow sense, since it is clearly the latter body to which he refers as "its own body"; and (2) that he **ought to** have done so, since the mind is the idea of that of which it perceives the nature, and that of which it is the idea is its body. The second corollary looks out of place. He has not argued for it; he claims to have explained it by examples in the appendix to Part I, but those examples concern the supposed subjectivity of sense experience, which can be **corrected** by science. Granted the correction, do we not come to perceive the nature of the "external objects" as adequately as that of "our own body" (or body-actual)? Do not the "external objects" thereby **come to be** as much part of our body (that of which our mind is the idea) as is "our own body", even if they were less so before?

But it will be said, even if Spinoza ought on his own premisses to have agreed that the old oak tree of which I have an idea is part of my body, is it not to his credit that he did not? For is it not a very strange doctrine? Strange it may be, but I believe the strangeness comes, not from the nature of the case, but from our Lockean habit of regarding ourselves as having exclusive property in our own bodies. For we can't have exclusive rights in our environment: the old oak tree stands on the Common, and belongs to all who frequent it. But then the idea that we can have exclusive rights over our own bodies (in the narrow sense) is also untenable. Of course, any civilised state will give people many rights over their own bodies-actual which it denies them over other people's. But this is not an absolute matter. My child has the right to my physical powers to help him do what he cannot do for himself. The aged and infirm have the right to a portion of the labour of the community to support them. Money, as Ruskin pointed out, just **is** a token of right over the labour of others (Ruskin 1985). And it would generally be recognised that in many extremities, the community may take far more far-reaching powers over the persons of its members.

As for extending the conception of our bodies beyond the limits of our bodies-actual, in some measure we do it every day. Prosthetics, spectacles, clothes being worn, are always treated as part of us. To a degree, tools being used and vehicles being ridden are too. Marx refers to tools of labour as extensions of our limbs; certainly, it is largely through them that our power to affect and be affected by many things is increased. Most people can think of some familiar objects which they habitually treat as having the same sort of intimacy as parts of their body-actual. Almost anyone who has ever suffered a burglary will feel the entry of their home as a violation. Nothing is so artificial as the distinction between crimes against the person and crimes against property: to take away my light is to take away my eyes; to take away my means of labour is to take away my hands; to take away my food is to take away my life.

To a lesser degree, people's feeling for their home town or village is of the same nature, and can generate the same anger at damage inflicted on it. As I have said, the intimacy of the inorganic body is a matter of degree, depending on the degree of interaction essential to us. But this degree is alterable, and is at stake in the specifically human **conatus**. This takes us to the question of human interests. What conception of interest follows from the conceptions of mind and body outlined here?

3 Interests

We have never understood what Spinoza says about anything until we have understood what he says about it under the attribute of extension, **and** under the attribute of thought, **and** how he can regard the two accounts as accounts of the same thing under two attributes. A misfit between the accounts under the two attributes is the sure sign of a misunderstanding. Let us apply this to the question what Spinoza means when he speaks of seeking ones own **advantage** (as Curley translates **utilitas**; "interest" is better, since "advantage" suggests one person's advantage over another).

Of course, in general terms, as an instance of the **conatus** theory, a person's interest is in their persistence in being, or self-preservation. But if this meant what is usually called self-preservation (i.e. the pursuit of survival), we would be at a loss to know how to unite such an interest under the attribute of extension with our interest under the attribute of thought, which is to understand clearly and distinctly (since that is "self-preservation" for the mind — see *E* 4P26D / *C* p. 559). One might end by assuming that understanding is a **means** to survival, or, slightly less implausibly, **vice versa**. But "No one strives to preserve his being for the sake of anything else" (*E* 4P25 / *C* p. 558), and talk of means-end relations between something under one attribute and the same thing under the other makes no sense.

But the case is quite different from this. For the features of an entity by virtue of which it persists in its being are quite different for different kinds of thing. In the case of a pile of stones, its **conatus** is simply a matter of the way they lie one on another; in an amoeba, it is a matter of consuming diatoms, or whatever; in an animal, it will be a complex interaction with its habitat and its kind (the "lust to propagate" is never adequately theorised by Spinoza, as Schopenhauer points out — see Schopenhauer 1958, vol. 2, ch. XLIV). In human beings, our power to persist in our being is our power to interact with (affect and be affected by) our environment in many ways. The more ways, the more power to persist in our being. This sort of "self-preservation" means venturing and augmenting ones talents, not burying them. As Hans Jonas puts it:

> "Only by being sensitive can life be active, only by being exposed can it be autonomous. And this in direct ratio: the more individuality is focused in a self, the wider is its periphery of communication with other things" (Jonas p. 278).

The tendency of our **conatus**, under both attributes, is expressed by Spinoza thus:

> "in proportion as a Body is more capable than others of doing many things at once, or being acted on in many ways at once, so its Mind is more capable than others of perceiving many things at once. And in proportion as the actions of a body depend more on itself alone, and as other bodies concur with it less in acting, so its mind is more capable of understanding distinctly" (*E* 2P13S / *C* p. 458).[2]

[2] The apparent paradox that greater interaction and greater independence are iden-

Now if what I have said about the inorganic body is correct, it would follow that insofar as we are successful in pursuing our fundamental interest — understanding/interacting with more of nature in more ways — we are actually extending our inorganic bodies, in one sense drawing more of nature into ourselves — but not necessarily in an appropriative way. For instance, we may acquire a clearer idea of the stars by looking at them through a telescope. This does not alter the stars, nor does my looking at them stop you doing so. But insofar as I acquire a clearer idea of the stars, that idea becomes part of my mind, and, to the same extent, the stars become part of my body. And this is clearly not achieved merely by keeping the organs of one's body-actual healthy; we mainly increase our active and passive powers exosomatically — with the help, in the present example, of Spinoza's trade of lens-grinding.

An analogy with Freud's formula for the undoing of repression, "where id was, there ego shall be",[3] may be illuminating here. Recently when I was defending Marx's notion of nature as our inorganic body, I claimed for it (and this goes for Spinoza as here interpreted too), that it combined a Baconian project of power over nature through understanding, with an ecologically sensitive conception of a caring, non-exploitive relation to nature. For the model for our power over nature is the power we have over our own bodies in the narrow sense. This phrase was pounced on as indicating a "repressive" attitude to our own bodies. Needless to say, nothing was further from my mind than victory for the "soul" in a monkish duel with the "body". I meant that power over our bodies which defines health as against sickness: the power that I lack over my arm if it is broken, my eye if I have got a cataract, etc. My point was that, as St. Paul puts it, "no man ever yet hated his own flesh; but nourisheth and

---

tified in this passage is resolved by my interpretation according to which greater interaction means that more is incorporated into the "self".

[3]The passage in Freud 1971, p. 544 reads:
"Its [i.e. Psychoanalysis's — A.C.] intention is, indeed, to strengthen the ego, to make it more independent of the super-ego, to widen its field of perception and enlarge its organisation, so that it can appropriate fresh portions of the id. Where id was, there ego shall be. It is a work of culture — not unlike the draining of the Zuider Zee."

cherisheth it" (*Eph.* 5.29) — and that our relation to nature, as well as to our fellows, can have something in common with this. Here, **Freud's** (as opposed to the popular) notion of repression may help.

Freud's account of repression is, of course, under the attribute of thought — it is **ideas** that are repressed (though we may well believe Wilhelm Reich that something corresponds to this under the attribute of extension — see Reich 1950). Repression is not an expansion of the ego's power but a contraction of it: a thrusting of the repressed ideas into the outer darkness, where they cannot be controlled, and whence they produce unknown and unwanted effects. By contrast the work of psychoanalysis in undoing repression is "a work of culture, not unlike the draining of the Zuider Zee" — i.e. not a work of fighting off, but of reclaiming and incorporating the erstwhile repressed ideas. As more and more of the person's ideas are drawn into the ego, that person is less and less at their mercy; understanding has replaced the blind effect of external causes.

Just as repressed ideas are in a sense outside us (even though we are what we are in large measure because of them), so nature insofar as it is not understood, and is interacted with randomly and so relatively little and with relatively unclear ideas of it, operates as a set of forces to which we are subject, and to which we react with indifference or hostility. But insofar as we understand/interact with nature, it becomes to a degree part of ourselves, our inorganic body, and comes to be treated more tenderly.

Now we can see just how far the "inorganic body" reading of Spinoza removes the **conatus** theory from Hobbesian egoism: the body which the **conatus** is striving to preserve is not the body-actual, but something indefinitely larger, and the more that that **conatus** succeeds, the larger the body that it pertains to.

On this account, the interest that a person has in the wellbeing, not only of other people, but even of inanimate parts of their world, is no more problematic than their interest in the wellbeing of their own physiological parts. And it seems to me that, if we cast off the blinkers of commercialist ideology,[4] human motivation really does look like that: phenomenologi-

[4] But this ideology, of course, is not "all in the mind"; it has its equivalent under the attribute of extension (see the following section).

cally, everything is as if our desires existed, not only in our bodies-actual, but in the bodies-actual of others, and elsewhere in our worlds.

The inorganic body thesis has the consequence that (under the attribute of extension) we have shared bodily parts. Whatever Spinoza would have made of this, he must surely have accepted the corresponding account under the attribute of thought. At any rate if the logicist reading of "mind" is correct, then if you and I have an adequate idea of the same thing, we have the **same idea** in the sense in which this implies partial identity (overlap) between our two minds. It is instructive to read *E* 4P34–37 with this in mind.

It is easy to see how this conception of interests transcending "ego boundaries" may be applied to those intellectual interests which were so dear to Spinoza. As researchers or teachers, we desire that the truth be known, indifferently to the questions "by whom?" or "thanks to whom?" But it does not apply only to intellectual life. Is it not essential to sexual desire that it aims at pleasure in the body-actual of the other?

4 The materiality of morals

This conception of both the good for human beings and its material basis can help to clear up one of the three supposed paradoxes which have dogged Marxist philosophy in the twentieth century. I refer to the supposed contradictions between

1. Marx's deterministic theory of history and his advocacy of active transformation of the world;

2. Marx's critique of ideologies as class-relative illusions, and his claim to have inaugurated an objective science;

3. Marx's disavowal of moral ideals and his condemnation of capitalism.

I think it is true to say that Marxist philosophy has been dominated by responses to these three "paradoxes", which most critics regard as damning contradictions in classical Marxism; and many Marxists have regarded them as at the very least, difficulties to be resolved by drastic revisions of Marxism. Yet the curious fact is that the classical Marxists themselves (i.e. Marx, Engels, the Marxists of the Second International, the Bolsheviks) did not regard them as paradoxical at all. That this is not a mere oversight on their part is shown by their responses to critics (e.g. Engels's response to Dühring, Plekhanov's and Lenin's responses to Mikhailovsky).

The most likely explanation is that they, unlike their critics, held a general philosophical position from which these combinations of beliefs did not appear paradoxical. I think that they did indeed hold such a position, and were right to do so, and that that position was — Spinozism. How far the influence of Spinoza was direct is doubtful. They had read Spinoza, no doubt, but the path of influence is probably via Hegel — i.e. (simplifying drastically) Hegel's philosophy is Spinoza with teleology added, Marx's is Hegel with teleology removed. Plekhanov is the most consciously Spinozist, and not only on the three issues referred to. He calls Marxism "a kind of Spinozism" (Plekhanov 1974ff., Vol. 3, pp. 71–3), and refers to his own conversation with Engels when he asked:

> "'So do you think ... old Spinoza was right when he said that **thought** and **extent** were nothing but two attributes of one and the same substance?' 'Of course,' Engels replied, 'old Spinoza was quite right'" (Plekhanov 1974ff., Vol. 2, p. 339).

In several places, he even praises Spinoza's supposed hylozoism (a mistaken reading, as I have argued).

But to return to the three "paradoxes": the two great consciously Spinozist Marxist philosophers — Plekhanov and Althusser — have each done fine work on one of these issues. Plekhanov gives what has always seemed to me the definitive response to the first of the three in "On the Question of the Individual's Role in History" (Plekhanov 1974ff., Vol. 2, pp. 283–315), and "Fundamental Problems of Marxism" (Plekhanov 1974ff., Vol. 3, pp. 117–183). Althusser has concentrated on the second "paradox"; his extremely important work on science and ideology doubtless needs correction at some points — but I have written enough about that elsewhere (Collier 1979, 1988, 1989). On the question of morality, however, the Spinozan heritage has not to my knowledge been explicitly retrieved in a Marxist context.

The essence of Marxist "anti-moralism" seems to me to be: we will not make people more moral by telling them to be more moral; we will not even make ourselves more moral by trying to be more moral. Better moral values (under the attribute of thought) cannot exist without a better set of causal relations between people, nature and the means of labour (under the attribute of extension). Thus Engels tells us:

> "Only very exceptionally, and by no means to his and other people's profit, can an individual satisfy his urge towards happiness by preoccupation with himself. Rather it requires preoccupation with the outside world, means to satisfy his needs, that is to say, food, an individual of the opposite sex, books, conversation, argument, activities, objects for use and working up. Feuerbach's morality either presupposes that these means and objects of satisfaction are given to every individual as a matter of course, or else it offers only inapplicable good advice and is, therefore, not worth a brass farthing to people who are without these means" (*Feuerbach and the End of Classical German Philosophy*, in Marx and Engels 1968, pp. 615–6).

And Plekhanov:

> "Virtue requires, not to be **preached**, but to be **prepared** by the reasonable arrangement of social relations" (Plekhanov 1974ff., Vol. 1, p. 493).

In an economy where survival depends on competitive success, one can no more proscribe egoism than command a lighted haystack not to burn. Egoism simply **is** that economic causal structure under the attribute of thought. But in an economy based on common ownership and production for use, ones "self-interest" would include the interest of others, not because "egoism" had been sacrificed to "altruism", but because the "self" would no longer be conceived narrowly as a Hobbesian social monad, but as the centre of a complex world in which others figured, in which we were "members of one another". And the self would be so **conceived**, not through a change of "attitude", but because it would **be** that way: the causal interaction between us would be such that each person's **conatus** included the wellbeing of others in the way that the **conatus** of any complex individual includes concern for the wellbeing of its parts. Insofar as our causal equilibrium with the civic and natural environment is transformed, so is our "attitude" to that environment, for what is that attitude but that equilibrium under the attribute of thought? Spinozism explains Marx's hostility to moralism, for what is moralism but the demand that we pass from lesser to greater perfection under the attribute of thought while our causal relations with the world under the attribute of extension remain unchanged? But it also sustains Marx's socialist advocacy (which always takes the form of exposing capitalism's inhibition of potential that is already present in it). Capitalism is a set of causal relations between people, nature and the means of labour, such that definite constraints (more constricting than the constraints inherent in nature — though those are real too) are placed on our power to affect and be affected by many things.

Let me give an utterly concrete example of the materiality of values — i.e. of the trans-attributive identity of any given set of values with a certain organisation of the material world with its causal structure. The example is transport in the modern capitalist world. (The issue is perhaps posed in a harsher way in England than elsewhere, due to the density of car-owning population, and the total absence of a transport policy during the Thatcher years.)

The dominant mode of transport in any community determines much of its material environment. The narrow lanes and steps which give a town like St. Ives in Cornwall its charm date from the time when the sea was the main highway for its communication with the rest of the world. The dramatic transformation (and desecration) which our cities have been undergoing in recent years is a result of the dominance of private motor transport over public transport and private non-motor transport. Roads are transformed for the benefit of through traffic and to the disadvantage of walkers and lingerers; provision of parking space dominates town planning; the city composed of streets is replaced by the city composed of motorways, car parks and shopping malls. This is not the effect of current technology by itself, but of a particular social organisation of that technology: **private** motor traffic. This transformation of our cities (and one could speak of the countryside too) is not simply a change in the "external" environment which could leave human relations, lifestyle and aesthetic sensibilities unaltered. It alters, in the first place, the way we experience space and time. We move from place to place, so to speak, by convection not conduction: we can get from A to Z much quicker, but it is more difficult — often impossible — to **pass** from A to Z through B, C, D, etc. It is in this sense that modern transport is said to "abolish space". As C. S. Lewis remarks:

> "The truest and most horrible claim for modern transport is that it 'annihilates space'. It does. It annihilates one of the most glorious gifts we have been given. It is a vile inflation which lowers the value of distance, so that a modern boy travels a hundred miles with less sense of liberation and pilgrimage and adventure than his grandfather got from travelling ten. Of course if a man hates space and wants it to be annihilated, that is another matter. Why not creep into his coffin at once? There is little enough space there" (Lewis 1977, p. 127).

In consequence, time too becomes fragmented into time-here and time-there, with the transition — the travel from here to there — treated as a mere wasted interval, to be shortened rather than enjoyed. The effect on human relationships of this "mobile privatisation" (as Raymond Williams called it) inevitably involves a decline of neighbourliness, a further move towards quasi-contractual relations with others, a "friendship market".

The modification of our lived aesthetic is perhaps even more striking. It is a feature of aesthetic appreciation of one's environment that it is in a sense "holistic": ones view can be spoilt by the presence of a gasworks behind one's back. But in the modern townscape or landscape, isolated objects of beauty are often "preserved" framed by ugly utilitarian objects. The thatched farmhouse is left standing alongside a motorway intersection which cuts it off from its farm and the surrounding woods; the fine city church is boxed in and overtopped by cuboid office blocks. We are often **compelled**, in seeking environmental beauty, to make a mental bracketing, an **epoché** of the context of an "aesthetic object", if we are to encounter it at all. But we do this at the price of the corruption of our aesthetic sense.

An atomistic "metaphysics", not as a philosophical position but as a **lived** metaphysics, is the necessary correlate of the material structure of our modern (i.e. car dominated) environment; and all our values — aesthetic, moral etc. — are structured by that metaphysics. That metaphysics is contrary to our nature as complex organisms capable of great and subtle interaction with our environment; it dislocates and fragments that interaction. But it does not hang in a vacuum: it **is** the modern environment under the attribute of thought. If we are to pass from this lesser to a possible greater perfection, the structure of our causal interaction with our environment must be changed.

The effect of transport on our material environment and therewith on our values is only one example, though obviously one about which I feel strongly. More central to ethics is the difference between forms of self-interest that are egoistic and forms that are not, each corresponding to a definite set of causal relations between people and the world. Let me conclude with one word of caution: we should not confuse the distinction between morality under the attribute of thought and the causal structure that corresponds to it under the attribute of extension, with the distinction between morality and politics, "the personal" and "the political"; the latter distinction was rightly recognised by Spinoza: the virtues of the rational person and the virtues of the good citizen may be complementary, but they are not the same, and can sometimes conflict (the rational person returns good for evil, the good citizen punishes wrongs). These two spheres must remain distinct because they are grounded in two mutually irreducible kinds of structure in both of which we find ourselves placed. These distinct structures correspond, under the attribute of extension, to the two sets of values under the attribute of thought: moral and political values.

* * *

At this point, I foresee two serious objections to my comments on the materiality of morals, given the argument of the earlier sections of this paper. Firstly, in talking about people's moral values as under the attribute of thought, I may appear to have slipped into a psychologistic way of speaking about that attribute, despite my logicist reading defended in the first section. Secondly, my treatment of the relation between values and their material embodiment in the causal interaction of people with their social environment **as a relation of trans-attributive identity** precludes any "relative autonomy" of one from the other. I may seem to be committed to the reductionism with which one of the great Spinozan Marxists (Plekhanov) is often charged, rather than the anti-reductive programme of the other (Althusser).

To acquit myself of these charges, I would need to make four points which I will set out here, though I have no space to defend them fully.

1. It is necessary to distinguish, and clarify the relation between an idea and the having of an idea. Ideas are logical entities belonging to the attribute of thought. An idea is as much an idea whatever it is of — just as, for Plato's Parmenides, there are Forms of Dirt and Mud as well as of Man, etc. (see Cornford 1939, p. 83). But the **having** of an idea is a function of the complexity of the individual that has it. Mud and dirt don't have ideas; we do; and since Spinoza recognises that animals are conscious, they must occupy some intermediate stage. Complexity of body enables us to have ideas because it enables us to interact in more ways with more of nature. "Mind" in the psychologistic sense just is extensive interaction. One might borrow a term from phenomenology, and say that the concept of **bodily intentionality** has replaced the concept of mind in this sense, except that the important thing about mind in the psychologistic sense, is that it is the having of **ideas** (in the logicist sense).

2. We must recognise the logical primacy of ideas over the having of ideas, i.e. of the attribute of thought over "subjective mind". For (a) the relations between ideas hold independently of the having of

those ideas; and (b) one cannot understand what is going on in any having of an idea without understanding the idea.

3. While of course there can be no **causal interaction** between ideas and bodies for Spinoza (since they are trans-attributively identical), the practical consequences of his theory are very like interactionism. For if event E^1 at time T^1 causes event E^2 at time T^2, then since both events must occur under both attributes, everything is as if (for example) E^1 under the attribute of thought caused E^2 under the attribute of extension. Effectively, an idea can, via my having of it, cause my actions and their physical consequences.

4. Finally, we need to show that values are ideas, and the having of values just is the having of (certain) ideas. The prejudices of non-cognitivism in ethics have to be overcome here. For ideas, according to Spinoza, are **true of** their ideata. Now it will be said: no ideas about what is true can move us to action (i.e. can be values) without some conative shove. And it is quite true (and Spinoza is clear about this) that the having of ideas could not be the having of values for any being were it not for that being's **conatus**. However, insofar as we have adequate ideas, those ideas will include ideas about our own conative nature. And such ideas can give rise, in Kant's phrase, to **assertoric imperatives** i.e. to hypothetical imperatives that can be asserted since the condition is known to hold:[5] "**since** you are a being with such and such needs, interests, etc., and the world is thus and so, do this". The having of an idea can be the having of a value, not because the haver supplies a "value-judgement" extraneous to the idea, but because the (complex) idea is the idea **of**, among other things, the haver's interests.[6] If these four points

[5] See the second section of the *Groundwork of the Metaphysic of Morals*. Cf. my discussions of the role of assertoric imperatives in naturalistic ethics in Collier 1981, 1990 and 1992.

[6] In support of this cognitivism: I think it can be shown that differences between moral theories rest on their different ontologies e.g. the difference between utilitarianism and the idealist ethics of Hegel or Bradley on the difference between atomist and organic social ontologies. See my writings cited in note 5.

are established, then the unavoidable psychologistic language can be interpreted non-psychologistically; and the reductionist pitfalls of "the materiality of morals" avoided.

BIBLIOGRAPHY

Works of Spinoza

Collected Works vol. 1, Ed. and trans. by Edwin Curley. Princeton N.J: Princeton University Press, 1985. (Cited as *C.*)

Balz, Albert: *Idea and Essence in the Philosophies of Hobbes and Spinoza.* New York: AMS Press, 1967.

Bennett, Jonathan: *A Study of Spinoza's "Ethics".* Cambridge: Cambridge University Press, 1984.

Collier, Andrew: "In Defence of Epistemology". In: *Issues in Marxist Philosophy*, vol. 3, John Mepham and David-Hillel Ruben (Eds.), Brighton: The Harvester Press, 1979, pp. 55–106.

Collier, Andrew: "Scientific Socialism and the Question of Socialist Values". *The Canadian Journal of Philosophy*, Supplementary Volume VII (1981), pp. 121–154.

Collier, Andrew: "Retrieving Structural Marxism". *Economy and Society* 17 (November 1988), pp. 543–552.

Collier, Andrew: *Scientific Realism and Socialist Thought.* Hemel Hempstead: The Harvester Press, 1989.

Collier, Andrew: *Socialist Reasoning.* London: Pluto Press, 1990.

Collier, Andrew: "The Inorganic Body and the Ambiguity of Freedom". *Radical Philosophy* 57 (Spring 1991), pp. 3–9.

Collier, Andrew: "Marxism and Universalism : Group Interests or a Shared World?". In: *International Justice and the Third World*, Barry Wilkins and Robin Attfield (Eds.), London: Routledge, 1992 (forthcoming).

Cornford, F.M.: *Plato and Parmenides.* London: Routledge and Kegan Paul, 1939.

Curley, Edwin: *Spinoza's Metaphysics : An Essay in Interpretation.* Cambridge, Mass.: Harvard University Press, 1969.

Freud, Sigmund: *The Complete Introductory Lectures on Psychoanalysis.* London: Allen and Unwin, 1971.

Jonas, Hans: "Spinoza and the Theory of Organism". In: *Spinoza : A collection of Critical Essays*, Marjorie Grene (Ed.), Garden City, N.Y: Anchor Books, 1973, pp. 259–278.

Lewis, C.S: *Surprised by Joy.* Fount Paperbacks, 1977.

Marx, Karl: *Early Writings.* Harmondsworth: Penguin Books, 1975.

Marx, Karl, and Engels, Frederick: *Selected Works.* London: Lawrence and Wishart, 1968.

Odegard, Douglas: "The Body Identical with the Human Mind : A Problem in Spinoza's Philosophy". In: *Spinoza: Essays in Interpretation*, Maurice Mandelbaum, Eugene Freeman (Eds.), La Salle, Illinois: Open Court, 1975, pp. 61–83.

Plekhanov, Georgi: *Selected Philosophical Works.* (5 volumes.) Moscow: Progress Publishers, 1974–81.

Reich, Wilhelm: *Character Analysis.* London: Vision Press, 1950.

Ruskin, John: *Unto This Last and Other Writings.* Harmondsworth: Penguin, 1985.

Schopenhauer, Arthur: *The World as Will and as Representation.* New York: Dover Publications, 1958.

Westlake, Michael: *The Utopian.* Manchester: Carcarnet, 1989.

[19]

SPINOZA ON THE KNOWLEDGE OF GOOD AND EVIL

WILLIAM K. FRANKENA

In Book IV Spinoza says a number of things about the knowledge of good and evil that appear to be significant for his argument as a whole, but which are by no means all as clear and distinct as they should be. The purpose of this essay is to examine these and certain related passages in order to say something more about Spinoza's meta-ethics than is said by C. D. Broad and others, in the hope that this will be of intrinsic interest to students of Spinoza, of Continental Rationalism, and perhaps even of meta-ethics, besides being of use in understanding and assessing Spinoza's view as a whole. It will therefore deal with those passages only insofar as they throw light on his meta-ethics, i.e. on his views about judgments of value . It will, moreover, be concerned only with the meta-ethics of judgments of good and evil, not with that of judgments of virtue and vice, which may be somewhat different.[1] To simplify matters I shall frequently talk only about the knowledge of good (KG) rather than about the knowledge of good and evil (KGE), assuming that what is said about KG will also apply, *mutatis mutandis*, to the knowledge of evil (KE), which is true except for one point to be dealt with separately in Section VII.

There are three main respects in which Spinoza's meta-ethics of good and evil is ambiguous and unclear. (1) He is ostensibly a naturalist in G.E. Moore's sense, or at least a

[1] Spinoza's definitions of virtue and vice are a bit different from those of good and evil. I believe, however, that he would say much the same things about the knowledge of virtue and vice that he does about KGE.

definist, for he explicitly gives definitions of good and evil (somewhat different ones in different places) that play a crucial role in his proofs of some of his own evaluative judgments, as Broad points out and as I assumed in an earlier paper on Spinoza's meta-ethics.[2] This suggests that he would say:

a. that judging x to be good = judging that x has DP (a certain defining property),

b. that knowing x to be good = knowing that x has DP.

However, this is not what he actually says when he tells us what KG is, as we shall see. What he does say sounds, on the face of it, not only like hedonism, but also more like emotivism than like naturalism. He also says other things that go more naturally with emotivism than with naturalism. (2) Usually, Spinoza's definitions of good and evil are not hedonistic, but, as we shall also see, there is one in which he seems to build hedonism into the very meaning of "good" and "evil". (3) He seems in some places to hold that we judge things to be good because we desire them, not vice versa, and in others that we desire them because we judge them to be good. These three problems will be touched on in due course and other unclarities and problems will turn up as we go along.

One might begin this review of Spinoza's meta-ethics with a discussion of his definitions of good and evil, but I have said something about them in the previous paper referred to and wish here to focus on what Spinoza says about KGE, referring to his definitions only as they become relevant. I shall therefore proceed by starting with his most important proposition about KGE, bringing in earlier and later passages that bear on it, and then going on to look at some (but not all) of the other things he says about KGE.

[2] See C.D. Broad, *Five Types of Ethical Theory*, New York: Harcourt, Brace and Co., 1930, pp. 257, 259; W.K. Frankena, "Spinoza's 'New Morality'," in *Spinoza*, ed. by E. Freeman and M. Mandelbaum, La Salle, Ill.: Open Court, 1975, pp. 61-84.

SPINOZA ON KNOWLEDGE OF GOOD

I

The proposition referred to is P8, IV:

Knowledge of good and evil is nothing but an affect of joy or sorrow insofar as we are conscious of it.[3]

Here we may read "pleasure" for "joy" and "pain" for "sorrow". As was indicated, this proposition is both more emotivistic and more hedonistic, on its face at least, than one would expect, given most of Spinoza's definitions of good. For, given those definitions, one would expect him to say that KG simply is the knowledge that something is useful to us, or contributes to the preservation of our being, or increases our perfection, or is a means by which we may come nearer to the model of human nature we set before us — depending on which definition one picks. Had he said this, and only this, his view would be clearly naturalistic and it would leave open the question whether pleasure is the good.[4] Actually, however, P8 seems to assert something more like one of the following views:

a. KG = having joy (or pleasure).

b. Judging x to be good = having an emotion of joy about x.

(a) would be gladly agreed to by a hedonist and (b) by an emotivist.

Three questions arise. (1) Just what does P8, IV, say? To answer this we must try to see what Spinoza had in mind, partly by looking at his proof, partly by looking at things he says earlier, and partly by looking at what he goes on to say. (2) How does his proof go, and does it establish what

[3] Here and elsewhere I quote from W.H. White's translation.

[4] Of course, he would be a naturalist only if the concepts used in his definitions can themselves be analyzed in naturalistic terms. Here I use "naturalist" for anyone who defines ethical concepts wholly in non-normative terms. I should note, however, that G.E. Moore took Spinoza to be a "metaphysical" definist, not a naturalistic one. I leave open the question whether he is right about this. See *Principia Ethica*, Cambridge University Press, 1929, pp. 110, 113. The assertion Moore cites as making Spinoza a metaphysical definist is not actually one of Spinoza's definitions proper.

P8, thus interpreted, asserts? (3) Why does he bring P8 in at all? Why does he not simply say that KGE = knowing whether things contribute to the preservation of our being (etc.) or not?

Since we must use Spinoza's proof in answering any of these questions, we must have it before us. I shall not quote it, however, but restate it in a series of numbered statements as follows, writing in the definitions and propositions Spinoza refers to:

1. We call a thing good which contributes to the preservation of our being. Defs. 1, IV.
2. The effort by which each thing endeavors to persevere in its own being is nothing but the actual essence of the thing itself. P7, III.
3. Therefore, we call a thing good which increases or helps our power of acting.
4. Joy is the passion by which the mind passes to a greater perfection. Def. of joy, Schol. PII, III.
5. Therefore, we call a thing good in so far as we perceive that it affects us with joy.
6. The mind not only perceives the affections of the body but also the ideas of these affections. P22, II.
7. Therefore, KG is nothing but the idea of joy that necessarily accompanies the affect of joy itself.
8. This idea [of joy] is united to the affect [of joy] in the same way that the mind is united to the body. P21, II.
9. Therefore, this idea is not really distinct from the affect itself. Schol. P21, II.
10. An affect is an affection of the body by which its power of acting is increased or diminished, helped or hindered, together with the idea of this affection of the body. Def. 3, III.
11. Therefore, the idea of joy [referred to in 7, 8, and 9] is not really distinct from the idea of the affection of the body, but only conceptually so.
12. Therefore, KG is nothing but the affect of joy itself in so far as we are conscious of it.

Filled out in this way, the proof gives us much of what we need to answer our questions.

SPINOZA ON KNOWLEDGE OF GOOD

It is obvious that P8 is making some kind of equivalence or identity claim. Just what is it claiming to be identical with what? And what kind of identity is it claiming to obtain between them?

(a) As was observed, P8 does not say that KG = knowing that x contributes to the preservation of our being, which is what one would expect a naturalist to say who accepts the definition of good used here by Spinoza. There is no reason to believe that Spinoza would not accept this equation; he would, in fact, be inconsistent if he did not. But this is not the equation he is concerned to assert explicitly. The equation he asserts in P8 is this: KG = having an affect of joy and being conscious of having it. But KG is rather more ambiguous than Spinoza recognizes, and so we are not yet sure just what is being identified with having an affect of joy and being conscious of it (or more simply, with being conscious of having joy). The phrase "the knowledge of good and evil" may refer to any of the following:

(1) Having concepts of goodness and badness,
(2) Being acquainted with goodness and badness,
(3) Having opinion about what is good or evil,
(4) Having true opinion about what is good or evil,
(5) Knowing that something is good or bad.

Prima facie these are distinct, and one may ask which of them Spinoza means to equate with being conscious of having joy. One is tempted to say that (3) and (4) presuppose (1) and that (1) presupposes (2), and Spinoza might agree, though the latter assertion smacks of empiricism rather than rationalism. It may be objected that Spinoza would not himself distinguish (1) and (2) from (3), (4), and (5), and this may be true; he does think that having an idea (image, concept, or percept) entails making a judgment.[5] It may also be objected that one should not put (3) and (4) under the heading of knowledge, as I have. But Spinoza does that himself when he speaks of "knowledge of the first kind" and when he says that KE is

[5] See e.g. Ps. 17, 28, 35, 43, 49, II.

inadequate knowledge and I am simply following his usage. Thus my question remains. I believe Spinoza would reply that KGE in all five senses is "nothing but" being conscious of having joy or sorrow. His proof makes it clear, however, that he has (3), (4), or (5) primarily in mind, since it centers on what we call or judge to be good. Thus, he means to insist, particularly, that the "knowledge" that x is good (which may be only opinion) = being conscious that x affects one with joy.

(b) As for the nature of the equation or identity claimed in P8, it is clear that it is not conceptual identity. Spinoza is not asserting anything that is true simply by definition, as he would be if he were to say that KG = knowing that x contributes to the preservation of one's being. Rather, the equation asserted in P8 is one that requires a proof using not only definitions, but also some propositions established earlier on the basis of axioms as well as definitions. Spinoza is asserting an identity that is necessarily, not just contingently, true, but not one that is true by definition alone.

We must remember in this connection that Spinoza holds that mind and body have precisely the kind of identity just described. They are the same in substance but differ in attribute or conceptually. In fact, Spinoza believes that there are three things:

(1) an affection of my body by which its power of acting is increased,

(2) an affect of joy in my mind, which is an idea or perception of that affection of my body, and

(3) an idea or perception of that affect of joy in my mind.

He also believes that these three things are substantially the same and differ only conceptually; by P21, II, (1) is united to (2) and (2) is united to (3) as the mind is united to the body. But, now, in P8, IV, he is asserting that KG is "nothing but" (3) and it is still not clear what kind of identity he means to ascribe to them. If, as I believe, he is not claiming them to be conceptually or definitionally identical, then he must hold that KG is conceptually a fourth thing besides (1), (2), and (3), and that it is "united to" (3) and is "nothing but" (3) in the same way in which the mind is "united to" or "nothing but" the body. I do not think Spinoza sees that he must hold this, but it follows from the fact that he defines the good as what is conducive to the preservation of one's being. For, on this definition, knowing the good is strictly identical with knowing what is conducive to the preservation of one's being and can be only synthetically identical with having an affect of joy of which one is conscious. On an older view of mathematics, which Spinoza would have shared, 2 + 2 and 2 × 2 were synthetically and necessarily equivalent or identical, and it seems to me that in P8 he must be maintaining that KG and (3) are equivalent or identical in a similar way.

II

Does Spinoza's demonstration, as stated, actually show that the equation or identity asserted in P8 is true? Looking back at his proof as reformulated earlier, I am not sure that it does. In the first place, (1) is not established by the definition cited, as it is stated by Spinoza. The definition reads, "By good, I understand that which we certainly know is useful to us."[6] But from this, even if we equate being useful with being conducive to our self-preservation (or perfection), it does not follow that we call a thing good which is conducive to our self-preservation (or perfection). To obtain this consequence Spinoza would have had to say, "By good, we understand that which . . .," where the "we" is not just editorial but refers to all of us, and it is doubtful that he thought we all do mean this by good.[7] Secondly, I do not see at all clearly how (2) gets us from (1) to (3) or even just how (5) follows from what precedes. Thirdly, the crucial step is (7), and it is not obvious that (7) follows from (5) and (6), especially since (7) contains a term, KG, which does not appear in (5) or (6). (5) must mean either of the following:

[6] Def. I, Bk. IV.

[7] See e.g. what he says about "good" in Appendix, Bk. I.

W.K. FRANKENA

5a. We call x good if we have an affect of joy in connection with it, i.e. an idea or perception of an affection of the body by which its power of acting is increased, or

5b. We call x good if we have an idea or perception of an affect of joy in connection with it, i.e. an idea or perception of an idea or perception of an affection of the body by which . . .

But (6)only tells us that, when our body has an affection of this kind, then we have not only an affect of joy but an idea of that affect, and how, given this and either (5a) or (5b) can we conclude that KG is nothing but this idea? Spinoza must be assuming two further unstated premises:

13. If we call x good insofar as we perceive that it affects us with joy, then *calling* x good is nothing but perceiving that it affects us with joy.

14. If calling x good is nothing but perceiving that it affects us with joy, then *knowing* x to be good is nothing but an idea of joy which necessarily accompanies an affect of joy.

No doubt, one can affirm some kind of equation, if the antecedents of (13) and (14) are true, but there are a number of possibilities here, and it is not clear one can assert the equation in P8.

However this may be, one wonders why Spinoza did not simply argue as follows, if he meant to start from his basic definition of good:

m. x is good = x is conducive to the preservation of one's being. By def.

n. Therefore judging x is good = judging x to be conducive to self-preservation.

o. And KG = knowing that x is conducive to self-preservation.

p. X is conducive to one's self-preservation = x increases one's power of acting = x affects one with joy.

q. Therefore judging that x is good = judging that x affects one with joy.

r. And KG = knowing that x affects one with joy.

So stated, his proof would comport with his previous theory and it would establish P8 if that previous theory is true.

III

However, the main question for us is not whether Spinoza proves P8 but what he has in mind in asserting it. About this we can become clearer if we consider P8 in another perspective, also provided by Spinoza. I mentioned earlier that it seems to say something like what a hedonist wants to say, viz. that knowing x to be good = finding it pleasant, being affected by it with pleasure, etc. For Spinoza, however, this is the conclusion of a longish argument beginning with a non-hedonistic definition of good and including a rather different conception of pleasure from that which is usual among hedonists. The good he defines as what is conducive to the preservation of our being, to an increase in our perfection, etc. Joy or pleasure he defines in several related but somewhat different ways:

(a) It is the passion by which the mind passes to a greater perfection.
(b) It is a man's passage from a lesser to a greater perfection.
(c) It is an affection by which the body's power of action is increased.
(d) It is an affection of the body by which its power of action is increased, together with the idea of this affection.[8]

Behind these definitions of pleasure or joy is the thought that what is basic is an affection of the body by which its power of acting is increased or aided, and that pleasure or joy is either the idea (perception) in the mind of that affection of the body by which its power of acting is increased or aided, and that pleasure or joy is either the idea (perception) in the mind of that affection or a combination of that affection with an idea of it (actually, for

[8] For these definitions see Schol. P 11, III; Schol. P59, IV; Def. III, Bk. III.

Spinoza, the affection and the idea of it are substantively identical though qualitatively different). This means that, on Spinoza's view, a joy or pleasure is or includes a kind of perception of good, since it is or includes a perception (perhaps confused, but possibly clear and distinct), of an affection of the body in which its power of acting, etc., are increased or helped, i.e. it is a kind of knowledge of good. It is or involves a cognition, however confused, of a certain fact, and is not simply a blind feeling, as emotivists and hedonists usually conceive it to be. This point is behind what Spinoza says in P8, but it is not what he says in P8. In P8 Spinoza identifiies the knowledge of good, not with the idea or perception of a modification of the body (the affect of joy), but with the higher level idea or perception of that first level idea or perception. But, of course, if the first level idea (the affect) is a kind of KG, then the second level idea will also be kind of KG. This is part of what P8 asserts, and it does follow from Spinoza's general position. Then, however, he should have said that there are in a sense *two* kinds of KG: that represented by the affect of joy itself and that represented by the consciousness of that affect; even though these are in some sense really identical for him, they are at least conceptually distinct. P8 does not make this point explicit. Indeed, it is a bit misleading in suggesting that only the higher level idea involves a cognition of good.

However, P8 is not saying only that consciousness of an affect of joy is a kind of KG. It says or seems to say that all KG is "nothing but" a consciousness of an affect of joy, and this does not follow from the position just sketched. I believe in fact that Spinoza cannot really be holding, as he here seems to, that P8 is true for all kinds of KG, and will try to explain why as we review other passages in which he touches on our subject.

Before looking at other passages dealing with KGE, however, let us consider briefly the third question raised earlier about P8, IV, viz. the question why, in the only place in which he says what KGE is, he gives us P8, instead of saying that KG = knowing that something contributes to the preservation (or perfection) of our being. As far as I can see, Spinoza's reply would simply be, (a) that there is this further equation which can be proved, and (b) that this further equation is important for later parts of his story. This may be so. It remains true, however, that, logically, KG just is basically the knowledge that something is conducive to one's self-preservation or perfection. It *is* consciousness of joy if an only if it can be shown that there is a sense in which every case of knowing x to be conducive to one's self-preservation or perfection = being conscious of joy in connection with x. But this equation we shall find reason to think Spinoza cannot establish, even on his own terms.

IV

Let us now consider other passages in the order of their appearance in the *Ethics.* (1) Then we must begin with an apparent definition of good — the first to appear — formulated in the Scholium to P39, III. In P39 itself he says that one who loves another will endeavor to do him good, and in the Scholium he explains that by good he understands "here" every kind of joy and everything that conduces to it, giving as his reason for so understanding it the fact, previously established, that we call a thing good because we desire it and do not desire it because we judge it good. Now, had he used this definition in demonstrating P8, IV, he would have had a much easier time. Why did he not use it? The answer, I think, is that this apparently hedonistic definition is not his most basic one. Most basic is the one used in proving P8, IV; this one is derivative from that one, since it embodies an equation that needs to be proved by the introduction of further premises. In fact, as I see it, this definition in effect presupposes Spinoza's point in P8 and so could not be used in proving it. Perhaps this is why this definition, if it may be called that, appears only in a Scholium, and why he says it indicates what he means by good "here". At any rate, his basic definition is not a hedonistic one.

Even so, there is another point to be noticed for our purposes. On this definition, KG would not necessarily be a consciousness of joy; it would be either that or a knowledge that x is conducive to joy, which is not the same thing. There would, in fact, be two kinds of KG, and so P8, IV, though it would be partly true, would not be true without qualification. There is no reason, even on Spinoza's own general theory, for thinking that knowing x to be good as a means is "nothing but" a consciousness of joy, even if knowing x to be good as an end is such a consciousness. Again we find reason to think, on Spinoza's own terms, that P8, although it tells a truth, does not tell the whole truth and nothing but the truth. Even if knowing that x is good as a means to one's end is always accompanied by joy, which is debatable, it does not follow that it is "nothing but" a consciousness of that joy.

It may be objected here that, according to Spinoza's basic official definitions, all judgments of value must be instrumental ones, none of them can be judgments of intrinsic value. In particular, Def. 1, IV, the definition usually used in Spinoza's proofs of P8, etc., says, "By good, I understand that which we certainly know is useful to us." Now it is true that he insists that the terms good and evil "indicate nothing positive in things considered in themselves," being purely relational. If x is good, it must be good to or for someone. Goodness is not an intrinsic property in G. E. Moore's sense for Spinoza; it is not non-relational. But its being relational is compatible with the making of judgments of intrinsic value. Even if "x is good" means "x is desired or enjoyed by y", it still may be that y desires or enjoys x for its own sake, and not just because of its consequences, i.e. that it has intrinsic value to or for y. Nevertheless, it remains a fact that, except for the secondary definition just discussed (which does allow for judgments of intrinsic value), Spinoza does word his definitions in such a way as to seem to equate value with instrumental value. We should not infer, however, that this is his real intention. For one thing, this would be incompatible with the thrust of P8 and of the definition just referred to. For another, as we shall see, Spinoza wants to say later that some things are "directly good" and 'indirectly" bad and vice versa. Furthermore, in P28, to which we will come,

SPINOZA ON KNOWLEDGE OF GOOD

Spinoza contends that the highest good of the mind is the knowledge of God, using Def. 1, IV, in the proof. Yet he cannot really mean that the knowledge of God is good simply because it is conducive to *something else*. It is true that, verbally, he would say that it is good because it is conducive to an increase in one's perfection. But his real point is that to be perfect *is* to have understanding and knowledge of his second and third kinds, not that it is something else to which these are only means. Thus he is asserting, in effect, that such understanding and knowledge are good as ends. In J. S. Mill's words, "x is conducive to perfection" covers three kinds of cases for Spinoza: the case in which x is simply a *means* to perfection, e.g. taking medicine or grinding lenses, the case in which x is a *part* of perfection (or perhaps even the whole of it), e.g. a knowledge of God, and the case in which it is *both* e.g. a knowledge of the physics of lenses.

(2) In the Scholium to P39, III, Spinoza also makes a very emotivistic sounding statement.

> Each person . . . according to his affect judges or estimates what is good and what is evil. . . . Thus the covetous man thinks plenty of money to be the best thing and poverty the worst.

He seems to think that this follows from his thesis that we judge a thing to be good because we desire it, not vice versa. Actually, it asserts something broader, viz. that what one judges good or bad is determined by one's affects or by the affect dominant in one at the time. This is not quite to say that I judge x good because I have a desire for *it*, but, rather, that I do so because I have some affect x bears on, perhaps a desire for y, or, to use Spinoza's examples, covetousness, ambition, or envy. The main question, however, is whether Spinoza can say this and also hold that judging x good = judging it to be conducive to self-perfection. It is compatible with P8, IV; indeed, he formulates this view in connection with the apparently hedonistic definition of good that presupposes P8. But is it compatible with the view that "good" means "conducive to perfection"? I believe it is, if we suppose that for Spinoza an affect like ambition or envy can determine, not only what we find joy and sorrow in, but also what we perceive as conducive to our perfection. Since he equates finding joy in a thing with a kind of perception of its conduciveness to one's perfection, he might well think that one's perception of this is affected by one's affects. Then, however, we may ask once more whether all of our judgments of value are wholly determined by our affects in this way, and especially by those that are passions; and the reply must be, it seems to me, (a) that judgments of extrinsic value may be at least partly determined by our knowledge of relevant facts, and (b) that such value judgments as P28, IV, are determined, not so much by our affects, as by our following a certain line of argument (even if our affects determine whether we follow this line of argument or not). Moreover, (c), even if, or insofar as our value judgments are determined by our emotions, they are not necessarily determined by passions, for there are also active emotions.

(3) P58, III, needs mention next. It says that, besides the joys which are passions, there are affects of joy which are related to us insofar as we act, a point which reappears in IV, P52 and the proof of P59. This means, I take it, that there are two kinds of joy, one a passion and the other an active emotion, and, if this is so, then, by P8, IV, there must in some sense be two corresponding and somewhat different kinds of KG. Moreover, by P1, III, the first of them must be inadequate and the second adequate. Of this more later.

(4) Note should also be taken of the fact that, both in passages already referred to and in some others, Spinoza stresses the relativity of our judgments of good and evil. Thus, in Preface, IV, he writes:

> . . . one and the same thing may at the same time be both good and evil or indifferent. Music, for example, is good to a melancholy person, bad to one mourning, while to a deaf man it is neither good nor bad.

Again, however, we must be careful what inference we draw. Spinoza is insisting that value is relational, but he is not a relativist, except perhaps about the value of things like music, money, glory, etc. He certainly is not thinking that the value of understanding in general or of the knowledge of God in particular is relative; such things are good in themselves, in the sense earlier explained, for all human beings whether they are melancholy, deaf, or whatever. It is true that, in his proof of P8, IV, Spinoza seems to emphasize the speaker-relativity of judgments of value; he seems there repeatedly to be saying that one calls good that which is or is perceived by him to be conducive to *his* perfection, enjoyable to him, etc. Back of this way of speaking is Def. 1, IV, quoted earlier. But it is ambiguous. It can mean either that the good is that which one knows is useful to *him* or that it is what humans know to be useful to everyone, and probably Spinoza intends it to cover both senses. At any rate, in the sequel to the above passage, Spinoza says that by good he understands "in the following pages everything which we are certain is a means by which we may approach nearer and nearer to the model of human nature we set before us" — and what is good in this sense is not relative to the individual, though it is, of course, relative to the species. In this respect, Spinoza is an objectivist about the value judgments he himself makes in the *Ethics* even though he remains a relationist throughout.

While this is so, however, Spinoza found it necessary to speak in the more speaker-relative way that is also permitted by his definition in order to establish P8, IV. For it would hardly be plausible for him to claim that a knowledge of what is beneficial to everyone can be equated with or derived from a consciousness of one's own joy at something. No doubt this is why he does not use P8 in his proofs of Ps. 27-28, etc., but only when he wishes to prove propositions like Ps. 19, 64, etc. If so, then we also see more clearly why he finds the equation in P8 interesting and does not stop with the one that follows from Def. 1, IV. However, even if Spinoza had held in an unqualified way that the meaning of a value judgment is wholly speaker-relative, he still would not necessarily be an emotivist. For he might maintain that "x is good" always means "I desire (or take a pro-attitude toward) x." In A.J.

Ayer's terms he would then be contending that when one says a thing is good one is *asserting* that one desires or favors it, one is not just *expressing* one's desire or liking.

V

(5) Ps. 14-17, IV, make reference to KGE, but we shall consider them only insofar as they bear on our inquiry, not as they related to the topics of human bondage or freedom. The first thing that strikes the eye then is the fact that Spinoza speaks here, not just of KGE, but of "true" KGE, making it clear that he means P8 to be true of such KGE. But now the question arises: Is there false KGE or is all KGE true? The fact that Spinoza calls it knowledge does not mean he believes it is all true, for not all knowledge of the first kind, as he calls it, is regarded by him as true. On the other hand, in P16 and its proof he talks of KGE and true KGE as it they were interchangeable, and, actually, I believe Spinoza must hold that any idea of joy, however confused, must contain some true knowledge of good (i.e. some true perception of an increase in one's perfection), except where the joy is excessive, as it may be. It is true that in his proof of P15 Spinoza says things which imply the view that true KGE can only be adequate knowledge, i.e. knowledge of the second or third kind, but he cannot really be holding this view, for he later contends, as we shall see, that KE is always inadequate knowledge (knowledge of the first kind). In any case, I think he is committee to the position that there are two kinds of KG, one adequate and the other inadequate, and both true, one confusedly and the other clearly and distinctly — of which also more later. It is true that he says falsity consists in the privation of knowledge which inadequate ideas involve, but this is consistent with holding that such ideas also contain truth.

In Ps. 14-15 and their proofs, both of which use P8, Spinoza distinguishes between KGE insofar as it is true and KGE insofar as it is an affect. It has then somehow two aspects. It is tempting to recall Spinoza's distinction between an affect of joy and the idea of that affect and to say that it is the latter that is true and the former that is an affect (the two being as identical as those two sides of a coin, the mind and the body, so that in a sense it is the same thing that is both true and an affect). But this will not do if, as Spinoza explicitly holds, it is only the latter that is a cognition of good. And, if we add that the former is also a kind of cognition of good, as I suggested earlier, then it will not do either. One must say, it seems to me, that both of them have both of the aspects in question. In any event, in insisting on the basis of P8 that KGE is an affect and not simply true, Spinoza seems to be implying that it is not just "an idea by which the mind affirms a greater or less power of existence for the body than it possessed before," but something with an affective quality as well.[9]

(6) This brings us to P19, IV, the proof of which also uses P8. It states that one necessarily desires that which he considers to be good. In fact, Ps. 15-17 already implied that a (true) KG generates desire for the objects involved. This seems to mean that Spinoza thinks that, when this happens, we desire what we desire *because* we judge or know it to be good, which is not what one would expect an emotivist to hold. However, as was indicated earlier, Spinoza elsewhere is very insistent that

> we neither strive for, wish, seek, nor desire anything because we think it to be good, but, on the contrary, we adjudge a thing to be good because we strive for, wish, seek, or desire it.[10]

He also seems to mean this without qualification, i.e. that we never desire a thing because we judge it good but always judge it good because we desire it. This is the sort of thing an emotivist would most naturally say. Also, if it is meant without qualification, it is contradicted by Ps. 14-19, IV, if these later propositions must be taken to mean that we sometimes desire a thing *because* we judge it to be good, as seems to be the case. This raises the third of the

[9] For the words quoted see Demonst. P 14, IV.

[10] Schol. P 9, III.

problems about Spinoza's meta-ethics mentioned in my introduction, besides bearing on the first. Before commenting on it, however, let us consider what Ps. 14-19 say. As I said, they seem to affirm that we sometimes desire a thing because we believe or know it to be good, but this still leaves two possibilities:

(a) that this desire is generated by KG insofar as it is true,

(b) that this desire is generated by KG only insofar as it is an affect of joy

On the first alternative, the mere fact that we know that x contributes to the preservation of our being generates a desire for x; on the second, it does not. What Spinoza says in P14 and in the proof of P19 suggests that he espouses (b), since it involves insisting that, by P8, KG is the affect of joy itself insofar as we are conscious of it; that it is true seems not to play a role. He could, however, agree that (a) is sometimes true, since he thinks of the essence of man as an endeavor to persevere in one's being.[11] For, given this *conatus*, one will necessarily desire x if one judges it to be useful in preserving or perfecting one's being. It may also be, of course, that Spinoza would regard the choice between (a) and (b) as academic, because for him a knowledge that x is useful to us and a consciousness of joy about x are two sides of the same coin. But, if this is so, why does he insist in P14 that no affect can be restrained by the true KGE insofar as it is true, but only insofar as it is an affect?

Perhaps a consideration of the proof of P19 will help us both with this question and with that of reconciling P19 with Schol. P9, III. For us the crucial part goes as follows:

1. KG = the affect of joy itself insofar as we are conscious of it. P8, IV.
2. We endeavor to bring into existence whatever we imagine conduces to joy. P28, III.

[11] P 7, II.

SPINOZA ON KNOWLEDGE OF GOOD

3. Therefore, one necessarily desires that which he considers to be good.

The clinching point seems to be that considering a thing to be good = imagining it to be conducive to joy = being conscious of it with joy. Now, here again, we may raise our earlier question whether P8 establishes the first of these two equations. Even if it establishes that judging or knowing x to be good in itself = being conscious of it with joy, it does not thereby establish that judging x to be good as a means = being conscious of it with joy. But, however this may be, we must observe that Spinoza is here in effect asserting only that, *if* we are conscious of x with joy (= judging or knowing it to be good), then we necessarily desire it. He is not quite saying that we then desire it because we are conscious of it with joy (or *because* we judge or know it to be good). Certainly he is not claiming that, whenever we desire a thing, we do so because we are conscious of it with joy or because we judge or know it to be good. Still, he does elsewhere seem to hold that we sometimes desire a thing *because* we judge or know it to be good or *because* we are conscious of it with joy.[12] We must simply conclude that Spinoza cannot consistently maintain that we *always* judge x to be good *because* we desire *it* or that we *never* desire it because we judge *it* to be good. Contrary to *Schol.* P9, III, he must be believing that we sometimes desire x because we judge or know *it* to be good (= being conscious of it with joy), even though we also sometimes judge or know it to be good (and are conscious of it with joy) because we desire it. This more qualified position is compatible with his also holding, as he certainly means to, that we would not judge or know *anything* to be good, or be conscious of *it* with joy, if we did not first have a desire to preserve or perfect ourselves. For, even if this is true, it still may be that we may not desire a certain x unless we judge *it* to be good or are conscious of *it* with joy — or even that we *never* desire a particular x unless we so judge or are conscious of it. We may also observe that, while this more qualified view is

[12] P 28, III; Ps. 14-17, IV.

compatible with an emotivism maintaining that when one makes a value judgment one is directly or indirectly expressing one's basic *conatus* to persevere in one's being, it is also consistent with a naturalism contending that "x is good" means "x is conducive to self-preservation."

(7) Next, we must examine Ps. 27-28, IV. In them Spinoza is not doing meta-ethics; he is making certain value judgments himself, and he is claiming, not only that they are true, but that he can prove them. This means that he believes them to constitute adequate knowledge of the second kind. The two value judgments are:

a. That which conduces to understanding is (certainly) good (and that alone).
b. The highest good of the mind is the knowledge of God.

What interests me now is not the content of these value judgments but their form and the implied claim that they constitute a kind of KG that is adequate or clearly and distinctly true. In form they are general or universal; what is said to be good is not some particular thing but a kind of thing. The question (once more) is whether Spinoza thinks that such general knowledge too is "nothing but" an affect of joy insofar as we are conscious of it. It is hard to see how he can believe this, even if he is talking only about knowledge of what is good in itself, and not about knowledge of what is good merely as a means. It is easier to believe that knowing a particular thing to be good in itself is a consciousness of joy in that thing than that knowing a general truth about what is good in itself is a consciousness of joy at a kind of thing. Is knowledge of the sort involved in Ps. 27-28 = consciousness of joy in getting understanding or a knowledge of God? Is it not rather a matter of grasping the proofs offered by Spinoza? We certainly seem there to be knowing these two truths, not via a consciousness of joy, but via a deductive argument, if we know them at all — a deductive argument, moreover, which contains no mention of any joy that accompanies the getting of understanding or of a knowledge of God, and also no use of P8. Even if our coming to know these two propositions about what is good is itself accompanied by a consciousness of joy, it cannot be *this* joy that constitutes a cognition of the goodness of understanding in general or of a knowledge of God in particular. Again, it appears that P8 cannot plausibly be true of all KG. If our KG in Ps. 27-28 comes to us through its being proved that understanding things and knowing God are profitable to the mind and not through our finding joy in such understanding and knowing, then it will hardly do to contend that all KG = finding joy in such activities of the mind. Even if such KG and such joy were somehow ontologically one, as mind and body are, they would not be in the same boat epistemically.

It may be that Spinoza would wish to reply here that there are two ways of knowing that the knowledge of God is good: a discursive way via grasping his proof and a more direct or intuitive way via finding joy in such knowledge. This is what I think he should say, but then again he must distinguish two kinds of KG and apply P8 — and perhaps also some of the other things he says about KG — to only one of them.

(8) Ps. 41-48 should be noticed here. In them again Spinoza is making and purporting to prove value judgments like:

(a) Joy is directly good.
(b) Sorrow is directly evil.
(c) Cheerfulness is always good.
(d) Pain may be indirectly good.
(e) Hatred can never be good.
(f) Hope and fear can be good only insofar as they are able to restrain excesses of joy.

Some of these are general judgments of intrinsic value and some general judgments of extrinsic value, and, once more, it seems to me that on Spinoza's own terms judgments of both kinds embody a sort of KG that is not nothing but an affect of joy insofar as we are conscious of it. In particular, I think that knowing certain joys to be excessive and knowing certain pains to be indirectly good cannot be equated, respectively, with being conscious of sorrow or joy in experiencing those joys or pains. In any case, we should observe that Spinoza is here distinguishing two cases of KG, viz. knowledge of direct goodness and knowledge of indirect goodness, and must be thinking that knowledge of both sorts may be adequate.

We may now usefully restate one result of our discussions in (1), (7), and (8) as follows Not every piece of KG comes via joy experienced at the time of the knowing or judging on Spinoza's view, though P8, IV, suggests it does. Even if it is accompanied by joy, this will be joy at having that piece of knowledge, i.e. this joy will be a cognition of the goodness of that bit of KG, not a cognition of the goodness of something not going on at the moment. But we do sometimes have knowledge of the value, direct or indirect, of something that is not going on at the time, e.g. when we know that exercise is conducive to health or that knowledge of God is the highest good. It still may be, however, that every joy or consciousness of joy is or includes a judgment of value that is at least partly true.

Also in Ps. 41ff, Spinoza says that joy may be excessive and/or indirectly bad, but maintains that cheerfulness, a certain kind of joy, can never be excessive and so is always good. His argument is that cheerfulness is joy which, insofar as it is related to the body (as, of course, it basically is), consists in the fact that all parts of the body are equally affected, so that the body's power of action is increased or assisted throughout, without any opposing tendency. Therefore it is good both directly and indirectly (or wholly good).

If this is so, then there are two kinds of joy: those that may be excessive and/or indirectly bad and those that can never be excessive or bad. But this seems to imply that there are two corresponding kinds of perception (or knowledge) of good: (a) those consciousnesses of joy that are or may be misleading because the joy is actually excessive or indirectly bad, and (b) those that are always entirely veridical. In the former, as was indicated earlier, there is always some truth (for joy is always directly good), but there is or may also be some error; there is error insofar as the outcome of what is reported as good via the feeling of joy is a net decrease in the body's power of action (or in one's perfection.) In the latter there is nothing but truth, though the truth may be confusedly perceived. It may be that this distinction collapses into that already made between joys that are passive and those that are active (and between corresponding kinds of KG). It does, if cheerfulness is an active joy, while those that may be excessive are passive ones. It is not clear, however, that Spinoza regards cheerfulness as an active emotion. The Scholium to P11, III, implies that it is a passion, but a later passage leaves this question open, and the proof of P59, IV, is ambiguous on the point.

VII

(9) In his Scholium to P62, IV, Spinoza reminds us that our knowledge of the duration of things is very inadequate and largely based on the imagination, so that our judgments about the order of things and the connection of causes are rather imaginary than real. Then he adds that, as a result, the true KGE we have is "only abstract or universal," and that this is why the desire to which it gives rise is so impotent against other desires. Here he seems to be thinking mainly of our knowledge of indirect good and evil, but he must mean what he is saying to apply to that of direct good and evil as well. He also seems to imply that we have no true knowledge of the value of particular things, as I have been assuming we do. I doubt, however, that he can mean this. In fact,. his premises and conclusions here will apply at least as much to our knowledge of the value of particular things as to that of the value of kinds of things. He might even be thinking that our knowledge of particular things is "abstract" too, in the sense of involving inadequate knowledge of their contexts. In any case, he is obviously speaking mainly about such KGE as is inadequate, again neglecting to mention that there is adequate KGE. Thus, he is not necessarily holding that the latter is "only abstract and universal," or that the desire it generates

is weak relatively to other desires. Indeed, if KGE can be knowledge of his third kind, i.e. intuitive knowledge, then such tenets cannot be true of it. It may be, however, that Spinoza is here implicitly denying that KGE can ever be intuitive in this way — that it is at best knowledge of the second kind.[15] But, even then, one may ask whether Spinoza intends what he says here to be true of the KG embodied in P28, IV. Though not based on imagination or inadequate knowledge, it is "only abstract and universal" in a way in which knowledge of the third kind is not, and so presumably a desire arising from it would not be as strong as one based on the third kind of knowledge. At the same time, P62 itself seems to imply that such a desire is not so impotent against desires for present things as desires for future ones are.

(10) In P64, IV, Spinoza says that the knowledge of *evil* is inadequate knowledge. This is the one point, mentioned earlier, on which what he says about KG does not apply to KE For, as we have seen, KG is sometimes adequate knowledge of Spinoza's second kind, though he does somewhat fail to emphasize this fact. In his proof of P64 he uses P8 to show that KE is sorrow insofar as we are conscious of it, and argues that, since sorrow is a passion resting on inadequate ideas, therefore the knowledge of sorrow (= KE) is inadequate. In my opinion, however, even if this conclusion is true for some KE, it cannot be true for all of it. In cases (b) and (e) listed under (5) we have pieces of KE that can be proved in Spinoza's system and therefore constitute adequate knowledge of the second kind. That the greatest evil of the mind is ignorance of God (the correlate of P28, IV) would be another instance of such KE. Either such KE = consciousness of sorrow or sorrow does not depend on having inadequate ideas. Surely Spinoza would prefer the first alternative, which again involves qualifying P8, but in either case he must also qualify P64.

(11) In the Corollary to P64 Spinoza adds that it follows the human mind would have no notion of evil if it had

[15] On this point see P 36, V.

none but adequate ideas. If I am correct in saying that not all KE is actually inadequate on his view, then he is mistaken in thinking this follows. Be this as it may, he elaborates his point in an interesting way in P68, IV:[14]

> If men were born free, they would form no conception of good and evil so long as they remained free.

For then they would have only adequate ideas and hence no concept of evil, and, since good and evil are correlative, no concept of good either — and therefore no KGE whatsoever. Here we may ask just which concepts and which knowledge we would not have if we were fully free. It is clear that we would not have any confused or inadequate ideas or knowledge and hence no KGE involving them, but we might still have KGE consisting of adequate ideas and adequate knowledge, since there may be such KGE. Against the background of Spinoza's system, there seem to be the following relevant concepts we might or might not have if we were fully free:

(a) the idea of an affection of the body in which its power of action is increased, in which it passes to a greater perfection, etc. (= the affect of joy itself).

(b) the higher level idea of that affect of joy itself (= a consciousness of that joy, which is itself a consciousness of an affection of the body).

(c) the concept of passing from a lesser to a greater perfection, etc.

(d) the concept of something's being conducive to a passage from a lesser to a greater perfection, to the preservation of one's being, etc.

Here (d) is the concept equated with the concept of the good in Spinoza's basic definition of the good, while (b) is the idea equated with KG in the derivative equation asserted in P8, IV, but (a) and (c) will also represent KGs of some sort. Now, what we could not have if we were fully free is anything that presupposes our actually having inadequate ideas or actually passing from a lesser to a

[14] See also his explanation of Def. III in Schol. P 59, III.

greater perfection, etc., as becomes clear in P17, V, where Spinoza argues that, since God has only adequate ideas and cannot pass either to a greater or to a lesser perfection, He must be free from passions and cannot be affected with any affect of joy or sorrow. In fact, this is not true, if there are such things as active joys and adequate kinds of KG (as we have seen, and will see again, Spinoza really believes there are, but tends to forget about) unless the having of these presupposes the actually having of inadequate ideas, passions, or passages from a lesser to a greater perfection. But this Spinoza never shows and, as was intimated earlier, may entail a more empiricist view than he is ready to espouse. Does Spinoza as a rationalist really want to say that God does not have the concepts involved in (c) and (d)? Even his saying that God can have no affects of joy seems to be inconsistent with P35, V, in which Spinoza says that God delights in His infinite perfection and hence loves Himself with an infinite intellectual love (which itself seems to contradict Cor. P17, V). Remember that love is joy attended by the idea of the object loved as its cause. But if God can know joy at all then by P8 He must have some kind of KG. In any case, we may suppose that He knows such things as P28, IV, and that ignorance of God is evil, by either the second or the third kind of knowledge, and, hence, that He would have both some KG and some KE even if he knew no joy. As far as I can see, all this would be true of the fully free person as well, if there were one (which Spinoza carefully points out to be contrary to actual or even possible fact).

Having said this, it occurs to me that Spinoza may be thinking that God and the fully free person would make no use of any evaluative or normative discourse. Their discourse would be wholly physical, psychological, mathematical, epistemological, or metaphysical, and they would simply act in certain ways. This view is somewhat suggested by what Spinoza says about God and the free or wise man in IV and V, e.g., in P17, V, and elsewhere. It would involve holding, much as Kant does, that evaluative and normative discourse presupposes the presence of contrary passions, imperfections, etc., and so would have no use for those in whom these are not present. It also implies, however, that judging x to be good is not simply the same thing as judging it to contribute to the preservation of one's being, as Spinoza's basic definition seems to say it is. For if these two judgments are the same, then there is no reason why a free man could not make them both, and also no reason why he should not use the term "good" in making his judgment, since he could hardly be prohibited from having two ways of saying the same thing, as Spinoza in effect has of asserting P28, IV. Another way of putting this is to say that this interpretation of Spinoza makes him something other than a pure naturalist about good. But to accept this interpretation is to suppose that the fully free person would not know P28, IV, in the form in which Spinoza states it, since he would know only that a knowledge of God is conducive to an increase in one's perfection, and even that, in some sense, he would not understand any of Spinoza's talk about good and evil in Books IV and V. He would have general understanding and a knowledge of God, but he would not know that these are good or that the opposite states (which he surely can conceive of) would be bad. Supposing this does not seem at all plausible, especially if "x is good" *means* "x is conducive to perfection," as Spinoza says it does.

As for the question which of the ideas listed in (a)–(d) above the fully free man would have — Spinoza would apparently *say* that he would have none of the four kinds of ideas, but, if I am right, he might at least have some of each kind.

VIII

In part, our results may be summarized by noting that Spinoza explicitly or implicitly recognizes several different ways in which KG (or KGE?) may be divided into two kinds. In the order of their appearance in our discussion these are:

(a) the distinction between the KG that = knowing what

is conducive to self-preservation or perfection and the KG that = having an affect of joy and being conscious of it;

(b) the distinction between the latter kind of KG and the kind of KG that simply = having an affect of joy [these are substantially identical but conceptually distinct];

(c) the distinction between knowing what is directly or in itself good and knowing what is indirectly or extrinsically good;

(d) the distinction between the KG (and KE, if I am right) that is adequate and the KGE that is inadequate;

(e) the distinction, under each of the kinds in (b), between the KG that = having a passive emotion of joy and being conscious of it and that which = having an active emotion of joy and being conscious of it [this distinction may coincide with that in (c)];

(f) the distinction between the KG that is particular, i.e. knowledge that a certain thing is good, and the KG that is general;

(g) the distinction between the KG that = a consciousness of excessive joy and KG of the kind involved in cheerfulness [This distinction may coincide with that in (e)].[15]

Besides distinguishing these, in most cases more explicitly than Spinoza does, I have argued that they cannot all be equated with having an affect of joy and being conscious of it, or, in other words, that some of them are not only conceptually but also more substantially distinct. In particular, I think this is true of the first kinds mentioned in (a) and (d) and the second kinds mentioned in (c), (d), and (f). In short, the equation asserted in P8, IV, is not the whole truth. It also is not Spinoza's most basic equation involving KGE, but a derivative one. The basic equation is that KGE = knowing what is and what is not conducive [in the broad sense indicated in (1) in Section IV above] to self-perfection, etc., which holds for all the kinds of KGE distinguished. The main point left of P8 is the assertion

[15] Perhaps one should add (h) the distinction between the KG that = judging or calling a thing good and the KG that = *knowing* that a thing is good.

that every affect of joy and sorrow and every idea of such an affect is a kind of KGE as defined by Spinoza's basic equation and always contains some truth, even when it is partly mistaken, as it sometimes is. This point by itself separates Spinoza from the hedonist and the emotivist, but this is done even more by the fact that Spinoza really holds, though he seems not to, that there are kinds of KGE which are not reducible to consciousnesses or expressions of joy and sorrow. As Broad puts it, pleasure (joy) is for Spinoza not the *ratio essendi* of the good, as it is for the hedonist, and in a way for the emotivist, but the *ratio cognoscendi.*[16] The *ratio essendi* of the good is conduciveness to self-perfection, i.e. to knowledge and understanding.

For the rest, we may summarize by referring to the three problems about Spinoza's meta-ethics listed in the introduction. About the third I have said what I have to say in (2) and (4) in Section IV and (6) in Section V; we saw that there is a consistent position behind the apparently conflicting things Spinoza says in the relevant passages, but that, to hold it, he must modify his claim about the priority of desiring x to judging it good.

The second problem, already touched on in this summary, was dealt with in Section III and under (1) in IV. In spite of one apparent definition that seems to do so, Spinoza does not make hedonism true by definition, his basic definition being non-hedonistic. He does not proceed from a hedonistic definition plus axioms and previous theorems. It is true that he could take his non-hedonistic definition of good as basic and yet equate pleasure (joy) and the good at a second or third remove, as he seems implicitly to do in P8. Really, however, even at second or third remove, he equates the good, not with pleasure, but with knowledge or understanding. He agrees with the hedonist that KG = knowledge of pleasure and of what is conducive to pleasure, but for him this does not mean that the good = pleasure; it means only that all consciousness of joy is a cognition of good and that all cognition of good is a

[16] *Op. cit.*, p. 52.

consciousness of joy — and we have found that he cannot really sustain the latter of these two claims.

As for the first problem — we have seen that P8, IV, does not commit Spinoza to emotivism. For him P8 is just a derivative equation that looks like something an emotivist would say but is not, for behind it is the view that an affect is a kind of cognition, confused or not, of a state of the body. It is true that Spinoza says emphatically that we judge a thing good because we desire it, not vice versa. But (a) this assertion must be qualified, since he also maintains the reverse, and (b) this dictum is compatible with his being a naturalist. Given his view that a person essentially *is* a *conatus* (endeavor) for perfection, Spinoza could hold either that a value judgment is an expression of this *conatus* on its finding that something is conducive to perfection or that it is simply an assertion *to* this *conatus* that something is conducive to perfection. In my previous paper I suggested that a Spinozist might choose the first alternative and be an emotivist, but Spinoza himself chose the latter and so is a naturalist. He does sometimes, however, say things that are, verbally at least, compatible with either alternative. He also allows, as we saw in (5) of Section V, that a piece of KG has a affective quality and is not just true (or false), but, if I am right in what I said at the end of (8), this may mean only that a knowledge of something's being conducive to one's perfection is accompanied by joy.

We also saw in (11), Section VII, that God and the fully free person are, on Spinoza's own premises, not so axiologically blind as he represents them as being. But, like our other findings, this only shows that all good things are as difficult as they are rare.

The University of Michigan
Ann Arbor, Michigan 48104
USA

Part IV
Salvation and the Mind's Eternity

[20]

KNOWLEDGE, ANTHROPOCENTRISM AND SALVATION

Herman De Dijn

1. "Many a man before Spinoza and since has found the secret of peace: but the singularity of Spinoza, at least in the modern world, was that he facilitated this moral victory (consisting in the reconciliation of one's heart with the truth) by no dubious postulates. He did not ask God to meet him half way; he did not whitewash the facts, as the facts appear to clear reason, or as they appeared to the science of his day. He solved the problem of the spiritual life after stating it in the hardest, sharpest, most cruel terms".[1]

These words of the famous Spinozist, George Santayana, I would like to take as my starting-point for a reflexion on Spinoza's Critique of Anthropocentrism. Spinoza's singularity, Santayana says, is that he discovered a new and unexpected sort of salvation: salvation obtained in facing the sort of truth present in the newly invented modern science. This truth was so unpalatable to almost everyone, that the new science had to be instantly disarmed: either by constructing a new metaphysical framework (be it Cartesian or Leibnizian), or by considering the new science as simply a kind of game without any real insight in things. The counterpart of Spinoza's discovery was his critique of anthropocentric illusion present in ordinary religion and traditional metaphysics. This critique, developed in the Appendix of *Ethics* I, shows how the ordinary world-view is steeped in falsehood and illusion, and how it leads to spiritual misery and slavery.

Immediately however we are confronted here with a problem: with what looks like a very *un*modern view of the relationship between knowled-

[1] Santayana 1923: 105. That Spinoza was not the only one in his days to provide a Critique of Anthropocentrism is argued in Roothaan 1991.

ge and salvation (or happiness). Since Hume we are used to the fundamental distinction between intellect and heart, between information and evaluation (and action). Science seems to have nothing to do with happiness or salvation. Perhaps Spinoza is not a modern thinker at all ? Santayana implicitly seems to recognize this problem. He slightly rephrases Spinoza: he speaks of a reconciliation of one's *heart* with the truth: these are not Spinozistic, but Humean terms. How can Spinoza say that knowledge of the *facts* brings real peace; why should peace not be found apart from knowledge?

The way Spinoza describes the central illusion of anthropocentrism in the *Appendix* already shows the complexity of the relationship between cognition and the conative-affective element. False knowledge based on ignorance is so closely intertwined with striving and affection that it must become *prejudice*, deeply influencing the whole way of living. False knowledge concerning oneself is analogically used by man to interpret the behaviour of other things and even of the origin of things itself. This illusory view of oneself and the world is inevitably and inextricably related to emotions like pride and jealousy, fear and hope. It develops into a religious worldview which reinforces the teleological conception with man as the centre of the world, and which is further characterized by magical behaviour and belief in miracles. Because this world-view includes belief in God's inscrutability, any way out of this anthropocentric illusion seems blocked in advance, "had not" - I quote Spinoza - "mathematics, which deals not in final causes, but in the essences and properties of figures, offered to men another standard of truth". But how possibly could mathematics, which after all is but a series of true propositions, help us overcome ingrained prejudices of such strength and magnitude? Must we not agree with Hume and P. F. Strawson that, in view of their natural condition men cannot give up ordinary human attitudes even if confronted with scientific information pointing to a deterministic view of things, because these attitudes are too deeply rooted in their minds (De Dijn 1985a: 207-208)? According to Strawson, scientific insights can, for a moment, shatter our ordinary view of things, but they cannot continue to get a hold on us. A fundamental change within our attitudes, like "the generalization of abnormal egocentricity" (the supposed result of a deterministic view), is inconceivable as a real possibility within human behaviour (Strawson 1974: 18).

Two questions present themselves to us here: 1) what is the role of mathematics in the fight against anthropocentric prejudice; 2) what to make of Spinoza's belief that true knowledge saves?

2. In the Appendix Spinoza does not say literally that it is mathematics itself which helps us to overcome prejudice. He says that mathematics "which deals not in final causes, but in the essences and properties of figures, offered to men another standard of truth". Mathematics refers here to something like Euclidean geometry (De Dijn 1986a: 68-69) showing us what real understanding means, and what it means to think truly. One can safely say that Spinoza saw this model at work in the new science of nature as developed by Copernicus, Galileo, Descartes and others, and that he wanted, following this model, to develop a true, rigourous metaphysics and science of man (and society). So, it is not mathematics as such which is important, but a rigorous, "scientific" metaphysics and science of man, which somehow must constitute an Ethics.[2]

The question then is: how can the truth about God, the world and man destroy prejudice and bring about "the new man"?

A first way to understand this, could be called the demystification-thesis. This thesis says that if man succeeds, with the aid of the new scientific view on man and nature, to undermine his natural prejudices, he cannot fail to see the truth about himself and to lead a new life. This may still mean different things. It may mean that man starts adapting his strivings to the new insights, thus succeeding to seek what is really useful to himself. A scientific understanding of ourselves and of the conditions of our real well-being allows us to obtain a neutral, objective view upon ourselves and upon the causal conditions governing our life. At the same time, it is recognized that man cannot stop persevering in his own being, and valuing things accordingly. Objective knowledge is then used in an instrumental way to promote one's conatus (cause-effect-insights are transformed into precepts for more efficient self-preservation). Such a view of the relationship between objective scientific knowledge and subjective striving might seem compatible with a Humean position, but is it Spinoza's view? This is very unlikely. It supposes that the possession of scientific knowledge would be sufficient by itself alone to break the impact on our lives of the complex of prejudices and emotions which seems to be our natural human condition. That this is not Spinoza, is indirectly clear from these comments he makes on the Stoics: "For we have not complete command over our emotions... The Stoics, however, were of the opinion that the emotions depend absolutely on our free will, and that we have absolute command over them. But they were

[2] De Dijn 1991: 119-131. That Spinoza's metaphysics is best seen as the result of a revision of Avicennan Aristotelianism under the pressure from the new science, is argued for convincingly by Carriero 1991: 47-96.

compelled by the outcry of experience, not by their principles, to confess that not little practice and zeal were required to restrain and moderate them" (E 5Praef). This kind of optimistic rationalism, typical for the Aufklärung-view, is not Spinoza.

The upshot of this optimistic rationalism vis à vis the force of prejudice (ideology) is almost inevitably cynicism. The scientific view allows us to see through prejudice, but is not effective enough to really eliminate it; on the contrary, attempts to better our condition on a scientific basis get caught within the web of prejudice and passion, perhaps making things even worse. So the only solution is a haughty withdrawal into a form of fatalistic moralism. A cynical view is often combined with a merciless egocentrism. Again it is clear that Spinoza does not accept this position: "... he who is moved neither by passion nor pity to help others is rightly called inhuman, for (E 2P27) he seems to be unlike a man" (E 4P50S); "But those who cavil at men and prefer rather to reprobate vices than to inculcate virtues, and who do not strengthen, but weaken the minds of men, - these, I say, are a nuisance both to themselves and to others. Wherefore many, owing to too great impatience of mind and a false zeal for religion, have preferred to live among beasts rather than among men: just as children or youths who cannot bear [etc.]." (E 4A13).

Both the optimistic rationalist and the cynic think that we can or should be able to cut ourselves loose from contingent, traditional values and ways of seeing, to identify ourselves with ideals which we as rational beings understand to be really desirable. Both the optimistic rationalist and the desillusioned cynic think they can escape the prejudices of anthropocentrism, but they only repeat them in a certain way. The rationalist by continuing the habit of seeing oneself as the center of things, who like a kind of causa sui can arrange everything in his favour; the cynic by failing to notice that precisely by haughtily dissociating himself from the human condition, he falls prey to a typically human conceit.

The rationalist - some would say scientistic - interpretation just discussed paradoxically sees reason only as a means to obtain a certain emotional state: it is not reason itself, but a certain effect obtained through it, which constitutes happiness. This renders unintelligible Spinoza's repeated reference to the importance of knowledge in itself, particularly knowledge of God.

Some recent interpretations of Spinoza concentrating on his political philosophy or seeing his philosophy as a whole as culminating in a kind of political practice,[3] make a similar mistake. In their view, thinking sub specie aeternitatis and the concomitant intellectual love of God should ultimately be interpreted as enlightened political activity, combined with love of the collective power of the multitude. Philosophical thinking is here reduced to a contribution to the self-reflexive auto-production of the free, democratic multitude and the free flow of its collective power in democratic discussion and decision-making. In my opinion, this is again to confuse what are necessary (political) conditions for the salvation of as many philosophers as possible, with the medium of salvation itself, i.e., intuition and intellectual love of God. Furthermore, the autoproduction of the free multitude is understood here as "the human conquest of substantiality, of adequate causality", as an activity in which the masses move in the direction of becoming collectively a causa sui (Tosel: 289, 296). However original and interesting this interpretation may be, is this not the unexpected return, through the backdoor of a kind of political religion, of the old anthropocentrism, now in the disguise of a collective activism?

Another way to understand the relation between truth and the elimination of prejudice, is contained in the view that truth reveals the real value or perfection of things, which will automatically make a strong appeal on us. This is the way Spinoza seems to be understood in "deep ecological thinking".[4] Man is only a part of the whole, which alone is intrinsic and ultimate value; all the parts are valuable only by degree, in function of the whole. As against the usual anthropocentric metaphysics, which is here called the main cause of the ecological crisis, we would need a new naturalistic metaphysics which is genuinely not-anthropocentric (Passmore 1975: 260). Some ecologists claim that this metaphysics is present at least in nucleo in Spinoza. It is true that Spinoza equates reality and perfection (E 2Def6), and that he sometimes speaks of degrees of perfection (e.g., E 4Praef). It is impossible to discuss this question here at length. But it seems to me that this deep-ecological interpretation is incompatible with what Santayana calls the hard truth realized by Spinoza. The equation between reality and perfection should not be read as "reality is attractive value", but "if perfection (degree of -) means anything, it means the same as reality (degree of -)". Spinoza's critique of anthropocentrism is inherently a critique of all talk which projects universal value upon things (be it beauty, goodness, order or perfecti-

[3] Especially Antonio Negri and André Tosel. See my critical discussions of their interpretations: De Dijn 1986b: 619-630; Review of Tosel in De Dijn 1985b: 417-422.

[4] Naess 1993. For further references on Spinoza and ecology, see e.g.: *Inquiry* 23 (1980): 293-325; and Schuyt 1993.

on). The anti-anthropocentric metaphysics, ecologists like Arne Naess would like to develop is really crypto-anthropocentric. We ought to be able, so they say, to develop a point of view which helps to transcend the spontaneous attitude to evaluate everything solely in function of man; a point of view which reveals that the whole and not the parts is intrinsic, absolute value. But why should we accept that an objective view of things has any reference at all to notions like value or perfection which, as Spinoza saw, clearly originate in human desire?[5]

3. Up to now, we have not been able to discover how exactly the new "mathematical" metaphysics leads to salvation. The straightforward answers don't seem to be Spinoza's. Before we engage in a new attempt, it may be fruitful to reinvestigate also the other thesis of Spinoza: that the ordinary religious view of things leads to spiritual misery and slavery. If we take into account the whole of Spinoza's work, and not only his Ethics, we get a much more nuanced view also on this point.

If we take the Tractatus Theologico-Politicus (= TTP) really seriously,[6] we get a much more sophisticated view of what Spinoza calls the imagination, the origin of knowledge of the first kind. (In the Ethics the imagination is seen chiefly as the source of illusions and prejudices; on the other hand in the Ethics we get a complex, sophisticated view of what Spinoza calls the intellect, which comprises reason as well as intuitive knowledge; on this last topic, the TTP presents a rather undifferentiated picture whereas (together with the *Tractatus Politicus*) it presents a much more differentiated picture of the imaginatio, which can also take the form of a kind of common sense).[7] To one's surprise, in the TTP Spinoza seems to fully accept the existence of a kind of religion which, although containing the usual anthropocentric ideas and beliefs (personal God, teleology, miracles, etc.), nevertheless can lead to real salvation (De Dijn 1990a: 241-251). This religion is the purified judaeo-christian religion. It is inevitably related to narratives, rites and dogmas; yet this form of religiousness results in a way of life comparable to the

[5] I owe this idea to my colleague Arnold Burms.

[6] And discard a Straussian interpretation of the TTP (For one good reason to do so, see review of Tosel in De Dijn 1985b: 419).

[7] The first to really argue for the presence of a more sophisticated idea of *imaginatio* or knowledge of the first kind, was: De Deugd 1966. My interpretation only confirms his views.

moral practice of the Spinozist philosopher. Spinoza seems to accept here a twofold way to salvation: one for the few, and one for the crowd.[8]

The way purified judaeo-christianity leads to salvation is not through scientific truth, but through the practice of obedience, charity and justice. The reason why this practice leads to salvation must have something to do with a kind of "forgetting oneself", "letting go of oneself" realized through the practice. If this interpretation of the TTP is right - which in spite of L. Strauss I have never doubted - , then we arrive at the following astonishing conclusion: what are obviously anthropocentric ideas, in fact can be combined with a sort of life, with a moral practice which is anti-anthropocentric. It must mean, of course, that these ideas and beliefs do not function here in the same way as the illusory prejudices Spinoza talks about in the Appendix of Ethics I. This is precisely what Spinoza claims in the TTP: the religious views, even dogmas, can be entertained - and this is crucial - in a non-dogmatic, non-illusory manner. Being part of a way of life which constitutes the right moral practice, these views are not opposed to, but are part of a kind of wisdom. What counts is not the surface, not what one says, or seems to know, but the way of life which proves itself in its result as salvation. From here one could argue how wrong are those who want everybody to display the same indiscriminate, enlightened dislike for all kinds of religion, and who think that salvation for the masses is to be obtained only through more contact with science.

4. It is time now to return to our central problem: how exactly is it that the hard truth of a scientific metaphysics, and the anti-anthropocentric view of things it contains, can have any relation with salvation. There is a double problem here: how can knowledge itself, and particularly knowledge sub specie aeternitatis, ever lead to real acquiescence of the mind ? It is already very difficult to see how any knowledge could constitute this, let alone the new sort of knowledge divorced from all tradition, cut off from all emotional and evaluative elements. The kernel of the problem is: how can a kind of knowledge which is so alienating and anti-anthropocentric constitute real human fulfilment?

The answer, if there is any, must be contained in the Ethics, especially Ethics IV and V. One solution has already been rejected: scientific know-

[8] This idea of a dual path to salvation can of course be found in several (religious) traditions. See e.g.: Nishitani 1991: 160: "The teachings of the Vedas of India, the Neo-Platonists, and the Holy Gate of Buddhism, speak of knowing, while Christianity and Pure Land Buddhism speak of love and reliance" (in Christianity itself there is also a dual way because of the mystical tradition).

ledge, particularly of ourselves, can be used to form a picture of what, in view of our human constitution, is really good or useful for it. In this interpretation the Ethics is seen as a form of optimistic practical rationalism, in which scientific insights in human emotionality (Ethics III) are transformed into therapeutic devices and precepts bringing us closer and closer to "the ideal man", "the rational life" (Ethics IV and V).[9] That this interpretation is wrong is not difficult to show. First, it does not explain how knowledge as such or in itself constitutes salvation. Secondly, it does not really understand the problematic of Ethics IV concerning a certain impotence of reason. Thirdly, it cannot account for the distinction between Ethics IV and Ethics V and certainly not for the presence of the second part of Ethics V (which in this interpretation is considered as an unfortunate relapse of Spinoza into old ways of thinking).

The second solution which, as we have seen before, is equally inadequate, simply says that a human being capable of rational knowledge can identify himself with the point of view from eternity, and so become powerful enough to change into a new man, the rational man. It is clear from Ethics IV and V that rational knowledge, even rational knowledge about oneself is not capable by itself to fundamentally change oneself. With Strawson, and against the Stoics, Spinoza is clever enough not to deny what is, as he says himself, quite obvious from experience. Ultimately speaking however, Strawson is also wrong, because he does not recognize that there is a kind of knowledge sub specie aeternitatis which does the trick, which makes us really free, saved and happy.

Spinoza's position then is very subtle: in one sense knowledge sub specie aeternitatis is not capable by itself to really free us from the bondage of our emotions and our anthropocentric illusions and prejudices; but in another sense it is capable to bring about a fundamental change in us, the start of a life consisting at least partially in real freedom. The difference indicated here is the difference between reason and intuitive knowledge. Although reason is a necessary condition of intuitive knowledge, it does not really succeed in making us free from bondage and misery. Only intuitive knowledge helps us to reach a way of life "beyond good and evil", a life of real freedom which should not be confused with the rationalistic or scientistic idea of freedom of the self.

That reason by itself is not making us really free can be gathered from a careful analysis of Ethics IV.[10] It has always puzzled commentators why

[9] For a clever version of such an interpretation, see Curley 1988.

[10] For such an analysis, see De Dijn 1992: 493-503.

Book IV of the Ethics should bear the title On Human Servitude. Many of the parts of Book IV seem to treat topics which would seem to deserve instead the title of Book V On Human Freedom. If we take the title of Book IV seriously, somehow the whole of Book IV must be understood from the perspective of human weakness. This perspective is clearly indicated by the only Axiom which dominates Book IV: man is inevitably a part of Nature which can always be overpowered by something stronger than himself. It is this weakness of man, more particularly of rational man, which is the topic of Book IV . The rational man is the sort of man who develops the sort of science of good and bad and of the ideal of freedom to be found in Ethics IV. It is precisely in so far as the rational man cannot help developing this science and longing for this ideal, that he is not really free.

I found a confirmation of this interpretation in this sentence of the distinguished French historian of philosophy, Victor Goldschmidt: "C'est en cette science du bien et du mal que réside, pour Spinoza, la servitude, et c'est pourquoi le code des "préceptes de la raison" se trouve nécessairement exposé dans le livre qui porte ce titre. La liberté, elle, ne consiste, ni à connaître les préceptes de la raison, ni à les mettre en pratique ... Elle consiste à n'en avoir pas besoin et, au juste, à les ignorer ..." (Goldschmidt 1978: 114).

I do not know of any place where Goldschmidt elaborates this fundamental insight, but I completely agree with it.[11] The rational man is the man who, while and after experiencing the value of rational thinking as an activity, remembers his weakness and vulnerability, and tries to secure what he sees as really desirable with the help of his science of what is really good and bad for man. But this implies that his life is still dominated by longing, by hope and fear. It means that he is not a purely rational being: such a being would not even form notions of good and bad (E 4P64C+P68) . It means that his life is inevitably characterized by disturbances, and by the very special sorrows which characterize such a life of reason (E 4P17S). One could say that the desires of the rational man, informed by his science of good and bad, are in a sense contradictory: they are desires for a state which is such that, if one is in it, one could not have such desires (e.g., in a state of real freedom, one no longer cares about the future) (E 4P62).

So there is still some kind of delusion even in an individual capable of knowledge sub specie aeternitatis: he still is under the spell of general pictures ("the ideal man") and of a longing for mastery of oneself. Is there an escape: is there a kind of knowledge and a relationship towards oneself which is beyond good and evil, which overcomes this longing ? Spinoza

[11] For a similar interpretation of *Ethics* IV, see Bartuschat 1992: 331-339.

thinks there is: it is intuitive knowledge. In advance one can indicate what must be its characteristics: 1) it must itself be knowledge sub specie aeternitatis (it cannot fall below the threshold of truth); 2) it must be knowledge and yet at the same time happiness, beatitudo, virtue: it must affirm the anti-anthropocentric truth and yet not lead to cynicism or self-rejection, but to acceptance of oneself as decentered; 3) it must be activity and yet not longing, therefore it must be a kind of non-activist activity; it must be philosophy without morality (in the sense of Ethics IV).

The problem we encounter here seems precisely the reverse of the problem of the first kind of knowledge: this kind of knowledge, shot through with emotion, is so dominating our lives that it becomes difficult to see how one could ever come to a kind of pure, neutral view of things. The problem we have now (with Strawson) is: how on earth can scientific (or metaphysical) knowledge ever become the kind of knowledge having the extravagant effects described? This is the problem of Spinoza's third kind of knowledge as discussed in Ethics V. It is the Socratic problem "knowledge saves" transposed into Modern Times.

It is clear that Spinoza saw such knowledge as a real possibility. It makes him, as Santayana saw, a most singular philosopher, the first in the modern world to attempt a reconciliation of what Pascal, with Strawson and most philosophers, would see as absolute opposites: the heart and the truth. How can the unpalatable truth about ourselves and our place in the scheme of things be accepted in such a way that it produces salvation?[12] One central condition seems to be that this truth does not remain "external" to us, that this knowledge really becomes self-knowledge (E 5P36S). This means that the science of metaphysical psychology (the science of man and of his emotions) has to be blended or fused with the particular awareness of oneself and of the concrete train of one's emotions (this process is described in E 5P1-20). What we have to arrive at, is like a sort of meditation in which our most individual emotions are constantly contemplated in the light of our psychological insights, which produce the same kind of detached look upon ourselves as the new science of physics produced vis-à-vis our contact with nature (E 3Praef). Metaphysical psychology has to operate in a special way: it has to become like a predominating mental outlook, without effort encapsulating everything which affects us in the crystal-clear, objectifying atmosphere of thinking, thereby changing radically the effect of our emotions on us. Even though this kind of meditative thinking (the five remedies mentioned in E 5P20S) cannot remove the passions entirely, it can bring it about

[12] Wetlesen 1979; Sprigge 1984, chapt. 8. See also De Dijn 1990b: 147-156.

that they constitute only a minimal part of our mind, that the flow of active thought and feelings grows into an almost unassailable stronghold. The intuitive kind of knowledge is not just scientific or metaphysical knowledge in general, it is a kind of meditative self-knowledge. But the right kind of self-knowledge is dependent on the right kind of metaphysical psychology. It is by meditatively confronting ourselves, in our awareness of our emotions, with the scientific truth about them and about ourselves, that we can really escape an illusory view upon ourselves.

But what has all this to do with salvation? The self-knowledge present in the meditative self-knowledge paradoxically combines an experience not of decentering and rejection, but of decentering and acceptance, thus producing moments of extreme happiness combined with love of the impersonal Whole to which we belong (these moments are the penetration of eternity in time as described in E 5P21-end). In the serene atmosphere of the scientific contemplation of our emotions, in which each affection is a new occasion for further contemplation, we experience not only a certain estrangement from the ordinary view and awareness of ourselves and of our emotions, but we also see ourselves and our emotions immediately as an activity which is part of the Whole produced by God as Thought. The blending of concrete self-awareness and scientific contemplation, must be "complemented" by another element: the realization that everything which happens to us and in us, according to inexorable psychological laws, are nothing but concrete modal expressions of an impersonal Power. Again, this should not remain a piece of external information: in contemplating ourselves as modes of God, the metaphysical truth about God-Nature is lived as a kind of personal relationship with the Impersonal Other, who is given the old, venerable proper name of God. This view of ourselves as God's modes constitutes a peculiar emotional experience, the intellectual love of God (E 5P35-36). This experience means at the same time a joyful acceptance of the truth about ourselves, a kind of paradoxical celebration (*gloria*) of ourselves as modes of God, as well as love towards this Impersonal God - a love which asks nothing in return (E 5P36S; P19). This intuitive knowledge of the self in God, together with the active emotions going with it, constitutes our blessedness or salvation (E 5P36S). All delusion concerning oneself and concerning God is gone here. We know and accept ourselves as being "nothing but clay in the hand of the potter",[13] we know and love God as the radically different

13. This expression can be found repeatedly in Spinoza, e.g. in: *Tractatus Politicus* 2/22; *Tractatus Theologico-Politicus*, Note 34 to Chapt. 16.

"depth" of our being (Natura Naturans), the absolute source even of this very knowing and loving activity.

5. The key to ethics is knowledge.[14] Anthropocentric knowledge means slavery; true knowledge, which looks unpalatable and cruel to the ordinary mind, paradoxically means real freedom. This is Spinoza. But it has to be understood carefully. Knowledge is the key to ethics if knowledge is understood as a particular way of seeing things and oneself, if it is understood as a relationship with things and with oneself. This is why, what superficially is anthropocentric knowing (as in purified religion), can nevertheless mean salvation. This is why, what is non-anthropocentric scientific knowledge, does not necessarily constitute real freedom. On the contrary, the problem is rather how this scientific knowledge ever could lead to salvation. As everybody knows today scientific activity can go together with typically anthropocentric longing and conceit. The problem is how scientific-metaphysical insights ever can produce the right sort of attention.[15] Spinoza's main problem is not the furthering of science as such, it is the ethical problem of salvation through knowledge; this knowledge is not simply some or other piece of scientific information, but an activity which is adequate to the kind of being we really are. Salvation does not lie in mastering external or even internal nature through thinking, salvation lies in the pure mastery of thinking nature in me in particular.

Spinoza's view of freedom as compatible with determinism is not simply an attempt to think human freedom in such a way that it agrees with the scientific discoveries of his day; it is also guided by a deep insight into the very nature of human freedom. This is not to be understood as self-reflexive activity which is the absolute origin of itself, but as the joyful experience of oneself as being the expression of the impersonal Life-force itself. To obtain this experience, the spontaneous anthropocentrism of man has to be broken, but in such a way that the activity of knowing becomes a way of seeing oneself as being "clay in the hands of the potter": a way of seeing which is at the same time self-distantiation and self-acquiescence in the Other, in God-Nature.

The paradox of decentration and acceptance concerning the self is matched in Spinoza by the paradox of God-Nature as absolutely Other and yet Source of ourselves. Spinoza's God-Nature is not the cosy Whole in which we can feel at home (and which modern man would foolishly betray and

[14] In this century, Wittgenstein has repeated this thesis in his own way. See Edwards 1982: 251.

[15] For an attempt to further tackle this problem see De Dijn 1990b.

spoil). Nature does not care about man; God is not a Father, but not a Mother either. At each moment Nature can be cruel and bring terrible disaster (that's why men have to stick together, and to use all their cleverness and power). Nature has no peculiar aims, neither to bring about mankind, nor especially to destroy it (even though it will, at some point, be destroyed). And yet, this very urge in us to live and to prosper, as human beings, is an expression of this formidable life-force itself: we are made of no other clay. The truth about us is that we are the not-wanted who cannot but want themselves and therefore cannot but want to be wanted. The problem then is how this wanting is related to the truth about itself: can man really face this truth and live with it? Or, is not to face it and live with it, the same as not to reach full life?

BIBLIOGRAPHY

Works of Spinoza

Spinoza: *Ethics*. Edited, with a revised translation by G. H. R. Parkinson (Everyman Classics). London: Dent & Sons, 1989. (Unexplained references in the text are to parts of the *Ethics*.)

Literature

Bartuschat, Wolfgang: "Die Theorie des Guten im 4. Teil der 'Ethik'", in: Atilano Dominguez (ed.), *La Etica de Spinoza: Fundamentos y Significado* (Actas del Congreso Internacional: Almagro, 24-26 de octubre, 1990), Ediciones de la Universidad de Castilla-La Mancha, 1992: 331-339.

Carriero, John: "Spinoza's Views on Necessity in Historical Perspective". *Philosophical Topics* 19(1991): 47-96.

Curley, Edwin: *Behind the Geometrical Method. A Reading of Spinoza's Ethics*. Princeton N.J.: Princeton University Press, 1988.

De Deugd, C.: *The Significance of Spinoza's First Kind of Knowledge*. Assen: Van Gorcum, 1966.

De Dijn, Herman: "The Compatibility of Determinism and Moral Attitudes", in: Giancotti, E. (ed.), *Spinoza nel 350° Anniversario della Nascita: Proceedings of the First Italian International Congress on Spinoza*. Napoli: Bibliopolis, 1985a: 205-219.

De Dijn, Herman: Review of André Tosel, "Spinoza ou le crépuscule de la servitude: Essai sur le Traité Théologico-Politique." *Studia Spinozana* 1(1985b): 417-422.

De Dijn, Herman: "Conceptions of Philosophical Method in Spinoza: Logica and Mos Geometricus". *The Review of Metaphysics* 40 (1986a): 55-78.

De Dijn, Herman: "Spinoza als bevrijdingsfilosoof: Omtrent Antonio Negri's Spinoza-interpretatie". *Tijdschrift voor Filosofie* 48(1986b): 619-630.

De Dijn, Herman: "Spinoza en de geopenbaarde godsdienst". *Algemeen Nederlands Tijdschrift voor Wijsbegeerte* 82(1990a): 241-251.

De Dijn, Herman: "Wisdom and Theoretical Knowledge in Spinoza". In: E. Curley & P.-F. Moreau (eds.), *Spinoza. Issues and Directions, The Proceedings of the Chicago Spinoza Conference*, Leiden (etc.): E. J. Brill, 1990b: 147-156.

De Dijn, Herman: "Metaphysics as Ethics". In: Y. Yovel (ed.), *God and Nature, Spinoza's Metaphysics*; Papers Presented at The First Jerusalem Conference (Ethica I), Leiden (etc.): E. J. Brill, 1991: 119-131.

De Dijn, Herman: "Spinoza's ethics: from the sorrows of reason to freedom and beyond." In: Atilano Dominguez (ed.), *La Etica de Spinoza: Fundamentos y Significado* (Actas del Congreso Internacional: Almagro, 24-26 de octubre, 1990), Ediciones de la Universidad de Castilla-La Mancha, 1992: 493-503.

Edwards, C.: *Ethics without Philosophy: Wittgenstein and the Moral Life*. Tampa (etc.): University Presses of Florida, 1982.

Goldschmidt, V.: "La place de la théorie politique dans la philosophie de Spinoza". *Manuscrito. Revista di Filosofia* 2 (1978): 114.

Naess, Arne: *Spinoza and the Deep Ecology Movement*. Delft: Eburon, 1993. (Mededelingen vanwege het Spinozahuis; 67.)

Nishitani, Keiji: *Nishida Kitaro*. Translated by Yamamoto Seisaku & James W. Heisig. Berkeley (etc.): University of California Press, 1991.

Passmore, J.: "Attitudes to Nature". In: R. S. Peters (ed.), *Nature and Conduct*, London: MacMillan, 1975: 251-264.

Roothaan, Angela: *Kritiek op het Antropocentrisme: De rol van ethiek bij Van Schurman en Spinoza*. Delft: Eburon, 1991. (Mededelingen vanwege het Spinozahuis; 66.)

Santayana, G.: "Ultimate Religion". In: *Septimana Spinozana*: Acta conventus oecumenici in memoriam B. D. S. diei natalis trecentissimi Hagae Comitis Habiti, Den Haag: M. Nijhoff, 1923: 105-115.

Schuyt, Kees: *Spinoza, het ecologisch denken en de verandering van levenssstijl.* Delft: Eburon, 1993. (Mededelingen vanwege het Spinozahuis; 68.)

Sprigge, T.L.S.: *Theories of Existence.* Harmondsworth: Penguin Books, 1984.

Strawson, P.F.: "Freedom and Resentment". In: Idem, *Freedom and Resentment and Other Essays.* London: Methuen & Co., 1974: 1-25.

Tosel, André: *Spinoza ou le crépuscule de la servitude: Essai sur le Traité Théologico-Politique.* Paris: Aubier, 1984

Wetlesen, Jon: *The Sage and the Way. Spinoza's Ethics of Freedom.* Assen: Van Gorcum, 1979.

[21]

DID SPINOZA LIE TO HIS LANDLADY?

J. Thomas Cook

In his biography of Spinoza, Colerus recounts the following exchange:

> It happened one day that his landlady asked him whether he believed that she could be saved in the religion she professed: He answered, 'Your Religion is a good one, you need not look for any other, nor doubt that you may be saved in it, provided, whilst you apply yourself to Piety, you live at the same time a peaceable and quiet life' (Colerus 1906: 41).

As biographical tales go, this one is pretty reliable. The biographer, Johannes Colerus, was a German Lutheran minister who took over pastoral duties at the local Lutheran church in the Hague some sixteen years after Spinoza's death. The Van der Spycks, Spinoza's landlord and landlady for the last six years of his life, were members of Colerus' congregation, and the pastor seems to have been intrigued by the contrast between the stories that they told of their quiet, pleasant, upright tenant and the execrable blasphemies and impieties that the preacher found in Spinoza's writings. In writing the biography, Colerus used those of his parishioners who had known Spinoza personally as sources for a number of details, and we can be quite sure that he got the report of the above exchange straight from mevrouw Van der Spyck herself.[1]

We cannot know, of course, how precisely mevrouw Van der Spyck remembered Spinoza's specific words, but it is likely that Colerus would have urged her to think carefully and to be as accurate as possible in her

[1] Colerus is quite explicit about the extent of his reliance on the Van der Spycks as sources. When discussing Bayle's account of a putative knife-assault earlier in Spinoza's life, Colerus writes, "But Spinoza's landlord and his wife, who are still living, give me quite another account of it" (Colerus 1906: 9).

recollection. He was something of a stickler when it came to precise wording, as evidenced by the very great lengths to which he went in his efforts to get the exact text of the pronouncement whereby Spinoza was banned from the synagogue in Amsterdam. Perhaps mevrouw Van der Spyck found the exchange sufficiently noteworthy that she wrote it down in her day-book when it happened. Or perhaps she told others about it at the time, and could thereafter draw on their memories as well. In truth, we do not know how reliable her account is, but we can be confident that she was the source of the report and that Colerus tried to get as accurate a version as possible.

However exact the report may be in the details of its wording, I find that the statement attributed to Spinoza has a strong ring of authenticity to it. In part this is a function of tone – the gentle tone of authority with which he responds to his landlady's query. But it is also a matter of the content of his reply. These words sound like precisely the sort of thing that we would expect Spinoza to say to mevrouw Van der Spyck in such an exchange.

The Problem

But why does it sound like just the sort of thing we would expect Spinoza to say? Is it (1) because we think that he believes it, and we would expect him naturally to tell his landlady what he honestly believes? Is it that (2) she seems to be seeking reassurance, and that we expect him generously (if patronizingly) to accommodate her by telling her what she wants to hear (even though he does not believe it)? Or is it that (3) we expect him to tell her what she wants to hear (although he does not believe it) in order not to anger or offend a person with some power over him (since she is, after all, his landlady)? Or is it perhaps (4) that we recognize in his words clever equivocation that allows him to say (and mean) one thing, knowing full well that she will understand him to have said something else – and that we expect this sort of equivocation from Spinoza when he talks with non-philosophers about their religions?

Spinoza scholars will recognize that the interpretive alternatives offered in the prior paragraph have application not only to the conversation with mevrouw Van der Spyck, but to Spinoza's overall treatment of "revealed religion" in the *Tractatus Theologico-Politicus*.[2] In that enigmatic

[2] Until very recently there has not been a good available English translation of the complete *Tractatus Theologico-politicus* (hereafter *TTP*). Fortunately that is changing. E.J. Brill has recently published a very readable and accurate translation by Samuel

work Spinoza often speaks very highly of revealed religions (especially Christianity) even as he appears to undermine the scriptural bases that have traditionally been thought to provide the chief warrant for belief in those religions. Cagily couched affirmations and carefully reserved denials follow on the heels of one another, leaving some reader with an impression of inconsistency, of confusion or of evasion. Commentators have charged Spinoza with naivete, with irony and with mendacity.[3]

Where the *TTP* seems evasive, it is sometimes possible to get clearer on Spinoza's views by focusing on his direct one-on-one exchanges with specific individuals.[4] In this paper I will approach some of the larger interpretive questions that surround the *TTP* by focusing on the modest exchange between the philosopher and his landlady. I do not assume that Spinoza spoke more openly to her than to his intended readership in the *TTP*. But I hope that the directness of her question might help to narrow the interpretive target, and, since we know whom he was addressing, perhaps we can dispense with some of the confusing questions about intended audience. He was not writing a treatise to influence the political future of the Netherlands; he was answering mevrouw Van der Spyck's question. He was not "addressing potential philosophers while the vulgar listened" (as Strauss describes Spinoza's project in the *TTP*); he was talking with his landlady.

My plan is to consider each part of Spinoza's statement in sequence, asking what it means, to what extent he might have believed it, and why. In each case, we will consider the prima facie grounds for thinking that Spinoza could not have meant what he said in a simple, straightforward way. In each case, we will assess the force of those considerations, and if need be will ask what else he might have meant by the words in question – and whether he in fact believed his own words, so interpreted. If all other inter-

Shirley. More importantly, E. M. Curley's full translation will be available relatively soon with the publication of the second volume of his translation of Spinoza's complete works. This edition will have a full index and glossary as well as the other scholarly notes and apparatus that have proved so valuable in Volume One. I am grateful to Professor Curley for making his translation of the *TTP* available prior to publication. Page references in this paper are in standard accordance with the Gebhardt edition.

[3] Leibniz charged irony (see Curley, forthcoming). See Strauss 1952 on mendacity charge.

[4] For example, if the *TTP* were to leave us in doubt about Spinoza's view of certain historically-based claims in favor of the Catholic Church, we can remove such doubts by attending to *Ep* 67 – Spinoza's blistering response to Burgh's efforts to proselytize him for the Mother Church. Again, if we find Spinoza evasive in the *TTP* regarding the Trinity and the Resurrection, we can find a clearer statement of his position in a letter responding to a direct question posed by Oldenburg (see *Ep* 74).

pretive efforts fail, we will regretfully conclude that he was simply lying to his landlady. It may come to that, but if so it will be the interpretive hypothesis of last resort.

Spinoza's response to mevrouw Van der Spyck falls naturally into three parts: (1) The positive assessment of her religion; (2) the assurance that she can be saved in her religion; (3) additional (behavioral) requirements for her salvation. The second and third of these cannot really be dealt with fully in isolation from each other, but we will begin by looking at each of the claims seriatim.

"Your religion is a good one, you need not look for any other..."

The Van der Spycks were Lutherans. Lutherans attribute considerable importance to two of the traditional sacraments; Spinoza believes that such religious ceremonies "contribute nothing to blessedness" and do not "have any holiness in them" (*TTP* 5: 76). Lutherans believe in the literal resurrection of Jesus Christ from the dead; Spinoza does not. Spinoza does not believe in salvation by faith alone.[5] Perhaps most importantly, Lutherans attribute anthropomorphic psychological states to God and tend to think of God's laws as similar to the edicts of a human prince. This final view is not only false, according to Spinoza, but likely to lead one to experience conduct in accordance with divine law as an onerous obligation rather than as active and free self-expression (*TTP* 2: 41).

This list could go on, but the point here is that Spinoza's endorsement of mevrouw Van der Spyck's religion is not the result of an overall agreement with the tenets and practices of the Lutheran faith. Indeed, Spinoza holds a number of these tenets and practices to be superstitious. Why then does he say that her religion is a good one and that she need not look for any other?

Spinoza plays down the importance of the specific beliefs of any given sect regarding sacraments and interpretation of scripture (*TTP* 14: 173). The main point and purpose of religion is to encourage piety, peace and loving kindness – and presumably the Lutheran church is, as a rule, as effective as any other in this regard. Spinoza also had specific knowledge of how the

---

[5] Though Spinoza's definition of faith is interesting and maybe designed in such a way as to bridge the gap between those who favor works and those who favor faith as the means to salvation. According to Spinoza's definition, obedience to God (= works) is a sufficient condition to guarantee the presence of faith (and faith is thereby a necessary condition for works). See also note 15 below.

gospel was preached in the Van der Spycks' own neighborhood church, for he occasionally attended services with the family. Apparently he thought very highly of Pastor Cordes' sermons and "... esteemed particularly [the pastor's] learned way of explaining the Scripture, and the solid applications he made of it"[6]. (Colerus 1906: 41) So, as institutionalized religions go, mevrouw Van der Spyck's was indeed, by Spinoza's lights, a good one, and she would have been unlikely to find a better faith had she gone looking for one. Moreover, for her to have undertaken such a search would have been a sign of the unfortunate inconstancy that so often attends superstitious belief in the hearts of the fearful (*TTP* Preface: 6).[7]

Of course mevrouw Van der Spyck was not interested primarily in Spinoza's general assessment of her religion. Rather, her question was quite specific and direct. She wanted to know if he believed that she could be saved in her faith. Spinoza's answer seems similarly direct: **"...nor [need you] doubt that you can be saved in [your faith]".** She asked; he answered affirmatively. What could be more straightforward?[8]

[6] All told, the relationship between the religiously suspect philosopher and the local churches seems to have been remarkably congenial. I find it quite surprising, for example, that he was buried in the churchyard of the (Reformed) Niewe Kirk at his death. Perhaps no one in the neighborhood of the Niewe Kirk had heard about the controversy surrounding the publication of the *TTP* and the views of its heretical author. Colerus suggests that they did **not** know "...what were the principles of Spinoza in point of religion" (Colerus 1906: 99). And yet, on the other hand, twelve years earlier Spinoza wrote to Oldenburg that "...ordinary people ... never stop accusing me of atheism" (*Ep* 30). Granted, "ordinary people" seems to be contrasted here with "theologians", and it is likely that Spinoza is not talking about people as ordinary as the neighbors and Pastor Cordes. Still, he lived among these people for upwards of six years, and it is hard to imagine that they knew nothing of his controversial status.

[7] I think that what has been said is an accurate account of what lies behind Spinoza's initial remarks to his landlady. But it might be mentioned that Spinoza may have had his own – primarily political – reasons for preferring the Lutheran over the Reformed (Calvinist) Church of his time. For one thing, Luther's "two kingdoms" doctrine is more congenial to Spinoza's political views in the *TTP* than Calvin's theocratic tendencies. Secondly, Spinoza wrote the *TTP* in part as an argument for "freedom of philosophizing and saying what we think" (*Ep* 30), and the Calvinist clergy was notoriously intolerant in this regard. (The pyre of Servetus still smoldered in the back of free-thinkers' minds). Finally, at the very moment that the *TTP* was being written, the Calvinist clergy in the Netherlands was agitating to do away with the Republican government and introduce monarchy under William of Orange. Spinoza rightly feared that such a development would be bad for the freedom of mind of the Dutch citizenry.

[8] I suppose that a skilled practitioner of the hermeneutics of suspicion might find readings of Spinoza's response that make it seem less than directly affirmative: for example, "You need not doubt that you can be saved {since such doubting will just upset

Unfortunately, we have reason to believe that insofar as Spinoza's answer really was affirmative, it could not have been an answer to the question she asked. We can be pretty sure what mevrouw Van der Spyck had in mind when she asked if she could be saved in her faith. She was concerned about being saved from damnation – salvation from an afterlife of everlasting punishment for her sins. As a general rule, an affirmative answer to a question suggests acceptance of the presuppositions that underlie the question.[9] But Spinoza's answer to his landlady violates this general rule, since he accepts almost none of what we may reasonably assume to be her presuppositions in the matter of salvation. Spinoza does not believe in an everlasting after-life in which we might suffer punishment. Nor does he accept the existence of the sort of God who, like a divine magistrate, would pass judgment and impose sentence upon the guilty. Insofar as Spinoza's affirmative answer suggests an acceptance, on his part, of the beliefs presupposed by the intended meaning of her question, his answer is misleading. Mevrouw Van der Spyck is led to believe that Spinoza holds that she can be saved (with all that that involves) whereas in fact Spinoza does not hold that, in her sense, she can be (nor, indeed, that she needs to be) saved.

But if Spinoza does not mean that she will be spared everlasting punishment, and if he is not just plain lying to the woman, just what does he mean when he assures mevrouw Van der Spyck that she can be "saved" in her faith? What does "salvation" mean for Spinoza? This, as it turns out, is a tricky and complex question. As indicated in Curley's Glossary-Index to the *TTP*, salus, salvare, and related terms require a variety of translational equivalents, depending upon context. On the whole it is possible to narrow in on those contexts in which the terms are being used in ways that are relevant to Spinoza's exchange with mevrouw Van der Spyck, but even then Spinoza seems to have a lot more to say about the "way to salvation" and what salvation depends on than about what salvation is.

---

you, and you don't need to get upset over things that you can do nothing about, such as the fact that there is no salvation of the sort that you are imagining}".

[9] If I reply "Yes" to your query about whether the present king of France is bald, I thereby indicate acceptance of the ontological assumption presupposed by the question. If I ask you whether angels have two wings (rather than four) and you reply, "You need not doubt that they have but two wings", you thereby indicate your acceptance of the existence of [wingéd] angels. The talk here of "indicating acceptance of the assumption" is intended to avoid any hard and fast Russellian claims about allegedly necessary truth-conditions. The claim is offered not as conceptual analysis, but as a claim about everyday informal semantics.

Some help can be found in Spinoza's tendency to equate salvation with blessedness,[10] and there is at least one passage in which Spinoza tells us what these are: "...true salvation and blessedness consist in true peace of mind...(vera animi acquiescentia)" (*TTP* 7: 111). The phrase animi acquiescentia appears at one other place in the *TTP*, but these words, coupled with "blessedness" (beatitudo) most readily remind us of Part 5 of the Ethics, where the phrase appears three times.11 Not surprisingly, it is here, in the *Ethics*, that Spinoza most fully develops his conception of salvation and blessedness.

Fortunately, our purposes do not require that we pursue the intricacies of that most difficult section of that most difficult work, for Spinoza provides a brief and clear outline of his view early in chapter 4 of the *TTP*. Using the phrases "our blessedness", "our highest good", and "our highest happiness" interchangeably (or so it seems to me), Spinoza presents an account according to which our blessedness depends entirely upon knowledge – specifically, knowledge of God:

> –"...our greatest good and perfection depend only on the knowledge of God" (*TTP* 4: 60).
>
> –"...our greatest good not only depends on the knowledge of God, but consists entirely in it" (*TTP* 4: 60).
>
> –"...the knowledge and love of God is the ultimate end toward which all our actions are to be directed" (*TTP* 4: 61).

This, I take it, is the doctrine (familiar from the *Ethics*) that our blessedness or salvation finds its consummation and culmination in the intellectual love of God.

This is what we should expect Spinoza to say, of course, but it does not bode well for our assessment of his veracity in conversation with his landlady. If salvation requires the sort of knowledge of God presupposed by *amor intellectualis Dei*, then (with all respect) mevrouw Van der Spyck is a lost soul. Such knowledge of God is an intellectual achievement of the highest order, and nothing suggests that she was up to such feats. Moreover, the theological instruction that she would have received in her Lutheran Church would have gone a long way toward undermining any chance that she oth-

---

[10] I.e. *TTP* 7: 111 and *TTP* Adn 9: 263, where the two are disjoined by *sive*. Likewise in *E* 5P36S and in *K*: 110.

[11] *E* 5P10S; P36S; P42S. Curley translates the last of these as "peace of mind", but the English reader might miss the connection to the *TTP* since the first two are rendered "satisfaction of mind".

erwise might have had of forming a true conception of God. If salvation "depends on and entirely consists in" a true knowledge of God, and if mevrouw Van der Spyck is incapable of achieving a true knowledge of God, then it seems to follow that salvation is beyond her reach.

Yet Spinoza assures her that she can be saved. If we are to avoid the conclusion that Spinoza is just plain lying (our interpretive hypothesis of last resort), then it seems to me that at least one of the following must be the case: either

(1) there is some other (non-cognitive, or at least non-intellectual) path to the salvation that we have been discussing; **or**

(2) there is some other sort of salvation available to those, such as mevrouw Van der Spyck, who lack the requisite intellectual wherewithal to achieve a true knowledge of God.

Both of these are possibilities; both are worth exploring.[12]

The Alternative Path to Salvation

In the *TTP* Spinoza explicitly acknowledges that "the intellectual, or [*sive*] accurate, knowledge of God is not a gift common to all the faithful..." (*TTP* 13: 168). From the closing sentences of the *Ethics* we know that such a saving knowledge of God is very rare indeed, and not only among the faithful.[13] This fact is unfortunate, of course, but in the *Ethics* Spinoza seems willing to accept the inevitable inference that salvation itself will be rare as well. In the *TTP*, on the other hand, we are told that scripture, rightly interpreted, indicates a way to salvation that does not require an "intellectual or accurate" knowledge of God. Spinoza claims that the prophets and the apostles are of one voice in teaching two very basic and very simple precepts – that we

[12] The best extended discussion of the issue here is to be found in Matheron 1971. I have benefitted immensely from his careful textual analysis, and have followed him in drawing distinctions between two kinds of salvation. If I have ignored the political dimension of this issue (while he emphasizes it) it is only because it seems to me less pertinent in the individual case of mevrouw Van der Spyck. If I understand him correctly, however, Matheron's solution to the issue before us requires that we attribute to Spinoza a belief in the possibility of multiple, temporally successive instantiations of a person's "individual essence" (Matheron 1971: 208). Such Spinozistic reincarnation is more than I can accept. If this is the only alternative, I will have to opt for the mendacity option (See note 26 below).

[13] "...if salvation were at hand, and could be found without great effort, how could nearly everyone neglect it? But all things are as difficult as they are rare" (*E* 5P42).

should love God and that we should obey Him by loving our neighbors as ourselves. Scripture tells us that we can be saved simply through adherence to these precepts.

According to this interpretation of scripture, accurate knowledge of God is not at all necessary for salvation. On the contrary, the truth or falsity of one's beliefs about God is of very little intrinsic importance.[14] The important test for one's beliefs is whether those beliefs are conducive to obedience to God via love of one's neighbor. If so, then one's cognitive states are pious and salutary, even if false. If not, one's cognitive states are impious and pernicious, even if true.[15](*TTP* 13: 172) Spinoza adduces seven basic "tenets of the universal faith" (*TTP* 14: 177), which he holds to be essential – not because of their truth, but because "...if any of these tenets is taken away, obedience is also destroyed".[16] (*TTP* 14: 178)

Thinking back to the exchange with mevrouw Van der Spyck, we remember that Spinoza's reassuring words were not unconditional. "...Nor need you doubt that you may be saved in [your faith], ***provided, whilst you apply yourself to Piety, you live at the same time a peaceable and quiet life***". Applying oneself to piety and living a peaceable and quiet life are not exactly the same as loving God and obeying Him through justice and lovingkindness, but we might charitably take the former as a shorthand (or vaguely-remembered) version of the latter. And if we do, we see that in assuring mevrouw Van der Spyck that she could be saved in her faith if she lived in the prescribed manner, Spinoza simply affirmed the doctrine that he

---

[14] Actually, the claim is even stronger than that. In many passages Spinoza suggests that the properties of God that we need to know are the ones that we can "imitate" in our own lives. Not surprisingly, that leads him to recommend belief in those properties characteristic of the most anthropomorphic (and hence false, from a Spinozistic point of view) conceptions of God. For example: "...God through the prophets asks no other knowledge of himself from men than knowledge of his divine Justice and Lovingkindness, i.e. such attributes of God as men can imitate in a certain way of life" (*TTP* 13: 170).

[15] It should be mentioned that Spinoza does not really think that "accurate or intellectual" knowledge of God could lead to disobedience. Such knowledge would necessarily lead to right conduct.

[16] Spinoza very cleverly threads the needle here between those Christians who favor the doctrine of justification through works (emphasizing the views in the New Testament book of James) and those who favor Paul's emphasis on justification by faith. Spinoza's sympathies are clearly on the side of James's doctrine, but since he claims that one cannot be obedient without believing the tenets of the universal faith, he can likewise claim that there is no salvation without faith.

finds in both Old and New Testaments – viz. that "...all and only those who obey God by living in this way are saved" (*TTP* 14: 177).

By providing an alternative path to salvation, this view resolves the difficulty with which this section began and provides an easy negative answer to the title question of the paper. But perhaps it has occurred to the reader that there is something suspicious in all this. The argument of the last three parts of the *Ethics*, as well as the passages quoted above from the *TTP* consistently reaffirm the importance of knowledge (accurate, intellectual knowledge) of God for human blessedness. This new doctrine would have it that one could be saved with quite false beliefs about the nature of God. This does not sound like Spinoza. Moreover, we have seen precious little in the way of argument for the truth of this doctrine. All Spinoza has provided is argument for the claim that the prophets and apostles believed and espoused the doctrine. Since when does Spinoza accept otherwise-unproven views based on the authority of ancient and philosophically unsophisticated writers?

To his credit, Spinoza is up-front about the epistemic status of this doctrine of salvation through obedience alone. Granting that it cannot be known by the light of nature alone, he urges that we can nonetheless accept it "with sound judgment" based on its pragmatic value and on "so many testimonies of the Prophets" (*TTP* 15: 187). The overall position of the *TTP* on the issue of how one can be saved can thus be summed up as follows: By the light of nature we can know with demonstrative certainty that "...the man ...who participates in the greatest blessedness is the one who loves above all else the intellectual knowledge of God..." (*TTP* 4: 60). This is the position argued in Chapter 4. In Chapters 12-15, on the other hand, Spinoza interpretively attributes to the prophets and apostles the view that one can be saved through obedience alone, without knowledge, and he urges that this view, though not demonstrable by the light of nature, should nonetheless be accepted.

Leo Strauss holds that a careful, philosophically sensitive reader will see that this position is inconsistent and unstable, and that Spinoza intended thereby to send a signal to the careful reader (Strauss 1952: 170).[17] According to Strauss, the first (knowledge-oriented) position is Spinoza's real view, and the later obedience-oriented doctrine is a sop to palliate the pious faithful and a smokescreen to confuse the persecutorially powerful. The careful reader, by noting the evident contradictions and by applying the interpretive

[17] Strauss calls attention to many more putative (and putatively intentional) contradictions in the *TTP*, but this one is the most basic.

method provided by Spinoza in Chapter 7, will be able to see through the smokescreen and dismiss the "exoteric" doctrine of salvation by obedience alone as artful deception. Errol Harris, on the other hand, rejects Strauss's position and argues that Spinoza means what he says when he tells us that we can accept the testimony of the prophets "with sound judgment" (Harris 1978). Mevrouw Van der Spyck's salvation and Spinoza's reputation for scrupulous honesty may both hang on this issue, so let us attend to it with some care.

Whether self-contradictory or not, the position that Spinoza espouses certainly is inherently precarious. This precariousness can be seen as an unavoidable consequence of the nature of the project he has undertaken. In a letter to Oldenburg Spinoza explained that one of his motives for writing the *TTP* was to defend "the freedom of philosophizing and saying what we think". As part of that effort, he seeks to draw a line of strict demarcation between philosophy and theology in order that he may maintain that theology is in no way threatened by philosophy. In the midst of a philosophical book about theology, he claims that what one holds regarding philosophical questions is just irrelevant to faith and theology. In a really remarkable passage, Spinoza juxtaposes a number of his own positions with more traditional religious views and claims that "as far as faith is concerned" it just does not matter what one believes about these things:

> Again, it does not matter, as far as faith is concerned, whether someone believes that God ...directs things from freedom or from a necessity of nature, or that he prescribes laws as a prince or teaches them as eternal truths, or that man obeys God from freedom of the will or from the necessity of the divine decree, or finally, that the reward of the good and the punishment of the evil is natural or supernatural... It's all the same whatever anyone maintains about these matters (*TTP* 14: 178).

These issues are all philosophical issues, he says, and philosophical issues have no bearing on the matters of justice, lovingkindness and obedience that are the proper concerns of faith and theology. „...[T]he goal of Philosophy is nothing but the truth, whereas the goal of Faith, as we have shown abundantly, is nothing but obedience and piety" (*TTP* 14: 179). This radical separation between philosophy and theology allows Spinoza to conclude that "Faith, therefore, grants everyone the greatest freedom to philosophize, so that without wickedness he can think whatever he wishes about anything..." (*TTP* 14: 180). A further consequence of this separation, however, is that philosophy (whose goal is truth) cannot be used to demonstrate the truth of "the foundation of theology – that men are saved only by obedience..." (*TTP* 15: 185). Theology, on the other hand, is equally incapable of demonstrating this important truth, since truth-demonstrations are not within theology's legitimate purview. So on what grounds are we to accept this important doctrine? Spinoza nicely sums up the epistemic/methodological dilemma in which he finds himself:

> ...since we cannot rationally demonstrate that the foundation of Theology - that men are saved only by obedience - is true or false, someone may raise against us too the question: why then do we believe it? If we accept it without reason, like blind men, then we too act foolishly and without judgment. On the other hand, if we want to maintain that we can demonstrate this foundation rationally, then Theology will be a part of Philosophy, and would not have to be separated from it (*TTP* 15: 185).

As mentioned above, Spinoza addresses this dilemma by offering two kinds of considerations in support of acceptance of the doctrine – pragmatic considerations and prophetic testimony. The first of these might best be thought of as incentives for acceptance, while the latter is best viewed as a complicated argument from inductively authenticated authority. Neither is without its difficulties.

> So even though this foundation of the whole of theology and scripture cannot be shown by a mathematical demonstration, we can still accept it with sound judgment. For it is indeed ignorance to be unwilling to accept what has been confirmed by so many testimonies of the prophets, what is a great source of comfort to those who cannot exert much power by reason, what brings no slight advantage to the state, and what we can believe with absolutely no risk or harm... (*TTP* 15: 187).

Consider, first, the pragmatic appeal being made here – an odd appeal by its very nature. The fact that a doctrine has been a great source of comfort to those lacking in reason is offered as grounds for accepting the doctrine – where "accepting the doctrine" presumably means "believing the doctrine to be true". But there is of course no logical relationship between a claim's being comforting to the weak-minded and that claim's being true. These considerations do not provide epistemic reasons in favor of the claim at all, though they do provide prudential motives that might induce one to believe – especially if one counts oneself among the weak-minded and is in need of comfort.[18] The mention of the advantage to the state raises pragmatic considerations of a different kind, but epistemically it is in the same boat. It

[18] In this context we might remember that it is a principle of Spinoza's psychology in the *Ethics* that a person endeavors to imagine those things which indicate an increase in our power to persevere in being. If the doctrine at issue seems to make salvation available to me, I will, in the absence of countervailing considerations, believe it.

seems to me most natural to read these passages that deal with the pragmatic considerations as encouraging the reader, with a conspiratorial wink, to play along with the (unsubstantiated) claim in the interests of political stability and the happiness of the feeble-minded multitude.[19] In addition, the note of paternalism in the mention of the comfort that this doctrine provides to the weak-of-mind seems to me further to support Strauss's claim that Spinoza does not want this doctrine to be taken seriously by serious-minded philosophers.

But there is more to Spinoza's position here than the pragmatic considerations discussed thus far. The doctrine of salvation by obedience alone is recommended not only because of the beneficial effects that belief in this doctrine has produced for individuals and for the state, but because it "has been confirmed by so many testimonies of the Prophets...". This is, I think, a stronger contention, but it too has its problems. The reader who has come this far in the *TTP* knows that Spinoza puts no credence in the claims of the prophets when they make pronouncements on speculative matters, but that true prophets are said to be reliable when they address "matters which concern integrity and morals" (*TTP* 2: 35): "...the Prophets could be ignorant of things that concern only speculation, but not those which concern love and how to conduct our lives" (*TTP* 2: 42). Presumably the doctrine of salvation through obedience alone is among those things about which they cannot be ignorant, so that the "many testimonies of the Prophets" regarding this doctrine can be accepted by us all without fear of error.

In fact, Spinoza does not recount a large number of specific cases in which specific prophets have embraced the doctrine in question, but I am prepared to accept his word that there are many such testimonies. More important is the way in which he seeks to establish the prophetic *bona fides* of those who have preached this doctrine.[20] First Spinoza reminds the reader of the conclusion of earlier discussions of prophecy – viz. that a true and reliable prophet is distinguished by his teachings and by the signs that accompanied the teachings (*TTP* 15: 186). Since biblical reports on the presence or absence of signs are unreliable (*TTP* 2: 31), we are forced to base our judg-

[19] My use of the term "conspiratorial" unavoidably suggests nefarious motives. This impression is further bolstered by the fact that Dostoevsky's Grand Inquisitor is modern literature's most articulate proponent of such a view. I (emphatically) do not want to suggest evil intent *vis a vis* the masses, though there is certainly an element of mendacity here.

[20] In this section, Spinoza is relying on the earlier extensive discussion of prophecy and the prophets (*TTP* 1 & 2: 15-44) That discussion deserves more thorough consideration than it receives here, but that is a topic for another day.

ments primarily on the teachings of the putative prophet. Regarding the doctrine of interest to us – salvation through obedience alone – Spinoza offers the following rather convoluted argument:

> Since we see that the Prophets commend Lovingkindness and Justice above all, and aim at nothing else, we conclude from this that they did not teach in bad faith, but from a true heart, that men become blessed by obedience and trust (*TTP* 15: 186).

It is not entirely clear to me how this argument is supposed to work. Even if we grant Spinoza his premise about what the Prophets commended and aimed at, how would the conclusion follow that they were not teaching in bad faith when they taught that obedience and trust are sufficient for salvation? Is the hidden assumption that the Prophets would not have commended anything that would not suffice to make men blessed? That does not sound convincing. Is the assumption that those who "commend" (*commendare*) Lovingkindness and Justice always speak and act "with a true heart?" That sounds even less convincing. Perhaps the claim is simply that prophets who teach Lovingkindness and Justice are in the moral mainstream of biblical teaching, so they qualify as *bona fide* prophets. That would in turn lend credence to their words when they spoke on the subject of salvation and obedience. At the very least we could conclude that when the prophets spoke about becoming blessed by obedience and trust, they "...did not say this rashly, and did not rave while they prophesied". This does lend some support to their words, but is hardly decisive.

Actually, Spinoza has another argument with which to follow this one, and I suspect that he might have thought it a better one.

> In this we are even more confirmed when we notice that they [the Prophets] taught no moral doctrine which does not agree most fully with reason. For it is not without reason that the word of God in the Prophets agrees completely with the word of God speaking in us (*TTP* 15: 186).

Presumably Spinoza is not including the doctrine that obedience suffices for blessedness among those doctrines that "agree most fully with reason", for (by hypothesis) reason cannot pronounce directly on that matter. So how is this argument supposed to work? As I see it, Spinoza wants to stamp the imprimatur of reason on a doctrine that he is forbidden to demonstrate rationally by deducing it from first principles. So he provides what amounts to an inductive argument in favor of the rational reliability of the prophets' pronouncements on the salvific power of obedience. In ninety-nine cases we are able to compare the moral teachings of the prophets with a rationally-deduced ethical system, and in all ninety-nine cases, Spinoza tells us, the positions match up. This then serves as an inductive argument for the claim that the hundredth teaching (that obedience suffices for salvation) will also be in accord with reason, even if we are not able to deduce it rationally (i.e. are not able to find its match in our rationally deduced ethical system).

Inductive arguments cannot be conclusive, but if they satisfy certain conditions they can lend persuasive support to a position – and persuasive support is all that Spinoza can expect here. The main conditions that must be satisfied in this case are the following: (1) the one teaching whose rationality we cannot directly assess must be relevantly similar to the ninety-nine teachings whose rationality we can directly assess; (2) there must not be significant independent grounds for believing that the teaching in question is *contrary* to reason. The first of these conditions seems *prima facie* to be met, since the hundredth teaching is a teaching of the prophets and is a teaching about the consequences (if not the dictates) of morality. There is one important *dis*similarity, of course, and that is the fact that, contrary to the other ninety-nine, the one teaching in question seems not to be directly deducible from first principles by the light of nature. That dissimilarity does not seem to me to be enough to undermine the argument unless it should turn out that the second condition is broken – i.e. unless it should turn out that the doctrine cannot be deduced directly *because* it is *contrary* to reason. Presumably Spinoza's *Ethics* provides the standard for what can be deductively derived from first principles using only the light of nature. So the test for whether the doctrine of salvation through obedience alone is contrary to reason is whether it is contrary to the truths deduced in the *Ethics*.

A moment's review is in order here. In our initial discussions of the meaning of salvation, we noted a tension between the emphasis upon high-level knowledge of God in the *Ethics* and the claim that mevrouw Van der Spyck could be saved in her faith. It was that tension that led us to search for another path (other than the cognitive/intellectual) by which one might attain salvation. We found Spinoza claiming that the path of obedience via justice and lovingkindness can also lead to salvation and citing the authority of the Prophets in support of that contention. Seeking to assess the epistemic force of Spinoza's appeal to authority, we found that the argument depends upon whether the doctrine of salvation via obedience is consonant with the doctrine of the *Ethics*. In light of our initial perception of a tension in that regard, this is not an encouraging finding.

And it gets worse. Spinoza emphasizes the agreement between the moral teachings of the prophets and rationally deduced ethical principles. In doing so, he calls attention to those parts of the *Ethics* in which the explicitly moral doctrine is developed. Among the most prominent propositions in the relevant section of Part 4 of the *Ethics* are the following:

> Proposition 26: What we strive for from reason is nothing but understanding, nor does the mind, insofar as it uses reason, judge anything to be useful to itself except what leads to understanding.
>
> Proposition 27: We know nothing to be certainly good or evil, except what really leads to understanding or what can prevent us from understanding (Curley translation).

It might seem, at first, as if these propositions are in direct conflict with the doctrine of salvation through obedience alone, since they seem to suggest that the only good, intrinsic or instrumental, is understanding. But in fact they are, at least technically, compatible with the doctrine. Having carefully qualified and circumscribed the claims ("from reason"; "certainly"), Spinoza can maintain (as he does maintain in the *TTP*) that it remains *possible* that obedience is a good even though we cannot know it via reason and hence cannot know it for sure. This resolves the surface tension between these propositions and the doctrine of salvation through obedience, but there is a deeper tension at work here that is not so easily resolved.

The two propositions just cited are absolutely central in the development of Spinoza's own ethical doctrine.[21] Of the forty-five propositions that come after these in Part 4 – propositions that articulate much of the normative ethical content of Spinoza's system – *twenty-four* are demonstrated by direct or indirect appeal to proposition 26.[22] For example, Proposition 37 tells us that "the good which everyone who seeks virtue wants for himself, he also desires for other men...". This sounds like the sort of proposition that Spinoza might have had in mind when he claimed that the moral teachings of the prophets "agree most fully with reason", for it seems like a descriptive version of the oft-quoted biblical prescription that we should love our neighbors as ourselves. But the proof of 37 relies heavily on Proposition 26 – on the claim that insofar as we use reason we want nothing but understanding. The proposition's plausibility depends entirely on the assumption

[21] Proposition 26 seems to me as important as any other in the *Ethics* for binding together the *ethical* doctrines with those of the metaphysics and epistemology. This also seems to me one of the most problematic and perplexing propositions in the work. Some of the problems and perplexities are discussed in Cook 1986: 191-198.

[22] When I say that proposition X makes "direct appeal" to proposition 26, I mean that proposition 26 is cited in the demonstration of X. When I say that proposition Y makes indirect appeal to proposition 26, I mean that the demonstration of proposition Y cites another proposition that makes a direct appeal to 26. I have seen this convention used elsewhere, but am unable to locate the source.

that the good that the individual wants for himself (understanding) is not a zero-sum good. So the proposition accurately tells us what a rational person would do whose desire is for understanding. But the biblical injunction was addressed to all – those whose highest good is to be found in rational understanding as well as those for whom that is not the case.

One of Spinoza's main arguments for the acceptability of the doctrine of salvation through obedience alone is the fact that many biblical prophets have taught that doctrine. These prophets are said to be reliable because other of their moral pronouncements agree with rationally deducible moral principles. But many of these rationally deducible moral principles, as it turns out, derive their validity from the claim that the one certain good for a person is understanding – and that, in turn, is a truth whose deduction requires the condition that the individual's striving is rationally based and that he/she "uses reason". But we have no grounds at all for thinking that what is good for a person insofar as he/she "uses reason" will be good for him/her *simpliciter*. On the contrary, since the plausibility of many of the later propositions of Part 4 of the *Ethics* derives from the fact that the one certain good for a person who uses reason is not a zero-sum good, it seems more likely that what is good for a person who "uses reason" will *not* be what is good for a person *simpliciter*. So it seems to me that our confidence in the moral pronouncements of the prophets (intended for everyone) should not be bolstered by the fact that they match the moral precepts of the *Ethics* (which, after 4, P26, apply to people insofar as they use reason). If the prescriptions of the prophets match the precepts of the *Ethics*, then the prophets are prescribing to people conduct whose goodness depends upon their not being who and what they in fact are. This should not increase our confidence in the moral teachings of the prophets.

In sum, Spinoza can legitimately claim that it is *possible* that salvation or blessedness (of the kind discussed in the final propositions of the *Ethics*) might be had through obedience alone, without knowledge. But try as he might, he has not, I think, provided any persuasive grounds for believing that it can. Pragmatic considerations about the comfort that the doctrine brings the weak-minded are epistemically irrelevant. And arguments based upon the fact that this doctrine was taught by the prophets fail for lack of a convincing case for the authority or reliability of the prophets. Perhaps there is an alternative, non-intellectual path to salvation. Perhaps mevrouw Van der Spyck *can* be saved in her faith, provided she devotes herself to piety and lives a quiet and peaceable life. But thus far we have not discovered good reason for believing it to be so. Since it is hard for me to believe that Spinoza would have accepted the doctrine on the basis of arguments that seem insufficient to convince, I suspect that Spinoza did not believe the doctrine either – at least not as it is presently being interpreted.

If it is difficult to construct a convincing case for the existence of an alternative path to salvation as initially understood, perhaps there is a different kind of salvation to which mevrouw Van der Spyck might aspire and which she might more likely attain. The *TTP* contains hints in this direction as well.

Salvation of a Lesser Kind

We have been trying to confirm the existence of an alternative (non-intellectual) path that might bring mevrouw Van der Spyck to the salvation enjoyed by those blessed with true knowledge and intellectual love of God. The effort has been laborious and has ended in failure. By contrast, if we are willing to settle for a lesser kind of salvation – for an *acquiescentia animi* less exalted, less blissful and more temporally circumscribed – our task will be quite easy. The path that leads mevrouw Van der Spyck from a life of peaceable piety to this kind of salvation is short and follows straightforwardly from the basic principles of Spinoza's psychology.

As developed in the *Ethics*, Spinoza's psychology is based on the primary affects of desire, pleasure and pain, where pleasure and pain are defined in terms of the increase or decrease in an individual's power to persevere in being. Other affects are then understood as modifications of these primary affects of pleasure and pain and thus inherit their connection to increases and decreases in our power to persevere in being.

Our basic *conatus* to persevere in being leads us to seek to maximize our power so to persevere, and hence to seek to maximize those affects which involve an increase in that power. The more we succeed in maximizing those affects, the greater our chances of perseverance in being and the greater degree of psychological and emotional stability we will enjoy. Chief among the affective states that further our power and stability is love, which is defined as pleasure (i.e. an increase in our power to persevere) accompanied by the idea of an external cause. Maximization of love is *eo ipso* maximization of one's power to persevere.

There are, however, other affective states that run counter to those of pleasure and of love. Chief among those are hatred and fear. Emotions of hatred and of fear are manifestations of pain and are thus by definition reductions in our vital powers. Conflict and contention with others are, of course, major sources of hatred and fear.

If one can maximize the place that love holds in one's affective life, and if one can minimize conflict and contention with other people, one can to that extent increase one's power to persevere in being and decrease the disabling effects of fear and hatred. The resulting peace of mind (*acquiescentia animi* of a lesser, but not inconsiderable kind) could, in a Stoic sense, save one from emotional upheaval – from being "tossed about on the waves of passion" as Spinoza puts it in the *Ethics*.[23]

Spinoza's reply to mevrouw van der Spyck can plausibly be read as a prescription for attaining this peace of mind. She is discouraged from undertaking a search for a better religion, for such a search would only confuse and unsettle her. To "apply herself to piety" is to focus on the love of God, and although her mistaken imaginative notion of God will infect her love with unfortunate traces of fear, she may nonetheless enjoy some of the benefits of love of the divine. To "live a peaceable and quiet life" is to avoid unsettling public controversies and private conflicts that foster fear and hatred. In these ways, and to this extent, mevrouw van der Spyck might indeed be "saved in her religion".

Fear, hatred and the instability of emotional vacillation are indeed bad things. If mevrouw Van der Spyck can avoid these by devoting herself to piety and by living a peaceable and quiet life, then she will have been saved from sure evils. On this reading, Spinoza's veracity and sincerity need not be questioned, for what he says is simply true. If we were to have doubts about the subtle details of Spinoza's systematic psychology, an inductive appeal could be made to everyday experience. Peaceable and piously faithful people very often impress us with their solid, calm confidence and dig-

[23] I am claiming that this *acquiescentia animi* is of a lesser kind than that experienced by those who have love of God based on true knowledge of God. E. Harris directly opposes this view. He writes, on the one hand, of "...the blessedness and peace of mind that supervene upon the true knowledge of God and his attributes, with the appropriate conduct which that knowledge prompts". On the other hand there are those believers who, lacking knowledge, are inspired to right conduct by imaginational or passional motives. Harris says, of these two: "The conduct is the same in both cases (for behavior prompted by passion may coincide with genuine action) and so likewise is the peace of mind" (Harris 1978: 12). Harris is surely right (spinozistically) in saying that the conduct might be the same in both cases, but he gives no reason for thinking that the supervening peace of mind would be the same. It seems to me, on the contrary, that there is every reason to think that it would not be the same. *Acquiescentia animi* represents heightened power and activity, and true knowledge of God is entirely unrivaled (in intensity, permanence, and certainty) as a source and locus of such power and activity.

nity.[24] Of course Spinoza need not rely on such empirical appeals, since the efficacy of love and quiet piety in bringing affective stability and thus peace of mind follows necessarily from first principles in the *Ethics*.

But this very fact should give us pause. Perhaps this is the sort of salvation that Spinoza had in mind when he spoke with his landlady. But if it is, it must be sharply distinguished from the salvation that is promised to obedient believers in the *TTP*. As our foregoing discussion noted, the *TTP* emphasizes repeatedly that the fact that one can obtain salvation through the practice of justice and lovingkindness can*not* be demonstrated philosophically. If it *can* be demonstrated that mevrouw van der Spyck can achieve a certain peace of mind through piety and a peaceable life, then this peace of mind cannot be the salvation that Spinoza is addressing in the *TTP*.[25]

Conclusion

Did Spinoza lie to his landlady? Well, it seems to me that he certainly misled her in answer to her question. He led her to believe that she could be saved (in *her* sense of "saved"), whereas in fact she could not. Not only could she not be saved in *that* sense, but in the sense of "salvation" that is most important to Spinoza, she cannot be saved "in her religion" at all. Her religion will foster in her false ideas about God and thereby preclude her attaining to the accurate knowledge of God required for the intellectual love of God wherein salvation consists. And while Spinoza may be correct in claiming that the prophets proclaim "with one voice" that salvation can be had through obedience alone, if we take "salvation" in Spinoza's own favored sense, he has not given us any reason to believe that what the prophets thereby proclaim is *true*. His suggestion that this doctrine has given great comfort to the weak-minded hardly gives us grounds to believe it, and his attempt to provide inductive support for the reliability of the prophets fails as well. So, if we take the term "save" in mevrouw Van der Spyck's intended sense or in the primary sense in which Spinoza uses it, the correct answer to her question is "no", and the affirmative answer that he gave was deceptive.

---

[24] The importance of this point was made clear to me (in conversation and correspondence) by H. De Dijn and by U. Goldenbaum.

[25] Matheron makes this point very clearly in his discussion of "salvation in the weak sense" and "obedience in the weak sense" (Matheron 1971: 191).

Mevrouw Van der Spyck can achieve an impressive peace of mind through Lutheran piety and a quiet life – and in so doing she will be saved from some of the greatest pains and miseries of human life. This is not to be taken lightly, and would be an attainment worthy of respect and emulation. But this was not what she had in mind when she asked her tenant about her chances of being saved in her religion, and it is not what Spinoza has in mind when he addresses the subject of salvation in either the *Ethics* or the *TTP*. Spinoza led his landlady to believe something which he himself held to be false. I conclude that to that extent he lied to her.

Why might Spinoza have lied to mevrouw Van der Spyck? I think that in this specific case we can reject the Straussian suggestion that Spinoza misrepresented his views "when addressing the vulgar" chiefly in order to avoid persecution. As his landlady she did have some power over him, but it is difficult to imagine that Spinoza would lie to mevrouw Van der Spyck in order to ward off hostility on her part. It is more likely, I think, that his misleading words were an accommodation to her level of understanding and to the level of freedom and fulfillment that she could realistically be expected to achieve in this life. Assuming that she is not going to scale the cognitive heights required for the intellectual love of God, the peace of mind referred to in the last paragraph is the most that she will be able to attain. Since a life of quiet piety is indeed conducive to this peace of mind, it is in mevrouw Van der Spyck's best interest that she lead such a life. Since her notion of salvation is an imaginational idea of a sempiternal afterlife of heavenly reward, she can be encouraged to do that which is in fact in her best interest by being told (misleadingly) that it will lead her to enjoy salvation of the kind that she imagines. Such mendacious accommodation to her imaginational way of thinking and her real capacities and prospects is a lie, but certainly not a malicious one. He is patronizing mevrouw Van der Spyck, but he does not do so with evil intent.

Should we be disturbed by this conclusion? I am not disturbed by it, though it seems that Mssrs. Matheron and Harris would be.[26] I am reassured by the fact that Spinoza attributes to the Apostle Paul and even to Jesus a similar kind of patronizing accommodation to the beliefs and capacities of

[26] Matheron asserts, as a fundamental given ("donnée) of his interpretive inquiry that "...Spinoza, conformément à l'idée qu'il se fait du philosophe, s'interdit de mentir; ...il ne dit jamais ce qu'il ne pense pas". (Matheron 1971: 149) If Harris's critique of Strauss's views can be extended to this context, he seems to think that "our judgment of Spinoza's intellectual integrity" would be compromised (Harris 1978: 2).

ordinary people.[27] Jesus, we are told, understood the truth about the divine law – viz. that it is written in our minds and in the laws of nature, rather than given prescriptively as by a lawgiver.

„...Christ perceived the things truly and adequately. If he ever prescribed them as laws, he did this because of the people's ignorance and stubbornness. So in this respect he acted in the manner of God, because he accommodated himself to the mentality of the people" (*TTP* 4: 65).

The phrase "If he ever prescribed them as laws..." suggests a hesitancy on Spinoza's part to attribute to Jesus the propagation of views which He knew to be false. One should indeed be hesitant in making such attributions. But the rest of the passage makes it clear that in the end Spinoza does make that attribution and is willing to accept the necessity of adapting one's teaching to the level of understanding of one's audience. Indeed, the passage itself exemplifies the very practice in question. The talk of God "...accommodating himself to the mentality of the people" unavoidably suggests an intentional, purposive act on God's part – a decision by God to be accommodating in this way. But in fact we know that when Spinoza writes in this way he is accommodating his own message to the mentality of those who think of God as a purposive agent.

Spinoza wanted to encourage mevrouw Van der Spyck in her life of quiet piety, and he judged it best mendaciously to reassure her that such a life would lead to her salvation. Spinoza does not condemn Jesus for misleading His followers in His desire to communicate important moral truths to them. Nor, I think, should we condemn Spinoza.[28]

---

[27] H. Gildin presents an interesting and convincing discussion of this issue (Gildin 1980: 165-166).

[28] For various kinds of input at various stages of this paper's development, I am grateful to: Ed. Curley and the members of his 1992 NEH Summer Seminar; H. De Dijn; U. Goldenbaum; Ch. Huenemann; the participants in the Southeastern Seminar in Early Modern Philosophy in Atlanta, 1994; and my departmental colleagues at Rollins College.

BIBLIOGRAPHY

Colerus, Johannes: *The Life of Benedict de Spinosa.* The Hague: Martinus Nijhoff, 1906. (Reprint of 1706 edition/translation with spelling modernized).

Cook, J. Thomas: "Self-Knowledge as Self-Preservation?" In: *Spinoza and the Sciences*, ed. M. Grene and Debra Nails, Dordrecht: Reidel, 1986 (Boston Studies in the Philosophy of Science; 91): 191-210.

Curley, Edwin M.: "Homo Audax". *Studia leibnitiana*, forthcoming.

Gildin, Hilail: "Notes on Spinoza's Critique of Religion". In: *The Philosophy of Baruch Spinoza*, ed. Richard Kennington, Washington, D.C.: The Catholic University of America Press, 1980 (Studies in Philosophy and the History of Philosophy; 7): 155-171.

Harris, Errol E.: "Is There an Esoteric Doctrine in the *Tractatus Theologico-Politicus*?". Leiden: E. J. Brill, 1978. (*Mededelingen vanwege het Spinozahuis*; 38).

Matheron, Alexandre: *Le Christ et la salut des ignorants chez Spinoza.* Paris: Aubier, 1971.

Strauss, Leo: "How to Study Spinoza's Theologico-Political Treatise". In: *Persecution and the Art of Writing.* New York: The Free Press, 1952: 142-201.

[22]

Errol E. Harris

SPINOZA'S THEORY OF HUMAN IMMORTALITY

I

There is, perhaps, no great philosopher who presents us, with so much confidence and assurance as Spinoza does, with such stark contradictions so rigorously deduced from indubitable first principles. Our first reaction is the conviction that something must have gone wrong with the reasoning at some obscure point; but more careful examination of his system and his explicit statements reveal that there is no actual inconsistency and that the conflicts in his doctrines are only apparent. Let us first notice briefly what they are.

Spinoza rejects in no uncertain terms the existence of a god conceived, as he traditionally is in the major religions, as an almighty, all-wise, compassionate ruler and judge.[1] Small wonder that, in his day, Spinoza was deemed an atheist.[2] But, on the other hand, he asserts, as indubitable and self-evident, the existence of an infinite, eternal God, in whom all things exist and have their being, and without whom nothing can either be or be understood; God, who is the universal, immanent cause of all things, whether infinite and complete in their kind, or finite and derivative. Small wonder, again, that later commentators called Spinoza a 'God-intoxicated man'.

In Spinoza's God there is nothing contingent.[3] He is a necessary cause of all things and nothing that He does could He have left undone—though this is no limitation upon His power, rather the contrary. Everything that is and occurs is therefore necessarily and inevitably produced by God. We are left in no doubt that Spinoza is a determinist. Free will, he says, is an illusion entertained by men because they are aware of their desires but unaware of the causes of their desires. Yet, on the other hand, Spinoza devotes a fifth

---

[1] Cf. *Korte Verhandling van God, de Mensch, en deszelfs Welstand, Deel* I, Chap. VII. Hereinafter cited as KV.

[2] Cf. Hume, *A Treatise of Human Nature,* I, Part IV, Sec. 5.

[3] Cf. KV, *Deel* I, Chap. IV.

part of his greatest work to the subject of human freedom and ends his *Korte Verhandeling* with a chapter on '*De Waare Vryheid*'.

The mind or soul of man, he tells us, is a mode (or idea) in the Attribute of Thought (God's attribute), of which the object is a mode of Extension (another of God's attributes) and nothing else.[4] The soul is the idea of the body, and as the body is altered so is the mind affected. If the body is destroyed the mind is likewise.[5] Nevertheless, the mind can be united not only with its body, but also, through its adequate ideas, with the infinite essence of God; and, through the third kind of knowledge, *Scientia Intuitiva,* with its corollary, the intellectual love of God, the mind of man may be united with God Himself and His infinite intellect, thus it may become immortal (*onsterfelijk*) and eternal (*aeterna*).

These apparent conflicts in Spinoza's philosophy are well known and there can be no dispute about the statements which seem to involve them. What is less apparent and more disputable is whether they can be resolved in the terms Spinoza offers, and, if so, how. My purpose in this paper is not to discuss all of them in detail, but to consider only the last; to ask whether Spinoza's account of the relation between body and mind is, or can be so understood as to be, compatible with any notion of immortality for the soul; to consider his view of the body-mind relation in particular; to examine Spinoza's conception of immortality, and to decide in what sense, if any, this notion of immortality is acceptable.

II

There have been, in the past, two predominant ways of conceiving human immortality. The first is to regard human personality as consisting of two separable parts, each in some sense complete in itself, a material body and an immaterial soul. On the death of the body, the soul is then thought of as being released, as continuing to live for an indefinitely long time, as it had in the past, but in a disembodied state. The second way of regarding immortality is that adopted in the Christian creeds, which envisage a soul separable from the body, but consider the resurrection of the body essential for the preservation in an after-life of the individuality of each par-

[4] Cf. *Ethics,* II, Prop. XIII.

[5] Cf. KV, II, *Voorreeden,* footnote; *Ethics,* Prop. XIV, *et seq.*

ticular person. The soul here does not die, as the body does, but presumably remains after bodily death in some state of suspended existence until such time as it can be reunited with the resurrected body, now somehow rendered indestructible so that the person can live on forever.

Spinoza's theory, at least prima facie, seems incompatible with both of these conceptions. He maintains first that the soul is the idea (which constitutes the human mind) of which the object "is a body, or a certain mode of extension actually existing, and nothing else" (*Ethics,* II, Prop. XIII). If and when the body ceases to exist, therefore, it should follow that its idea ceases to exist likewise—except in the rather special sense that every mode of Extension and of Thought exists eternally in God. This latter existence, however, cannot be equated with human immortality in particular, because it applies to all finite entities without exception, whether they are modes of Extension or of Thought. The eternal existence of the modes in God's eternal and infinite attributes pertains to God. It is not the *continued* existence in time of any one of them beyond its allotted span, and thus by no means is it the continued existence of a human person enjoying successive experiences after the death of his body. It seems to follow, further from the Scholium to Proposition XIII, that body and mind are inseparable and co-terminous. There Spinoza asserts that the relation between them is no different from that between any other *idea* and its *ideatum:* "For of everything there must necessarily be given an idea in God, of which God is the cause, in the same manner as there is an idea of the human body." The only difference between man's mind and the ideas in God of other things is what follows from the difference between the human body and other bodies, the former being, for the most part, more complex. As the human body is more apt than others for doing and suffering many things at the same time, so the human mind is more apt to perceive many things at the same time.

Similarly, we are told in the *Short Treatise* that as the body suffers changes caused by other bodies, so these changes are felt by the soul as sensations (Part II, Introductory footnote) and also that the soul, being an idea of the existence of some thing (*Zaak*) in Nature, endures and changes according as that thing endures and changes (Part II, Chapter XXIII). But in both of these contexts Spinoza qualifies his statement in a way which he then uses to prove human immortality. To this qualification I shall return presently. What is to be noted here is that the doctrine of body-mind relationship is so conceived that the mind (the idea of the body) cannot be separated from the body and can exist only if and when the body exists.

It should follow that the immortality of the soul, if it is to be affirmed at all, cannot be conceived, as it was by Plato in the *Phaedo*, as the continued existence of a disembodied soul after the death and dissolution of the body.

Nor is the notion of a resurrection of the body anywhere suggested by Spinoza and it would seem impossible in his system. For the existence of each finite thing (according to him) is necessarily caused in the succession of finite causes which flow endlessly from God's infinite being; and no cause is mentioned, or seems to be contemplated, which could reconstitute exactly so complex a body as the human organism, once its dissolution has been brought about by other necessary causes. That such reconstitutive causes might be included in God's infinite reality is conceivable, but Spinoza never considers the possibility, nor is there anything in the nature of things as we experience them that requires us to assume the actuality of such causes. However that may be, Spinoza's account of human immortality does not include or require the resurrection of the body as a particular mode of extended Substance, though it does, as we shall see, require a special interpretation of the nature and status of individual bodies and a transformed idea of them.

So far, however, we can say that Spinoza's theory of body-mind relationship is in conflict with both the most usual traditional forms of the notion of immortality, and would seem to exclude the possibility of any kind of 'life after death'. Yet Spinoza, in the Scholium to Proposition XX of *Ethics*, V, says: "And with these [demonstrations] I have dealt with everything that concerns this present life. . . . It is, therefore, time now that I should pass to those matters which pertain to the duration of the mind without relation to the body." And in Proposition XXIII he declares: "It is not possible for the human mind to be absolutely destroyed with the body, but something of it remains which is eternal." The first statement is significant, not simply because no existence of the mind without relation to the body should be possible consistently with what we have been told earlier, but also because reference is made to it specifically as 'duration'. It would, however, be quite consistent with this statement subsequently to deny that the mind endures at all without relation to the body; and this is precisely what Spinoza does say in the Scholium to Proposition XXIII: "Our mind . . . can be said to endure, and its existence can be defined by a certain time, only so far as it involves the actual existence of the body." Nevertheless, he maintains in the same place that "we feel and know (*experimur*) ourselves to be eternal." How is all this to be understood so that it is self-consistent?

III

Most commentators resolve the difficulties first by contending that the reference to 'this present life' in the Scholium to Proposition XX, in Part V of the *Ethics,* is a concession to ordinary parlance or a 'momentary slip',[6] and to the mind's duration without relation to the body as a use of 'duration' different from that in which he distinguished it from 'eternity'. Joachim explains[7] that Spinoza is not quite consistent in his use of this word. While, as a rule, he refuses to predicate duration of what is eternal and identifies the former with persistence in time, he also regards duration as that of which time is the measure. Time results from the subdivision of duration into periods. But this subdivision is the work of *Imaginatio,* for strictly duration is indivisible (like Extension generally);[8] and when adequately conceived all its 'parts' are seen as coeval. It then becomes identical with eternity, or, as Joachim puts it, is "the general term of which eternal existence and temporal existence are forms."[9]

Spinoza's otherwise clear distinction between duration and eternity can then be taken seriously, and his attribution of 'immortality' (a term he uses himself only once in the *Ethics*) to the human mind is explained as the identity of its adequate ideas with the eternal

[6] Cf. H. H. Joachim, *A Study of the Ethics of Spinoza* (Oxford: Oxford University Press, 1901), p. 296. Joachim, nevertheless, points out that for Spinoza 'this present life' would naturally mean 'our life so far as we are imaginative', i.e. the life of *Imaginatio,* associative thinking, passion and illusion.

[7] *Op. cit.,* pp. 294 ff.

[8] Cf. KV, I, Chap. II.

[9] *Loc. cit.*

and infinite intellect of God. The adequate ideas in human knowledge are thus eternal. In *Scientia Intuitiva* man achieves an adequate knowledge of the essence of things through an adequate idea of the formal essence of certain of God's attributes,[10] and such adequate ideas as constitute this third kind of knowledge (as well as such adequate ideas as constitute the second, scientific kind—*Ratio*) are eternal and timeless. "In eternity there is no *when, before* nor *after*." (*Ethics,* I, Prop. XXXIII, Sch. 2). And as the ideas in which these forms of knowledge consist are part of the complex idea which is the human mind, there must be some part of it 'that remains eternal' when the human body dies and passes away. The use of the word 'remains' here is again regarded as a slip or a concession to ordinary language. It is repeated by Spinoza in subsequent propositions, but there is no need to understand it otherwise than as meaning that there is something eternal in the human mind besides what ceases to 'endure' when its body dies (as we say in arithmetic: $15 \div 6 = 2$, remainder 3). The eternity of the 'immortal' part of the human mind or soul is thus not a continued duration after the death of the body, but a quality of being.

What corresponds in extension to the eternal part of the human mind is then not the human body as a finite mode merely, or as imagined in the first kind of knowledge (*Imaginatio*), but the human body as adequately understood in the second and third kinds of knowledge. As so understood it is an eternal consequence of God's infinite, immanent causality, and must be conceived in its proper place in the total scheme of things, through an adequate idea of the formal essence of (presumably) the attribute of Extension. Inasmuch as adequate ideas are identical with God's, their counterpart in extension will not be any merely finite mode but the whole extended world—*facies totius universi,* for all modes of Extension are mutually connected in the infinite series of causes which flow of necessity from God's nature; and Spinoza explains at some length how simple bodies become included in more complex bodies in a hierarchy that eventually includes the whole of nature.[11]

---

[10] *Ethics,* II, Prop. XL, Sch. 2.

[11] Cf. *Ethics,* II, Lem. VII, Sch.: *"Et si sic porro in infinitum pergamus, facile concipiemus, totam Naturam unum esse Individuum, cujus partes, hoc est omnia corpora, infinitis modis variant, absque ulla totius individui mutatione."* "And if we proceed still further to infinity, we can easily conceive that the whole of

This interpretation of Spinoza's doctrine, or something very similar to it, is accepted by all the best known commentators. Joachim expounds it with admirable clarity; Leon Roth and Ruth Saw do not depart from it; Stuart Hampshire adheres to it; and much the same view, with minor variations, is taken by John Caird and A. E. Taylor.[12] Indeed, there is a great deal in the Spinoza text to support it. However, I shall not discuss the views of these commentators individually here, nor shall I consider at this point the implications for morality and religion of the conception of human immortality involved. It would be better first to discuss the conception itself, for it leaves difficulties unresolved and significant implications in Spinoza's theory undeveloped.

IV

In the first place, with all due respect to Joachim, as one of the most accurate and careful scholars who has commented on the *Ethics*, it is not wholly correct to attribute the idea of the human body as a finite mode of Extension in interaction with other finite modes, merely to the confusion of *Imaginatio*. "The modes are not 'parts' of Substance," writes Joachim. "The oneness of the modes in God is more intimate than the oneness of parts in a whole. It is the separation of the modes from God (as if they were 'parts' of a whole) which causes the inadequate understanding of the imaginative consciousness, for which Reality becomes a world of finite things."[13] Joachim is certainly right to insist that for Spinoza the finite modes are not separable from God, as material 'parts' may be imagined as separable from a material whole—that conception of whole and parts is certainly inadequate to Spinoza's Substance. And Joachim, too, is fully aware that the unity of the modes in Substance does not preclude or cancel out their multi-

nature is one individual whose parts, that is, all bodies, vary in infinite ways without any change of the individual as a whole."

[12] Cf. H. H. Joachim, *op. cit.*, Book III, Chap. IV; Leon Roth, *Spinoza* (London, 1945), pp. 140-63; Ruth Saw, *The Vindication of Metaphysics* (London, 1951), pp. 128-36; Stuart Hampshire, *Spinoza* (London, 1946), pp. 126-32; John Caird, *Spinoza* (London, 1910), Chap. XVI; A. E. Taylor, "Spinoza's Conception of Immortality," *Mind*, 5, No. 18, New Series.

[13] *Op. cit.*, p. 299.

plicity.[14] But if it is an imaginative error to think of the modes as separable parts, it is not simply an error to think of them as finite and as distinguishable within the unity of Substance. (Joachim, of course, is not suggesting that it is—but *we* should not fall into the error of thinking it so.) And though it may well be only for imaginative consciousness that Reality is *merely* a world of finite things, there really is in a legitimate sense a world of finite things contained in the infinite reality of Substance.

The finite modes really are finite insofar as they are determined by other finite modes and in whatever other ways they may be determined (e.g. by their specific attribute). And these determinations follow necessarily from God's essence and are produced, as are all changes and events in time, by His immanent causation. It follows that the finiteness of man's body is not a figment of the imagination or simply a misconception, nor are the facts of its coming into existence at his birth and passing away at his death. The same applies to his mind as the idea of his (finite) body. It is really finite and its limitations follow necessarily from the nature of things. His finiteness is no misconception, though we can (if only for that reason) entertain misconceptions about it. We cannot, therefore, simply dismiss the notion of the body as a finite mode, nor of the mind as dependent on, or correlative to, it, as inadequate ideas typical of *Imaginatio*. The finite body and the idea of it, which according to Spinoza constitutes the human mind, are actual factors in the nature of Substance. They are parts of Nature, and, even though their essence cannot be adequately understood in isolation from the rest of Nature, and can be adequately understood only in their essential and inseparable relation to the other parts and to Nature as a whole, yet they cannot be dissolved away into Substance as a single, seamless, undifferentiated unity excluding finiteness and temporal change. The character of the finite, as finite, remains in contrast and in opposition to the eternal and infinite, and the problem of reconciling them in the divine nature cannot be sidestepped.

Secondly, it does not help us to understand what Spinoza means by human immortality to assert that the idea of the finite human body exists eternally in God and follows necessarily from His

[14] *Op. cit.*, p. 300.

essence, just as the existence of the body itself as a finite mode of Extension follows necessarily from God's infinite essence. For this is true of all finite bodies and all the ideas of them. They are all eternally in God's infinite understanding. If this sort of eternal being is all that is involved in man's immortality he is no more immortal than a fish and rather less so than a piece of granite.

Similarly, it does not help to insist that an adequate idea of the body involves its relations to all other bodies and so involves an adequate idea of the whole of Nature; and that such an adequate idea of Nature would be itself an infinite and eternal idea. Because such an infinite and eternal idea would be the intellect of God and not of man, which, for all his accomplishments, remains finite. True, Spinoza does assert, and with reason on his principles, that insofar as man achieves adequate knowledge his mind becomes united with that of God; but this is surely only to a limited extent, and it is not easy to understand how one can enjoy immortality only to a limited extent, or be eternal only in a limited sense. This surely would involve us in flat self-contradiction. Yet, as Spinoza's very philosophy exemplifies, according to its own principles, man does frame more or less adequate ideas of Nature and of the essence of God. Again we are faced with the problem: How are we to reconcile the finite with the infinite?—and Spinoza's doctrine of human immortality is very largely his answer to the question.

An objector may point out that the difficulty I have raised is spurious, for Spinoza repeatedly asserts that things vary in degree according to the amount of reality their essence includes, and the essence of the human mind is intelligence, and it embraces more of reality the more it conceives things *sub specie aeternitatis*. It could therefore be eternal in a degree, or to the extent to which its ideas were adequate, and to that extent it would be immortal. In short, as Plato and Aristotle alleged, not the whole human soul, but only its thinking part is immortal—in Spinoza's terminology the part that consists of adequate ideas.

But now we are involved in another difficulty (or perhaps the same one over again, seen from a new angle). The personality of man is a single whole and does not consist of separable parts, even though it includes many distinguishable functions and traits. The category of whole and part is inappropriate to the mind, just as Joachim explained it to be inadequate to express the nature of

Spinoza's Substance. We cannot, therefore, intelligibly maintain that the human soul is immortal only in bits and pieces, and that other parts die with the body. If it lives at all it must live as a whole. Thus though it is legitimate to hold that we can attain to the truth or to goodness in varying degrees, it hardly seems proper to maintain that we can, *qua* individuals, enjoy eternal life only in some degree of eternity inferior to the eternity of Substance, or only in certain parts of our conscious being.

V

How far is it possible to extract from Spinoza's writings a solution to this problem? The first clue to follow is the one suggested in the previous paragraph. Spinoza has told us that the human mind is the idea of the body; and in Epistle XXXII he says that "the human body is a part of Nature" and "the human mind . . . also is a part of Nature: since I maintain that there exists in Nature an infinite power of thinking, which, so far as it is infinite contains in itself objectively the whole of Nature and whose thoughts proceed in the same way as Nature which, to be sure, is its *ideatum.*" This power of thinking is undoubtedly the intellect of God—a system of ideal modes, or ideas, corresponding to the extended modes, or bodies, which constitute the physical world. None of these is a mere collection or congeries of separate or separable items (though, the modes in each attribute are distinguishable and ordered as a system of interrelated terms). The world of extended things, we learn from this letter, as from the *Korte Verhandling* (and elsewhere), is an indivisible whole, and the order and connection of things is the same as the order and connection of ideas.

In the same letter that I have quoted above Spinoza goes on to say "the human mind is the same power [of thinking] not insofar as it is infinite and perceives the whole of Nature, but insofar as it is finite and perceives only the human body, and in this way I declare that the human mind is part of a certain infinite intellect." But the contents of the divine intellect, as follows both from what Spinoza says in this epistle and from his general position stated repeatedly in various works, is an indivisible and systematic whole; and from what he says further about ideas it becomes apparent that any adequate idea belonging to this system will, in a significant sense, comprehend the whole of it.

For ideas are not static pictures or replicas, in some different medium, of the extended things; they are their 'minds' or consciousness, just as the idea of the human body is its mind and consciousness. But the idea of a relatively simple body, like a stone or a coprolite, taken simply in and by itself, is so rudimentary and inadequate that it hardly merits the name 'mind' at all, though Spinoza is prepared to concede it in some degree.[15] The human body, however, is far more complex and is "apt to do and suffer many things," hence its mind (or idea) is equally complex and is highly developed.

Sir Frederick Pollock complains that Spinoza uses the word 'idea' ambiguously [16] to mean both concept (our more usual sense of the word) and also the counterpart of a physical thing or physiological process (its *'ideatum'*), and that he fails to distinguish between them. But I believe Spinoza did this deliberately and committed no inadvertent confusion. An idea in the sense of 'concept' is abstract, and can be entertained only by a developed mind such as man's. It is one of the complex products (or 'parts') of the 'idea' (in the other sense) of a complex body such as the human organism. And Spinoza did not hold that abstract and general ideas corresponded to anything actually existent. He was no 'realist' (like Plato), in the Mediaeval sense of that word, but was nominalist in his doctrine of general ideas. He regarded them as no more than convenient devices for description and classification—part of a technique or activity of thinking peculiar to men. Such concepts are possible, nevertheless, only for a conscious being, and consciousness is always 'idea' in the other sense—the idea of the body. This we should understand as the direct awareness of the body such as we have in sensation, which is the origin and basis of all 'higher' forms of consciousness. I shall return to this point in a moment; here let us note that concepts and general 'ideas' are part of and are derivative from this more basic consciousness, a fact which Spinoza never forgets. This is why he refuses to distinguish between the two senses of 'idea' and treats one simply as a special form or elaboration of the other.

We must not, however, as I said above, think of ideas or minds

---

[15] *Ethics,* II, Prop. XIII, Sch.: "Individual things . . . are all, though in varying degrees, beminded (*animata*). For of each thing there is of necessity an idea in God . . ."

[16] Cf. *Spinoza, His Life and Philosophy* (London, 1912), Chaps. VI and IX.

as mere reproductions, in the attribute of Thought, of the essences of their corresponding bodies. They are awarenesses, self-conscious and self-illuminating. Every idea implies an *idea ideae*; every knowing a knowledge that one knows, and though each such *idea* is, according to Spinoza, formally distinct from every other (as an act of consciousness), the *idea* and the *idea ideae* are 'objectively' identical and are both substantially one and the same with their (physical) *ideatum*.[17] Each is a mode under a different attribute of the same identical substance. The nature of this dual unity we shall consider more closely below, but first I wish to draw attention to the fact that consciousness, the basic sense of Spinoza's *idea*, even at its most primitive level, is never, *qua* consciousness, confined or limitable to a bare particular of any sort. Its most elementary object is and must be at least a particular distinguished from a background, or set in a context. An entity's idea or consciousness of itself must and can only be its awareness of itself in distinction from something else, and so must any awareness of anything whatsoever. Consciousness, therefore, is always not merely self-consciousness (*idea ideatum*) but is also self-transcendent, and tends to comprehension of some whole to which its object belongs. Spinoza never makes this point explicitly but I am sure he was fully aware of it, and if it is taken along with what I have cited above about the wholeness and indivisibility in essence of Nature, of Extension, of the intellect of God, and with what follows from that, very significant consequences ensue for the human mind which will illuminate Spinoza's doctrine of its capacity for eternity. Let us now return to the detail of his view of the relation of body and mind.

First, our ideas are the awareness of our bodies and nothing else. We feel and are conscious of the activities, passivities, and functionings of our bodies, and of other bodies only as and through our own (*Ethics*, II, Props. XIX-XXVI). In becoming conscious of external things we are commonly unaware (because we suppress the awareness) of the mediation of our own sense organs ("the eye sees not itself"), but that mediation is nonetheless an essential condition—as it were, the matrix—of such awareness. Our awareness of the world is thus at the same time our awareness of the body, and it is only by making distinctions, in an appropriate and systematic way,

[17] *Ethics*, II, Prop. VII, Sch., Prop. XXI, Sch., and Prop. XLIII.

within this total self-awareness, that we distinguish what belongs to our own bodies from what pertains to others. According to Spinoza, we make these distinctions well or ill according as we think adequately or inadequately; and we do the former insofar as we are active (and vice versa), and the latter so far as we are passive—so far as our body suffers effects from interaction with other bodies. Though Spinoza writes as if action depended only on adequate ideas, and as if inadequate ideas resulted only from bodily passivity, to understand him aright we must realize that activity and passivity are each one and the same in both body and mind.

Next we must notice that the body is strictly and in fact related in some way to everything else in the extended world, and all these relations must, therefore, be registered in idea either as passions or otherwise. In *Imaginatio* ideas are confused and *in se* false but they are nevertheless consequences of the impingement of external things upon the body. Thus even so far as its confused ideas are concerned, the mind embraces and reflects the whole of Nature.

But each thing endeavours, so far as it is in itself, to persevere in its own being (*Ethics,* III, Prop. VI), and this *conatus* is the actual essence of the thing itself (*Ibid.,* Prop. VII) and involves not finite but indefinite time (*Ibid.,* Prop. VIII). The mind, both in its confused and in its adequate ideas, is aware of this *conatus* (*Ibid.,* Prop. IX), and the *conatus* is nothing less, in the last resort, than the power of God Himself (cf. *Ethics,* IV, Prop. IV, dem. and I, Prop. XXIV, Cor.). It is by virtue of the *conatus* that the mind advances from *Imaginatio* and confused ideas to *Ratio* and adequate ideas and thence to *Scientia Intuitiva.*

The human mind, therefore, as the idea of the body embraces within its consciousness all the affects of the body and all its relations with the rest of the world, and so is all-inclusive even in its passions and its confused ideas. Its *conatus* is towards action and so impels it to develop (in ways the detail of which we need not here rehearse) [18] towards the perfection of its own being in the intellectual love of God. The result of this development is not just the supplementation of confused ideas by a new set of adequate ideas; it is the transformation of the entire personality on its emotional and

[18] See *Ethics,* V, Props. I-XVII and XXVIII.

practical side, as well as in its theoretical content. Adequate knowledge is not simply part of the mind beside and separable from inadequate ideas still harboured within it. Adequate knowledge is an order of activity, different from the passivity of *Imaginatio,* which annuls all cognate confused ideas, transforms the passions into healthy, positive and beneficient emotions, and divisive and turbulent motives of action into love and compassion. All this, moreover, is the product of the *conatus in suo esse perseverare* of the individual, which is the very power of God (what in Christian doctrine would be called the Holy Spirit) working within him.

This transformation is of mind and body in one. Spinoza quite explicitly says that "he who has a body capable of many things, has a mind of which the greater part is eternal" (*Ethics,* V, Prop. XXXIX), having already stated that the attainment of the third kind of knowledge implies the mind's knowledge of itself and of its body *sub specie aeternitatis* (*Ibid.,* Prop. XXX). This knowledge of the body is surely no mere way of viewing it, as might be some inadequate imagination of it. It must be a conception of the body as it really is in Nature, the more so as the initial feeling of the body in *Imaginatio* was already a confused awareness of the *facies totius universi,* which has become progressively clarified into a true and adequate knowledge of God.

The body is thus revealed as the vehicle of God's own self-revelation in and through the mind of man. The power of God, causing the infinite system of modes which is Nature, and working immanently throughout that system produces man's body, as it produces all others, and *pari passu* produces man's mind, the idea of his body, which it then urges through a process of internal development from *Imaginatio* to *Scientia Intuitiva* and to a revelation of His own infinite and eternal nature, the supreme object of perfect and unadulterable love. It is the love of God Himself, by which in one and the same act He loves Himself and His creation, including man. In short, it is a union or self-identification of man with God.

This knowledge and love of God, the eternal and infinite Substance, transforms all man's emotions and actions, and constitutes the life of blessedness, which is the highest virtue, the final object of rational desire, as well as the completest and most adequate knowledge of Nature. This is man's immortality, and its special character follows from the nature of self-consciousness, which, as was said above, is no inert replica of the extended world, but is an active self-awareness of the body, that embraces in its purview and comprehension all its relations to other bodies and the infinite, eternal character of the essence of Substance to which it belongs.[19] Consciousness is self-transcendent. It cannot exist simply in space and time and at the same time be aware, as it is, of the interrelations of all parts of space and time. It cannot be limited within a restricted space or period, and also know the relations of that restricted space and period to the rest of the extended world. In order to enjoy that knowledge it must somehow be identical with its object(s), as it is impossible for the body as a finite mode of Extension to be identical with its causes and effects. Although the body is related to all other bodies and registers within itself their effects, it can do so only if it is distinguished from them as the mind cannot be distinguished from them if it is to be conscious of them. Spinoza insists that *idea* and *ideatum* are identical, but a finite mode of Extension cannot be similarly identical with other finite extended modes with which it is in spatio-temporal and causal relations. The mind, on the other hand, in knowing these relations and their systematic integration into one individual as the whole of Nature (*totius facies universi*), transcends all spatial and temporal limits and is one with the whole eternal structure.

Time and space, the coming to be and passing away of finite things in the material world, are then seen as partial features or aspects of a single, indivisible, infinite and absolute totality—the eternal being of God or Substance. They are not unreal, for they do proceed necessarily from the divine essence. They are actual elements within the reality of Nature and do constitute a real and necessary attribute of God. Only for the imagination are they merely fleeting episodes of ephemeral significance. Their finiteness is not illusory, for their mutual determinations are essential to the multiplex unity of Substance. But their existence in itself is not of ultimate significance, for it can neither be nor be conceived except through the infinite being of God, in which they all live and move. To understand the world in this way is to conceive things adequately and

[19] "Proceeds from an adequate idea of the formal essence of certain attributes of God to the adequate knowledge of the essence of things." *Ethics,* II, Prop. XL, Sch. 2.

is the part of *Scientia Intuitiva*; and any mind that has perfected its knowledge to this degree has transcended the finite nature of its body, and the transient existence of things in space and time in a synoptic awareness of "all time and all existence."

It follows, as Spinoza says, that "the human mind cannot be absolutely destroyed with the body, but there is something besides (*aliquid remanet*) which is eternal." It transcends, in its consciousness, the temporal as well as the spatial limits of the body, while yet, paradoxically, being identical in substance with the body, as its idea. It is its ideal, its conscious, character that is transcendent, and this transcendence is typical of all consciousness, even of *Imaginatio*. There is a valid sense, therefore, in which every mind is in some degree eternal. All consciousness is in some degree transcendent of its immediate object in time and space; and to the extent that it is not limited, as its object is limited, it participates (if we may, like Plato, use a term which is not wholly appropriate) in eternity. For to be aware of spatio-temporal relations is, of necessity, not to be merely one term in any such relation. To be conscious of space and time is to transcend space and time—to be eternal. To be adequately conscious of the whole structure of Nature as it really is, to know it as God knows it, would be to be united with God and to share (if, again, so inappropriate a word may be permitted) in His eternal being.

Obviously this eternal reality cannot be an extended duration. If this is the nature of man's immortality it cannot possibly be an "after-life" or temporal existence prolonged beyond the temporal existence of the body. For the mind transcends the body just because its consciousness is *not* in time and so far as it does *not* endure. It transcends time and space in the sense that time and space are *for* it and it is not in them. Consciousness embraces time and space as orders or wholes, and so is aware of the place within these orders of the body, of which the human mind is the conscious life. The human mind is thus both finite and potentially infinite, both the idea of a finite mode of Extension (and thus itself a finite mode of Thought) and, nevertheless, in being idea, capable of adequate knowledge of the total scheme of things. It is thereby self-transcendent and eternal.

The objection may be raised that this conception of immortality dissolves away human individuality, which, in accordance with this view, cannot survive the body. What 'remains', if we follow Spinoza, is the impersonal eternal essence of Substance with which the mind in *Scientia Intuitiva* has become identified. This must inevitably be the same for all minds, for there is only one infinite being, as Spinoza spares no effort to demonstrate; and to become identified with it must be to become wholly absorbed into it and as a distinct individual to be obliterated by it.

This objection, however, is misconceived. Immortality, we have seen, is not a matter of the "survival" of his body by any personality. Nothing "remains" in the sense of continuing in time. What remains when we have accounted for the mortal form of experience is an eternal awareness of God's essence and the complex system of the world. It is an awareness enjoyed by each and every individual who attains to adequate knowledge, and it transforms and perfects his personality by developing it to its fullest moral capacity. So far from being lost or swallowed up in the boundless ocean of Substance, the individual personality becomes whole, internally harmonious and perfectly self-determined. The limitations of bodily life are clearly understood in their relation to the rest of Nature—nothing is blurred or obliterated—but these limitations are no restriction to the mind's self-awareness as a necessary and intimate pulse in the total life of the universe. Temporal transience is not felt as an irremediable handicap or an inescapable confine.

All that Spinoza writes in the final propositions of the *Ethics* about human blessedness and the eternal nature of man's mind follows from what has here been set out. "The greatest endeavour of the mind and its greatest virtue is to understand things by the third kind of knowledge." This kind of knowledge gives the greatest satisfaction. It is the highest good and to enjoy it (blessedness) is its own reward. It would be ridiculous to imagine that any gratification of the appetites or indulgence of lesser desires, for however long a period, could be more satisfactory and could thus serve as a reward of virtue. Equally absurd is the assumption that men can become truly virtuous through fear of eternal punishment. Nothing is more ludicrous than to speak as if virtue were a hardship to be endured for the sake of carnal pleasures promised later, or to escape greater pains threatened in an everlasting after-life.

Further, a mind that understands the eternal nature of things and its own place in the total scheme, which thus enjoys God's knowledge of Himself and is eternal in the knowledge of His eternal essence, will be unconcerned about the temporal limits of the body. Its concern for the life of the body will be to understand its place in nature adequately and truly, and to act accordingly. Its longevity or otherwise will therefore become of minor importance. The actual time and eventuality of its death will have little significance, once the third kind of knowledge and the intellectual love of God have been attained. "So far as human bodies are capable of many [acts], there is no doubt but that their natures can be referred to minds which have a great awareness (*cognitio*) of themselves and of God, and of which the greater part and the most important (*praecipua*) is eternal, and that therefore they should scarcely fear death." (*Ethics,* V, Prop. XXXIX, Sch.) For death will have been transcended in the sublime contemplation of the totality of being—the infinite and eternal reality of God.

This interpretation of Spinoza's theory does not equate human immortality with the idea, eternal in God's intellect, of the finite mode of Extension which is the human body. Nor does it identify, as the 'body' of that part of the human mind which is eternal, the whole face of nature conceived *sub specie aeternitatis.* This indeed is the *ideatum* of the eternal idea—or, at least, the *ideatum* is nothing less than this. But for man it is so only through the registration in the finite human body of the effects and relations to that body of the rest of Nature. Immortality, then, while being no extended duration beyond the temporal life of the body, consists in the mind's transcendence of the body's finite limits, because of the inherently transcendent character of consciousness—a transcendence characteristic of idea, as such. The accepted interpretation of Spinoza seems to me to overlook, or to fail to stress sufficiently, this element of transcendence in the nature of consciousness. Thus it fails to draw out its implications, and so leaves unresolved the difficulties which I outlined in Section III.

ERROL E. HARRIS

NORTHWESTERN UNIVERSITY

[23]

CANADIAN JOURNAL OF PHILOSOPHY
Volume XI, Number 1, March 1981

Spinoza's Theory of the Eternity of the Mind

DIANE STEINBERG, University of Illinois, Urbana-Champaign

In part I of this paper I argue that on his theory of the mind as the idea of an actually existing body Spinoza is unable to account for the ability of the mind to have adequate knowledge, and I suggest that his theory of the eternity of the mind can be viewed as his solution to this problem.[1] In part II I deal with the question of the meaning of 'eternity' in Spinoza (i.e., whether eternity involves timelessness or not, and why), in regard both to God and the human mind, and I sketch a line of thought which I believe may have provided him with further motivation for his theory that a part of the mind is eternal.

---

1 The material in this paper is based on a portion of the author's doctoral dissertation, "Spinoza's Theory of the Mind," University of Illinois at Urbana-Champaign, 1977. An earlier version of it was presented at a colloquium at the University of Toledo in November, 1977.

Diane Steinberg

I

According to Spinoza the whole of reality is properly conceived as a single absolutely infinite substance (God) consisting of an infinite number of attributes, each one of which is infinite in its kind (unlimited by anything else of a similar nature). The attributes constitute the ultimate divisions of reality and each one must be conceived through itself. Thought and extension are the only two attributes known to man. What Spinoza called the infinite modes are held by him to follow necessarily from each attribute considered absolutely or as such; these are the universal features of reality conceived under a given attribute, for example, motion and rest in the case of extension. Each attribute is also modified by an infinite number of finite modifications — individual things as we know them. These do not follow from the attribute considered absolutely, but each one has for its cause another finite mode, which in turn has for its cause another, and so on *ad infinitum*. Because the attribtues are really distinct (each one is conceived through itself), the causal chain in each is complete; that is, there is no interaction between modes of different attributes. There is, however, a correspondence between the modes of different attributes, which Spinoza expounded in E, II, 7, when he wrote that 'The order and connection of ideas is the same as the order and connection of things.'[2] In other words, for every extended thing there exists in God an idea of that thing, or as Spinoza put it in the corollary to E II, 7, 'God's power of thinking is equal to his actual power of acting; that is to say, whatever follows *formally* from the infinite nature of God, follows from the idea of God [*idea Dei*] in the same order and in the same connection *objectively* in God.'

According to the presentation of Book II of the *Ethics*, the human mind is the finite mode of thought or idea which corresponds to a certain mode of extension, the human body, and mental functions and events are understood as being the parallel of bodily processes and events. To perceive a pain in one's finger is to have the idea (mental event) in one's mind which corresponds to a certain bodily

2 In this paper references to the *Ethics* are indicated by 'E' followed by a capital Roman numeral designating the appropriate part of that work. Following this, an Arabic numeral standing alone designates a proposition, 'dem.' stands for 'demonstration,' 'dfn.' for "definition,' 'ax.' for 'axiom,' 'cor.' for 'corollary,' 'lem.' for 'lemma,' 'post.' for 'postulate,' and 'sch.' for 'scholium.' All references are to Benedict de Spinoza, *Ethic*, trans. W. Hale White, trans. rev. Amelia Hutchison Stirling, 4th ed., rev. (London: Oxford University Press, 1927). All Latin quotations are from the *Ethica* in *Spinoza Opera*, ed. Carl Gebhardt, 4 vols. (Heidelburg: Carl Winter, 1925), 2.

modification, namely, an injury to the finger. The occurrence of such a perception in a mind *m* which corresponds to body *b*, is explained by the perceived event's occurrence in *b*, in conjunction with E II, 7. The latter proposition thus allows Spinoza to avoid Cartesian interactionism.

E II, 7, along with its explication in its corollary and scholium, has two important functions in Spinoza's metaphysics. In the first place it explains what Spinoza called the 'union of mind and body,' as we have shown above. In the second place it helps to bolster Spinoza's rather precarious monism; it explains the unity of the attributes in God. Spinoza's conception of reality as a single substance consisting of infinite attributes is problematic because, since each of the attributes must be conceived through itself, that is, as really distinct from every other, it is difficult to see in what the unity of the absolutely infinite substance consists. Spinoza addressed himself specifically to this problem in the scholium to E II, 7, when he wrote

> Before we go any farther, we must here recall to our memory what we have already demonstrated, that everything which can be perceived by the infinite intellect as constituting the essence of substance pertains entirely to the one sole substance only, and consequently that substance thinking and substance extended are one and the same substance, which is now comprehended under this attribute and now under that. Thus, also, a mode of extension and the idea of that mode are one and the same thing expressed in two different ways . . .. For example, the circle existing in nature and the idea that is in God of an existing circle are one and the same thing, which is manifested through different attributes; and, therefore, whether we think of nature under the attribute of extension, or under the attribute of thought, or under any other attribute whatever, we shall discover one and the same order, or one and the same connection of causes; that is to say, in every case the same sequence of things.

The two functions of E II, 7 in Spinoza's metaphysics are mutually supportive: insofar as that proposition is taken to explain the relation of the attributes to one another, and of the modes of one attribute to those of another, it provides the framework within which Spinoza can explain the nature of the mind and its relation to the body; and, insofar as it explains the union of mind and body, it provides him with a concrete illustration of what it means to say that a mode of one attribute is identical with or the same as a mode of another.[3]

---

3 I have characterized the relation which Spinoza holds to obtain between a thing and the idea in God of the thing (the relation between modes of different attributes which is dealt with in E II, 7 and its scholium) as correspondence rather than identity (the term by which Spinoza designates that relation in E II, 7 and its corollary and scholium) because to me the claim that two things which must be conceived as really distinct are strictly identical is obscure. Nothing, however,

Diane Steinberg

It should be evident now that Spinoza's assertion in Book V of the *Ethics* that 'The human mind cannot be absolutely destroyed with the body, but something of it remains which is eternal,' is problematic because it appears to contradict E II, 7. The human mind is the idea of the human body; if it is such that it is not destroyed with the body, then it would appear to be false that the order and connection of ideas is the same as the order and connection of things. At best the mind itself becomes an exception to the thesis which was intended to explain its nature; and the monistic thesis is rendered less intelligible insofar as E II, 7 is weakened.

With these difficulties in mind, many commentators have tried to explain Spinoza's attribution of eternity to the mind in such a way as to reconcile it with E II, 7. Errol Harris interprets Spinoza on this point as meaning simply that the mind *qua* thought has the ability to transcend the barriers of space and time.[4] Similarly, Joachim and Pollock hold that the eternity of the mind consists not in unlimited continued survival after the destruction of the body, but rather in its actual knowing of itself in a special way, namely, as a necessary part of the whole order of nature.[5] Since E II, 7 requires only that for every mode of extension there exist a corresponding ('identical') mode of thought, and *vice versa*, not that the corresponding modes of thought and extension have identical properties, to say therefore that the mind possesses a certain sort of knowledge while the body does not is no more problematic than to say that the body possesses a certain degree of motion while the mind does not. What is wrong with solutions of this sort, however, is that to say that the mind is able to transcend in thought the barriers of space and time, or to know itself as a necessary consequence of the divine nature, does not imply that the mind 'cannot be absolutely destroyed with the body, but something of it remains which is eternal' (E V, 23).

---

which follows in this paper turns on this interpretation, and I intend my use of 'correspondence' here to be taken in such a way as not to preclude the possibility that corresponding modes of distinct attributes might somehow be understood as identical in the strict sense. That is, if a thing and the idea of that thing in God be strictly identical, then they still may be said to correspond, although not *vice-versa*.

4 Errol E. Harris, "Spinoza's Theory of Human Immortality," *Monist*, 55 (1971), pp. 668-85.

5 Harold H. Joachim, *A Study of the Ethics of Spinoza* (1901; rpt. New York: Russell & Russell, 1964), pp. 308-9; Frederick Pollock, *Spinoza: His Life and Philosophy* (1880; rpt. Dubuque: Wm. C. Brown Reprint Library, n. d.), pp. 293-96.

In contrast to these and most other commentators who have interpreted E II, 7 as requiring Spinoza to hold 'the same' for the body as the mind, and who have hence construed the eternity of the mind as something other than its continued existence in time, Alan Donagan has maintained that Spinoza's conception of eternity is best understood as necessarily omnitemporal existence and that

> astonishingly . . . the very theorem that the order and connection of ideas is the same as the order and connection of things, entails that what, in the attribute of thought, corresponds to a mere possibility in the attribute of extension, must be more than a mere possibility.[6]

Donagan interprets Spinoza as holding (see E II, 8, which Spinoza held to follow from E II, 7) that at all times there exists in the infinite idea of God an idea of each possible individual, irrespective of whether or when that particular individual may actually exist. (Whether or not a given possible individual does become actual depends on the order of nature.) The eternal part of the human mind is the idea of one such individual; it is actual at all times as a part of the infinite idea of God even though its object (the human body) is actual only for a particular limited time. Donagan's interpretation is interesting because instead of trying to explain away the eternity of the mind he has (as he says) accepted it at its face value and reinterpreted E II, 7 in such a way that the latter is shown not only to be consistent with, but to imply, the eternity of the mind.

Donagan's approach seems basically sound, for the text of Book V indicates strongly that Spinoza did in fact mean to attribute an actual eternal existence to the mind and not to the body: whereas the body, at some point in its existence, is 'absolutely destroyed' (Latin: *absolute destrui*), something of the mind is not (E V, 23). And while my own version of the reasoning behind Spinoza's 'proof' for immortality, viz., the demonstration of E V, 23, would differ from his in a number of details, it is clear that Spinoza did justify his theory that a part of the mind is eternal on the basis of the requirements of divine omniscience. E V, 23 rests primarily on the preceding proposition which states that 'In God, nevertheless, there necessarily exists an idea which expresses the essence of this or that human body under the form of eternity.'

The main objection to any interpretation along the lines taken by Donagan is that, if it is correct, then the correspondence or identity of modes of different attributes, which we took to be a fundamental

6 Alan Donagan, "Spinoza's Proof of Immortality," in *Spinoza: A Collection of Critical Essays*," ed. Marjorie Grene (Garden City: Doubleday, 1973), p. 254.

theorem of Spinoza's metaphysics, is destroyed. That is, on Donagan's interpretation, there are countless numbers of ideas in God of *possible* extended things which do not and may never exist. Such ideas have no physical correlates, but rather are modes of thought which are not and may never be, expressed in the attribute of extension. Donagan's answer to this objection is that the ideas of non-existent things are not themselves finite modes of thought, but are merely parts of the infinite idea of God (*Dei infinita idea*), which is an infinite mode of the attribute thought.[7] Donagan, however, has not provided us with an explanation of how these ideas, which surely are modes not substances, and finite rather than infinite, are not finite modes of thought, or how a things's being a part of an infinite mode precludes its being a finite mode. (In general the latter must be false, since the mediate infinite mode of extension — the face of the whole universe, which though it varies in infinite ways always remains the same — would appear to be made up of the entire totality of finite modes of extension.[8]) Indeed, Spinoza himself is guilty of obscurity in regard to the infinite modes and their relation to finite modes, and Donagan sheds no light on this question.

I believe that Spinoza did reject what might be called the simple parallelism of E II, 7, but not merely — as the proof for E V, 23 seems to imply — on the basis of the theological truism that God must be omniscient. In what follows I shall try to show that it was a certain deficiency in his conception of the mind as the idea of or mode of thought corresponding to an actually existing body which led Spinoza to his theory that a part of the mind must be eternal.

According to the presentation of Book II of the *Ethics*, the human mind is the idea of (the mode of thought corresponding to) the human body (E II, 11, 13). Ideas differ from one another according as their objects vary. Because 'one [idea] is more excellent and contains more reality than another, just as the object of one idea is more excellent and contains more reality than another,' it follows that 'in proportion as one body is better adapted than another to do or suffer many things, in the same proportion will the mind at the same time be better adapted to perceive many things, and the more the actions of a body depend upon itself alone, and the less other bodies co-operate with it in action, the

---

7 Donagan, p. 255.

8 See Letter 64 in *The Correspondence of Spinoza*, trans. and ed. A. Wolf (1928; rpt. New York: Russell & Russell, 1966); and E II, lem. 7, sch.
Subsequent references to Spinoza's correspondence will be to Wolf's edition.

better adapted will the mind be for distinctly understanding' (E II, 13, cor., sch.). Since the abilities and limitations of the mind are coincident with, if not dependent on, the abilities and limitations of the body, Spinoza includes in Book II a short account of the nature of bodies.

According to that account all bodies are either in motion or at rest, and all are capable of varying degrees of motion (Ax. 1 and 2, following E II, 13). No body can set itself in motion or bring itself to rest, but rather 'a body in motion will continue in motion until it be determined to a state of rest by another body, and . . . a body at rest will continue at rest until it be determined to a state of motion by another body' (E II, lem. 3, cor.). Thus, while a body's resultant state of motion or rest depends partly upon its state of motion or rest prior to its being acted on by another body (Ax. 1, following lem. 3), no body can move itself or induce changes in its own motion. The most simple bodies, Spinoza tells us, 'are distinguished from one another by motion and rest, speed and slowness alone.' Composite bodies are formed 'when a number of bodies of the same or of different magnitudes are pressed together by others, so that they lie one upon the other, or if they are in motion with the same or with different degrees of speed, so that they communicate their motion to one another in a certain fixed proportion' (E II, dfn., following Ax. 2, following lem. 3). All the properties of a composite body are ultimately reducible to the motions (and rest) of its component parts (E II, ax. 3, following lem. 3; lemmata 4-7). Spinoza thus gives a thoroughly mechanical account of the nature of bodies, and the mind, as the idea which corresponds to the body, is necessarily a kind of 'immaterial automaton.'[9]

Spinoza's theory of perception, as set forth in Book II of the *Ethics* is as follows. Since the mind is the idea which corresponds to the body, there necessarily exists in the mind an idea of anything which happens in the body, or the mind perceives whatever happens in the body (E II, 12). Because (by post. 3 and 6, following lem. 7) the human body both is affected by and is disposed to affect external bodies, it follows that 'The human mind is adapted to the perception of many things, and its aptitude increases in proportion to the number of ways in which its body can be disposed' (E II, 14), In other words, the greater the number of events which happen in the body, or states which the body is capable of assuming, the greater the number of events which are or can be perceived by the mind.

9 A description used by Spinoza in *On the Improvement of the Understanding* in *The Chief Works of Benedict de Spinoza,* trans. R. H. M. Elwes, 2 vols. (1883; rpt. New York: Dover Publications, 1955), II, 32.

The mind perceives external bodies, as well as its own body, only through its perception of the states of its own body (E II, 16, cor. 1; 19; 26). This is possible because (from E I, ax. 4, and E II, ax. 1, following cor. lem. 3) 'The idea of every way in which the human body is affected by external bodies must involve the nature of the human body and at the same time the nature of the external body' (E II, 16). Spinoza accepts the principle that nothing can be contained in or belong to a thing which it does not derive from its cause; in his own (Cartesian) terminology, that whatever properties a thing has must be contained either formally or eminently in its cause. From this it follows that to know something about a thing or event is to know something about its cause. Since the states of the human body result from the action of external bodies on the human body, the knowledge (perception) of these states involves knowledge of the external bodies and of the human body itself.

Since an external object is only perceived through an idea whose physical correlate is a certain bodily state, it follows that if the body is affected in the way in which it is generally affected by a given external object, the mind will contemplate the external object as present to it (whether in fact it is present or not), until the body is affected in a way which excludes the presence of the external object (E II, 17, and dem.). Thus, Spinoza is able to account for the mind's ability to imagine things which are not present, as well as dreams, hallucinations, and memory.

Besides corporeal nature the mind is able to perceive its own ideas (E II, 22), and through these ideas, itself (E II, 23). Spinoza characterized the knowledge we have of all these things — the human body itself, its modifications, external bodies, the ideas of the bodily modifications, and the mind itself — as *inadequate*.

Adequate knowledge or an adequate idea of a thing *x* is such that (i) it is complete in the sense that it contains at least implicitly all the properties of *x* (that is, all the properties of *x* can be deduced from it); and (ii) it involves the idea of the ultimate and complete cause of *x*, and *x* is understood as following from its cause.[10] Spinoza referred to ideas which fail to satisfy the first condition as 'mutilated,' and to those which fail to satisfy the second as 'confused'. Ideas which do satisfy the second condition, that is, which are clear and distinct, necessarily satisfy the first. This is because just as the complete and ultimate cause of a thing is sufficient to produce the thing, so the knowledge or idea of the complete and ultimate cause of the thing is sufficient to deduce the thing and all its properties.[11] Because clearness and distinctness is a sufficient

10 Steinberg, pp. 82-85.

11 Letter 60.

condition for the adequacy of ideas, Spinoza often uses 'clear and distinct' as synonymous with 'adequate' in referring to ideas.

From the corollary to E II, 7 that 'whatever follows *formally* from the infinite nature of God, follows from the idea of God [idea Dei], in the same order and in the same connection *objectively* in God,' it follows that all ideas are adequate in God. That is, in God (a) there is a complete idea of each thing; (b) there is a complete idea of the cause of each thing; and (c) the idea of each thing follows from the idea of its cause. These three conditions taken together insure that in the divine mind the idea or knowledge of each thing is adequate. God, insofar as He constitutes the human mind, has ideas or knowledge only of the modifications of the human body, or equivalently, the human mind perceives *directly* nothing but the modifications of its body (E II, 9, cor.; 11; 12; 13). Hence, whatever the human mind is able to know adequately must be such that a complete idea of it, and of its cause, and of the cause of its cause, etc., be involved in the ideas of the body's modifications.

The ideas by which the mind perceives the states of the body are the ideas which correspond to the bodily modifications. Consequently, these ideas completely represent to the mind the modifications of the body. Since the bodily modifications result from the interaction between the body and external bodies, adequate knowledge of the bodily modifications would require (among other things) complete knowledge of the body itself and of the external bodies with which it interacts. But because the body itself and external bodies are known to the human mind only through the ideas which correspond to the modifications of the body, and because they are not completely represented in these ideas, it is impossible for the mind to have complete or adequate knowledge either of its own body (E II, 27), or of external bodies (E II, 25). Hence the mind's knowledge of the body's modifications themselves is also confused or inadequate (E II, 28). Since 'the order and connection of ideas is the same as the order and connection of things' (E II, 7), it follows by similar reasoning that the mind has only inadequate knowledge of itself (E II, 29) and of its ideas of the bodily modifications. By E I, 28 every finite thing or event — the parts of the human body, the human body itself, external bodies, interactions between the human body and external bodies — is the result of other things and/or events, which in turn are themselves the result of yet others, and so on, *ad infinitum*. Because the human mind is itself finite and can know other finite things only insofar as its (finite) body is affected by them, its ideas of all finite things necessarily are like 'conclusions without premises' — inadequate.

Despite its finitude the human mind is able to have some adequate ideas. According to E II, 38, 'Those things which are common to everything, and which are equally in the part and in the whole, can only be adequately conceived.' What is common to all things (all the modes of a single attribute) is the attribute under which they are conceived and the infinite modes of that attribute. Thus, the nature of extension (substance as extended) and motion and rest, must be adequately conceived by all men (E II, 38, cor.). The reason for this is that the ideas of the 'causes' of motion and rest and extension are not beyond the reach of the human mind. The cause of motion and rest — which is present in everything — is nothing else than the nature of extension itself which is likewise omnipresent. And extension itself is understood through itself: it has no prior cause, and is involved in the idea of every extended thing. Further, because no individual thing can be conceived except through one or the other of the divine attributes, it follows that an adequate knowledge of the divine essence itself must be involved in every idea which the mind has (E II, 45, 46, 47).

Spinoza held that when the mind has adequate ideas it *acts* in his technical sense (E III, 1, 3). That sense is defined at the beginning of Book III where he tells us first that

> I call that an *adequate cause* whose effect can be clearly and distinctly perceived by means of the cause. I call that an *inadequate* or *partial cause* whose effect cannot be understood by means of the cause alone (E III, dfn. 1; my italics).

And then:

> I say that we *act* when anything is done, either within us or without us, of which we are the adequate cause, that is to say (by the preceding Def.), when from our nature anything follows, either within us or without us, which by that nature alone can be clearly and distinctly understood. On the other hand, I say that we *suffer* when anything is done within us, or when anything follows from our nature, of which we are not the cause excepting partially (E III, dfn. 2; my italics).

Clearly the notions of an adequate cause and adequate knowledge are reciprocal and parallel. Our earlier formulation of the notion of adequate knowledge of a thing might be simplified by characterizing such knowledge as knowledge of a thing as following from its adequate cause. And Spinoza could have defined an adequate cause as one whose effect could be adequately known by means of (knowledge of) the cause. Further, just as the adequate cause of *x* adequately causes *x* in the world of things, the adequate knowledge of the cause of *x* adequately causes the knowledge of *x* in the realm of ideas.

The definition of an adequate cause, taken in conjunction with E I, 28, that

> An individual thing, or a thing which is finite and which has a determinate existence, cannot exist nor be determined to action unless it be determined to existence and action by another cause which is also finite and has a determinate existence; and again, this cause cannot exist nor be determined to action unless by another cause which is also finite and determined to existence and action, and so on *ad infinitum*

implies that no finite thing or event and no finite set of finite things or events is able to be the adequate cause of any other. Consider, for example, the case in which a ball *A*, is struck by a second ball *B* moving with velocity v_1, in such a way that *A* is impelled by *B* to move with velocity v_2 from a point x_1 to x_2.

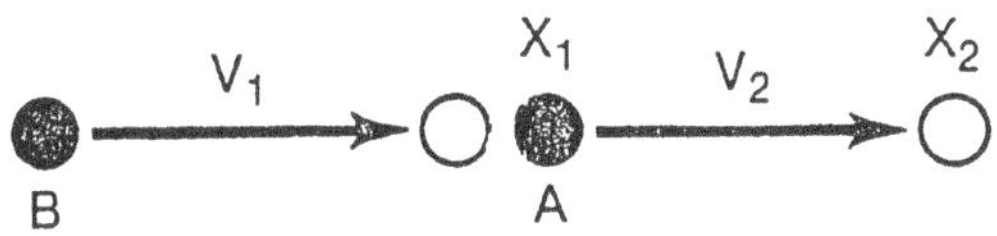

The striking of *A* by *B* is not the adequate cause of *A*'s movement because *A*'s movement cannot be clearly and distinctly (adequately) perceived merely through the striking of *A* by *B*. In order clearly and distinctly to perceive (or understand) *A*'s movement, we must understand how *B* came to have the particular movement it had when it struck *A*, and how some other body *C*, whose impulse caused *B* to move in turn received its motion, and so on *ad infinitum*. We must also know how *A* came to be in the particular position and state it was in when it was struck by *B*, which involves another infinite chain of events.

According to the account of the nature of bodies which Spinoza gives in Book II, all events in the physical realm are properly understood on the model of our illustration. That is, ultimately, any change of state in a simple or complex body must be understood as a change in its motion or rest (or the motion or rest of some of its constituent parts) which has been caused by some external force. Thus, Spinoza's account of the nature of bodies is quite consistent with the implication of E I, 28 that there can be no finite agents, i.e., things which act in his technical sense (see E III, dfn. 2, above). (The principle of inertia, which Spinoza states in E II, lem. 3, and cor., follows from a direct application of E I, 28, to the realm of extension. See E II, lem. 3, dem.)

By contrast, Spinoza maintained in Book III of the *Ethics* that the mind acts whenever it engages in adequate thinking (E III, 1, 3). He wrote: '. . . of that effect of which God is the cause, in so far as He is affected by an idea which is adequate in any mind, that same mind is the adequate cause . . .. Our mind, therefore . . . in so far as it has adequate ideas, necessarily at times acts . . .' (E III, 1, dem.). Despite E I, 28, the text is clear: the mind, unlike the body, is able to act (be the adequate cause of its own states) and does so whenever it engages in adequate thought. The difficulties involved in the assertion that the mind acts in so far as it has adequate ideas do not end with E I, 28. Although few commentators seem to have been aware of it, Spinoza's contention that the mind acts whenever it engages in adequate thinking is as much in conflict with the parallelism of extension and thought asserted by E II, 7, as is his thesis that a part of the mind is eternal.[12]

What lies behind what might be called the thesis of the agency of the mind I believe is Spinoza's implicit assumption that the cause of knowledge of a thing is (must be) the knowledge of the cause of the thing. A genetic account of how I came to know *s*, that is, a description of the series of states of mind which preceded the state describable as 'knowledge of *s*' cannot count as the causal explanation of my knowing *s* unless that series includes states of mind describable as 'knowledge of the cause of *s*.' In other words, no state of mind can be described as knowledge in the full-fledged sense unless it is completely determined by a rational justification. (One's knowledge that *s* is completely determined by a rational justification either when *s* itself is a self-evident truth, or when *s* is seen to follow deductively from other truths which are self-evident.)

The knowledge of individuals through sense perception which is in the human mind is inadequate because knowledge of the adequate causes of individual things is beyond the reach of the human mind. The human mind is a part of the infinite intellect of God, and such knowledge exists in Him not insofar as He constitutes an individual human mind, but only insofar as He constitutes the infinite intellect (E II, 11, cor.). Since Spinoza held that the cause of an idea (knowledge) of a thing is the idea (knowledge) of its cause, it follows that when the human mind has inadequate knowledge it is determined from without, that is, the adequate cause of its ideas (the adequate knowledge of the causes of their objects) is not present in the mind. Hence, the mind suffers insofar as it has inadequate knowledge or ideas. By contrast, since the mind has adequate knowledge of a thing only insofar as it has adequate knowledge of the cause of the thing, it follows that *when the mind has adequate knowledge of a thing the cause of its knowledge* (adequate knowledge of the cause of the thing) *is completely internal*. Spinoza's notion of adequate knowledge seems to dictate that any mind which has such knowledge must contain within itself the cause of its

12 David Bidney is an exception. See his *Psychology and Ethics of Spinoza* (New Haven: Yale University Press, 1940), pp. 61 ff.

having that knowledge. But insofar as according to that notion a mind (human or divine) can be said to have adequate knowledge of a thing only insofar as it also has knowledge of the complete and ultimate cause of the thing, and understands the thing as following from its cause, it has built into it the condition of any genuine knowledge: that it be completely determined by a rational justification.

The very conception of knowledge as a state of mind completely determined by a rational justification entails that insofar as the mind has knowledge it acts in Spinoza's strict sense, that the mental process of adequate thinking cannot be a passive, mechanical one. But as Spinoza characterized the physical world in Book II of the *Ethics*, there can be no physical finite agents (things which act in the technical sense), but rather only an infinite chain of moved movers. Thus the body is not able to act or be the adequate cause of its own states (or of those of any other body).[13] It seems evident then that to the extent that Spinoza maintained the parallelism of mind and body he was unable to give an account of the (active) mental function of adequate thinking, for there is no *active* physical process to which adequate thinking might correspond.

It might be objected in reply to this criticism that Spinoza's notion of action is not absolute, but rather admits of degrees, that everything is to some degree a positive factor in whatever situation it may be involved in, and that to the extent which it influences any situation it is in, each thing acts. It is true, I think, that Spinoza does have such a notion as I have described, a notion of agency which admits of degrees. It is nothing other than his concept of the *conatus* or endeavor of a thing: each thing is always to some degree a positive factor in whatever situation it may be involved in, and to the extent that it is such, it might be said to act. In this sense the body as well as the mind is necessarily (to some degree) an agent. But the definition of action which Spinoza sets forth explicitly at the beginning of Book III (dfn. 2) is not meant to explicate this notion. Rather it defines a notion of action which does not admit of degrees. Moreover, it is clear from the context of the statement that when Spinoza says in the demonstration of E III, 1 that insofar as the mind thinks adequately it acts, he means 'acts' in the absolute sense set forth in definition 2. Finally, it is the case that, viewed from the standpoint of its content, a finite sequence of ideas may be self-contained in a way no finite sequence of physical events can ever be. It is this viewpoint which Spinoza took in so far as he maintained that the mind,

13 It should be noted that the account of 'human' action in Books III and IV of the *Ethics* is an account of mental action only.

in adequate thinking, acts. I believe that Hampshire is wrong when he states that corresponding to the mind's power to think adequately there is a 'power of the body to maintain itself in a normal state in spite of external disturbances,' which power is to be understood in terms of the complexity of the bodily organism.[14] For no matter how complex a mode of extension the human body may be (and consequently, no matter how stable it may be), it is still a part of nature, interconnected with the whole, which at every moment in every aspect of its existence is dependent on other parts of extended nature outside itself.[15]

J. H. Randall seems to have been aware of the problem under discussion here, for he has written that

> the difficulties [in Spinoza's conception of knowledge] are not merely those of detail, for they go back ultimately to his fundamental assumption that the logical order of ideas is identical with the mathematical order of nature, that knowledge grasps the basic structure of the universe directly. It is hard to see how, if the logical order of ideas in reasoning merely reflects the mechanical order of the effects of other bodies upon our body, we could ever escape the dominion of inadequate and confused ideas. As Spinoza himself proves, "inadequate and confused ideas follow by the same necessity as adequate or clear and distinct ideas" (II, prop. 36), and consequently there seems to be no intelligible ground for believing that the mind can lift itself to a vision of truth under the form of eternity. That it can so know must be accepted as Spinoza accepts it, as a bare fact, as a basic assumption. . . . It is easy to be caught by Spinoza's vision and intoxicated by it, to be convinced that he has worked out a consistent system. It is easy to overlook the fact that he ultimately failed, and that however inspiring or true the last three books of the *Ethics* may appear, they are not adequately grounded in the first two.[16]

Randall has pinpointed the difficulty quite nicely, but what he fails to see is that Spinoza was aware of this problem and offered a solution for it. That solution was his theory of the eternity of the mind. According to the presentation of Book II, the mind is the idea of, or the mode of thought corresponding to, an actually existing body. On this account all mental events and processes must be understood as being coincident with corresponding bodily events and processes. But in Book V Spinoza gives a different account. There, after having established that there is an

---

14 Stuart Hampshire, "Spinoza's Theory of Human Freedom," *Monist*, 55 (1971), p. 562; *Two Theories of Morality* (Oxford: Oxford University Press, 1977), sec. 17.

15 See Letter 32.

16 John Herman Randall, *The Career of Philosophy*, vol. 1'; *From the Middle Ages to the Enlightenment* (New York: Columbia University Press, 1962), pp. 445-46.

eternal part of the mind, namely, the idea of the essence of the body 'under the form of eternity' (E V, 23), and following a number of propositions dealing with the third kind of knowledge (25-28), Spinoza tells us that 'Everything which the mind understands under the form of eternity [i.e., all adequate knowledge] it understands not because it conceives the present actual existence of the body, but because it conceives the essence of the body under the form of eternity' (E V, 29). What Spinoza is saying in this proposition (and its demonstration, at which we shall look more closely later on) is: the mind has knowledge of things under the form of eternity; but insofar as it is merely the idea of an actually existing body it cannot have such knowledge; hence, it has such knowledge in so far as it is something other than (more than) the idea of an actually existing body, that is, insofar as it is the idea of the essence of the body under the form of eternity.

Spinoza proceeds in the second part of Book V as follows: first, he shows that certain mental functions, namely, all those which are properly understood as the mental correlates of certain bodily processes (perception, memory, imagination in general), cease with the destruction of the body (E V, 21). Second, he shows that there exists in God an idea which expresses the essence of the body 'under the form of eternity' (E V, 22). Third, he identifies this idea in God, which is eternal and independent of the body, as a part of the mind (E V, 23). Fourth, he attributes the function of reason or adequate thinking to this part of the mind (E V, 29, 31, 40). Essentially Spinoza solves the problem we have presented here, that of the inability of the theory of the mind as the idea of an actually existing body to account for the mind's ability to engage in adequate thinking, by presenting in Book V a new theory for the origin of the intellect. The intellect or reason is not a function of the idea of an actually existing body; rather, it is a function of the idea of the essence of the body 'conceived under the form of eternity.'

But what is the 'idea which expresses the essence of this or that human body under the form of eternity' which is first introduced by Spinoza in E V, 22, then identified by him with a part of the human mind in the following proposition? In contrast to the conception of the mind as the idea or mental correlate of an actually existing body which Spinoza introduced in Book II, the former is at best not well-defined within his metaphysical system. An outline of an answer to the question, however, is suggested by the demonstration of E V, 22:

> Prop. XXII. — In God, nevertheless, there necessarily exists an idea which expresses the essence of this or that human body under the form of eternity.
> *Demonst*. God is not only the cause of the existence of this or that human body, but also of its essence (Prop. 25, pt. 1), which must necessarily be conceived through the essence of God itself (Ax. 4, pt. 1) and by a certain eternal

> necessity (Prop. 16, pt. 1). This conception, moreover, must necessarily exist in God (Prop. 3, pt. 2). — Q.E.D.

Here, for an idea to express an essence 'under the form of eternity' seems to mean for an essence to be conceived 'through the essence of God *itself* [*per* ipsam *Dei essentiam*] . . . and by a certain eternal necessity [*idque aeterna quadam necessitate*] (italics mine). I take the phrase 'through the essence of God *itself*' to mean 'through the essence of God considered absolutely' or 'through the absolute nature of one or the other of the divine attributes,' as opposed to 'through the essence of God considered as modified by some finite mode' or 'through some attribute in so far as it is considered to be affected by some finite mode.' Things which are conceived through the essence of God considered absolutely are conceived as following immediately or directly from the divine essence; that is, they can be inferred from (understood through) the conception of the divine essence (one or the other of the divine attributes) alone. Things which are conceived only through the essence of God (one or the other of the divine attributes) considered as modified by some finite mode require for their (adequate) conception the idea of their finite cause, which in turn requires the idea of its cause, and so on. What Spinoza seems to mean then by the idea of the essence of the body 'under the form of eternity' is an idea of the essence of the body conceived as following necessarily from the divine essence (the attribute of extension) considered absolutely.

The distinction between the divine essence or one of the divine attributes 'considered absolutely' and the divine essence or one of the attributes 'considered as affected by some finite mode' is employed by Spinoza in E I, 21, 23, 28, and 29 to explicate the relation of cause and effect which obtains between God and the infinite modes and between Him and the finite modes respectively, Thus, there is an obvious difficulty involved in holding that the essence of a finite thing — the human body — may be conceived as following from the divine essence considered absolutely ('through the essence of God itself'). According to Spinoza's metaphysical doctrine finite things do not follow from the divine essence or from any attribute considered absolutely, but rather only from that essence considered as modified by some other finite mode, which in turn follows from it considered as affected by some other mode, and so on (E I, 28); how then can something finite (such as the human body) be truly conceived as following necessarily from the divine essence *itself*?[17]

---

17 That the phrase 'through the essence of God itself' ('*per ipsam Dei essentiam*') is not an accident is indicated by its recurrence in E V, 23, dem.

In order to understand how this is possible it is necessary to take into consideration Spinoza's metaphysical doctrine that all possible essences must be actual. He stated this doctrine in E I, 16 (which is cited by Spinoza in the proof of E V, 22, in support of the assertion that the essence of the body 'must necessarily be conceived through the essence of God itself . . . *and by a certain eternal necessity*' — italics mine). That proposition states that

> From the necessity of the divine nature infinite numbers of things in infinite ways (that is to say, all things which can be conceived by the infinite intellect) must follow.

Prima facie, E I, 16, looks false. Given, for example, the axioms and lemmas concerning the nature of bodies which Spinoza states in Book II, it would seem to be equally conceivable by the infinite intellect that a certain body *b* be in motion at time *t* and that *b* be at rest at *t*. Yet both possibilities cannot be the case. If it is replied that the infinite intellect has knowledge not only of all the laws of nature in terms of which all possibilities must be conceived, but of the infinite chain of events prior to a certain event e, and hence knows 'ahead of time' how things will come out, and only conceives of things as in fact they will come out, then E I, 16, is trivial. That is, if the fact that a thing will happen is the condition of its being conceived by the infinite intellect, then in telling us that whatever is conceivable by the infinite intellect must follow from the divine nature, Spinoza is merely telling us that what will be will be.

I believe that E I, 16 is not trivial, and that in it Spinoza is telling us something important about his notion of God as the most perfect being. God, by definition, is 'Being absolutely infinite, that is to say, substance consisting of infinite attributes, each one of which expresses eternal and infinite essence' (E I, dfn. 6). As he stated in a letter to Hudde:

> . . . since the nature of God [who is absolutely perfect] does not consist of a certain kind of being but of absolutely unlimited being, His nature also requires all that perfectly expresses *being*; otherwise His nature would be limited and deficient (Letter 36).

In other words, all possible attributes must in fact belong to God or he would not be perfect, hence would not be God. What is true at the level of the attributes is also true of the modes of each attribute: God consists of infinite attributes *each one of which expresses eternal and infinite essence*. That each attribute must produce every possible mode (i.e., every mode which is conceivable in terms of that attribute) is what Spinoza is saying in E I, 16. If it were otherwise God would not be perfect.

But in the face of the blatant contradictions to which it appears to give rise, how are we to make sense of such a doctrine? The answer I think is that Spinoza meant by E I, 16 simply that every possible (i.e., conceivable in terms of the notions and laws, knowledge of which constitutes knowledge of an attribute and its infinite modes) essence must at some time be actual. Since whether or when a thing exists depends not on its essence, for otherwise it would be self-caused, but rather on the order of nature as a whole, it follows that a proper conception of the essence of a thing does not involve its existing at any particular time. Hence, given infinite time, there is no contradiction in supposing that every possible essence or individual at some time exists. If two individuals are such that the existence of one precludes the existence of the other, then they cannot exist at the same time. But since the essence of no individual involves its existing at some particular time, it is impossible that the actualization of one possible individual prevent absolutely that of another. And E I, 16 is meant to tell us that all possible individuals do in fact follow from the divine nature. (Whatever description is given of the essence of a certain body *b*, 'moving at *t*' or 'at rest at *t*' is not part of that description since both of these imply that *b* exists at *t*, and it cannot follow from *b*'s essence that it exists at any particular time.)

Because a true conception of the divine essence involves the knowledge that each attribute expresses eternal and infinite essence, i.e., must produce every mode which is conceivable in terms of it, it follows that every essence can be truly conceived as following necessarily from the divine nature itself, or equivalently, can be conceived 'under the form of eternity'.

Regarding human knowledge Spinoza wrote in E V, 29, sch., that

> Things are conceived by us as actual in two ways: either in so far as we conceive them to exist with relation to a fixed time and place, or in so far as we conceive them to be contained in God, and to follow from the necessity of the divine nature. But those things which are conceived in this second way as true or real we conceive under the form of eternity, and their ideas involve the eternal and infinite essence of God ...

and he seems to have held that there was a similar duality in the divine intellect: the conception of the essence of the body under the form of eternity 'must necessarily exist in God' (E V, 22, dem.).

We are now able to say what God's idea of a thing 'under the form of eternity' is, and how it differs from his idea of it as actually existing. The idea of a thing actually existing comes into being and passes away when the thing itself comes into being and passes away, reflecting during its existence all the states and changes which occur in the thing itself (E II, 7, 8, 9 cor.). Further, insofar as its object is a thing actually existing (essence + existence), such an idea derives from the idea of the cause of the thing, which in turn derives from the idea of its cause, and so on (E II, 9). By contrast, an idea in God of a thing 'under the form of eternity' (a) is eternal, i.e., neither comes into being nor passes away (E V, 23, and dem. and sch.; 31 dem.);[18] (b) is of the essence of the thing alone and involves no reference as to whether or when the thing exists (see above, pp. 39ff.; and also E I, 23; E V, 22, 29); and (c) derives directly from the divine essence itself.

So far I have argued that (a) Spinoza's conception of the physical world is such that no extended thing can ever *act* in his technical sense; while (b) his conception of adequate knowledge implies that any mind which has such knowledge or engages in adequate thinking, to that extent, acts. Thus, (c) on his theory of the mind as the idea of (corresponding to) an actually existing body, Spinoza is unable to give an account of the ability of the mind to engage in adequate thinking, for there can be no physical process which corresponds to the mental function of adequate thought. I have suggested (d) that in order to remedy this difficulty Spinoza identified the active part of the mind with the idea of the essence of the body 'under the form of eternity'. That is, his theory of the eternity of a part of the mind can be seen as his attempt to provide an account of the origin of the intellect. Finally, (e) I have attempted to explicate the notion of the idea of the essence of a thing 'under the form of eternity' and to say how it differs from the idea of a thing actually existing.

Insofar as my interpretation involves the assertion of the existence of a whole realm of ideas — the ideas of the essences of things 'under the form of eternity' — which have no physical correlates, it (like Donagan's) is vulnerable to the criticism that it ignores Spinoza's basic metaphysical guidelines (specifically, E II, 7). To this I have no answer but that the text of Book V seems to indicate that Spinoza himself ignored them, and that my primary aim in this paper has been to ask why he might have done so.[19] To the extent that I have provided a good

---

18 The question of whether to construe the eternity of the mind (and other ideas of things 'under the form of eternity') as timeless or not is dealt with in the second part of this paper.

19 If, as I maintain in the second part of this paper, the eternity which Spinoza attributes to the human mind and, *mutatis mutandis*, to all ideas of the essences of things 'under the form of eternity,' is timeless, then it seems to me that the criticism is somewhat mitigated: for, in this case, E II, 7 (and 8) can be understood as applying only to the temporal realm, and hence, as exceptionless. And while E II, 7 is no longer able to fulfill its apparent function of explaining the unity of the attributes in God, it still plays a significant role in explaining the nature of the passive (temporal) part of the mind — the part which perishes with the body.

explanation, I have argued indirectly for the former conclusion. In part II of this paper I shall discuss the nature of the eternity which Spinoza attributes to a part of the human mind and sketch a line of thought, based on the text of Book V, which I believe explains why Spinoza was led to attribute *timeless* eternity to the human mind. Again, if my reading of the text is plausible, and my arguments are good ones, I shall have succeeded in providing further support for the thesis that Spinoza himself rejected — or never held — the theory that to every idea there must correspond an actually existing thing.

II

The question of the meaning of 'eternity' in Spinoza's philosophy has recently received fresh treatment from two writers, Donagan and Martha Kneale.[20] Donagan has maintained (contrary to most of the previous commentary on the subject) that there is no passage in the *Ethics* which 'is not consistent with the conception [of eternity] . . . as necessarily omnitemporal existence.'[21] Kneale has argued that the notion of eternity involves two components, necessity and timelessness, that the latter is obscure unless it is understood as necessity or sempiternity, and finally that necessity must entail sempiternity, but not vice versa. Thus whatever is eternal must also be sempiternal. Kneale attributes these insights to Spinoza in Part V of the *Ethics*, holding that there 'He thought that eternity was essentially necessity, that he can prove the necessity of the human mind and from this the sempiternity of the human mind.'[22] Even if Kneale's analysis of the notion of eternity and its relation to sempiternity is correct, it seems unlikely to me that Spinoza would have agreed with it. I agree with Kneale that the conception of eternity as excluding all temporal relations or of timelessness as something other than sempiternity is very obscure; yet it seems to me that there are very strong reasons (apparently overlooked by Kneale) for thinking that, with the exception of the *Short Treatise on God, Man and*

20 Donagan, pp. 241-46; Martha Kneale, "Eternity and Sempiternity," *Proceedings of the Aristotelian Society*, **69** (1968-69), pp. 223-38 (reprinted in M. Grene's *Spinoza: A Collection of Critical Essays*).

21 Donagan, p. 246.

22 Kneale, p. 236.

His Well-Being, his first philosophical work, Spinoza held throughout all his writings that God must be eternal in a sense which excludes all temporal relations, a sense which is incompatible with sempiternity. In what follows I shall try to make clear what these reasons are and why I think Spinoza did not share Kneale's views *simpliciter* concerning the relation between eternity and sempiternity either as regards God or the human mind.

At the beginning of Book I of the *Ethics* Spinoza gives the definition of eternity:

> By eternity, I understand existence itself, so far as it is conceived necessarily to follow from the definition alone of the eternal thing.
>
> *Explanation.* — For such existence, like the essence of the thing, is conceived as an eternal truth. It cannot therefore be explained by duration or time, even if the duration be conceived without beginning or end (E I, dfn. 8).

The definition of duration, which is here opposed to eternity, is given at the beginning of Book II. It reads

> Duration is the indefinite continuation of existence.
>
> *Explanation.* — I call it indefinite because it cannot be determined by the nature itself of the existing thing nor by the efficient cause, which necessarily posits the existence of the thing but does not take it away (E II, dfn. 5).

According to the definition of eternity, God or absolutely infinite substance is the sole eternal existent, since God alone is the only being from whose definition existence follows. But Spinoza also attributed eternity to the infinite modes (E I, 21) and to at least a part of the human mind. Besides this problem, that Spinoza defined eternity as necessary existence, then attributed it to things which presumably do not exist necessarily, there is another. That is that while at times Spinoza spoke as though eternity were properly conceived as excluding all temporal relations, at other times he seemed to use the term as equivalent to omnitemporality. Thus E I, 21 states that 'All things which follow from the absolute nature of any attribute of God must for ever exist, and must be infinite; that is to say, through that same attribute they are eternal and infinite.' Here 'eternal' seems to mean 'must for ever exist.' In E I, 33, sch. 2, however, Spinoza tells us that 'in eternity there is no when nor before nor after.' And the definition of eternity itself (E I, dfn. 8, above) explicitly states that eternity is not to be understood as unlimited duration.

In all of Spinoza's writings, the most complete discussions of the notion of eternity and the related notions of duration and time occur in the *Thoughts on Metaphysics*. There, in the chapter entitled "Of Duration and Time," he tells us that

Diane Steinberg

> From the fact that we have divided being into being whose essence involves existence and into being whose essence does not involve existence except as possible, there arises a distinction between eternity and duration. Of eternity we shall speak more fully below.
>
> 1. *What eternity is.* Here we simply observe that eternity is an attribute under which we conceive the infinite existence of God.
>
> *What duration is.* Duration is an attribute under which we conceive the existence of created things according as they endure in their own actuality. From this it clearly follows that duration is not to be distinguished from the complete existence of a particular thing except in thought. For as the duration of a thing diminishes, its existence also necessarily diminishes. To determine duration, we compare the duration of one thing with the duration of others which have a fixed and determinate motion, and this comparison is called "time."
>
> 3. *What time is.* Thus time is not an affection of things but rather a mode of thought or, as we have said, a logical being; for it is a mode of thought serving to explain duration. One should notice here, regarding duration, something which will be useful later when we speak of eternity: that duration is regarded as greater or less, as though composed of parts, and also that it is attributed only to existence and not to essence.[23]

'Duration' is simply Spinoza's name for that property of the existence of created things in virtue of which it can be said to be longer or shorter. The fuller discussion of eternity which Spinoza promises occurs in Part II, chapter 1, section ii of the *Thoughts*. There he writes that

> The outstanding attribute [of God] which must be considered before all the rest is the eternity of God, by which we explain his duration; or rather, since we do not attribute duration to God, we say that he is eternal. For as we noted in Part One, duration is an affection of the existence of things and not of their essence; but we cannot attribute duration to God, whose existence belongs to this essence. Those who do attribute duration to God separate his existence from his essence. Indeed there are those who ask whether God has existed any longer now than at the time when he created Adam; and since this seems sufficiently clear to them, they accordingly decide that duration must in no way be removed from God. Truly, such people beg the question, because they suppose that the essence of God is distinct from his existence. For they ask whether God, who existed up to the time of Adam, has not from the time of Adam's creation until now added additional time to his existence. On this basis, they attribute to God a duration increased by each passing day, and suppose that he is, as it were, continually created by himself. If they did not distinguish the existence of God from his essence, they would not attribute duration to him, since duration absolutely cannot pertain to the essence of

23 *Thoughts on Metaphysics* in *Earlier Philosophical Writings: The Cartesian Principles and Thoughts on Metaphysics*, trans. Frank A. Hayes (Indianapolis: Bobbs-Merrill, 1963), Pt I, ch. 4. Subsequent references to the *Thoughts* will be to Hayes' translation.

things. Surely, no one will ever say that the essence of a circle or triangle, as an eternal truth, has endured longer now than in the time of Adam.

The last sentence of this passage appears to be the basis for M. Kneale's contention that in this passage Spinoza uses an argument 'exploded by Searle' in order to distinguish the notion of eternity from that of duration.[24] Kneale argues plausibly that

> all that is meant by calling mathematical truths "timeless" is that there *is no point* in asking when two and two are four in the way that there is point in asking when the daffodils are in bloom. But this does not mean that it is not the case that two and two are four today, that they were four yesterday, and that they will be four tomorrow. These statements are not meaningless or untrue, but simply so obvious as to be pointless. I would go so far as to say that it is true that today two and two have been four for a day longer than they were yesterday and similarly that, if God exists, he has existed longer today than he had existed yesterday. I admit that such remarks seem paradoxical, but they are neither meaningless nor untrue.[25]

Kneale goes on to state that

> The argument which led Spinoza and has led many others to deny that an eternal object exists at a given time is of a type which has seemed to some very modern. It was much favored by the late Professor Austin and its general form is: "There are no conceivable circumstances in which it would be pointful to utter the sentence *S*. Therefore the sentence *S* is meaningless." Using this kind of argument, Austin denies in effect that "He sat down intentionally" in normal circumstances expresses any proposition at all, let alone a true proposition. It seems to me that this type of argument has been sufficiently dealt with by Professor J. R. Searle in his paper "Aberrations and Assertions". As he remarks, it depends on confusing the conditions under which it is correct (conventionally or socially) to assert that-*P* with the conditions under which it is true that-*P*. It is very rarely, if ever, socially correct to assert that two and two are four on Wednesdays, because this suggests that they might be something else on other days. Nevertheless it is perfectly true.[26]

I agree with Kneale that Professor Searle succeeds in refuting the argument in question (the Austin argument), but I do not believe that this is the argument used by Spinoza in the above passage in the

24 Kneale, p. 235.

25 Kneale, pp. 227-28.

26 Kneale, pp. 228-29. Kneale gives the following references for the papers of Austin and Searle to which she refers: (Austin) "A Plea for Excuses", *P.A.S.*, 57 (1956-57), pp. 1-30, repr. in *Philosophical Papers*; (Searle) *British Analytical Philosophy*, ed. B. Williams and A. Montefiore, pp. 41-54.

Thoughts on Metaphysics (II, 1, ii) to show that God is eternal in the sense of the term which entails timelessness. What Spinoza is saying there is: to attribute duration to God implies that his existence is separate from his essence, but this is false, hence duration cannot belong to God. Why is it the case that God's having duration implies that his essence must be separate from his existence? Suppose the implication did not hold. Then we would have both

(1) God's existence is identical with his essence; and

(2) God has duration.

Since it is the case that

(3) As the duration of a thing increases or diminishes, its existence increases or diminishes;

it follow that

(4) God has existed longer today than he had when he created Adam.

Therefore, by (1)

(5) God has a different essence today than he had when he created Adam.

But (5), of course, is false, since the essence of a thing — be it God or creature — cannot change. Thus the assumption that God has duration leads to a disjunction, that God's essence is separate from his existence or that God's essence can change, both disjuncts of which are false.

The argument, depending as it does on the distinction between essence and existence, and the concept of existence as a property, shows just how medieval Spinoza could be and has little appeal for modern philosophers. Nevertheless, it seems to be a valid argument. One might attempt to attack it by objecting to Spinoza's notion of duration as an inseparable property of created existence. Spinoza lacked, or at least never articulated, any notion of temporal passage as prior to things which exist.[27] But even if we substitute for Spinoza's notion of duration the notion of temporal passage, or give the notion of

27 See Donagan, p. 242.

duration a sense in which it is prior rather than posterior to things which endure, the argument seems to go through. For if God is in time (even if his existence is omnitemporal) then each day he has existed a day longer than the day before. Existence is longer or shorter whether duration or time is prior or posterior to things which exist. Hence again, either God's existence is not identical with His essence, or His essence itself is continuously changing. It is true too that the *Thoughts on Metaphysics* is an early work of Spinoza, and that in it his intention was to express the views of Descartes and not his own; yet in this case it appears that Spinoza held the fundamental premises of the argument to be true at the time he wrote the *Ethics*. Thus E I, 20 states that 'The existence of God and His essence are one and the same thing.' He also retained the notion of essence as immutable. It seems then that Spinoza, at the time of writing the *Ethics*, would have accepted the argument found in the *Thoughts*, II, I, ii that duration cannot be attributed to God, but rather He is eternal in a timeless sense. I conclude, therefore, that at least as regards the divine essence, Spinoza could not have accepted Kneale's view that timelessness is either identical with sempiternity or mutually entails it.

We are able to see too what Spinoza would have replied to Kneale's argument that a necessary object must also be sempiternal or that necessity entails sempiternity. Kneale writes that

> if there is a necessary object, e.g., God, then there is a true proposition expressible by the sentence "God necessarily exists". Now suppose God not to be sempiternal; obviously there will be a true proposition expressible, according to the time of utterance, by "God did not exist", "God does not exist", or "God will not exist". It follows that "Possibly God does not exist" expresses a true proposition, so that . . . we have a contradiction. Therefore, if God is eternal, in the sense of "necessary", he is also sempiternal.[28]

Spinoza would have objected (quite reasonably) that this argument simply begs the question of whether the existence of God is temporal or non-temporal, or of whether or not there can be such a thing as non-temporal existence. The supposition that God is not sempiternal does entail that there is a time at which God does not exist, which in turn entails not 'Possibly God does not exist' where 'exist' is taken in an unqualified sense, but rather 'Possibly God does not exist in time,' or 'Possibly God does not $exist_t$' where '$exist_t$' means 'exists in time.' Spinoza held not only that the latter was compatible with 'God necessarily exists' (and hence that Kneale's contradiction does not ac-

28 Kneale, p. 232.

tually occur), but that it was entailed by it since, according to him, 'God necessarily exists' would entail 'Necessarily God does not exist$_t$' which in turn entails 'Possibly God does not exist$_t$.'

In Book I of the *Ethics* the distinction between eternity in the timeless sense and duration seems to be applied along the same lines as were drawn in the *Thoughts on Metaphysics*. Thus, in the passages in which eternity is opposed to duration or time Spinoza is talking about a thing whose essence involves existence, i.e., God or substance (E I, dfn. 8; EI, 33, sch. 2); but when eternity is equated with everlasting existence — unlimited duration — he is talking about things whose essence does not involve existence, the infinite modes (EI, 21). Book V, however, is another story. The human mind and any parts it may have, is surely a created thing, or a thing produced by God, and hence its essence does not involve existence. It seems appropriate then that Spinoza refers to the 'duration of the mind without relation to the body' in contrast to 'this present life' (E V, 20, sch.), and speaks of the eternal part of the mind as *remaining* when the body is destroyed (E V, 23). But in a number of passages dealing with the eternal part of the human mind — passages which are at once puzzling and highly suggestive — Spinoza speaks of eternity as being opposed to duration and time (especially E V, 23, dem., and sch.; E V, 29, dem. and sch.). Here I would like to offer a possible interpretation of those passages which suggests a theory as to why Spinoza might have held a part of the human mind itself to be eternal in the timeless sense.

I have already suggested that Spinoza's attribution of eternity to the mind can be viewed as his solution to the problems arising in his account of the ability of the mind as the idea of an actually existing body to have adequate knowledge. What, I think, prompted him to choose this particular solution was that he came to hold that in so far as adequate knowledge is knowledge of a thing 'under the form of eternity' it can only be had by a being which is itself timelessly eternal.

When the mind has knowledge of a thing 'under the form of eternity' it knows it as following necessarily from the divine essence (E II, 44, cor. 2, sch.; E V, 29, sch.). All adequate knowledge is knowledge 'under the form of eternity.' One type of adequate knowledge, which Spinoza called reason or knowledge of the second kind, includes the common notions (adequate ideas of the attributes and their infinite modes) and the adequate ideas of properties common to the human body and other bodies by which it is generally affected, and whatever can be deduced from these (E II, 38, 39, 40). The other type of adequate knowledge, which he called the third kind of knowledge or intuitive science, he described as advancing 'from an adequate idea of the formal essence of certain attributes of God to the adequate knowledge of the essence of things' (E II, 40, sch. 2). It is generally agreed among commentators that by 'reason' Spinoza means scientific knowledge or the mind itself insofar as it has scientific knowledge. Reason is general or abstract and provides knowledge of individuals only insofar as they are subsumed under general laws. By contrast, the description quoted above of intuitive science seems to imply that that form of knowledge is a kind of knowledge of individuals, for to Spinoza essences are individual. Exactly how it should be construed has been a problem for commentators, both because Spinoza said little about it, and because of the difficulties which seem to arise if it is held, within the context of Spinoza's system, that adequate knowledge of individuals is possible for man. The scope of this paper precludes a long discussion of Spinoza's third kind of knowledge, but I believe that what he had in mind was a kind of knowledge identical in form to the characterization I have given of God's knowledge of things 'under the form of eternity.'[29]

I think that Spinoza meant to say that we have intuitive knowledge of a thing when we conceive its essence completely and correctly in terms of the fundamental notions and in accordance with the laws, knowledge of which constitutes knowledge of an attribute. By E I, 16 a thing so conceived is understood as following necessarily from the divine nature or equivalently, is conceived 'under the form of eternity.' While it is not conceived as existing here and now or at any other particular time (since it does not follow from the divine nature considered absolutely, or from the essence of a thing considered alone, that it exists at a particular time) because (by E I, 16) it is conceived as a necessary consequence of the divine nature, it is conceived in a certain sense as actual. Thus Spinoza writes in the demonstration to E V, 30 that 'To conceive things . . . under the form of eternity, is to conceive them in so far as they are conceived through the essence of God as actually existing things, or in so far as through the essence of God they involve existence.' When the mind has intuitive knowledge of an individual essence it sees that essence as a possible mode of an attribute, hence as a necessary consequence of that attribute (God as conceived under the given attribute).

In Book V Spinoza repeatedly connects the mind's ability to know things 'under the form of eternity' with its own being as an eternal entity, and its ability to know things 'in relation to time' or as 'actually existing' with its durational existence (E V, 21, and dem.; 23, sch.; 29 and dem. and sch.; 32 and dem.). I believe he was led to hold that only insofar as it was eternal could the mind know things under the form of

29 See above, pp. 49-53. Spinoza's conception of *scientia intuitiva* is explored at greater length in the author's dissertation, pp. 177-93.

eternity, and only insofar as it possessed duration could it know things in relation to time because he was influenced by certain insights into the logic of knowledge of the timelessly eternal versus that of knowledge of things in time. That is to say, he came to hold that there were logical reasons why a temporal being could not apprehend the timelessly eternal and an eternal being could not know things in relation to time.

What Rescher has called the 'temporal equivocality of is' is partially exhibited in the following two examples:

(a) Seven is a prime number;

(b) It is raining.[30]

Whereas the 'is' of (a) is generally taken to be atemporal and tenseless, that of (b) means 'is now' and its meaning is expressed in regard to future and past events by 'will be' and 'was' respectively. It seems that at least some other verbs may be similarly equivocal: Spinoza held that in the sentence 'God exists,' 'exists' must be understood in an atemporal sense, i.e., as tenseless, while in a sentence asserting that a thing produced by God exists, 'exists' should generally be understood in its more usual temporal sense.

The difference between the temporal and atemporal senses of 'is' (we shall indicate them by 'is_t' and 'is_a' respectively) is perhaps best brought out if we try to translate sentences containing 'is_t' into the timeless mode. Clearly (b) cannot be rendered by

(b′) It is_a raining

because raining occurs at or during a certain time and the use of 'is' ('is_t') in (b), as opposed to 'was' or 'will be,' indicates that the time of the raining is now.

It might seem that (b) could be rendered in the timeless mode by using a system of dates. Thus if *now* is 9:00 P.M., CST, on June 15, 1956, we could say

(b″) It is_a raining on June 15, 1956, at 9:00 P.M., CST

thus expressing timelessly the fact that something occured at a certain time. But (b″) lacks something which (b) has, namely, an indication of

30 Nicholas Rescher, *Topics in Philosophical Logic* (New York: Humanities Press, and Dordrecht-Holland: D. Riedel, 1968), p. 198.

its time of assertion relative to the fact asserted. As Rescher remarks, when we shift from a tensed to an atemporal 'is' what we have

> is always rather a *transformation* than a *translation*: something is always lost in the process — to wit, the temporal placement of the event at issue with respect to the time of assertion, i.e., the actual present. When a tensed copula is used the statement asserted is itself made from within the temporal framework; when the atemporal *is* is used, the statement may well be about something that happens within the temporal framework, but the assertion itself does not have a stance within the temporal framework. Even when "the same fact" is viewed, there is a crucial difference in perspective here — a difference so wide that there is no way to bridge it over.[31]

I think that Spinoza had this difference in perspective in mind when he stated in E V, 23, sch., that

> Only in so far . . . as it involves the actual existence of its body can the mind be said to possess duration, and its existence be limited by a fixed time, and so far only has it the power of determining the existence of things in time, and of conceiving them under the form of duration.

In other words, only insofar as the mind is itself in time does it have the temporal viewpoint or the ability to conceive of things as standing in temporal relations to itself. He expressed the same point again in the demonstration of E V, 29:

> In so far as the mind conceives the present existence of its body does it conceive duration which can be determined in time, and so far only has it the power of conceiving things in relation to time (Prop. 21, pt. 5, and Prop. 26, pt. 2).

For example: only a being which is itself in time or whose existence is temporal can meaningfully say 'It's raining now,' or 'Thank goodness, that rain is over,' because only such a being has the temporal viewpoint.

Continuing the demonstration of E V, 29, Spinoza tells us that as a temporal being the mind cannot know things under the form of eternity.

> But eternity cannot be manifested through duration (Def. 8, pt. 1, and its explanation); therefore the mind so far has not the power of conceiving things under the form of eternity . . . .

Why is this? First of all, I think Spinoza held that insofar as the mind is a temporal being it must know or conceive things from a temporal point

31 Rescher, p. 200.

of view. That is, it must conceive of things as standing in some temporal relation to itself here and now, as being now, or in the past, or in the future (or possibly all three). A being which is itself in time necessarily views all reality from this perspective, i.e., as standing in some temporal relation to itself. He also evidently held that insofar as there are timeless objects such objects cannot be adequately known or understood from a temporal point of view. The reason for this is that from a temporal point of view a timeless object doesn't look timeless, but appears to stand in temporal relations. From a temporal point of view — expressed in our language by the use of tenses — it seems to be the case that God exists now, that He existed at the time of Adam, that He will exist in the year 2000, and so forth. From the temporal viewpoint the timelessly eternal looks like the omnitemporal.

In Spinoza's ontology there is at least one object which is timelessly eternal: God. He concluded therefore, that if the mind were to have true or adequate knowledge of God, it must itself be timelessly eternal; otherwise it could not apprehend the timeless eternity of God.

In the proof of E V, 29, however, Spinoza seems not to be concerned to explain the mind's knowledge of God, but rather its knowledge of things under the form of eternity. Such knowledge of a thing is knowledge of its essence insofar as it follows from or is contained in the divine essence. While it is knowledge of a thing whose nature (unlike that of God) is to exist in time, there are two reasons why Spinoza held that such knowledge must be from the eternal point of view (hence, could not be possessed by the human mind insofar as it was in time). The first is that insofar as an essence is conceived as following from or as contained in the divine essence, its conception involves knowledge of the divine essence itself. The second is that considered in itself, i.e., as contained in the divine essence, an individual essence does not stand in any temporal relations, hence it cannot be accurately thought as being now or future or past.

Spinoza, of course, did not argue a *posteriori* from the fact that we have knowledge which is adequate and which involves God's eternal essence to the conclusion that the mind is eternal; rather his method was to prove first that the mind is eternal (E V, 23), and then to show that its knowledge of things under the form of eternity could be accounted for by the fact that it is eternal. Thus, EV, 29 concludes

> but because it is the nature of reason to conceive things under the form of eternity (Corol. 2, Prop. 44, pt. 2), and because it also pertains to the nature of the mind to conceive the essence of the body under the form of eternity (Prop. 23, pt. 5), and excepting these two things nothing else pertains to the nature of the mind (Prop. 13, pt. 2), therefore this power of conceiving things under the form of eternity does not pertain to the mind except in so far as it conceives the essence of the body under the form of eternity. — Q.E.D.

It seems to me, however, that the desire to account for adequate knowledge — knowledge of things 'under the form of eternity' — was a powerful motivating factor in Spinoza's coming to hold that the mind is eternal.

I have argued that Spinoza came to hold that a part of the mind is eternal in order to account for the mind's ability to have adequate knowledge and knowledge under the form of eternity. In part I of this paper we saw that the conception of the mind as the idea of an actually existing body was inadequate to account for its ability to have adequate knowledge because on Spinoza's account when the mind has adequate knowledge it acts. It was shown that insofar as the mind was conceived as the idea of an actually existing body it could only be understood as having passive functions. As a solution to this problem Spinoza in Book V offers what is essentially a new theory for the origin of the intellect. According to this new theory the mind consists of two distinct parts: one, the idea of the body actually existing, is the mental correlate of the body and exists just so long as the body itself exists; the other, the idea of the essence of the body conceived under the form of eternity, is virtually independent of the body and is timelessly eternal. To engage in adequate thinking or to know adequately is a function of the latter part.

But has Spinoza offered a real solution to the problem? The question which arises now is how can the mind's having a part which is timelessly eternal explain its ability to act in time? Presumably the mind *comes* to understand things, i.e., understanding is a process or state which takes place or comes to be in time. How can that be done by an eternal thing? The only reply which Spinoza could make to this is that understanding is not a temporal process. The mind's ability to act which is manifested in its having adequate knowledge can be understood in non-temporal terms. The mind itself, of course, is ontologically prior to its knowledge (E V, 31), but it need not be temporally prior. When it has adequate knowledge of a thing its knowledge follows from other adequate knowledge which it has, but such 'following' may be strictly logical, not temporal. Insofar as the mind's ability to act can be understood in non-temporal terms there is no conflict between the conception of the mind as able to act and as timelessly eternal.

In some passages of Book V Spinoza seems clearly to be aware that the thesis of the eternity of a part of the mind entails that that part cannot change in any respect. Thus, in E V, 33 he states that

> The intellectual love of God which arises from the third kind of knowledge is eternal.

> *Demonst.* — The third kind of knowledge (Prop. 31, pt. 5, and Ax. 3, pt. 1) is eternal, and therefore (by the same axiom) the love which springs from it is necessarily eternal. — Q.E.D.
>
> *Schol.* — Although this love to God has no beginning (Prop. 33, pt. 5), it nevertheless has all the perfections of love, just as if it had orignated; — as we supposed in the corollary of Prop. 32, pt. 5. Nor is there here any difference, excepting that the mind has eternally possessed these same perfections which we imagined as now accruing to it, and has possessed them with the accompanying idea of God as the eternal cause. And if joy consist in the passage to a greater perfection, blessedness must indeed consist in this, that the mind is endowed with perfection itself.

The intellectual love of God does not come to be, and of course, cannot be destroyed:

> Prop. XXXVII. — There is nothing in nature which is contrary to this intellectual love, or which can negate it.
>
> This intellectual love necessarily follows from the nature of the mind, in so far as it is considered, through the nature of God, as an eternal truth (Props. 33 and 29, pt. 5). If there were anything, therefore, contrary to this love, it would be contrary to the truth, and consequently whatever might be able to negate this love would be able to make the true false, which (as is self-evident) is absurd. There exists, therefore, nothing in nature, &c. — Q.E.D.
>
> *Schol.* — The axiom of the Fourth Part refers only to individual objects, in so far as they are considered in relation to a fixed time and place. This, I believe, no one can doubt.

(The axiom to which Spinoza refers here states that 'There is no individual thing in nature which is not surpassed in strength and power by some other thing, but any individual thing being given, another and a stronger is also given, by which the former can be destroyed.')

In light of these passages Spinoza's remarks in EV, 38 and 39, and their demonstrations and scholia, which imply that what we do here and now in this life determines the degree to which the mind participates in eternity, are puzzling. EV, 39, states that 'He who possesses a body fit for many things possesses a mind of which the greater part is eternal.' Spinoza's line of thought in the proof for this proposition is that he who possesses a body fit for many things is least agitated by affects which are contrary to his nature, hence is able to cause the ideas of all the affects to be related to the idea of God, and consequently is affected to a greater degree with the intellectual love of God. The demonstration is problematic both because the eternal part of the mind cannot suffer any evil (E V, 37, and dem., and sch.), and because no matter how the affects or bodily affections are arranged or controlled, there can be no increase in the intellectual love of God (E V, 33, and dem.). Equally puzzling is his declaration in the scholium to E V, 39, that 'In this life, therefore, it is our chief endeavor to change the body of infancy, so far as its nature permits and is conducive thereto, into another body which is fitted for many things, and which is related to a mind conscious as much as possible of itself, of God, and of objects . . . .'

These remarks, it seems to me, can only be understood in light of Spinoza's statement in E V, 31, sch., that 'although we are now certain that the mind is eternal in so far as it conceives things under the form of eternity, yet, in order that what we wish to prove may be more easily explained and better understood, we shall consider the mind, as we have hitherto done, as if it had just begun to be, and had just begun to understand things under the form of eternity.' In other words, Spinoza must be understood as saying in E V, 39 and its scholium, not that a mind which is connected with a body fit for many things is capable of *developing* a greater love for God (since by E V, 33 the intellectual love of God cannot come to be), but that a mind which is connected with such a body simply *has* a greater love for God, hence participates to a greater degree in eternity.

In other words, Spinoza cannot consistently be understood in Book V of the *Ethics* to be explaining the connection between a good mind and a good body in terms of the latter's being better adapted to facilitate the acquisition of knowledge or the development of the intellectual love of God. According to the conception of the mind which Spinoza expounded in Book II, all mental functions and states including the acquisition and possession of knowledge were explained as being the parallel in the mind of certain bodily events or states (E II, 13, cor., sch.; E II, 14). But Spinoza revised his conception of the mind and came to hold that it consisted of two parts: one, the idea of an actually existing body, and the other the idea of the essence of the body under the form of eternity. The latter is virtually independent of the body (E V, 23). According to his new theory, knowledge is a function of the idea of the essence of the body under the form of eternity, the eternal part of the mind, and Spinoza is committed to the thesis that whatever knowledge the mind has, it has eternally, i.e., it is not acquired at a certain time and place. The connection between having a body fit for many things and having a mind which knows a great deal (hence has a greater love for God and participates to a greater degree in eternity) can only be understood in terms of Spinoza's theories that the eternal part of any mind is the idea of the essence of a particular body, and that ideas vary as their objects vary (E II, 13, sch.).

On Spinoza's new theory there can be no tension in the mind between the two parts, no struggle for dominion on the part of one or the other, for there simply cannot be any effect of one part on the other. The passions cannot overpower the intellectual love of God (E V, 37), nor can the intellectual love of God, which is eternal (E V, 33) have any effect in time. The eternal part of the mind is not in the realm where struggle takes place (E V, 37, sch.). The salvation which Spinoza mentions at the very end of the *Ethics* is not merely difficult, but does not exist in the sense of something one might strive for even though the result of one's striving were preordained. Whatever knowledge, whatever intellectual love of God a mind has, it has eternally.

Spinoza all but severed the tie between one part of the mind — the intellect — and the body, and between the two parts of the mind. Perhaps it was in part dissatisfaction with this aspect of his theory which led him to remark near the end of Book V, as though there were still a question in his own mind, that 'Even if we did not know that our mind is eternal, we should still consider as of primary importance Piety and Religion, and absolutely everything which in the Fourth Part we have shown to be related to strength of mind and generosity' (E V, 41).

December 1979

Part V
Religion and Politics

[24]

Yirmiyahu Yovel

SPINOZA: THE PSYCHOLOGY OF THE MULTITUDE AND THE USES OF LANGUAGE

I

The multitude was a major philosophical concern for Spinoza. The philosopher lives among the multitude, surrounded on all sides by its powerful presence. He cannot attain his goal by ascetic isolation nor can he realistically expect that all men and women will rise to the life of reason. In Spinoza, therefore, the question how to deal with the multitude and what strategy to develop about it, becomes a major philosophiical question in itself - even, indeed, part of the general question, what is the good life, what is the life of philosophy.

Spinoza regards the multitude as a special category in itself. Individuals could rise above the *imaginatio* and attain *ratio*, even *scientia intuitiva*; but the great majority is incapable of doing this - and the concept of the multitude is defined by this majority. Therefore, by definition, the multitude will always be there, guided as it is by the powers of the imagination and by the special psychology of the masses to which it gives rise. In the usual course of things, the psychology of the *imaginatio* breeds conflict, discord, violence and war, as well as fanaticism and the various forms of intolerance; it is the source of social instability and insecurity in life. Yet even if we must admit that the multitude will remain governed by the imagination must we also accept these characteristics of its behavior and their catastrophic consequences? Or, perhaps, is there a natural way to reshape and re-channel the effects that *imaginatio* has on the multitude, so as to neutralize its destructive results and engender behavior that is socially beneficial, because it conforms with what reason itself would have recommended?

This question, as I have argued elsewhere (Yovel 1973), defines the philosophical program of Spinoza's *Theological-Political Treatise* (TTP) and underlies it throughout. The overall aim of this work is to establish mental and institutional mechanisms that would transform the imagination into an external imitation of reason, using state-power and the purified popular religion as vehicles of a semi-rational civilizing process. This program is carried out in both parts of the TTP, the *theological* and the *political*, and confers systematic unity upon them. It contains Spinoza 's main answer to the question, what is to be done about the multitude - a question which, under the interpretation I should like to defend, dominates the TTP 's philosophical concerns.

Spinoza 's answer is not to raise the multitude to the level of reason, a task he considers impossible. Rather, accepting that the majority of people will inevitably remain in the domain of the *imaginatio*, Spinoza seeks to transform and institutionalize the latter 's effects in semi-rational patterns. This civilizing intervention in the crude course of nature will be carried out by a perfectly natural mechanism, based upon scientific knowledge of the passions and their effects; indeed, there will be two such mechanisms, one religious and the other political.

The man of the *imaginatio* will not act from adequate ideas or purely rational motivation. He will continue believing in external authorities (God 's will, the Bible, the transcendental status of moral and legal precepts, etc.) and will require the coercive power of the state to restrain his socially destructive passions by still more powerful passions. In direct contrast to the philosopher, obedience to authority and fear of punishment will be dominant in the man of the multitude 's motivation. And yet, these two radically different human types will manifest little or no difference in their external conduct, for each in his way will act in accordance with the rules of justice and mutual social benefit. Purified religion and the rationalized state are thus designed to engender in the multitude the same form of conduct that the rational model requires, even though it will be motivated by non-rational powers and by inadequate ideas.

Hence the close link between the problem of the multitude and the problem of language. Spinoza 's program for the multitude as we have just summed it up cannot use clear and distinct ideas and their verbal correspondents as its vehicle. To have the desired effect upon its target-group, it must be suited to the latter 's mental powers and tendencies. This requires the philosopher to use language also rhetorically, so that his discourse could trigger the desired effects in the intended audience. The rhetorical use of language has therefore a social and cultural role which, far from being regarded an undignified corrolary, acquires philosophical import in Spinoza as part of his general theory of discourse. Just as people are divided into those led by the imagination and those guided by reason, so there must be different types of discourse suitable to each group - and also discourse that will fit an eventual transition between the groups or allow the rational model to externally inspire and reshape the imagination.

The Psychology of the Multitude

Given that the multitude is the underlying problem of the TTP, it is not surprising that this work opens with a succinct discussion of the psychology of the multitude. The main characteristic of the multitude is **uncertainty**, resulting from ignorance of true causes and the lack of adequate ideas. Uncertainty makes the person of the multitude prey to the alteration of fear and hope, between which he or she vascillates back and forth without sufficient reason. Controlled as he is by his passions, he also pursues goals which are unstable and uncertain in themselves (riches, pleasure, fame). As a result, his fundamental state of mind is *fluctuatio animi*, the fluctuation of the soul,[1] which explains the notorious unreliability of the multitude, its quick reversals of sides and positions, and the fierce, intolerant way in which it takes up any of these accidental stands.

Uncertainty also breeds superstition, the essence of historical religion according to Spinoza. The greater the lack of certainty and of rational self-confidence, the stronger the temptation to invest one's faith in superstitious explanations. *Fluctuatio animi* and the constant vascillation between fear and hope also account for the multitude's inflammable nature, its outbursts of cruelty and violence and its inclination to intolerance and to fanaticism - which is a false form of self-confidence. For although the multitude is liable to constantly change its positions, always it clings to its current position with absolute passion. Paradoxically, the basic uncertainty of the person of the *imaginatio* does not result in tolerance, but in a permanent revolving of absolute stands.

This picture which is drawn from the TTP should be complemented by what the *Ethics* has to say about the psychology of the association and of anthropomorphic explanations. The world-image of the multitude is based upon contingent connections of ideas to which nothing constant and objective corresponds in reality (associations). This produces cognitive instability which both underlies and compounds the mind's emotional instability. In lacking necessary order and regularity, a world-picture based upon association again invites superstition. This is enhanced by the anthropomorphic nature of the imagination, its tendency to explain everything in terms of human-like purposes and intentions. Together, they provoke an irresistible appeal to occult human-like forces which are believed to act arbitrarily behind the natural phenomena and which must be appeased or otherwise influenced by flattery, submission, sacrifice, and spurious irrational acts.

Vana religio as Alienation

Spinoza thus carries his analysis of cognitive superstition into the domain of historical religion, the one based upon cult and prayer and upon some particular kind of revelation. All historical religions are *vana religio* to Spinoza, a spurious religion of the imagination. Their dominion over the lives and minds of their adherents is a form of repression - indeed, of *self*-repression, because the fear which makes this dominion possible arises from the masses' own weaknesses and expresses their psychological needs. This mental state of servitude is exploited by and readily enhances a political form of servitude: religious superstition becomes an instrument of tyranny. The supreme secret of monarchy, Spinoza says (momentarily denouncing all monarchic governments alike), is to wrap up the fear which it instills in its subjects in a nice religious mantle, so that the subjects *will fight for their servitude as if it were their salvation*. Here Spinoza seems to anticipate the concept of self-alienation which became prominent in later philosophical and psychological discussions. The psychology of the masses, by its own dynamic, and as long as it is not reformed in the semi-rational manner the TTP recommends, is liable to produce self-alienating religious and political institutions, in which the individual is made to turn against himself, viewing his bondage as freedom and a value worth of sacrifice, and thus deepening his bondage in the second degree.

The reformed religious and political institutions suggested by the TTP are designed to alleviate this alienation even without rising to the level of true rationality. This program is to be carried out in two stages which can be summarized as follows.

Stage I: The Universal Popular Religion

In the first stage, historical religion as *vana religio* must be undermined. This is the negative, indeed subversive side of the critique of religion; its role is to clear the ground for genuine rational life in those capable of it - and for a semi-rational substitute in the general populace. The substitute is the *religio catholica* (universal religion) discussed in the theological part of the TTP; it is a popular version of the religion of reason which remains rooted in passion and the imagination and is rational only externally or by imitation. Designed as it is for the multitude, not for the true philosophers, the *universal* character of this religion lies in its stressing the same patterns of conduct which reason, too, will recommend; but it does not presuppose any true knowledge, only obedience. Hence its universality is not self-constituted but borrowed: in order to know its own content and set the model it has to follow, the reformed religion cannot rely upon itself but must turn to what is an external authority for it - namely, to reason. This makes the use of reason not truly rational, since contrary to its nature, reason is used here without understanding, as yet another external authority to which one shows mere obedience. Herein lies the **semi**-rational nature of this kind of religion, its being a mere imitation of reason within the realm of the imagination.

Moreover, this form of religion has a second master as well: the authority of Revelation, or the Bible, which the person of the imagination is disposed to obey and by which, on Spinoza's plan, the multitude could be moved to comply with the semi-rational model. But this requires that the content of the Bible be re-interpreted to suit the message of the new universal religion. Although Spinoza insists that Biblical hermeneutics must become an objective and autonomous science, he tacitly also expects it to serve as a means for reforming historical religion by reducing the content of the *word of God* and the *true meaning of the Prophets* to a nuclear set of general (and rather secular) principles such as justice, solidarity and mutual help.

Thus, while the authority of reason supplies the paradigm of conduct to be followed, the motivation for doing so is still drawn from Revelation, or more precisely, from the multitude's persisting attachment to it. Spinoza treats this attachment as a real psycho-cultural force by which the mechanism of the reformed imagination can be set up in a purely naturalistic manner, diverting the passions of the multitude from their normally destructive course to produce benign and socially stabilizing effects.

Stage II: The Secular State spells out the Word of God

The second, political stage in Spinoza's program for the multitude is required both for its own sake and as a complement to the theological stage. The nuclear doctrine of justice and mutual help to which revealed religion and its *word of God* have been reduced is far too general and vague to serve as a basis for action. The principles of justice and solidarity must be spelled out in specific legislation and adapted to the social context in which they are meant to apply.

Nothing of this can be determined a priori, either by mere philosophizing or by consulting the Scripturer. The Bible had not been meant as a concrete political constitution except for one people, the ancient Hebrews, whose state has long been destroyed. Therefore, whatever specific commands the Bible contains are now outdated and irrelevant for both gentiles and modern Jews alike. What the Bible goes on teaching validly is its abstract moral nucleus, the general principles of social ethics which are non-denominational and therefore must be interpreted, that is, translated into actual legislation, by the secular authorities of each state.

By making the political authorities the sole *interpreters of the word of God*, Spinoza grants the secular government a monopoly over the normative domain as a whole, that is, over right and wrong, justice and injustice in all their valid applications. Since no normative concept can validly exist in Spinoza except by virtue of the enforceable law of some actual state, his move undercuts the superior authority which priests, rabbis, and church organs claim for themselves over and above the state, and denies their right to construe religion as a *realm within a realm*. The clergy may well participate in the battle of ideas, but they should not enjoy institutional privileges of any kind. By a similar token, Spinoza will have to dismiss as meaningless the moral claims of secular intellectuals who assume a **super**-natural or transcendental point of view in criticizing the laws passed by government, although he will of course not deny their right to be heard: logical and political legitimacy are sharply distinguished concepts.

In so far as the laws of the state, which specify how justice and mutual help are to be practiced within a given community, are seen as interpreting the *true word of God*, the political system employs religious sentiments and authority as an auxiliary. But the state, of course, also has its own independent system of obedience and authority. In monopolizing both might and right, power and justice, in possessing undivided sovereignity and the exclusive capacity to punish, deter, and enforce its laws, the state is the most potent mechanism by which the power of the imagination can be reshaped in the multitude as an imitation of reason.

This is particularly true of a free state whose laws come close to a rational paradigm; but Spinoza holds that even arbitrary laws are preferable to the fierce dangers of anarchy. The civil state is a realistic middle term between two hypothetical extremes: the state of universal strife where no government is available and the state of universal rationality where no government is necessary. Had all people been acting out of reason alone then, Spinoza holds, justice and solidarity would have reigned of themselves and state-power would have become superfluous. Although no rational person is altruistic in his motives, rational people understand what is truly beneficial and thereby are driven to act toward others in ways that in most cases have the same effects as altruism.

The multitude, however, which lacks such rational understanding must have another power, another natural mechanism to re-direct its passions toward socially beneficial conduct. This mechanism is supplied by the authority and legislation of the state. Using obedience, fear, and other non-rational powers, the state can elicit the desired conduct from its citizens, although it cannot change their fundamental motivation or reform the inner quality of their lives. The state, moreover, can attain this goal not through fear and intimidation only but by the routinized practice of the laws themselves and by remolding the associative functions of the populace in habitual patterns that echo or imitate a rational paradigm.

The state thus becomes in Spinoza not a mechanism of power only but equally a civilizing agent, an instrument of education. In both ways it uses the typical psychology of the multitude to create an institutionalized form of semi-rationality, imposed by obedience and enhanced by habit, repetition, and the memory of its advantageous results. This does not mean that the state should indoctrinate its citizens with any particular opinion; on the contrary, the state as such must be tolerant of all conflicting views and ideologies. Yet the very functioning of the state as an institution can mold the psychology and habits of the citizens and thereby, in turn, reinforce itself.

Such a program of course has its dangers; Spinoza left the difference between educating and manipulating the multitude much too broadly defined; he has given far too little thought to the need for checks and balances that would disperse the concentrated power of the state without compromising its authority or dividing its sovereinity; and, more generally, he paid little heed to the danger of a *despotism of reason*, a concept Spinoza must have deemed incoherent but which historical experience has since validated and to which Spinoza's theory is not sufficiently immune.

These are all serious lacunae that deserve separate discussion. But the above sketch of Spinoza's program in the TTP already indicates **the importance that non-philosophical discourse must have in its execution**. Using metaphor, allusion, equivocation and other rhetorical devices is essential to Spinoza's program no less than strict philosophical discourse and as an indispensable complement to it. This is even more so considering the special nature of the TTP which is not a theoretical work merely but a form of **philosophic engagée**. As I have tried to elaborate elsewhere,[2] the TTP was not written as a pure philosophical tract on religion, the Bible, the state etc. but was meant to intervene in the social and cultural processes of its time and, by its use of discourse, **to trigger and enhance the execution of its own recommendations in the very act of putting them forth**. In having this practical and even *performative* side, the TTP must use several levels of discourse simultaneously, address different audiences in the same text, exploit the connotations of old words to serve radically new meanings and, in general, use language in other ways than for pure description and philosophical reasoning. Significantly for Spinoza, these rhetorical forms acquire philosophical import in themselves, because they figure in a general theory of discourse which the philosopher must work out in response to the genuinely philosophical problem of how to live among the multitude - that is, in simpler terms, of how to live.

Let me now turn to Spinoza's uses of language. Although my main interest lies in the nonscientific kind of discourse, I shall have to consider the scientific or **geometrical** kind as well. The link between the two will come out more clearly in the last part of this essay.

II

The Scientific-Theoretical Context of Discourse

This is, above all, the universe of discourse of the *Ethics*. Language in this domain is to serve adequate ideas only; it takes its model from the formal-mathematical calculus, construed *ordine geometrico* and ideally requiring an absolute degree of transparency. According to this model, everything must be fully defined and conclusively established at the outset, without an opaque margin or residuum of any sort. Definitions are considered stipulative and exhaustive, they are supposed to create the full semantic scope of each term, barring any additional input, tacit connotations, or other satelite elements of meaning which might hover around the term in its natural use; and the same clarity which is attained at the outset must be transmitted to all further steps without loss or distortion. Thus, at every stage of the explication, each unit of discourse will be subject to the same principle of absolute **distinctness** with no blur or 'halo' at its semantic bounderies.

This also means that the theoretical language is fully synchronic. What the language means depends exclusively upon the inner and simultaneous relations between its components. To understand a term we are neither allowed nor required to transcend the given linguistic system to something else - natural language, the history and etymology of words, the linguistic habits of actual speakers - or to consider the role of metaphor, connotation, and other semantic inputs and accompaniments. The entire information necessary and relevant to fully understand this language is supposed to reside in the system itself, as if it were a formal-deductive calculus.

I say *as if it were*, because, as we shall see, this ideal is neither actually implemented by Spinoza **nor can it be carried out where substantive philosophical issues are at stake**. But before considering Spinoza's practice I should like to indicate a difficulty that he seems to encounter even in his ideal model, and how he may have responded to it.

The problem arises from the use of primitive terms and axioms and from the purely linear, one-directional character of the progression. This may seem to reflect the finitude of the philosopher as a particular mode; but the philosopher is also thought capable of rising to the standpoint of eternity, where he grasps things according to the order and connection of their ideas in the *infinite intellect of God*; if so, how can he allow for the opacity and arbitrariness that seem to be incurred by the use of axioms and primitive terms, or be satisfied with a linear and quantitative progression, which is bound to miss the true nature of the totality as a qualitative infinity and as a system of mutual relations?

Spinoza's response consists, I think, in the intuitive mode of knowledge by which he flanks the linear deduction on **both** its sides.

1. First, axioms as Spinoza understands them are not arbitrary but necessary and self-justifying truths. The absence of anything beyond them which could ground them does not mean they are groundless but that their grounds lie in themselves. This offers an analogy in the domain of discourse to Spinoza's ontological concept of the *causa sui*. The *lumen naturalis* is supposed to give us beliefs which are justified in the very act of grasping their meaning; it thus supplies self-verifying axioms which are not arbitrary but rational - not opaque but *clear and distinct*. As for the **terms** by which such ideas are conveyed, they must count as linguistic expressions of rationally necessary first concepts, not of mere conventions or arbitrary notions.[3]

In this respect, the only opacity remaining in the system is trivial - that of the sign-conventions chosen by the language in use; and these, Spinoza seems to hold, may be perfectly functional in invoking the intended words and ideas, and can be replaced by the signs of another language without loss of meaning or efficiency.

But signs, of course, can also be misleading - as Spinoza takes them to be most of the time. The natural attitude of the mind is to draw its beliefs from the crude and direct interpretation of signs, without the mediation of causal-rational knowledge; hence knowledge from mere signs (linguistic and non-linguistic) is classified by Spinoza among the forms of the *imaginatio*, the lower cognitive degree. At the same time, linguistic signs can also be made to trigger the exact ideas we intend, and non-linguistic signs can be severed from their alleged interpretation and given a proper scientific one. Thus, although we depend upon signs, they do not necessarily curtail the rationality of our knowledge.

Does the dependence upon signs indicate a remaining difference between the finite philosopher and the *infinite intellect of God*? Must God also think in some natural language? Does God speak Hebrew, or Latin, or perhaps 17th century Dutch? These questions can happily be spared in Spinoza's system, where God does not think or speak at all because he is not a person, and where the *infinite intellect of God* does not mean a subjective process but the sum total of all objective truths about the universe, taken in their mutual interrelations. The components of this intellect are not words or sentences but ideas, understood as pure logical entities, propositional contents etc. Of course we must assign words to designate these ideas (and signs to invoke the words) but Spinoza, as we have seen, postulates that this can be done adequately, without submitting this postulate to any serious scrutiny: he is greatly concerned with the adequacy of ideas to things, not so much of words to ideas.

2. Just as the linear deduction is preceded by intuitive reason in the form of *lumen naturalis*, so it culminates in *scientia intuitiva*, another form of rational intuition. The philosopher can reach the authentic point of view of God, or eternity, only in the third degree of knowledge which supersedes *ratio* and its merely linear progression. Although the precise nature of intuitive reason is notoriously vague in Spinoza, it is undoubtedly supposed to transcend the step-by-step march of the deduction and grasp the relevant field of connected items of knowledge in one synoptic review. As the network of logical relations becomes manifest to him as a single totality, the philosopher can start seeing the components of the system in their mutual grounding

and no longer as a one-directional sequence. Instead of a linear form and a false quantitative infinite he now reaches a true, qualitative infinite, that is a closed and comprehensive system where everything depends on everything else in an eternal movement of circular grounding (Ep 12 and in E 1P15S). This model provides a fuller and more adequate interpretation of the concept of *cause of itself*. Everything in the system depends on everything else, except for the totality which depends on itself alone. This implies in turn that the totality must appear as a self-**particularizing** unity, one that is cause of itself by being the cause of an infinite diversity with which it is identical. Substance as *natura naturans* grounds itself qua *natura naturata*. The totality seen as unity is not separate from the totality seen as diversity but exists within and through this diversity alone. This deeper view of the system breaks its seemingly linear rigidity and presents God, or Substance, as the circular and qualitative infinity it is.[4]

This deeper insight of the totality (and of the place of the actual particular within it) cannot be attained in the merely discursive stage of *ratio* but requires the synoptic grasp of *scientia intuitiva*. This is why Part I of the Ethics is only grasped in full when revisited from the standpoint of Part V. Spinoza's book, no less than his system, has this tacit circular (or spiral) form which overshadows its apparent linearity. The linear progression is a necessary condition for attaining the third kind of knowledge but is transformed and superseded in its new, holistic and intuitive grasp.

That such an intuitive experience should come together with an overpowering emotion is plausible, and a major reason why Spinoza maintains that *scientia intuitiva* alone, and not *ratio*, is capable of dominating the other emotions and bringing about a moral and qualitative revolution in the inner person.

More geometrico as Metaphor

So much for Spinoza's possible response to the problem of opacity and arbitrariness in his ideal model. But the fact remains that Spinoza did not implement this model in his actual philosophical discourse. The two political tracts are not written *more geometrico*, and one of them, the TTP, is saturated with non-theoretical language - metaphors, rhetorical devices etc. Moreover, even the *Ethics*, where Spinoza speaks his purest and most direct and literal philosophical language, approaches the deductive model only from afar.

The definitions in the *Ethics* do not create the full semantic scope of their objects. Spinoza's seemingly stipulative definitions are in fact theoretical ones; they offer a new, sometimes revolutionary interpretation of a traditional philosophical issue which the term chosen as the *definitum* invokes. Thus, when defining *God*, *substance*, or *freedom*, Spinoza is not just laying down a convention: he exploits the existence of these terms in traditional philosophical discourse in order to put forth his own theory about their objects, stating what God, substance and freedom actually are. His definitions thus provide nuclear philosophical doctrines which Spinoza introduces at the very outset. Indeed, some of these definitions contain no less than Spinoza's major revolutionary insights of which a good deal of the rest is only explicatory. Moreover, as readers of the *Ethics* often realize, the order in which these insights are spelled out *more geometrico* could frequently be reversed, placing a theorem in lieu of a definition or a definition in lieu of an axiom and vice versa.

This may lead to a further consideration: the linear progression serves to spell out a fundamentally closed, circular system whose contours precede and implicitly guide the deduction (although they require the deductive process in order to take explicit shape). The full scope of the system emerges only after two necessary conditions have been fulfilled: first, the actual deductive process must be performed and secondly, its linear form must eventually be transcended toward the synoptic overview of *scientia intuitiva*.

This implies a distinction between the system of truth as it is in itself (in the *infinite intellect of God*) and the cognitive order in which it is reconstructed and known by the mind. As a rough analogy we can think of a geographical map whose various landmarks stand in constant mutual relations but can be approached in a variety of back and forth routes. Thus in metaphysics, the *ordo cognoscendi* may be open to variation, as long as its paths follow the same logical map (and as long, of course, as it respects the logical rules of inference which alone can serve as its vehicle). But at the final stage, when the progression is supposed to be replaced by a holistic and simultaneous grasp of the map, the question of cognitive order becomes obsolete.

These considerations should modify the sometimes exaggerated role which commentators of the *Ethics* attribute to the geometrical order and to its specific layout in Spinoza's text. We may recall that Spinoza had published a geometrical exposition of Descartes' philosophy while disagreeing with much of its content, and he avoided the geometrical form in his own political tracts; which serves to prove that the geometrical model is not as sacrosanct to Spinoza as is sometimes supposed for it neither guarantees nor is indispensable to the attainment of truth. The geometrical model is more a matter of philosophical form and mood; it stresses the need for rigor, clarity, and step-by-step consequentiality as necessary conditions of rationality, and it also calls for philo-sophical detachment in dealing with the most passionate issues. Thus it dramatizes certain main features of Spinoza's philosophical message. But Spinoza makes no claim that the infinite intellect of God takes the same cognitive route as traced in the *Ethics* - nor does he actually shape his book as a formal-deductive calculus. As any reader of the work will recognize, the *Ethics* draws heavily from natural language and the traditional vocabulary of philosophy; thus it contains elements of content, allusion, connotation etc. which an actual formal model will have to reject as opaque; and if the *Ethics* admits them, it is because its language is **not** truly **geometrical** and cannot be so by the nature of its subject-matter. Spinoza's actual discourse, even in pure philosophy, is fundamentally non-formal but depends on history, natural language and the accumulated human experience, although it is organized in weak analogy to a formal system. Thus the geometrical model, the antithesis of metaphorical discourse in Spinoza, is in this respect itself a kind of metaphor.

In a succinct and crucially important remark, Spinoza explains his use of philosophical terms (and its dependence on cultural connotations):

> *I know that these words have a different meaning in ordinary use. But my aim is to clarify not the meaning of words, but the nature of things, and to designate them by those names whose meaning in ordinary use does not in the main contradict the meaning I wish to assign to them* (E 3AD20Ex; my translation).

Although Spinoza makes this remark with reference to moral and psychological terms, it applies to the *Ethics* throughout and has the role of a major statement. It clearly states Spinoza's preference for historical terms that have a relevance, or affinity, to the subject-matter he explicates.[5] Whether it be God, substance, freedom - or love, envy, error etc. - the subject matter is invoked by the proper term, then re-defined or reinterpreted by Spinoza. His aim, as he indicates, is to give theoretical rather than lexical definitions, conveying not what a term signifies in ordinary use but the true nature of the thing it invokes and is used to designate.

This presupposes, of course, that philosophical issues have their historical continuity and that traditional terms are capable of properly designating such an issue (or subject-matter) even if the term is given a false or distorted interpretation. For example, we may well have God as the reference of our talk even if we miscontrue his nature. Designating a subject-matter and correctly interpreting its objects are two distinct acts, one logically independent of the other; the first is best served by using traditional terms, while the second requires that they be given new definitions that sometimes amount to true semantic revolutions. Only the new definition is supposed to capture the actual object which other definitions have missed, but in order to know where our target lies (and to render the talk about its being 'missed' or 'captured' meaningful), we must first delineate its general domain, be it in vague contours only; and this, Spinoza holds, is best done by traditional terms which are retained even after their meaning has submitted to a radical shift.

A further role of using the historical vocabulary of philosophy is to exploit the emotive and connotative halo of a traditional word and, by re-definition, transport it to a new object to which it is believed to properly belong. To this major use of pilosophical terms I shall refer later, when discussing the *hermeneutic* function of language.

In summary, as long as philosophy must deal with substantive and perennial questions it cannot, from Spinoza's own standpoint, be reduced to a purely formal and synchronic form of discourse. This is why the geometrical ideal could neither be implemented nor have been intended by Spinoza in any other way than as a weak analogy. Even so, there is a valid concept of **strict** philosophizing which requires rigor, clarity, and a form of discourse that is plain, literal, and unambiguous, free of metaphor and equivocation, with no attempt to suggest, persuade or otherwise affect the reader rhetorically. The philosopher will adhere to this kind of discourse when reasoning with himself or with others who share his standpoint; however, there are other contexts in which, for **philosophical** reasons, he will have to use different types of discourse as well. To these other uses we shall now turn.

III

The Rhetorical Context

Spinoza was a master not only of clarity and rigor but equally of equivocation and double language. His rhetorical use of language was elevated to the level of art. Here Spinoza brought to perfection a style and skill in which his ancestors, the Marranos, had excelled for generations (Yovel 1982 and forthcoming Part One, Ch. 4), adapting it to his own situation, that of an esoteric thinker enmeshed in the world and living among the multitude.

The rhetorical use of language is not, however, only a personal talent in Spinoza. We have seen that it also has a philosophical import in itself, as part of a general theory of discourse which the philosopher must evolve in view of the problem of the multitude.

Inevitably using inadequate ideas, the rhetorical use of language adapts itself to the mind of its target-audience. Its aim is to affect the imagination in ways that will produce desirable (that is, semi-rational) effects in the audience's perception, emotions, and especially conduct. Spinoza assigns three major functions to the rhetorical use of language: one passive, or defensive; another active, even aggressive; and the third I shall call constructive (or hermeneutical).

The Passive or Defensive Function: Prudence

Caute was not merely an inscription on Spinoza's ring. Prudence to him was far more than a personal strategy or temperament: it was a philosophical issue in itself, what *ratio* must recommend to any philosopher as part of the life of reason. Given the all-embracing presence of the multitude, the question of how to handle it acquires a major ethical importance. We have seen that the philosopher cannot attain his goal in isolation, nor can he expect all people to rise to his own level. At the same time he has to engage in public affairs (or public discourse) since part of his goal is to help others, who are capable of it, to cast away their superstitions and share in the life of reason. Yet this goal has nothing of a missionary zeal in it; motivated as it is by rational motives, it is held in check by a sense of realism and also by a kind of paternalistic benevolence on the part of the philosopher: knowing that most common people are incapable of truly rational attitudes, he will not seek to provoke and shatter their lives in vain. He is also well aware of the dangers involved, which makes his benevolence join forces with his prudence, both nourished by purely rational motives. Yet some people are capable of rising over their situation, and the philosopher must seek and pick them out by a differentiated discourse which speaks in different voices to different sub-groups, even though officially it is addressed to all of them alike.

In other words, the philosopher must trace the delicate and narrow path between two conflicting imperatives: provoking a rational conversion in those capable of it and concealing his true message from those whom it will not benefit and might endanger for having expressed it. Both aims are dictated to him by reason; as a rational person he must both help to educate the first group and refrain from provoking the second; and since his major instrument is public discourse (which officially addresses itself to everyone alike), the rhetorical and equivocal use of language becomes rationally necessary for the philosopher.

Barred from revealing his true mind to everyone, the philosopher cannot be fully trained for a life of reason unless he learns how to use language defensively, masking his true aim and intentions and passing tacit messages to some while in the same text or phrase misleading others. One of the most efficient masks is the use of pious phrases, images, formulae, borrowed from the Scriptures or from accepted religious beliefs.

For example, Spinoza in the TTP repeatedly mentions the will of God, his decisions, thoughts, and commands; he cites biblical verses, alludes to Christian dogma, to Providence, to God's Son, to events where God had acted as a particular agent, and he otherwise cloaks himself in language which his strict philosophy would reject as superstitious. Spinoza even puts forth a list of *articles of faith* which, taken literally, flatly contradict his metaphysical doctrine (they refer to a personal God, moved by justice and love and not by necessity etc.), although almost every article contains a nucleus which can be translated into a strict Spinozistic thesis and be seen as its metaphoric mask.

In his correspondence Spinoza used a similar technique. Many of his statements have a fasade that sounds acceptable to the current theological idiom, and an inner meaning that is bluntly heretical. That his prudence did not spare him the charge of atheism may testify to the limits of linguistic camouflage. After all, Spinoza was not just a reformer of revealed religion but its adamant enemy; his philosophy of immanence (or so called pantheism) stood out in dissenting affront not to the established religions alone but to all the other philosophers of reason as well, who accepted the transcendent status of God and the duality between God and His world. Spinoza was in this respect a loner even among the daring heterodox minority; and with such a gulf separating him from his contemporaries, he could not sufficiently mask his thoughts without compromising his intellectual identity and goals beyond what he deemed tolerable. At the same time Spinoza did practice concealement in varying degrees and lived - even in the relatively tolerant Netherlands - as his own kind of Marrano, a **Marrano of Reason** (as I think we may call him) whose hidden esoteric truth, the one that leads to true salvation, is not Judaism in opposition to Christianity but the immanent religion of reason in opposition to all historical religion alike (Yovel 1982 and forthcoming Part One, Ch. 4).

The Offensive Function: Persuasion and Polemics

Spinoza uses language rhetorically also in order to arouse doubts in his audience and combat the malignant effects of *vana religio*. This offensive function of language is independent of the defensive and would have existed even if there were no problem of prudence. It expressed the philosopher's need to reach out to those who may be capable of rational life and subvert their entrenched beliefs as preparation for philosophy. This is a kind of mental *purification* similar to the one which the method of the doubt and the free will behind it were supposed to achieve in Descartes. Spinoza recognizes no free will, only causal processes, and he uses language rhetorically to induce doubts and queries in his audience through a natural course of cause and effect. Whereas Maimonides (Spinoza's chief mentor in the matter of rhetoric and the multitude) had written a *Guide of the Perplexed*, Spinoza wished to start by provoking perplexities in order, first, to loosen the grip of religious superstition over the multitude and make it prone to accept new, rationally guided authorities (including the secular state), and also, for a smaller group, in order to clear the ground for a genuine life of reason.

Neither goal can be reached by merely arguing *more geometrico*. In order for rational arguments to be effective against the entrenched religious beliefs, the audience must already recognize the superior authority of reason over Revelation. But this is precisely the iron bar which Spinoza confronts, it is a deep chasm over which no particular usage of reason can bridge because nothing less than an overall conversion will do. Vicious circle? Perhaps, but Spinoza believes that it can be broken with the assistance of less-than-rational factors which will mediate the ascent of rationality even before it had totally vanquished. Thus the rhetorical use of language can build an imaginary or semi-rational bridge where a genuinely rational one could not have been erected. Since the person of the imagination cannot share the philosopher's rational standpoint, the philosopher will have to assume his interlocutor's language and standards in order to transform their meaning and eventually turn their authority against itself. This requires a dialectical skill and ample use of metaphor, allegory and equivocation. Using dual language, the philosopher can adjust his talk to the minds of the multitude while always retaining a sense in which his discourse is true also philosophically. The latter point seems important to Spinoza in order to avoid being (or appearing) totally deceptive; in order to pass a tacit message also to the initiated; and perhaps above all, in order to prefigure the rational outcome which the allegory should serve or help emerge, and thus give the whole exercise a semi-rational and educational character, its only justification from Spinoza's standpoint.

Thus Spinoza, the enemy of allegoric reading of the Bible, falls back upon allegoric and metaphorical language when the text concerned is neither the Holy Scripture nor strict philosophical reasoning but discourse that is to help the latter undermine the authority of the former and eventually usurp it.

Another familiar technique is to undermine the Bible's authority by finding contradictions in it. But this again presupposes that some credence is given to the force of reason, for although the contradictions are materially drawn from Biblical statements, it is reason which passes the verdict, judging these contradictions as flaws (rather than, say, as signs of God's imponderable wisdom). This may remind us that the dichotomy between reason and revealed authority is frequently not as neatly defined as an abstract model presumes; religious believers who respect their rationality are often deeply disturbed when having to admit the irrationality of the Scriptures, even if this will not necessarily shatter their belief. On the whole, the more inquisitive or doubtful one already is, the more effective this technique will prove to be in his or her case. Here again we are dealing (at most) with a tool of gradual change, not with a sweeping one-time instrument of conversion.

As in other clashes between reason and revelation, this kind of polemics manifests that reason is incapable of proving its superiority, but must in the last analysis affirm it. A rationalist standpoint such as Spinoza's or Descartes' involves an existential stand which no proof can provide or guarantee. Of course, these two philosophers will not admit of any such description; to them, reason is its own light or provides immediately its own justification. But even on their theory, rationality does not emerge from argument and proof but from the kind of intellectual intuition they call the *lumen naturalis*: a secular kind of revelation which is supposed to be open to everyone and entails no religious connotations. This rational illumination either occurs, taking complete possession of the mind, or it does not. What if it does not? What should be done while the mind is still possessed by false ideas and is not ready for clear and distinct ones to emerge? The inevitable answer is: one must pre-figure rationality by less-than-rational means and suggest the horizons of reason as a temptation rather than as a self-grounding truth. Rationality itself must in this inevitable stage be made an object of persuasion.

Although it is customary to view Spinoza's thought as utterly unhistorical, his theory of language and the multitude may serve as a counter-example. Here Spinoza reasons in terms of a process rather than in those of an abstract dichotomic model. He does not envisage a radical, one-time revolution but a gradual growth of rationality from within the domain of *imaginatio*, and he thinks it is the philosopher's task to provide tools for dealing with the various forms of this transition - as he himself does in his theory of allegory, metaphor, and non-scientific discourse generally.

The Constructive Hermeneutical Function

While part of the multitude whose dogmatic beliefs had been undermined by the pre-philosophical discourse will eventually move on to genuine rationality, the majority will remain in the realm of the passions and the imagination, which, on Spinoza's program, must be reorganized as an external imitation of reason. In both cases rhetorical discourse does not have defensive and subversive uses only but serves a constructive hermeneutical function: in the case of the first group it helps preserve the semi-religious dimension which Spinoza considers essential to rational truth as well; and for the second and larger group, the hermeneutical use of metaphor and equivocation should help establish the semi-rational imagination which is Spinoza's answer to the problem of the multitude.

I use the term 'hermeneutical' in the general sense of adding or extracting a new structure of meaning within a body of discourse which carries a cultural or traditional import. Spinoza's constructive use of hermeneutics is seen in two major ways:

a. In his reading of the Bible Spinoza extracts a rational message from what seems otherwise to be, and serves as, a tool of the imagination;[6] and

b. In his systematic works he exploits the affective and connotative halo of select traditional terms, especially 'God' and its various derivatives, in order to give the philosophy of reason its import as a substitute to religion.

Metaphoric-Systematic Equivalence

There is a whole series of terms which serve Spinoza as metaphors but are perfectly translatable into strict philosophical language. By redefining these traditional terms Spinoza carries them over from the realm of the imagination into that of reason. Although the literal sense of the term may be very misleading (e.g. *the will of God*), there is another, philosophical sense into which it can be translated and which constitutes its tacit new meaning.

An example of such a translation is given in the following quotation:

> *To those who ask why God did not create all men so that they should be governed only by reason I give no answer but this: because matter was not lacking to him for the creation of every degree of perfection from highest to lowest; or, speaking more properly (**vel magis proprie loquendo**), because the laws of his nature are so vast as to suffice for the production of everything conceivable by an infinite intellect* (end of E 1A; emphasis added).

Spinoza sees a tacit equivalence between the first statement, which uses the common theological idiom (God as omnipotent creator) and the second statement which translates this imprecise metaphor into proper philosophical discourse (that of E 1P16 et al.). The second statement uses adequate ideas to pass the same message which the first had couched in inept words and inadequate ideas. The philosopher using or reading such discourse will know that the true meaning of the first statement is actually given in the second, although for rhetorical reasons the equivalence is not allowed to appear on the surface.[7]

We may designate the relation between this type of metaphor and its rational equivalence by the symbol:

$$P \overset{(met)}{=} P'$$

This is of course a special kind of equivalence; P and P′ are not interchangeable in their literal senses, although each in its own idiom translates essentially the same idea. Literally taken, the two expressions have different meanings and (when they make a statement) opposing truth values; but as the modifier *(met)* serves to indicate, only one of them is autonomous and meant to be taken literally; the other draws its meaning from the first and serves as a rhetorical envelope for it.

Given this relation, we can offer a few translations of metaphoric expressions of this kind into their systematic equivalents:

(*) *God's intellect*:	The totality of adequate ideas (including all true propositions and theories about the universe) taken in their interrelations.
(*) *God's will*:	The totality of facts, events and processes in the universe taken in their necessary causal connections.
(*) *God's power*:	The same as God's will (with a subjective emphasis on factuality).
(*) *Creation*:	The inner particularization of the substance in accordance with the logical laws of its nature (E 1P16).
(*) *Salvation*:	Knowledge of the third kind coupled with intellectual love of the nature-God.
(*) *God's omnipresence*:	The fact that all modes are in the substance.
(*) *God loves justice and benevolence*:	Justice and mutual help are models for imitation in conduct (for non-philosophers).

Depending on context, the expressions on the left (and others like them) can be used either in their metaphoric capacity or as direct substitutes for the expressions on the right. Spinoza uses them both ways, to suggest or prefigure rational ideas and, in parti-cular, to embody and encapsulate some of the major principles by which the imagination is to be reshaped as a practical imitation of reason. Metaphoric discourse serves him here in a constructive capacity, as building-blocks of the semi-rational imagination.

This is illustrated, perhaps without refinement, in the *articles of faith* which Spinoza designs for the multitude as mere postulates for action. Although Spinoza says they need not be true in themselves as long as the agent believes them to be true, and indeed couches these beliefs in a particulary heavy theological idiom, at least six of the seven principles in Spinoza's list can also be given some general philosophical interpretation. That God exists; that he is one and unique, omnipresent, and possessing supreme right (=power) and dominion over everything; that his worship consists exclusively in leading a life of justice and loving one's neighbor; and that only those (and all those) who obey God in leading such a life will be *saved* - all these statements are compatible with both superstition and philosophical truth, depending on the mode of interpretation.[8]

Spinoza himself explains how this duality works. For example, he says in another revealing aside, the dictum that God provides the model of true life can be taken to mean either that

a. God is a person animated by justice and mercy (superstition), or that

b. God is the natural ground of all things and ideas, including the philosopher's ideas about the good life (Spinozistic truth).

As we move to interpret each dictum in more detail, the gulf that separates their true from their superstitious construal will become more apparent; but the multitude, Spinoza insists, need not engage in a detailed interpretation in order to use these beliefs as postulates for action. In these examples Spinoza again exposes his technique of metaphoric equivalence and lets us see the constructive role he assigns it in building a semi-rational imagination.

However, not all metaphoric expressions can be translated into neutral equivalents without incurring significant loss from Spinoza's philosophical standpoint. This warning should apply above all to the word 'God' and some of its derivatives, but also to such terms as 'piety', 'beatitude', 'love', etc. which maintain the affinity of rational philosophy with religious sensibilities and concerns. Although Spinoza secularizes religion, he equally creates a system of naturalist rationalism which preserves some of the supreme goals of the historical religions it rejects. It is therefore essential to Spinoza that the semi-religious dimension of philosophy will not be shed along with the many errors and superstition that had governed it in the past. Spinoza's philosophy is also concerned with God, or the absolute, and the soul's identification with it; and ultimately it aspires to wisdom and a new form of salvation. Here a major difference lies between Spinoza and naturalist philosophers after him, most of whom renounced such exulted aspirations and made do with much more modest spiritual results. (Freud, in certain ways a modern Spinozist, is an interesting case in point).

In Spinoza, although the life and ethics of reason are marked by naturalism, this does not make them banal and prosaic. On the contrary, some of the major drives that had in the past animated religion and even mysticism are still partly at work within Spinoza's naturalist rationalism; and to preserve this essential feature it is indispensable to continue using some of the old vocabulary, above all the word *God* in its new and daring association with *nature*. Neither of these twin terms can be abandoned, since it is their conjunction in *deus sive natura* which passes Spinoza's actual message - both naturalizing God and sacralizing nature.

Hence the need not only for the word *God* but for other charged terms, whose re-definition transports their high spiritual tension to the correct object. From a strictly formal or *geometrical* viewpoint, the terms *deus* and *natura* are interchangeable; yet *natura* alone, because of its historical connotations, cannot express the immanence and inherence of the divine in everything natural. The semantic halo of this term, as used for centuries and as embedded in every European mind since childhood, has negative shades which suggest something quite antithetical to the divine and inherently devoid of it. Spinoza, however, seeks to attach the elevated states of mind which traditionally have been associated with God and with things divine to their true, natural objects; hence he must employ select theological terms even within his strictly philosophical discourse while giving them revolutionaryly new interpretations. (This, as we have seen in another context, is also a reason why a fully transparent and synchronic language is impossible in Spinoza).

Like 'God', the terms 'piety' and 'religion' as used e.g. in E 4P37S1 have similar function. But let us consider another example which is not directly connected with theology. In discussing 'natural right' Spinoza uses the term *jus* as interchangeable with *potentia* (power or capacity). However, only power is the strictly adequate term; in the state of nature, before any political authority has been constituted, the concept of right is meaningless under any interpretation. Man's so called natural right extends to anything that his passions covet and his power can attain, and he will seize this thing by natural necessity alone: no normative concept whatsover is or can be involved here. Yet

Spinoza, who insists upon calling this natural power by the name of 'right' (TTP 16 et al.), performs thereby another rhetorical exercise. He exploits the dignified sense which *jus* has in traditional, non-Spinozistic discourse (as something transcendental and higher than mere nature), in order to attach this dignity to the natural power itself. But strictly speaking, no right can yet be involved here, and the term *jus* is used parasitically to convey a halo of favorable connotations to something other than itself.[9]

There may be several degrees in which exploiting such traditional connotations is important to Spinoza. We may consider this practice less vital in political discourse than in metaphysics or in the higher form of ethics. But whatever the degree involved, we have seen how an affective or metaphoric structure of discourse must often be retained from tradition and embedded in Spinoza's philosophical language in order for the true import of his thought to emerge and be grasped.

Conclusion

The constructive role of hermeneutics is seen in broader perspective as Spinoza re-defines traditional terms into building blocks of his own innovative system. While this often produces semantic revolutions, Spinoza always preserves certain elements of the historical term which he reinterprets, drawn from either its kernel or from its semantic margins and halo; and he addresses himself to what is to him a real philosophical issue or object which the historical terms have vaguely envisaged but always missed because of false or prejudiced definitions.

In this sense Spinoza, the official proponent of the geometrical method, is more historically oriented than meets the eye. Writing as he does within a given set of conventions and vocabularies which for long have articulated the philosophical concerns of his culture, Spinoza cannot start philosophizing from the linguistic *tabula rasa* which Descartes' method may seem to suggest. Actually, even in his 'geometrical' train of reasoning Spinoza performs a good deal of hermeneutical work, using traditional concepts to reinterpret others while preserving some of their denotative and much of their connotative import and, in general, remains within the broad confines of a philosophical quest which former philosophers as Spinoza sees them had both delineated and misconstrued.

Of course, the task to re-define the true objects of this quest amounts in Spinoza to no less than a momentuous revolution. His departure from the dominant trends in philosophy and theology is most radical. Yet even this revolution is inevitably articulated by means of the historic and linguistic context which at the same time it transforms. It also has other, social and strategic aims which require that language be used in a versatile manner, adapted to its various functions as we have examined them in this essay. Historical terms, metaphor, and equivocation have all their indispensable place within this scheme: we have seen how they are used defensively as a means of prudence; offensively as a weapon to subvert superstition; and constructively, as a hermeneutical tool which helps set up the reformed, semi-rational imagination which is Spinoza's positive program for the multitude. In addition, a constructive hermeneutical function applies to the level of rational discourse proper, where historical terms and/or their connotative halo are used to draw some vital elements from the old philosophy and theology, especially to maintain an affective and semi-religious dimension in Spinoza's philosophy of reason, which does not seek knowledge only but a change of heart and mind and eventually a new form of salvation.

NOTES

1 I use this term in slightly broader sense than used in E 3P17S, where it is defined as opposition between two contrary emotions. Such opposition, as between love and hatred, joy and sorrow, fear and hope and their derivatives, is not to be understood as just one emotional state among others in the list, but as the **underlying structure** of the emotional life of the person of the imagination. As such, the fluctuation of the soul is the most characteristic trait of the multitude, from which most of its other irrational and destructive outbursts derive.

The centrality of this unstable mental state in Spinoza's theory, as the major mental disturbance to be overcome by reason, is also indicated by the opening pages of TIE where Spinoza 's philosophical program is laid down. (Incidentally, a play by Spinoza 's tutor, Van den Enden, in which Spinoza may even have acted, dealt with the fluctuation of the soul as a theme; - I learnt this point from Mr. Bedjai in Paris, at a meeting of the *Association des Amis de Spinoza*).

2 Yovel 1979, shortly to be reproduced with several changes in my forthcoming book, *Spinoza and other Heretics*.

3 This, of course, raises the general question of the relation of words to their ideas, to which Spinoza did not devote much reflection. He seems to view words as independent designators which in proper strings and contexts may invoke their designated ideas in precise and adequate ways, and which can moreover be translated from one linguistic system (say, a natural language) into another without losing this function. His declaration that he is interested in the nature of things rather than the meaning of words (see below) may imply a belief that words are neutral vehicles which can be dissociated from their ideas and therefore discounted. Ideas, on the other hand, cannot be dissociated from their objects because they are ontologically the same as their object, and the order and connection of ideas is the same as that of things.

4 Under this interpretation, Spinoza is much closer to Hegel than the latter realized or was willing to admit. Spinoza lacks (or rather would reject) the dialectical logic by which Hegel sustains this metaphysical picture, and the anthropomorphic ontology it implies.

5 There is some understatement in Spinoza 's saying that the term he chooses must not *contradict* his new interpretation: this condition should not be construed as sufficient (most words in the dictionary will trivially pass this test) but as an added condition, the first condition being, as stated, that the term have a natural or historical link to the matter under consideration.

6 That Spinoza succeeds in doing this on the basis on an anti-allegorical method of interpretation is almost a *tour de force*. Although he is scientific and predominantly literal in his approach to concrete issues in the Biblical text, Spinoza reverts to tacit a priori generalizations (motivated by his strategic aims) when declaring that nothing in the Bible contradicts reason or claiming that the Bible teaches nothing except justice and mutual help as practical precepts.

7 It seems surprising that on this occasion Spinoza undercuts his own metaphor by making its rational translation immediately explicit; but at least he gives us here, even if inadvertently or by implication, a statement of his usual rules of the game.

8 Only the seventh principle, that God forgives the sins of the repentent, seems impossible to translate into philosophical language and to reconcile with Spinoza 's rejection of guilt and remorse. But we should remember that this rejection applies to the rational person, whereas the *articles of faith* are formulated for the multitude, which should remain affected by guilt no less than by fear and by respect for authority if a

semi-rational pattern is to govern its conduct (and thus benefit or *save* it). Also, Spinoza's *articles of faith* can be translated into philosophical truths only in their general metaphoric outline, not in all the details which accompany them. This is why I spoke above of a 'particulary heavy theological idiom'.

9 Spinoza capitalizes here on equivocation in the concept of *having a right to x*. This concept can be construed as meaning that there is no law or norm prohibiting me to do or to possess x. But why is there no such norm? Two totallity different reasons can be given:

a. we live in a state of nature where no law exists and hence no norm can have a meaning;

b. we live in a legal system which does not prohibit me to do or to possess x.

Only the second answer can make Spinozistic sense of the concept of right, which presupposes the existence of a political state and its enforceable legislation. Yet Spinoza uses *right* also in the first context where strictly speaking it cannot yet have any sense. Why? One plausible answer is that Spinoza tries to capitalize on the edifying resonance which the word *right* invokes in his audience because of the latter's false beliefs in order to facilitate the ushering in of something very different indeed, namely, his own naturalistic revolution of the concept of norm. In addition, as the positive halo of the term *right* radiates over the notion of natural power, nature and all things natural are being further highlighted as expressions of the divine. (Here we connect again with the theological context).

[25]

WHY SPINOZA CHOSE THE HEBREWS: THE EXEMPLARY FUNCTION OF PROPHECY IN THE *THEOLOGICAL-POLITICAL TREATISE**

Michael A. Rosenthal[1]

Introduction

In 1664 Ferdinand Bol produced a large painting for the Magistrate's Chambers (*schepenkamer*) in the Amsterdam Town Hall entitled 'Moses Descends from Mount Sinai with the Ten Commandments'.[2] Bol depicts Moses as he is described in *Exodus* 34:29–30: face shining and eyes wide with awe, descending with the tablets of the Law. We can infer that this is Moses' second descent from Sinai from the position of the tablets. The first time Moses descended Sinai with the Law, he was greeted by the sight of the Hebrews dancing around the Golden Calf, the idol they had erected in his absence (*Exodus* 32:19). Other contemporary artists, among them Rembrandt, depicted Moses at this moment holding the tablets high above his head in anger, ready to smash them in his disgust with the scene. But in Bol's painting, we see Moses holding the tablets in front of his chest, preparing to show a far more reverent crowd the Law God has just given them.[3] Below Moses are the people of Israel, who are attentive

* Throughout this paper passages from the *Theological-Political Treatise* (*TTP*) are referred to by chapter, paragraph number and reference to Gebhardt's edition of Spinoza's *Opera*. All translations are taken from a typescript copy of the second volume of *The Collected Works of Spinoza*, translated and edited by Edwin Curley. I would like to thank Professor Curley for allowing me to use his work before it has been published.

[1] For their comments on an earlier draft of this paper I would like to thank Ann Jensen Adams, Daniel Brudney, Edwin Curley, Daniel Garber, Joseph Stern and Janelle Taylor.

[2] See photograph between pages 216 and 217. For my discussion of this painting I draw upon two works of Albert Blankert, *Kunst als regeringszaak in Amsterdam in de 17e eeuw: Rondom schilderijen van Ferdinand Bol* (with summary in English) (Lochem 1975), and *Ferdinand Bol* (Doornspijk, 1982); and the doctoral dissertation by Barbara Joyce Buchbinder-Green, *The Painted Decorations of the Town Hall of Amsterdam* (Evanston, 1974). There is some uncertainty as to the actual date of this painting. It was probably commissioned in 1659, though it was not actually hung until 1664, at the earliest. Bol did not receive payment for it until 1666. See Blankert, *Ferdinand Bol*, pp. 109–10, and Buchbinder-Green, *Painted Decorations*, pp. 152–3.

[3] According to Buchbinder-Green (*Painted Decorations*, pp. 153–4), it was Jan van Eyk in 1758 who first suggested this interpretation of the painting in *Kunst en historie-*

and bathed in his heavenly light. Having already been brutally punished for their idolatry (*Exodus* 32:26–35), all the figures below Moses look up to him in awe, except for one who looks down at the ground. If the painting itself does not provide certain evidence for the identity of this figure, we can turn to the marble frieze commissioned by the Amsterdam burgomasters to be set underneath the painting.[4] At its centre, the frieze depicts the frenzied scene around the Golden Calf, and at its periphery, the increasingly debase forms of behaviour (singing, dancing, gambling, fighting, fornicating) that occurred as a result. The draped and bearded figure who leads the worship of the idol at the focal point of the frieze is most likely the same as the figure bowing down in the painting. It is Aaron the high priest, who, after having succumbed to the demands of the people and fashioned the idol (*Exodus* 32:4), now contritely avoids looking directly at Moses. Indeed, the drama of this painting is contained in the tension between these two figures, counterposed along a diagonal axis of light: Moses bearing the law, and Aaron the high priest bowing down before him. As we shall see in more detail below, this painting was commissioned by the Magistrates in order to make a specific political point: the welfare of the people requires that the priests, the interpreters of the law, be subservient to the authority of the state and its founding principles.

We know, from two letters, that Spinoza was working on something resembling the *Theological-Political Treatise* in 1665, a year after Bol's painting was finished and hung in the Town Hall; he had probably begun work before then.[5] Although the medium is obviously different in each case, Spinoza's *Theological-Political Treatise* shares certain crucial features with Bol's painting. First and foremost they share a common subject, the Hebrews, and more specifically the position of Moses the lawgiver. As we shall see, while Spinoza treats a great range of Biblical events in the *Theological-Political Treatise*, that of Moses in his capacity as founder of the state is clearly the most important.

---

kundige beschryving en aanmerkingen over alle de schilderyen op het stadhuis te Amsterdam (Amsterdam, 1758), p. 139. See also Blankert, *Ferdinand Bol*, p. 110. On this point, both Buchbinder-Green and Blankert refer to an article by Christian Tümpel, 'Studien zur Ikonographie der Historien Rembrandts', *Nederlands Kunsthistorisch Jaarboek*, XX (1969), pp. 169–75, which compares Bol's Moses with that of Rembrandt.

[4] See Katherine Fremantle, *The Baroque Town Hall of Amsterdam* (Utrecht, 1959), p. 77; and also Blankert, *Kunst als regeringszaak, pp. 32–3, 65.*

[5] See letters 29 and 30 (IV/165–6), respectively from and to Oldenburg. For more on the structure and composition of the *Theological-Political Treatise*, see Edwin Curley, 'The *Theological-Political Treatise* as a Prolegomenon to the *Ethics*', in *Central Themes in Early Modern Philosophy*, ed. J.A. Clover and Mark Kulstad (Hackett, 1990). Since Bayle's *Dictionary* entry, most scholars agree that some of the *Theological-Political Treatise* was drawn from an earlier (and now lost) defence Spinoza had written for himself and his principles at the time of his ban (*herem*) from the Amsterdam Jewish community in 1656 (see Henry E. Allison, *Benedict de Spinoza: An Introduction* (Yale, 1987), p. 7).

Second, Bol's painting and Spinoza's text share a particular socio-historical context: not only were they composed at roughly the same time and place, but they address the same fundamental issues: the ongoing struggle between the Dutch Reformed Church and the state, and also, by extension, the related struggle between the republican States party and the monarchist Orange party. Third, Spinoza and Bol (or at least the authorities who commissioned his work) both understand that to be effective, political argument must go beyond the realm of purely rational discourse. For this reason they attempt to appeal to the masses through an interpretation of an imaginative example.[6]

But why should this example taken from Scripture be relevant to Spinoza's situation, much less to our own? While scholars have all noted Spinoza's use of the Hebrews in the *Theological-Political Treatise*, their interpretations of it have varied. Some have argued that Spinoza used the ancient Hebrews only as an example of 'a rude and barbarous way of life'; they are not models to be imitated but 'antagonists of the Moderns'.[7] But while Spinoza is trying to overturn a sacred view of history and replace it with one more secular, he does not reject the use of the Hebrew example altogether. Important work has also been done tracing Spinoza's relation to his medieval predecessors, particularly Maimonides, and to his liberal successors in the Enlightenment.[8] But while the broad historical perspective is crucial to a complete understanding of Spinoza's argument in the *Theological-Political Treatise*, here I would like to concentrate

[6] Although Spinoza says in the preface to the *Theological-Political Treatise* that 'I do not ask the common people to read these things' (*TTP*, Preface, 34: GIII/12), in a letter to Oldenburg he had said that one reason why he was undertaking the project that culminated in the *Theological-Political Treatise* was to avert the accusation held by the 'common people' [*vulgus*] that he was an atheist (Letter XXX to Oldenburg, September or October 1665, in *The Correspondence of Spinoza*, trans. and ed. A. Wolf (New York, 1928), p. 206, GV/166). Moreover, although Spinoza did not permit the translation of the *Theological-Political Treatise* from Latin into Dutch, at that time it was perhaps due more to the unexpectedly strong reaction that the book had provoked rather than an original plan to limit its audience (Letter XLIV to Jarig Jelles, 17 February 1671, *Correspondence of Spinoza*, ed. Wolf, p. 260, GV/227). I will return to this issue again in the conclusion. All that we need assume at this point is that the audience is composed of persons other than those led exclusively by reason, which is most people.

[7] André Tosel, 'Y-a-t-il une philosophie du progrès historique chez Spinoza?', in *Spinoza: Issues and Directions*, ed. Moreau and Curley (Leiden, 1990), pp. 306–26. A. Matheron also points to passages in which Spinoza emphasizes the primitive aspect of the Hebrews (see e.g. Alexandre Matheron, *Le Christ et le salut des ignorants chez Spinoza, (Paris, 1971), pp. 14 ff), but he would not, I think, subscribe to all other aspects of Tosel's interpretation.*

[8] See Shlomo Pines, 'Spinoza's *Tractatus Theologico-Politicus*, Maimonides and Kant', in *Scripta Hierosolymitana*, Vol. 1, ed. Ora Siegel (Jerusalem, 1968), and more recently, Steven B. Smith, 'Spinoza's Paradox: Judaism and the Construction of Liberal Identity in the *Theologico-Political Treatise*', *The Journal of Jewish Thought and Philosophy*, Vol. 4 (1995), pp. 203–25.

on a more immediate context: the politics and political theory of seventeenth-century Netherlands. Others have insisted that Spinoza adopted this example solely for external reasons, because it belonged to his ostensible political enemies — the Calvinist clergy — and he simply wanted to subvert their use of it.[9] Yet, as I hope to show, that approach pays too little attention to the constructive use Spinoza makes of the example. Finally, there are those who are sensitive both to the context in which Spinoza writes and the use he makes of this example to illustrate his own theoretical principles.[10] In what follows I will argue along with those in this last group who think that Spinoza uses his analysis of the function of prophecy and religion among the ancient Hebrews to illustrate and solve a deep theoretical problem in the Hobbesian social contract theory as he sets it forth in the *Theological-Political Treatise*: the need to explain how self-interested individuals in the state of nature are willing to surrender their natural rights and create a sovereign power to which they become subject in civil society. The history of the Hebrews is an example of just how that transition is managed through the prophetic appeal to divine authority and the institutionalization of religion in the life of the state.

But we need to delve farther into the question. For even if prophecy solved the Hebrews' dilemma, how can the Hebrew theocracy be relevant to the Dutch republic, or to the modern secular age?[11] The deeper relevance of the Hebrews must transcend the specificity of their historical situation; their story must tell us something about the constitution of society and the nature of political language itself.[12] I suggest that what makes the Hebrews exemplary in this deeper sense is the function of prophetic language: the appeal to an imaginative narrative example that claims universal authority in order to justify a particular set of institutions.[13] In other words, when the prophets called the Hebrews

[9] Lewis S. Feuer, *Spinoza and the Rise of Liberalism* (Boston, 1958), p. 119.

[10] Sylvain Zac, 'Spinoza et l'etat des Hébreux', in *Philosophie, théologie, politique dans l'oeuvre de Spinoza* (Paris, 1979), pp. 145–89. Etienne Balibar, in *Spinoza et la politique* (Paris, 1986), also places the *Theological-Political Treatise* within the political struggles of the time.

[11] P.F.-Moreau, in 'Les Principes de la lecture de l'Ecriture Sainte dans le T.T.P.', *Groupes des Recherches Spinozistes: Travaux et Documents*, No. 4 (Paris, 1992), pp. 119–31, compares Spinoza's use of the Hebrews to that of his contemporaries (e.g. Koerbagh, Meyer) and concludes that 'Il est donc possible de reprendre comme exemple [i.e. the Hebrew state] ce que l'on refuse comme modèle' (p. 127). In the same volume are useful articles on other uses of the Hebrew state: F. LaPlanche, 'L'érudition Chrétienne aux XVIe et XVIIe siècles et l'état des hébreux' (pp. 133–47); and C.R. Ligota, 'Histoire à fondement théologique: la republique des hébreux' (pp. 149–67).

[12] Norman O. Brown, in his article 'Philosophy and Prophecy: Spinoza's Hermeneutics', *Political Theory*, Vol. 14 no. 2 (May 1986), pp. 195–213, also considers the function of prophetic language in the political argument of the *Theological-Political Treatise*, though he focuses on its transformative rather than its exemplary effect.

[13] The general sense in which I use the term *exemplar* is indebted to two recent discussions of this rhetorical concept: Timothy Hampton, *Writing from History: The*

'chosen', they were performing a function essential to any society: using the imagination to transcend individual interests and to create a common standard of judgment and behaviour. When Spinoza and Bol chose the Hebrews as their subjects, they were performing a function analogous to that of the prophets, appealing to and interpreting the collective imagination of a people in order to exhort them to right conduct in society. In order to see this further point we must turn to Spinoza's own discussion of *exemplaria* and the use of such models in the contemporary historical discourse.

In what follows, then, I will make four basic points. First, I will take what Spinoza says in the *Ethics* about an exemplar of human nature as a clear and basic indication of what the purpose of an exemplar is: to transform value from an individual and subjective utility to a universal and objective standard. Second, I will argue that the function of prophecy in the foundation of the state is essentially to fulfil the role of an exemplar, but on a political (not ethical) level; that is, to persuade the individual that his or her interest is only fulfilled through submission to the state's authority. Third, I will show how the history of the Hebrew state exemplifies the tension inherent in an exemplar between its particular imaginative origins and its universal pretensions. Fourth, I will claim that the narrative of the Hebrews' use and misuse of prophecy spoke directly to the Dutch of Spinoza's time and speaks indirectly to the political theorists of our own time.

The Origin and Function of an *Exemplar* in the *Ethics*

Spinoza defines 'good' and 'evil' in the opening definitions of Part 4 of the *Ethics*: 'By good I shall understand what we certainly know to be useful to us';[14] 'By evil, however, I shall understand what we certainly know prevents us from being masters of some good'.[15] The striking feature of these definitions is their apparent subjectivity. What is useful and therefore good to one person may not be useful and good to another. This, of course, raises an immediate problem about the use of these value terms; for if they are to be meaningful they must apply not only to the experience of a single individual, but to others

Rhetoric of Exemplarity in Renaissance Literature (Ithaca, 1990), and John D. Lyons, *Exemplum: The Rhetoric of Example in Early Modern France and Italy* (Princeton, 1989).

[14] *E*4d1. References to the *Ethics* are made in a format used by Jonathan Bennett in his book *A Study of Spinoza's Ethics*: *E* refers to the *Ethics*; the following number refers to the Part; the subsequent letter and number refers to either an axiom (a), definition (d) or proposition (p); which may be followed by another letter and number which refers to a specific demonstration (dem), corollary (c) or scholium (s); a preface is indicated by Part number and 'pref.' (e.g. *E*4pref.); an appendix by Part number and 'app.'. For example, *E*4p4c refers to the *Ethics*, Part 4, proposition 4, corollary. All translations are those of Curley in his *Collected Works of Spinoza*, Vol. 1. References to the Latin are made, when necessary, to the standard Gebhardt edition of the *Opera* (G), by page and line number.

[15] *E*4d2.

as well. Spinoza is aware of this difficulty and points the reader back for further explanation to the preface of Part 4, in which he discusses the idea of an exemplar of human nature,[16] whose very purpose is to bridge the gap between 'subjective' value and 'objective' value.[17] The structure and function of such an exemplar is exactly parallel to the structure and function of prophetic language in the *Theological-Political Treatise*: it provides a bridge for the transition from personal interest or utility to general interest. Let us therefore first examine how Spinoza conceives an exemplar in the *Ethics*, and then explore the extent to which it is able to fulfil its stated function.

An exemplar of human nature is itself an example of a broader category of exemplary ideas, each of which creates a standard on the basis of which value judgments can be made. Let us, for example, take a look at the origin of a judgment about something's perfection. In order to judge the perfection of a thing, men must have some idea of the end for which it was made. Spinoza thinks that he can discern the origin of the notion of perfection in the Latin etymology [*perficere* from *per* + *facere*] of the word itself: 'If someone has decided to make something, and has finished it, then he will call his thing perfect — and so will anyone who rightly knows, or thinks he knows, the mind and purpose of the Author of the work'.[18] The idea is that when something is complete, and the end of the author or maker of the work is realized, then it can be said to be perfect. Thus, in an example Spinoza himself gives, when a man desires to build a house, he will consider it perfect when it is complete according to the initial plan he had of it. Of course, if another person does not know the intentions of the author, then he will be unable to judge whether the work is imperfect or perfect. In this sense, a judgment of value is not tied to any broad exemplar, but to a specific work and its relative stage of completion.

As soon as people wish to make judgments about other things whose author's intentions remain unknown, the need arises for broader categories that can subsume specific judgments under more general ideas. That is precisely the structure of an exemplary idea: it is a universal idea that functions as a model, on the basis of which a value judgment may be made. Although the idea of an 'end' is no longer readily apparent, it is nonetheless still very much present as the universal idea towards which things do or should tend. As Spinoza writes:

> But after men began to form universal ideas, and devise *exemplaria* of houses, buildings, towers, etc., and to prefer some *exemplaria* of things to others, it came about that each one called perfect what he saw agreed with

[16] In the preface to Part 4, Curley renders the Latin *exemplar* as 'model'. I prefer to use the cognate 'exemplar' in order to maintain the dual sense of 'example' and 'model' in the original. My reasons should become clearer in what follows.

[17] I am indebted for this way of putting the problem of value in Part 4 of the *Ethics* to an article by Edwin Curley, 'Spinoza's Moral Philosophy', in *Spinoza*, ed. Marjorie Grene (Notre Dame, 1979), pp. 354–76.

[18] *E*4pref.; GII/205.

> the universal idea he had formed of this kind of thing, and imperfect, what he saw agreed less with the *exemplar* he had conceived, even though its maker thought he had entirely finished it.[19]

In this way, the meaning of 'perfect' changes from a literal judgment of whether a thing is complete or not, to a more abstract judgment of whether and to what extent the thing conforms to the universal idea of which it is supposed to be exemplary. This move allows an individual to judge the perfection of a thing, whether or not he actually knows the intention of its maker; for now the judgment is not whether a specific intention has been realized but whether the thing realizes an abstract degree of exemplification, under which the particular intentions have been subsumed. One man can thus judge the house of another, not on the basis of the owner's original plan and whether it is complete or not, but on the basis of an idea of what a house as such should be like — in other words, on the basis of a universal idea of a house that serves as an exemplar. Thus, at the same time that a universal idea groups together diverse things under a single category, it also functions as the ground for a judgment of their relative 'perfection'. An exemplar is the abstract 'end' towards which things of a certain kind should tend if they were to become perfect; it is the standard which they should imitate and by which they should be judged.

The same process occurs in a judgment about whether something is 'good' or 'bad'. As we have seen, Spinoza originally defines these terms from the point of view of an individual's utility. But obviously what is useful to one person at any given time and place may not be to another. Spinoza notes that '. . . one and the same thing can, at the same time, be good, and bad, and also indifferent. For example, Music is good for one who is melancholy, bad for one who is mourning, and neither good nor bad to one who is deaf'.[20] So if the value terms 'good' and 'bad' are to mean something more than the mere subjective utility of an individual, then it is necessary to find a basis for this kind of judgment that transcends the particular judgment of the individual. That is the specific purpose of an exemplar of human nature. So, after the apparently 'subjective' definitions of good and evil in 4d1 and d2, Spinoza points the reader back to the preface, where he says, 'I shall understand by good what we know certainly is a means by which we may approach nearer and nearer to the model of human nature that we set before ourselves. By evil, what we certainly know prevents us from becoming like that model'.[21] Exemplars can serve this purpose, as a bridge from 'subjective' to 'objective' value, because they are formed not on the basis of an individual judgment but on the basis of a universal idea. In this

[19] *E*4pref.; GII/206. I have retained the Latin *exemplaria* here instead of Curley's uniform translation of it as 'models' because I want to argue that the word means both 'model' and 'example'. I think the first instance of the word in this passage could be translated as 'model' while the second could be translated as 'example'.

[20] *E*4pref.

[21] *Ibid.*

light, when an individual judges something (or someone's actions) as good, he is making a judgment that assumes the existence of a model that applies to all things or actions of that kind, and therefore has some injunctive force.

However, the so-called 'universal idea' upon which an exemplar is constructed is itself questionable. Spinoza distinguishes between ideas that he calls 'common notions', which 'are the foundations of our reasoning', and 'axioms, *or* notions, [that] result from other causes', which are 'ill-founded'.[22] Universal ideas belong to the second category. Whereas 'common notions' are 'equally in the part and in the whole, [but do] not constitute the essence of any singular thing',[23] 'universal ideas' (despite their name) are the confused product of a single mind that has been affected by the ideas of many external bodies.[24] Thus a 'universal idea' is derived not from the reflective comparison of the ideas we have of different bodies but from 'fortuitous encounters with things'[25] that are superimposed one upon the other through the equally haphazard procedures of memory.[26] It is not an 'adequate idea', the clear and distinct foundation of reason, but an 'inadequate idea', nothing more than a mutilated and confused *image* of the external world.[27]

As an image, a 'universal idea' is, at bottom, always the product of a subjective experience of the external world.[28] For this reason it can never attain

[22] *E*2p40s1.

[23] *E*2p37.

[24] As Spinoza writes, 'Those notions they call *Universal*, like Man, Horse, Dog, etc., have arisen . . . because so many images (e.g, of men) are formed at one time in the human Body that they surpass the power of imagining — not entirely, of course, but still to the point where the Mind can imagine neither slight differences of the singular [men] (such as the color and size of each one, etc.) nor their determinate number, and imagines distinctly only what they agree in, insofar as they affect the body. For the body has been affected most [forcefully] by [what is common], since each singular has affected it [by this property]. And [the mind] expresses this by the word *man*, and predicates it of infinitely many singulars.' (*E*2p40s1.)

[25] *E*2p29s.

[26] Memory is nothing other than the association of two or more images following the order of a body's interactions with the external world. See *E*2p18s.

[27] Spinoza sums up the difference between adequate and inadequate ideas in *E*2p29s, where he writes: 'I say expressly that the Mind has, not an adequate, but only a confused [and mutilated] knowledge, of itself, of its own Body, and of external bodies, so long as it perceives things from the common order of nature, i.e. so long as it is determined externally, from fortuitous encounters with things, to regard this or that, and not so long as it is determined internally, from the fact that it regards a number of things at once, to understand their agreements, differences, and oppositions. For so often as it is disposed internally, in this or another way, then it regards things clearly and distinctly . . .'.

[28] '. . . to retain the customary words, the affections of the human Body, whose ideas present external bodies as present to us, we shall call images of things, even if they do not reproduce the [external] figures of things. And when the Mind regards bodies in this way, we shall say that it imagines' (*E*2p17s).

the truly universal status to which, in name at least, it aspires. Certainly some images are more frequently repeated or are more vivid than others. But that only explains why some images become the central features of these general ideas, not why they ought to be described as 'universal'. The fact is, as Spinoza emphasizes, that 'these notions are not formed by all [men] in the same way, but vary one to another, in accordance with what the body has more often been affected by, and what the imagination recollects more easily'.[29] So, although someone might think that the term 'Man' refers to a single, universal genus, in fact, when the actual description of that genus is given, its extension will vary depending upon which features of bodies have most vigorously or frequently affected the knower. One person will understand by 'Man' 'an animal of erect stature', another, struck by another feature, will understand it as 'an animal capable of laughter', and so on.[30] These definitions may be useful as an aid to memory, but as a means towards scientific truth or as a way to overcome the subjectivity of value judgments they are flawed.[31] The problem is that the very appearance of universality in which such terms are draped obscures their origins as judgments made by individuals in particular circumstances. These particular judgments appear as natural categories, which are discovered rather than invented, but like teleological explanations of nature, they are really impositions of human categories, produced by the imagination, upon the natural world. Thus, as Spinoza himself realized, since exemplars are not common notions derived from adequate ideas of the natural world, but rather ill-founded constructs of the imagination, these so-called 'beings of reason' are perhaps more aptly named 'beings of the imagination'.[32] For the sake of gaining an appearance of objectivity, we can easily confound exemplars based upon universal ideas with real natural categories, but we can never really escape their origins in the subjective domain of imaginative experience.

[29] *E*2p40s1.

[30] *Ibid.*

[31] See *Appendix Containing Metaphysical Thoughts*, Part 1, ch. 1: 'So when Plato said that man is a featherless biped, he erred no more than those who said that man is a rational animal. For Plato was no less aware than anyone else that man is a rational animal. But he referred man to a certain class so that, when he wished to think about man, he would immediately fall into the thought of man by recalling that class, which he could easily remember. Indeed Aristotle erred very seriously if he thought that he had adequately explained the human essence by that definition of his. Whether, indeed, Plato did well, one can only ask.' (*The Collected Works of Spinoza*, p. 301; GI/235.)

[32] 'We see, therefore, that all the notions by which ordinary people are accustomed to explain nature are only modes of imagining, and do not indicate the nature of anything, only the constitution of the imagination. And because they have names, as if they were [notions] of beings existing outside the imagination, I call them beings, not of reason, but of imagination.' (*E*1app.; GII 83/11–17.) See also Martial Gueroult, *Spinoza II (L'Âme)* (Hildesheim, 1974), Vol. 2, p. 415.

The need for exemplars should now be clear. They are 'notions we are accustomed to feign'[33] in order to compare and judge individuals of the same kind in some certain manner. But it should also be clear that insofar as they are based upon universal ideas that are ultimately derived from the imagination, they fail in their purpose of defining an 'objective' and certain standard of value. It remains an open question to what extent these exemplars, 'beings' more of the 'imagination' than of 'reason', can lead us to an ethical life in Spinoza's sense — tnat is, a life governed by reason.[34] But in another guise, that of prophecy, exemplars will play a critically important role in the political sphere, where, according to Spinoza, the imagination and the passions necessarily dominate.

The Origin and Function of Prophecy in the *Theological-Political Treatise*

As Spinoza analyses it, prophecy has both the structure and the function of an exemplar: it is an inadequately conceived universal idea that serves to guide the judgment and conduct of men. Although prophecy originates in the particular situation of the prophet, and uses highly imaginative language to express its views, it claims universal authority because it appeals to God as its source. Moreover, despite the theologians who 'have accommodated Scripture to [Aristotelian and Platonic] speculations',[35] the true purpose of prophecy is not speculative but ultimately practical: it exhorts men to 'right conduct' in their affairs. In other words, it creates a standard that is used to judge other people's actions and that also serves as a model to be imitated.

Spinoza explains the idea of the Hebrews' 'chosenness' or 'election' by God in the context of the exemplary function of prophecy. The prophets called their people 'chosen' in order to exhort them to conduct that the people would believe was approved by God, a universal authority; and, to the extent that the people followed the exhortations of their prophets and adhered to a single standard of 'right' conduct, they saw visible proof of their 'chosenness' in the positive results of social cohesion, which in turn became a further incentive to follow the prophets, thus reinforcing the original standard of conduct. As we shall now

[33] *E*4pref.

[34] Edwin Curley, in 'Spinoza's Moral Philosophy', thinks that Spinoza 'must be saying that we can form an idea of human nature, which is rightly used as a standard of judgment' (p. 364). Curley holds that there are adequate ideas of human nature which, because an individual will necessarily desire to realize them (as a function of his or her *conatus*), are objective. For other views on the relation of an adequate idea of human nature and the ideal of the 'free man', see Dan Garber's recent paper delivered to the Jerusalem Spinoza conference, 'Dr. Fischelson's Dilemma: Spinoza on Freedom and Sociability' (Leiden, forthcoming); and Lee Rice's '*Tanquam Naturae Humanae Exemplar*: Spinoza on Human Nature', in *The Modern Schoolman*, LXVIII (May 1991), pp. 291–303.

[35] *TTP*, Preface, 18; GIII/9.

see, the exemplary structure of prophecy uniquely suits it to solve the problem that arose for the ancient Hebrews in the Sinai desert, a problem equivalent to that found in the theoretical 'state of nature'. Let us look at this problem and how Spinoza thinks that prophecy was able to address it.

In the *Theological-Political Treatise*, Spinoza examines in great detail how Moses was able to found the Hebrew state. When the Hebrews were cast out of Egypt into the desert they had no constitution of their own: 'they were no longer bound by the legislation of any other nation, so that they were permitted to enact new laws as they wished'.[36] But the problem was that 'they were quite incapable of ordaining legislation wisely and retaining the dominion in their own hands, as a body', for the reason that '[a]lmost all of them were crude in their understanding and weakened by wretched bondage'.[37] Moses, aware that his people were unable to take control of their own destiny, realized that 'the dominion had to remain in the hands of one person only, who would command the others and compel them by force, and who would prescribe laws and afterwards interpret them'.[38] But how could Moses alone compel his people, against their divergent individual interests, to obey him? He realized that even if he were able to gather enough support to use physical force he would not be successful in his long-terms aims; for the 'obstinate temperament of the people . . . would not allow itself to be compelled by force'.[39] Peaceful means were necessary if the state would endure. Thus, Moses had recourse to the invocation of divine power and authority, first in order to ensure his own power;[40] then, to institute the state.[41] Moses overcame the lawless condition of the Hebrews by instituting a state based on religious practices and ceremonies, a successful theocracy whose nominal king was God.[42] That alone, for Spinoza, was the purpose of Sinai, and was what entitled the ancient Hebrews to be called a 'chosen people'.

We have already noted the situation of the ancient Hebrew nation once they had left the slavery of Egypt under Pharoah. They were no longer subject to the laws of any other nation and were free to choose a new set of laws and

[36] *TTP*, V, 26; GIII/74.
[37] *TTP*, V, 27; GIII/75.
[38] *Ibid.*
[39] *TTP*, V, 28; GIII/75.
[40] 'Moses was easily able to retain this dominion, because he excelled the others in divine power, persuaded the people that he had it, and showed this by many testimonies' (*ibid.*).
[41] 'Moses, by the divine power and command, introduced religion into the Body politic, so that the people would do their duty not so much from fear as from devotion' (*TTP*, V, 29; GIII/75).
[42] In the *Discourses on Livy*, Machiavelli makes a very similar argument. Numa was able to transform Rome from a city based on force to a city founded on principles of law and obedience, through the use of religion (*Discourses*, Book 1, §11). On this parallel see also Brown, 'Philosophy and Prophecy: Spinoza's Hermeneutics', pp. 208–9.

institutions for themselves.[43] Spinoza describes their situation in these juridical terms:

> For after they had been freed from the intolerable oppression of the Egyptians, and had not attached themselves to any mortal by any contract, they again acquired their natural right to do everything they could, and each of them could decide anew whether he wanted to keep it, or to surrender it and transfer it to someone else.[44]

But why could they not just decide their future rationally among themselves? Using concepts developed in his analysis of 'natural and civil right' in the previous chapter, Spinoza suggests that the difficulties the Hebrews faced in their task were juridically and psychologically equivalent to those faced in the 'state of nature'.[45] The dilemma they faced in their lawless condition was that, although each person had the natural right to make his or her own decision, the very exercise of that right prevented the institution of civil society.

The problem results from the fact that most people are led not by reason but by their appetites and passions. In the state of nature, an individual lives according to his 'natural right', that is, an individual has 'the supreme right . . . to exist and act as it is naturally determined to do'.[46] But most people are not determined to live according to the dictates of reason; 'on the contrary', Spinoza emphasizes, 'they are all born ignorant of everything'.[47] Consequently, the exercise of natural right in the state of nature is determined 'not by sound reason, but by desire and power'.[48] If individuals were led by reason, they would quickly realize that mutual aid and cooperation would be the most beneficial course of action; but since they are led by their desires, they are instead brought to division and conflict.[49] For it is a fact that 'according to the laws of appetite each person is drawn in a different direction';[50] and since all persons have a natural right to act according to their individual appetite, no one has any obvious reason to compromise the effort to satisfy their desires. Indeed, in the state of nature, each person 'is permitted, by supreme natural right, to want and to take, in whatever way, whether by force, by deception, by entreaties, or by whatever other way is, in the end, easier'.[51] The 'obstinacy' of the Hebrews was anchored

[43] See also *TTP*, XVII, 26; GIII/205.

[44] *Ibid.*

[45] See Matheron, *Le Christ et le salut des ignorants chez Spinoza*, p. 10. Douglas Den Uyl argues that the Hebrews were not in the 'state of nature' because they already had the rudiments of social life (D. Den Uyl, *Power, State, and Freedom: An Interpretation of Spinoza's Political Philosophy* (Assen, 1983), Ch. 2.

[46] *TTP*, XVI, 4; GIII/189.

[47] *TTP*, XVI, 7; GIII/190.

[48] *Ibid.*

[49] See also *TTP*, XVII, 13–16; GIII/203.

[50] *TTP*, XVI, 14; GIII/191.

[51] *TTP*, XVI, 8; GIII/190.

in the exercise of each individual's right to live as he or she wished, and without the intervention of Moses' divine revelation the inevitable result would have been conflict and distrust.

In Spinoza's view, Moses understood what needed to be done in order to forestall the disintegration of the state of nature into conflict and chaos: he needed 'to so establish a political authority that there is no room left for deception, indeed, to so establish everything that everyone, no matter what his mentality, prefers the public right to the private advantage'.[52] Moses also realized that an appeal to disinterested reason would not work with the Hebrew masses, who were led by desire and appetite. Spinoza, in fact, denies the possibility of the masses ever acting on the basis of anything but personal advantage: individuals must therefore somehow be led to act for the public good through their own self-interest. Since 'no contract can have any force except by reason of its utility' it is entirely rational for an individual to break a contract (whether private or social) when it no longer serves his purpose.[53] Hence, if someone is going to try to persuade the unruly masses that cooperation is in their best interest, he or she must be aware of what Spinoza calls the 'universal law of human nature', namely that 'between two goods, each person chooses the one he judges to be the greater, and between two evils, the one which seems to him lesser'.[54] To this iron law of self-interest, though, Spinoza adds one important rider: the individual chooses 'the one which seems to the person choosing to be greater or lesser, and not that things necessarily are as he judges them to be'.[55] According to Spinoza, not only did Moses understand the underlying logic of self-interest that governs human behaviour in the state of nature, but he also understood that the appearance of self-interest was as important as its reality. His strategy therefore was to manipulate the appearance of self-interest until it was identical with real public interest. Moses appealed to revelation as the source of his views because prophetic revelation spoke to the situation of the people in language that they could readily understand and which directly appealed to their individual self-interest while using it in service of the community.

Spinoza defines prophecy as vivid imaginary language used by a person (a prophet) to exhort the ignorant masses to right conduct. The Prophets 'perceived God's revelations only with the aid of the imagination, i.e. by the mediation of

[52] *TTP*, XVII, 16; GIII/203.

[53] *TTP*, XVI, 20; GIII/192. Spinoza gives the example of a person who deceptively contracts with a robber to deliver goods to him at a later date. Spinoza thinks that when the person is later released he has no obligation whatsoever to make good on the promise; indeed he is fully rational to renege on it (*TTP*, XVI, 16–19; GIII/192). As Curley notes, this contrasts starkly with Hobbes's view of the same example (see Hobbes, *De cive*, ii, 16; Hobbes, *Leviathan*, xiv).

[54] *TTP*, XVI, 15; GIII/192.

[55] *Ibid.*

words or of images, the latter of which might be either true or imaginary'.[56] God communicated to the prophets through either a 'true voice', as he did to Moses alone, or 'visible forms and symbols', that often appear in a dream, such as Joshua's vision of an angel with a sword. But Spinoza does not think that the prophets had any super-human capacity, only a more vivid imagination than others: 'the Prophets were endowed, not with a more perfect mind, but instead with a power of imagining unusually vividly'.[57] However, since the people were ignorant of its causes, they misinterpreted the richness of prophetic language as an indication of its divine, supernatural origin. Thus, 'the Prophets were said to have the Spirit of God because men were ignorant of the causes of Prophetic knowledge, wondered at it, and on that account, were accustomed to refer it to God — as they did all other abnormal phenomena — and to call it the knowledge of God'.[58] It is this very misunderstanding of prophecy which paradoxically explains its success in guiding conduct. For although the prophet speaks in language that addresses a particular set of circumtances it is understood by its audience to have universal authority. Only then can it function effectively to define 'right' conduct, not only for the individual but for everyone.

We can see how this misunderstanding works if we continue our examination of interpersonal dynamics in the state of nature. Because there are no prior rules of conduct in the state of nature, each interaction between individuals has an uncertain outcome. A person can never be sure whether he will be deceived by the other or not, or even whether he will live or die. This uncertainty produces a constant affective vacillation in the individual between hope that things will turn out in his self-interest and fear that they will not. Indeed, the wretchedness of the state of nature is such that there is good reason among people to presume strongly that things will not turn out in their favour. As Spinoza remarks in the opening sentences of the preface to the *Theological-Political Treatise*, people 'are often reduced to such straits that they can bring no plan into operation, and since they generally vacillate wretchedly between hope and fear, from an immoderate desire for the uncertain goods of fortune, for the most part their hearts are ready to believe anything at all'.[59] This is the fertile ground upon which both prophecy and superstition take root. Despite their differences, which we shall examine in some detail in the next section, both prophecy and superstition address the same problem: the uncertain relations an individual has with external persons and things (fortune),[60] and the wrenching vacillation between hope and fear that stems from this human condition.

[56] *TTP*, I, 43; GIII/27.
[57] *TTP*, II, 1; GIII/29.
[58] *TTP*, I, 42; GIII/27–8.
[59] *TTP*, Pref., 1; GIII/5.
[60] It is not a coincidence that both the opening sentences of the *Theological-Political Treatise* and the preface to Part 4 of the *Ethics* both begin with a reference to fortune —

Prophecy is able to address a person's self-interest, not by way of the abstract principles of reason, but by way of his or her own uncertainties and passions. Since, 'by natural right each person can act deceptively, and is bound to stand by the contract only by the hope of a greater good and fear of a greater evil',[61] each person is interested in arriving at what concrete expectations (i.e. whether his hopes will be realized or not) he has for an outcome that would justify his abiding by the contract. A person will listen to prophecy because he or she desires above all to have a consistent set of expectations about the results of his or her actions, and prophecy appears to provide that certainty through its status as the word of God. In other words, it regulates a person's hopes and fears by locating them in a larger scheme of things, which is itself governed by an omnipotent being who expresses various passions (anger, care, etc.) — God. In this way, what works for one individual can also work for another, and a pattern of stable interactions emerges.[62] Thus, while prophecy satisfies a person's narrowly conceived self-interest in the state of nature — i.e. to end or at least regulate the vacillation between hope and fear — it also serves the general interest by establishing a standard of conduct which all can be expected to, and indeed do, imitate.

Therefore, when Moses called the Hebrews 'chosen' he was not merely describing their relation to God but actively exhorting them to a certain way of life that would mould them into a nation. In fact, the Hebrews were a lawless band of people cast out into the desert, thrown back into the 'state of nature'. But Moses, in order to give them a law and found a nation, placed their particular experience into the framework of a divine plan in which they played a special role. He took what had been their bad fortune and recast it as the founding principle of their society.[63] This idea appealed to the particular experience of each person and yet provided a broad basis upon which to build institutions and practices that would coordinate interactions between self-interested individuals. Although in effect the people were transferring their

---

the power of external things over human affairs. While here I assume the theoretical construct of the state of nature, I think Spinoza does provide a deeper metaphysical reason for its actual existence, which he calls the power of 'fortune'. For the role of 'fortune' in the *Theological-Political Treatise*, see F. Mignini, 'Theology as the Work and Instrument of Fortune', in *Spinoza's Political and Theological Thought*, ed. C. De Deugd, pp. 127–36.

[61] *TTP*, XVI, 23; GIII/193.

[62] Of course there still may be people who refuse to accept the prophet's teaching and continue to deceive others. But Spinoza's claim is that most men will desire to accept it. From that point on, the few who do not could be compelled by force.

[63] Machiavelli admires Moses — along with Romulus, Cyrus and others — for this same reason: 'They owed nothing to fortune but the opportunity which gave them matter to be shaped into what form they thought fit; and without that opportunity their powers would have been wasted, and without their powers the opportunity would have come to nothing' (Machiavelli, *The Prince*, Ch. 6).

natural right to the authority of Moses, they believed they were giving it to God. 'God alone, then, held political authority over the Hebrews, and this authority alone, by the force of the covenant, was rightly called the Kingdom of God, and God was rightly called also the King of the Hebrews.'[64] This anthropomorphic concept of God, a ruler who made a covenant with and guided his people in history, made it easy for individuals to transform their hopes and fears in relation to other people to hopes and fears in relation to God himself. Moses was not interested in teaching the people, with their 'childish understanding', about the nature of God as such, but only about His plan for their conduct, their rights and responsibilities to Him. Thus, he ritualized the people's relationship with their God into ceremonies and practices, and in this way the worship of God meant the establishment of institutions through which the founding principles of the state were made to endure.

Spinoza is quick to point out that the sense in which the ancient Hebrews were chosen was unique to them; 'the Laws of the Old Testament were revealed and prescribed only to the Jews'.[65] Moreover, we might think that this story would only be useful to other people who themselves are of 'primitive' understanding and predisposed to believe in an anthropomorphic God, miracles and the like.[66] But there is an important point to be made before we dismiss the history of the ancient Hebrews out of hand: it is not necessarily the content of prophecy itself that is exemplary to others — for as Spinoza emphasizes, it addresses the particular situation of the Hebrews alone — but the structure of prophecy and its function in the state. Spinoza is not trying to eliminate the use of prophetic language, for he realizes that it would be impossible as long as people are driven by their appetites and passions and think through the medium of the imagination alone. Rather, he is attempting to attack and undermine a misuse of exemplary language, a misuse which is also illustrated in the narrative of the Hebrews.

Prophecy and Superstition

Although, as we have seen, prophecy performs a crucial function in the creation and maintenance of the state, a danger lurks in the exemplary language of prophecy itself: people can easily mistake the appeal to divine authority for an appeal to the truth. That is precisely the difference between religion and

64 *TTP*, XVII, 30; GIII/206.

65 *TTP*, III. 22; GIII/48.

66 We see a parallel case in Machiavelli's *Discourses* where, after having described the men who were persuaded by Numa's invocation of the gods as 'rude' and 'uncultivated', he ironically undercuts the distance between ancient Roman times and his own by pointing out that '[it] did not seem to the people of Florence that they were either ignorant or rude, yet they were persuaded by Friar Girolamo Savonarola that he had converse with God' (Machiavelli, *Discourses*, I.11).

superstition, in Spinoza's view.[67] Both religion and superstition originate under the same set of circumstances: the uncertainty and affective vacillation characteristic of the state of nature; and both have the same immediate purpose: to guide wretched, self-interested individuals in their behaviour; but while religion uses the prophetic language of revelation for the sake of teaching knowledge about right conduct alone, superstition uses prophetic language to lay claim to the truth about nature itself. While religion stays within the epistemological limits of prophetic language, superstition oversteps them. The result, at least in the scriptural history of the ancient Hebrews, is that while the prescriptions of religion are based upon devotion to God and produce a relatively stable society, the prescriptions of superstition are almost invariably based upon the fear of an earthly King and produce conflict. To see why this is so, we must first explain in more detail the tension within prophetic language itself and then turn to how this tension is exemplified in the narrative of the Hebrews.

We can see how superstition develops out of prophetic language if we briefly recall what Spinoza says about exemplars in the *Ethics*. The purpose of an exemplar is to establish an 'objective standard' of value, which it does by an appeal to a 'universal' idea. But this universal idea of something can itself often

[67] One problem here is that Spinoza is not entirely consistent in the terms he uses for this distinction between the proper and improper use of prophecy. In the preface to the *Theological-Political Treatise*, for instance, he refers to the difference between religion and superstition (*TTP*, Pref., 3). But we must also keep in mind the difference between what Spinoza calls 'true religion' (e.g. see *TTP*, XII, 22–3; GIII/162–3) based on divine law, and ceremonial religion based on rituals, which is something like the distinction Wernham, in his introduction to *Spinoza: The Political Works*, makes between 'inward' and 'outward' religion (A.G. Wernham, *Spinoza: The Political Works* (Oxford, 1958), pp. 34–5), though Wernham does not insist that inward religion must be true only that it cannot be controlled by the sovereign. Curley also interprets Spinoza's reference to *superstitio animi* (*TTP*, Pref., 5; GIII/6, literally, superstition of the mind) as 'true superstition' in contrast to 'false' superstition, which is 'a manipulative use of seers which takes advantage of the people's belief in them' (see his footnote 2). I identify this 'true' use of superstition with the proper function of 'outward' religion. That is, it directs the populace to act in conformity with the precepts of inward religion (such as acts of charity, justice and loving kindness) even though they are not necessarily motivated in these acts by reason (see *TTP*, XII, 34–7; GIII/165 for the foundations of Scriptural religion). In the 'false' use of superstition, we find the confusion of speculative knowledge with the realm of the imagination, a confusion common (albeit in different ways) to both those theologians who would interpret Scripture on the basis of Platonic or Aristotelian categories and to soothsayers who would interpret events with the auguries of birds (*TTP*, Pref., 18; GIII/9, and Pref., 5; GIII/6). In both cases, the 'seer' is using some notion of divine authority not to inspire 'right conduct' but rather to 'prop up his own inventions' and authority (*TTP*, XII, 40; GIII/166). I would like to standardize these different terms in line with my interpretation: that is, 'religion' does not refer to the knowledge of the divine law (for that is 'philosophy') but to the proper function of cermonial law (i.e. to exhort to right conduct as true superstition does), while 'superstition' refers to the perversion of the function of ceremonial law itself.

prove to be epistemologically inadequate. That is, although it claims to represent the ideal or model of all things of a certain kind, in fact it might represent only a particular image or a series of particular images blurred together. For example, an exemplar of human nature formed in this manner looks as if it is a rational concept common to all, though in fact it is an imaginative construct that reflects the unique situation and experience of its creator. For it to fulfil its function — to create a standard of value and conduct — an exemplar of human nature trades on its *appearance* of universality in order for it to be accepted in practice by the multitude. Thus, even as Spinoza underlines the need and desire for such an exemplar, he is also careful to point out the possible errors to which its underlying epistemological ambiguity can potentially lead. Most importantly, Spinoza warns the reader against imagining that exemplars and the values attendant on them are found in, or products of, nature itself. They are just human constructs made in order to compare things, to judge relative value, and to emulate in one's actions. As soon as those who use an exemplar violate its intrinsic epistemological and practical limitations — using it to explain nature itself, rather than simply as a guide to conduct — it tends to lose its ability to serve its original purpose. Moreover, certain people, realizing that the masses are ignorant and easily manipulated, are eager purposely to misuse these exemplars to gain power for themselves.

What the misuse of an ethical exemplar and the misuse of prophetic language have in common is a misapplication of teleological thinking to nature. In the preface to Part 4 of the *Ethics*, Spinoza reiterates his critique of cosmic theology articulated in the appendix to Part 1: 'Nature does nothing on account of an end . . . What is called a final cause is nothing but a human appetite insofar as it is considered as a principle, or primary cause, of some thing.' But, since human beings imagine nature to act as they do, on account of an end, 'when they see something happen in nature which does not agree with the model they have conceived of this kind of thing, they believe that Nature itself has failed or sinned, and left the thing imperfect.[68] It is legitimate to speak of 'ends' in human affairs, which are conceived primarily through the imagination, but it is a profound error to read such an idea back into nature, which is governed by eternal and necessary laws conceived by reason alone. Teleological explanation has an inevitable role in the explanation of human affairs, but must be eschewed in the explanation of nature itself.[69] The problem is that some people want to

[68] *E*4pref.

[69] If man were able to live by reason alone, which would be to live and act in accordance with his determinate nature, then there would be no need for concepts of good and evil at all (*E*4p67). For these concepts arise because man is dependent on external things in order to preserve his existence. Since an individual desires what he lacks, to the extent that external things prevent a person from living in full accordance with his nature, he will imagine and desire a more perfect nature, one that lacks nothing, as his 'end'.

claim that their exemplar of human nature, which they set before men as an end that ought to govern their action, is itself an end of nature, thus confusing the realm of the imagination, which for the most part determines human values, with that of reason, which legitimately determines the truths of nature.[70]

We see this problem most acutely when we look at how Moses' prophetic pronouncement, that the ancient Hebrews were 'chosen' by God, was interpreted and used in the course of their history. As we saw above, the prophets used this notion in order to bridge the gap between personal interest in the state of nature and common interest in a civil society. It worked because it appealed directly to the Hebrews while locating them within a story of universal significance; it explained what had happened and what they ought to do in the future. But although this narrative was teleological — i.e. it established an end (a model of conduct) that the Hebrews were obliged to realize (or imitate) in their actions — it did not necessarily violate the epistemological limits of prophetic language. Moses himself, as Spinoza emphasizes,[71] knew that the Hebrews were 'chosen' to be distinct from other nations, not for their wisdom or blessedness, but for their success in temporal matters. He spoke 'according to the Hebrews' power of understanding' in order 'to warn [them] in this way . . . so that he might bind them more to the worship of God'.[72] In other words, Moses used prophetic language for its effect, which was to strengthen those very institutions for which the Hebrews had been 'chosen' in the first place, not because it taught anything about the workings of nature itself. The notion of a 'calling' allowed the Hebrews to preserve themselves by placing the self-interested desires of each individual within the larger framework of God's plan that involved choosing them, the Hebrew nation, as the main protagonist. On this level the teleological explanation of their actions was perfectly legitimate; there was no attempt to claim that the Jews were chosen for anything other than a certain form of conduct.

Moses was able to translate the prophetic idea of a 'calling' directly into concrete social practices that preserved the state and protected its own ideological (i.e. prophetic) foundation. In other words, he established a religion whose purpose was to maintain the social pact (in the form of a theocracy, in which power and right was symbolically transferred from the masses to God) through the institution of law. But at no point did Moses confuse the laws of the Hebrew theocracy, which were created by humankind and particular to it, with the eternal laws of nature decreed by God. Certainly, piety was identified with obeying civil law, and civil law was not distinguished from religious ceremonies and rites: 'So, in this state civil law and Religion (which as we have shown,

[70] For more on this point, see the appendix to Part I of the *Ethics*, especially from II/81/25 to the end.

[71] *TTP*, III, 3; GIII/44. Spinoza refers to Deuteronomy 9:6–7.

[72] *TTP*, III, 5; GIII/44.

consists only in obedience to God) were one and the same thing.'[73] Yet there was no confusion of this religious law, which encompassed both ceremonial and civil law, with the 'divine law', which teaches the eternal truths of nature through the 'natural light' (reason) innate in every person.[74] Moses himself claimed no true knowledge of God[75] but only exhorted the Hebrews to right conduct: 'The tenets of the religion were not teachings, but laws and commands; piety was thought to be justice, and impiety a crime and an injustice'.[76] For the individual, obeying the ceremonial laws of religion was identical not only with his moral duty[77] but also with his political life: for '[a]nyone who failed in his Religious duties ceased to be a citizen, and for this alone was considered an enemy; anyone who died for Religion was considered to have died for his Country, and absolutely no distinction was made between civil law and religion'.[78] Therefore, for the state, the practice of these ceremonial laws by its citizens meant its continued existence. In this light, it was no accident that most of the religious ceremonies of the Jews could not be performed by an individual alone but only by a group.[79] Moses identified civil law and religious law in order to inspire obedience to the state, not through fear of an earthly authority, but through devotion to a divine being who had ordered the life of His chosen people in this particular manner, that is, through the practice of these laws and rituals.

Moses also showed that he was aware of the epistemological and practical limits of the prophetic foundation of the Hebrew state in the way he organized its constitution. Most importantly, he separated the powers of the political authorities, whom he charged with administering and enforcing the law, from those of the priestly caste, charged with interpreting the law. In order to foster the belief that God was the actual monarch of the state, Moses constructed (with common resources) a 'palace of God' that would serve as a visible reminder of the presence of divine authority.[80] He also appointed one of the tribes, the Levites, to be 'the courtiers and administrators of this divine palace'.[81] The chief

[73] *TTP*, XVII, 31; GIII/205.

[74] 'By human law I understand a manner of living which serves only to protect life and the state; by a divine law, one which aims only at the greatest good, i.e. the true knowledge and love of God' (*TTP*, IV, 8: GIII/59). Spinoza describes the universal character of divine law and the way in which the 'natural light' (reason) that discovers it is common to all men in *TTP*, I, 1–5; GIII/16.

[75] See *TTP*, IV, 29; GIII/64.

[76] *TTP*, XVII, 31; GIII/205.

[77] Even general moral precepts were not taught to the Jews as true for all men but were accommodated to their particular understanding and practice (*TTP*, V, 7; GIII/70).

[78] *TTP*, XVII, 31; GIII/205.

[79] See *TTP*, V, 2; GIII/69. Conversely, the divine law does not require any ceremony at all (*TTP*, IV, 20; GIII/62).

[80] *TTP*, XVII, 42; GIII/208.

[81] *TTP*, XVII, 43; GIII/208.

of the Levites, Moses' brother Aaron, was 'the supreme interpreter of the divine laws, the one who gave the people the replies of the divine oracle, and finally, the one who petitioned God on behalf of the people'.[82] But Moses made sure that although they had the power of interpretation of God's law the priests did not have the authority to command that they be obeyed; this he left exclusively in his own control. The priests perpetuated the illusion that God had ordained the state by performing the religious rites that were commanded by God and interpreting the law, and in this way the actual leaders of the state, beginning with Moses himself, enjoyed the venerated status of being 'the ministers of God's state and . . . God's agents' without giving up any of their political authority.[83] This separation of powers set limits on the leaders' authority, since they had to act within the law, and preserved the integrity of the priestly caste, since they could only gain honour by their interpretations.[84] Moses used the priests to maintain the fiction that preserved his authority. But he also realized the dangers involved. When he kept for himself the exclusive right of enforcing the law, he tried to make sure that the priests would not overstep their bounds and think that their privileged status as the interpreters of God's law gave them the power to enforce it, a step that would effectively undermine his own authority.[85]

However, the very fiction that had founded the state could easily be perverted by the priests for their own ends. Spinoza sees the prophetic foundation of the Hebrew state as both its greatest strength and its fatal weakness. For although prophecy could serve as the basis of a religion uniquely suited to bind the Hebrews together into a cohesive political entity, it could also be distorted into a set of superstitious beliefs that might tear them apart. Spinoza argues that the Levites were the cause of the ruin of the state for at least two reasons.

First, their privileged status led to grumbling and division among the other tribes. Already, before Moses' death, there were signs of discontent among the élite. The very success of the state constituted at Sinai had given men the leisure to question the divine authority Moses had claimed, and his motives in establishing the Levites (with his brother at their head) as a special caste.[86] Once the

[82] *Ibid.*

[83] *TTP*, XVII, 65; GIII/212.

[84] *TTP*, XVII, 63–4; GIII/212.

[85] I neglect other features of the Hebrew state that Spinoza cites in *TTP*, XVII, as important reasons for its success: its citizen army (para. 69); the fact that the leaders themselves were bound together through a common religion (para. 70); the equal share in the ownership of all lands and fields (para. 84); the feasts and rituals that inspired a joyful devotion (paras. 89–91); the reverence for the temple (para. 92), etc.

[86] Compare to *TTP*, Preface, 2; GIII/5: 'For no one has lived among men without seeing that, when they are prospering, even those who are quite inexperienced are generally so overflowing with wisdom that they believe themselves to be wronged if anyone wants to give them advice.'

veil of appearances that was so crucial to the maintenance of the state had been pierced, rebellion became so widespread among the people that it was only ended through some 'great calamity or plague';[87] this was a sign of things to come.

Second, and more importantly, the special role reserved for the Levites led them to abuse their own power and thereby create turmoil in the state. Especially after Moses died and no one assumed the role of supreme commander, there was little to prevent the Levites from claiming not only the right to interpret laws but also the right to enforce them.[88] They found the justification for their actions in a new interpretation of God's word, provided by the 'many relentless and foolish Theologians' among them.[89] These men were not mindful of the limits of prophetic knowledge and interpreted the laws of right conduct as if they were the laws of nature itself. But this move, although useful in justifying the priests' claim to political authority, had catastrophic consequences for the state. Religious authority, originally instituted by Moses in order to benefit all, became the means of a small group to dominate the rest.

Spinoza thinks that in order to gain power the priests claimed that they had not only the right to interpret the laws but the right to enforce them as well. To justify this claim they used prophecy not to exhort the masses to right conduct, as had originally been intended, but in order to manipulate the masses to their own ends. The masses' very ignorance and fear which made prophecy effective in the first place, paradoxically also enabled the theologians to transform their religion into superstition.[90] As we have seen above, since the masses were ignorant of the divine law of nature, they were quick to take anything extraordinary as a sign of God's will. The vivid nature of prophetic language itself was inexplicable to them, and was consequently considered as a higher form of knowledge.[91] But the willingness of the masses to accept the word of prophecy allowed them to be easily misled. For example, whereas Moses and the other prophets had used 'miracles' not to teach about the nature of God but to guide the people,[92] the priests used them much as soothsayers used the entrails of

[87] *TTP*, XVII, 103–4; GIII/219.

[88] *TTP*, XVII, 100; GIII/218.

[89] *TTP*, XVII, 99; GIII/218.

[90] See *TTP*, Preface, 4; GIII/5: 'Because reason cannot show a certain way to the hollow things they desire, they call it blind and human wisdom hollow. The delusions of the imagination, on the other hand, and dreams and childish follies they believe to be divine answers.'

[91] '[T]he Prophets were said to have the Spirit of God because men were ignorant of the causes of Prophetic knowledge, wondered at it, and on that account, were accustomed to refer it to God — as they did all other abnormal phenomena — and to call it the knowledge of God' (*TTP*, I, 42; GIII/27–8).

[92] The true prophets did not think miracles were exceptions to the law of nature (*TTP*, VI, 6; GIII/83), nor did they think they could be used to learn about the nature of God (*TTP*, VI, 30; GIII/87). For a detailed analysis of Spinoza's discussion of miracles in the

animals to predict the future.[93] What at first had been used to exhort the masses to right conduct, later became a tool to inspire fear in them. But the very hollowness of such interpretations of religious law had an unfortunate effect: they served the immediate interests of the priests but not the long-term interests of the masses.

Introducing elements of superstition into the religious structure of the Hebrew state served the priests well, at first. They were not interested in maintaining the order of the state through a stable practice of religious rites which inspired devotion among the masses; rather, they ruled by introducing into the state practices that encouraged the very passions found in the state of nature. In fact, the power of superstition is such that it plays even more intensely on the hopes and fears of the individual than does religion, but without resolving them in the way religion does. Superstition, Spinoza argues, does not have any interest in allaying the fears of an individual; it does not seek to transform fear into devotion but uses this fear to maintain control of the state. 'Fear then, the reason why superstition arises, is preserved and is encouraged.'[94] In this manner, instead of maintaining a stable set of laws, the Levites began to constantly increase them, 'determining everything by priestly authority and daily issuing, concerning ceremonies, the faith, and everything else, new decrees, which they wanted to be no less sacred and to have no less authority than the laws of Moses'.[95] But the paradoxical effect of these decrees and the theological disputes that accompanied them was to undermine the faith of the masses in the religious order of the state. Since the epistemological ground of the priests' authority was ambiguous, and they had strayed from its proper domains, it was easy for others to dispute them. The signs that once had meant one thing could easily be twisted by the imagination of someone else to mean something different.[96] The doubt that a few individuals had already displayed in the time of Moses grew unchecked under the rule of the Levites. In response, the priests only redoubled their attempt to manipulate the desires of the people, allowing 'great license' and agreeing to whatever the common people desired when fear

Theological-Political Treatise, see the paper by Edwin Curley, 'Spinoza on Miracles', in *Proceedings of the First Italian International Congress on Spinoza*, ed. Emilia Giancotti (Naples, 1985).

93 'Indeed, [men who immoderately desire uncertain things] believe God rejects the wise and writes his decrees not in the mind, but in the entrails of animals, or that fools, madmen and birds foretell his decrees by divine inspiration and prompting' (*TTP*, Preface, 4; GIII/5). Hobbes also distinguishes between two senses of 'prophecy': (i) as God speaking to the people; and (ii) as prediction of future events (see Hobbes, *Leviathan*, Ch. 36, pp. 456–8).

94 *TTP*, Preface, 5; GIII/5.

95 *TTP*, XVIII, 8; GIII/221.

96 Compare this to what Spinoza says about the conflicting interpretations of 'universal ideas' in *E*2p40s1 and *Metaphysical Thoughts*, Part 1, ch. 1. These passages are quoted in notes 24 and 31, above.

was not effective.[97] The vacillation between hope and fear which had been endemic to the state of nature was reintroduced into the state by the very agents who had been appointed to preserve its stability. Thus weakened, the Hebrew state easily succumbed to the invasion of foreign armies.

Once the mechanisms for inspiring devotion among the people had broken down or proven inefficacious, something else was needed to keep the people in check. If the state was to be reconstituted, the only possible solution was to introduce an earthly sovereign to rule the masses in conjunction with the priests; and that, as Spinoza tells us, is what the Hebrews did.[98] This finally broke the political effectiveness of the covenant with God that Moses had instituted. If Moses' technique for ruling had been to gain the people's assent to his authority through the perpetuation of a fictive covenant with God, the technique of the kings was by necessity more direct — the threat of force. What had been a theocracy thus devolved into something resembling Hobbes's 'Leviathan'[99] but with annoying and dangerous relics of the past regime still extant. The élite of the old order had survived and were content with their lot, but their successors were not and were continually trying to usurp command. The institution of the priesthood persisted and continued to treat the rights of the ruling authorities with disdain, looking always to re-establish its own prophetic rule. Thus, the Hebrew kings 'had a state within a state, and ruled precariously'.[100] The constant strife within the state between the priests and the kings produced conditions from which only tyranny could result. Once 'religion declined into a pernicious superstition and . . . the true meaning and interpretation of the laws was corrupted', dissension reigned and only force could settle matters.[101] The eventual total collapse of the Hebrew state was caused not from without but from within.[102]

[97] See *TTP*, XVII, 106; GIII/219, and XVIII, 9; GIII/222.

[98] After the Hebrews had failed in their loyalty to divine worship, 'there were great changes, and a great license to do anything, and extravagant living, and negligence, with everything going from bad to worse, until, after they had often been conquered, they completely broke away from the divine law, and wanted a mortal king, so that the royal house of the state would not be the Temple, but the course, and so that the tribes would all remain fellow citizens, not any longer in virtue of divine law and the priesthood, but in virtue of the Kings' (*TTP*, XVII, 106; GIII/219).

[99] See Hobbes, *Leviathan*, Ch. 17.

[100] *TTP*, XVII, 108; GIII/220.

[101] *TTP*, XVIII, 9; GIII/222. See also *TTP*, XVIII, 12; GIII/223: 'we cannot have nay doubt that the flattery of the Priests, and the corruption of religion and the laws, and the incredible increase in the number of the laws gave a very great and frequent opportunity for arguments and disputes, which could never be settled. For where men begin to argue with the fierce heat of superstition, and the magistrate aids one of the other side, they can never be calmed, but must be divided into sects.'

[102] Spinoza argues that 'the necessity of solving this problem [i.e. of how to establish a regime in which people prefer public good to private advantage] has, indeed, compelled people to invent many solutions, still we have never reached the point where a state is

The significance of this story is manifold. Whether it accurately depicts and explains the events in the Old Testament is far less important than the use that Spinoza makes of it. On one level it appears simply to relate the narrative of the rise and fall of the ancient Hebrew state. On another level, this story proves to have a moral: the institutionalization of the priesthood leads to the state's corruption and collapse. Indeed, Spinoza writes that 'the more I consider this change [in the priesthood] the more it compels me to burst out in the words of Tacitus: at that time God's concern was not with their security, but with vengeance'.[103] Going yet one level further, I think that this story is a political allegory which translates the awful mystery of 'God's vengeance' on the Hebrews into a meaningful philosophical point about politics. In other words, Spinoza wants to show that the imaginative means which are necessary to produce political community can just as easily cause its downfall. But all of these elements are buried in yet another level of significance, which is Spinoza's use of the story itself as an exemplar. Spinoza is not only asking his readers to understand the underlying significance of the narrative of the Hebrew state, but exhorting others to regard it as an example to imitate. It is to this that we turn in the next section.

The Hebrews and the Dutch

What the prophets were to the ancient Hebrews, the Holy Scriptures were to the Dutch. It was for this cultural and historical reason that Spinoza chose the ancient Hebrews to be his example in the *Theological-Political Treatise*, and this is also what allows this example to function as an exemplar, or guide to action. On this score Spinoza followed his own advice:

> If someone wishes to teach some doctrine to a whole nation, not to mention the whole human race, and wishes it to be understood in every respect by everyone, he is bound to confirm his teaching solely by experience, and for the most part to accommodate his arguments and the definitions of things to be taught to the power of understanding of the common people, who form the greatest part of the human race.[104]

Just as an appeal to reason would not have been effective with the Hebrews, Spinoza thought that he must appeal to the experience of the Dutch if he was to persuade them of anything. But the 'experience' of a people is a broad and complex notion. As Simon Schama has so admirably shown, the seventeenth-

---

not in more danger from its own citizens than from its enemies, and where those who have command do not fear the former more than the latter' (*TTP*, XVII, 17; GIII/203–4). He cites the example of Rome and refers to Tacitus (*Histories*, Bk. iv).

[103] *TTP*, XVII, 97; GIII/218. See also *TTP*, XVII, 112; GIII/220: 'With this we see how Religion was introduced into the Hebrew state, and how its political authority could have been everlasting, if the just anger of the lawgiver had permitted it to stay the same. But because this could not happen, it had to perish in the end.'

[104] *TTP*, V, 37; GIII/77.

century Dutch saw themselves in the light of three important sets of historical narratives or myths: the story of their own heroic struggle against the Spanish, which retroactively assumed a strongly nationalist bent; the story of the Batavians depicted in the writings of Tacitus (and other Roman historians) and drawn upon by Grotius (amongst others) to explain the ancient origins of the nation; and the story of the ancient Israelites.[105] It is this last founding myth in which Spinoza is obviously interested in the *Theological-Political Treatise*.[106] As we shall now see, his interpretation of the rise and fall of the Hebrew state is not just a random example of his political philosophy, but one chosen for very concrete reasons. Spinoza hoped to use the Hebrews to show that tolerance of opinion and belief was necessary for the welfare of the state.

Although it would be far beyond the scope of this paper to explicate the full range of the meaning the story of the ancient Hebrews had for the seventeenth-century Dutch, we can focus on the aspects most relevant to Spinoza by returning to the painting described in the introduction, Ferdinand Bol's 'Moses Descends from Sinai with the Ten Commandments', hanging in the Magistrates' Chamber (*Schepenkamer*) of the Amsterdam Town Hall. As the art historian Albert Blankert has pointed out, there are undoubtedly two reasons why this somewhat unusual subject was selected for such a prominent place.[107] The first and more general one is that the Calvinist influence on the Dutch state at this time was strong, and from the earliest days of their sect the Calvinists

[105] Simon Schama, *The Embarrassment of Riches* (Berkeley, 1988), Part 1, ch. 2, *passim*. Grotius' account of the Batavian origins of the Dutch Republic, which includes a description of their manners and mores, is *Liber de Antiquitate Republicae Batavicorum* (1610). Schama's list of national origin myths (what I would call 'exemplars') obviously should not be understood to exclude all other sources that may have contributed to the formation of Dutch character, political institutions and intellectual life in the 'Golden Age'. In his book, *The Myth of Venice and Dutch Republican Thought in the Seventeenth-Century* (Assen, 1980), E.O.G. Haitsma-Mulier has stressed the importance of what he calls the 'myth of Venice' in Dutch thought during this time and discusses its influence on Spinoza via the work of Pieter de la Court. Although the influence of this myth may well have been felt in higher intellectual and political circles, it does not seem to have had equal resonance in popular political culture. So while Venice may have served Spinoza in a variety of ways, it was not useful to him as a common imaginative referent, an *exemplum*. That might explain why there are so few actual references to Venice in Spinoza's work, at least in the *Theological-Political Treatise*.

[106] In several places Spinoza directly compares the situation of the Hebrews to that of the Netherlands — see, for instance, *TTP*, XVII, 54; GIII/210 — but I am claiming a far more extensive parallel than these explicit comparisons alone would justify.

[107] According to Blankert this was only the second treatment of this subject by a Dutch master (the other one was by Rembrandt) and the first to be hung in a courtroom anywhere in Europe. See the English summary of Blankert's *Kunst als regeringszaak*, p. 65. What follows is based on this source; on Blankert's other monograph, *Ferdinand Bol*, pp. 50–1; and on Schama, *The Embarrassment of Riches*, pp. 116–25.

had identified with the exodus of the ancient Hebrews.[108] The second, more specific reason has to do with the troubled relations between the Calvinist clergy and the Amsterdam government. Let us look first at the general reason why this painting was commissioned before we look at its specific place in debates of the time. We need to know how the Calvinist identification with the Israelites was adopted by the Dutch before we examine the struggle over its interpretation.

The painter's use of this scene from Exodus to make his point was not a mere conceit or mannered allusion. By 1665 the Dutch had already repeatedly used the Old Testament to describe their own trials and tribulations, especially against the Spanish. In order to break the siege of Leiden by the Duke of Alva in 1574, William of Orange had taken the desperate measure of breaking the great dykes that protected the city and surrounding land from the ocean, with the result that the Spanish, panicked by the rising waters, had fled. Local historians and city fathers described this event in providential terms as the hand of God intervening on the side of his 'chosen people' against their oppressors, who were compared to the Midianites. The parallel with the trials of the ancient Israelites was not expressed in speeches alone: the chronicle of this event was recited thereafter during a commemorative meal with food (bread and herring) that was supposed to remind the celebrants of the Leiden citizens' heroic sacrifice and providential salvation.[109] This 'commingling of scripture and historical chronicle' helped foster the emerging Dutch nation's sense of separateness, creating a vision of destiny which transformed the profane events such as battles, trade, and even household chores, into the unfolding of a sacred plan.[110] The Calvinist preachers produced a theology in which the Dutch

[108] G. Groenhuis, in his article 'Calvinism and National Consciousness: The Dutch Republic as the New Israel' (in *Britain and the Netherlands*, Volume VII, ed. A.C. Duke and C.A. Tamse (The Hague, 1981), pp. 118–33) contests the view of some eminent Dutch historians (among them E.H. Kossman) who hold that, while the Calvinists frequently made the analogy with ancient Israel they did not see themselves as the 'inhabitants of a New Israel, as an elect nation'. See also Dr Groenhuis's work *De Predikanten* (Groningen, 1977) (with English summary), which Simon Schama draws upon.

[109] Schama discusses this incident and its contemporary interpretation on pp. 26–7 of *The Embarrassment of Riches*. A fuller description of the military and political aspects of the siege can be found in Geoffrey Parker, *The Dutch Revolt* (Harmondsworth, 1988). Groenhuis also notes many instances of the use of biblical imagery in the struggles of the early Dutch republic, as well as seventeenth-century uses, such as the poet Joost van den Vondel's comparison, in his play *Het Pascha* (The Passover), of the deliverance of the children of Israel with the liberation of the United Provinces (see Groenhuis, 'Calvinism and National Consciousness', pp. 120–1 and *passim*).

[110] Schama, *The Embarrassment of Riches*, p. 103. He also points to the curious way in which the Dutch penchant for cleanliness, frequently noted by foreigners, was a 'militan[t] . . . affirmation of separateness': 'What was cleansed was the dirt of the world that had obscured the special meaning of Dutch history and the providential selection of its people' (*ibid.*, p. 380).

Republic was the 'New Canaan,' and they saw their own position 'as equal to the prophets of ancient Israel'.[111] On this basis, the 'more demanding Calvinists . . . attempt[ed] to establish godly norms of social behaviour that were laid down as fitting for the new Chosen People. Conformity to these norms guaranteed perpetuation of the covenant; deviation from them, the downfall of the commonwealth.'[112] In other words, both politicians and preachers constantly exhorted their people to 'right conduct' through the terms of this scriptural identification. Thus, close to a century after the events in Leiden, when Bol was commissioned to paint Moses for the town hall, and when Spinoza was composing the *Theological-Political Treatise*, the identification with Israel had been well sown in the imaginative and moral landscape of the nation.

The scriptural narrative of the ancient Hebrews, then, was exemplary for the Dutch in both structure and function: in structure, because it could appeal to the imagination of the individual in the guise of an end that appeared to have divine, and therefore universal significance; and in function, because it served to exhort people to act in relation to an 'objective' set of values and in prescribed ways. But even if the narrative of the Hebrews was exemplary for the Dutch, the precise sense in which it was exemplary was not always construed in the same way by all people. A particular exemplary figure, like Moses, could be identified with different messages, religious or political. The Catholic tradition viewed Moses as a 'prophetic harbinger of the messianic return', while the Dutch identified him at different times with William of Orange, the noble liberator, and with Oldenbarneveld, the magistrate who was 'the first great patriot of the law' and a proponent of religious toleration.[113] As we have seen, Spinoza argues that the philosophical basis of this variety in interpretation and use of an exemplar stems from its foundation in an inadequately conceived universal. So what appears to be universal really reflects some more particular interpretation of experience. The fact that the scriptures were 'revealed' hides their human origins. This is both an exemplar's strength and its weakness; that which allows it to appeal to a large group of people separated by different interests and ideas can itself be used in service of those different interests. It is certain that the Dutch drew upon the scriptural history of the Hebrews to understand and shape their own experience in a common way, but the possibility of more than one interpretation of this identification opened the way for conflict over the exemplary authority of one use or another of the universal idea (scripture) that was at its foundation. The image of Moses holding the tablets of the law that he brought down from the top of Mount Sinai was indisputably an image of authority for the Dutch nation in the seventeenth century, but it 'was when individual Dutch men and women tried to discern what exactly it

[111] Groenhuis, *De Predikanten*, p. 183.
[112] Schama, *The Embarrassment of Riches*, p. 381.
[113] *Ibid.*, pp. 112–13.

was that those tablets commanded . . . that the scriptural analogy began to lose clarity and coherence'.[114]

The dispute over who had the ultimate right to interpret scripture, and who had the authority to enforce its commands, was the specific reason behind the commission and conception of Ferdinand Bol's painting of Moses. In 1665, although the clergy was officially under the rule of the state's authority, nonetheless 'a bitter controversy arose on the issue of the primacy of the church or the state'.[115] Since the revolt against the Spanish in the sixteenth-century the head of the House of Orange-Nassau had led the army and, for a good measure of that time, the country, through the office of the stadholder.[116] But after the sudden death of William II in 1650, the States party, led by Jan de Witt, formed a republican government without a stadholder.[117] The majority of the Calvinist clergy was stridently opposed to the government of the States party, not least because of the relative tolerance it showed to other religions.[118] Although they bowed (voluntarily or forcibly) to the power of de Witt's government, nonetheless their natural and historical sympathies were with the Orangists.[119] The Orangists in turn sought to use the clergy's influence over the populace to serve its own end, which was to re-establish itself at the head of the government. It is not surprising then to find the Calvinist clergy using Biblical analogies in order to argue for the reinstitution of the stadholderate.[120]

Blankert argues that the Amsterdam burgomasters chose Bol's painting and specified the depiction of its subject-matter in order to counter the claims of the Calvinist clergy — and also by extension those of the Orangist party — and thereby strengthen their own position and that of the States party within Amsterdam.[121] The painting stresses the role of Moses as Lawgiver and depicts his brother Aaron, the high priest, as subordinate to him and the Law. As we

[114] *Ibid.*, p. 114.

[115] Blankert, *Kunst als regeringszaak*, p. 65.

[116] See H.H. Rowen, 'Neither Fish nor Fowl: The Stadholderate in the Dutch Republic', in *Political Ideas and Institutions in the Dutch Republic* (W.A. Clark Memorial Library, UCLA, 1985).

[117] For a succinct discussion of Jan de Witt, see H.H. Rowen, *John De Witt: Statesman of the 'True Freedom'* (Cambridge, 1986). For general background, see P. Geyl, *The Netherlands in the Seventeenth Century* (London, 1964), Vol. 2.

[118] See Groenhuis, 'Calvinism and National Consciousness', p. 129, though P. Geyl, in *History of the Low Countries* (London, 1964), p. 165, notes that '. . . it would be wrong to conceive of Reformed and Statist as necessarily making a contrast. Most of the regents of the True Liberty brand were as good Reformed as any, and many Reformed ministers accepted them with befitting respect and submission as their lawful governors.'

[119] Groenhuis, 'Calvinism and National Consciousness', pp. 129–30.

[120] Rowen, 'Neither Fish nor Fowl', pp. 19–20.

[121] For a detailed discussion of the political debates among the Amsterdam regents and their role in the choice of subjects in the decoration of the Town Hall, see Margaret Deutsch Carroll, 'Civic Ideology and its Subversion in Rembrandt's *Oath of Claudius Civilis*', in *Art History*, Vol. 9 (1986), pp. 12–35.

have already noted, the burgomasters, as if worried that the point of the painting would be missed, also commissioned a marble frieze underneath it, detailing the frenzied scene around the Golden Calf. 'The relief as a whole', Blankert comments, 'served as a warning of the chaos that results when the leadership of the people falls into the hands of the clergy'.[122] In addition, just in case the audience did not immediately grasp the visual evidence, a poem by Joost van den Vondel was also commissioned, apparently sometime before the painting itself was completed:

> The Hebrew Moses has received the Law from God, with which he returns from above to the people, who greet him reverently and welcome him eagerly. The free State begins to flourish when the people respect the laws.[123]

Using images drawn from the Old Testament, the town authorities hoped to make a statement to the clergy (in their own language) and to the populace about the proper order of the state.[124]

Spinoza, like Bol, emphasized the image of Moses as lawgiver because he wanted to argue for the priority of the state in all matters, whether secular or religious. It is Moses as the prophetic founder of the state who was able to lead the people from the state of nature into civil society; and it is Moses as lawgiver who organized the state so that it would endure. But, as we saw in the previous section, the means by which Moses founded the state could easily be perverted by others for their own interests. This is precisely the point where Spinoza's analysis of the causes and effects of the corruption of religion into superstition

[122] Blankert, *Kunst als regeringszaak*, p. 65.

[123] Quoted in Katherine Fremantle, *The Baroque Town Hall of Amsterdam* (Utrecht, 1959), p. 77. It derives from a poem by Vondel published in 1659. Vondel himself had been a target of the Calvinist clergy in 1654 when his play *Lucifer* was banned (Schama, *The Embarrassment of Riches*, pp. 119–21). While Fremantle thinks that the poem was actually inscribed beneath the painting, both Buchbinder-Green (*The Painted Decorations of the Town Hall of Amsterdam*, p. 153) and Blankert (*Ferdinand Bol*, p. 52) disagree. However, other expository poems by Vondel *were* inscribed under other paintings in the town hall; for example, under Van Helt Stocade's *Joseph Distributing the Grain in Egypt*, which was in the Treasury Room (*Schoorsteenstuk*) (Buchbinder-Green, *The Painted Decorations of the Town Hall of Amsterdam*, p. 110).

[124] It is worth noting that a few years before the Magistrates commissioned Bol's painting to assert their own authority over and against the Calvinist ministers, they had also commissioned for another part of the town hall a large painting of the Batavians, depicting Claudius Civilis who had led the Batavians in a revolt against the Romans. Claudius was readily identified with the first William of Orange who had led the Dutch in their revolt against the Spanish in the sixteenth century. The commission was a sign of respect made as a result of the re-emergence of the Orangist faction, who claimed the stadholderate for the young William of Orange, later to be William III. For a detailed discussion of this episode, see the article by Carroll, 'Civic Ideology and its Subversion', *passim*, and Blankert, *Kunst als regeringszaak*, p. 64. This example shows how different exemplars can be used in different (complementary and conflicting) ways at the same time.

on the Hebrew state is deeply related to his argument for toleration in the contemporary United Provinces.

Spinoza bases his argument for toleration on the distinction between the proper and improper use of exemplary language (like prophecy or scripture) in the state. To see this let us go back to the preface of the *Theological-Political Treatise* where Spinoza tells the reader that he has a twofold purpose in writing the work. His overarching aim is, as he declares in the Preface, to show 'not only that . . . freedom [of opinion and worship] can be granted without harm to piety and the peace of the State, but also that it cannot be abolished unless piety and the Peace of the State are abolished with it'.[125] But in order to achieve this end Spinoza thinks he must first 'indicate the main prejudices regarding religion, i.e. the traces of our former bondage, and then also the prejudices regarding the right of the supreme powers'.[126] For '[m]any, with the most impudent licence, are eager to take away the greater part of that right, and to turn the heart of the masses (who are still liable to pagan superstition) away from the supreme powers, so that all may rush again into slavery'.[127] Spinoza intends to expose these prejudices and their causes through a thorough critique of Scripture, thereby undermining the position of those who seek to exploit the masses through the manipulation of superstitious beliefs.

But under the political and social conditions of the United Provinces of his time, Spinoza did not think it was possible to deny any and all roles to religious belief and practice. As he writes:

> For everyone knows how highly the people value the right and authority regarding sacred matters, and how much everyone hangs on the utterances of the one who has it, so much so that we can rightly say that the person who has this authority has the most powerful control over their hearts.[128]

Given these conditions, Spinoza found it necessary to appeal to these religious beliefs and practices through his own interpretation. In this light, the use of the ancient Hebrews was not just an analogy or illustration of a theoretical point, but an exemplar, something to be imitated. Spinoza realized that the same underlying dynamics involved in Moses' appeal to revelation in the foundation of the state were at play in the Dutch identification with the scriptural narration of the Hebrews. If Moses called his people 'chosen' in order to exhort them to 'right' conduct, i.e. to conduct that would best maintain the state, then the Dutch Calvinist belief that their nation had been 'chosen' to endure and triumph over their travails could also be used to maintain a certain political order. But Spinoza also thought it imperative that the priestly caste, the Calvinist preachers, not be allowed to overstep the bounds of their proper role in the state. In this sense,

[125] *TTP*, Preface, 12; GIII/7.
[126] *TTP*, Preface, 13; GIII/7.
[127] *Ibid.*
[128] *TTP*, XIX, 40; GIII/235.

the painting compresses the exemplary meaning of the narrative of the founding of the Hebrew state, as Spinoza has reconstructed it, in one highly symbolic moment: the priests must kneel below Moses, who embodies the proper relation of prophecy to the law. In other words, the right to interpret the law must be distinct from the right to enforce the law. Only under this constitutional arrangement could the state persist.

In the Netherlands of Spinoza's time, however, the priests threatened to cross this boundary, claiming that they had the right not only to interpret the law but also to enforce their interpretation of it.[129] They pursued this claim just as the ancient Levites had, not directly but indirectly through the establishment of a king whom they believed they would control. In the period during which the struggle between church and state, republicans and Orangists, was most intense, '[t]he idea of the Dutch Israel was given its most elaborate treatment' by Calvinist ministers for whom the 'State existed for the sake of the Church'.[130] Moreover, the rhetorical strategy the Calvinists adopted to further their own ends was that of the 'relentless and foolish' theologians whom Spinoza had identified among the Levites of the young Hebrew nation: they claimed that their interpretation of the laws was identical with philosophical truth.[131] Once in power, Spinoza believed, the theologians would attempt to suppress all interpretations of the law that differed from theirs; for to allow a diversity of beliefs would effectively undermine their very claim to power as exclusive arbiters of true knowledge and belief. But any such attempt would be illegitimate, impractical and ill-advised: illegitimate, because each person has the natural right to maintain whatever idea or passion they have;[132] impractical because it is impossible for any individual (or group of individuals) to control the minds of the multitude;[133] and ill-advised, because the imposition of pro-

[129] Spinoza is not opposed to the various interpretations of scripture but only to the attempt by one sect to prohibit the interpretation of another (see *TTP*, XIV, 3; GIII/173). For earlier disputes, see D. Nobbs, *Theocracy and Toleration. A Study of the Disputes in Dutch Calvinism from 1600–1650* (Cambridge, 1938).

[130] Groenhuis, 'Calvinism and National Consciousness', pp. 122–3. The books Groenhuis discusses are: Abraham van de Velde, *De Wonderen des Allerhoogsten* (The Miracles of the Almighty), published in 1668, and Herman Witsius, *De Twist des Heeren met syn Wyngaert* (The Quarrel of the Lord with his Vineyard), published in 1669.

[131] We see Spinoza's opposition to this attitude reflected in the correspondence Spinoza had with Willem van Blijenbergh who, although himself not a Calvinist priest, believed that when the truths of reason and revelation were in conflict, the truth of revelation superseded that of reason. See letter 20, in *The Collected Works of Spinoza*, especially pp. 361–2 (GIV/97–8), and Spinoza's reply, letter 21, pp. 375 ff. (GIV/126f.).

[132] '... it cannot happen that a mind should be absolutely subject to the control of someone else. Indeed, no one can transfer to another person his natural right, *or* faculty of reasoning freely, and of judging concerning anything whatsoever, nor can he be compelled to do so' (*TTP*, XX, 2; GIII/239).

[133] '... however much the supreme powers are believed to have a right over all things, and to be the interpreters of right and religious duty, they will still never be able to bring

phetic (i.e. exemplary) language into the realm of either intellect or belief would not succeed but only be called tyranny (*imperium violentum*) and be met with resistance.[134] What the priests and preachers had forgotten or purposefully denied was that their role is to perpetuate language (and institutions that maintain this language) whose goal is to inspire conduct commensurate with the aims of the state — peace and liberty. What Spinoza did in the *Theological-Political Treatise*, just as Bol attempted to do in his painting, was remind interested parties of this fact. The toleration of belief and ideas, if not of action, was central to the maintenance of the social contract.

Obviously, Spinoza was reading contemporary Dutch concerns back into the scriptures. He was not concerned with the 'historical truth' of the scriptures, even if one were available,[135] but only with the use of its contents for his own purposes. Like others who came before him in the tradition of writers who employed the rhetoric of exemplarity, Spinoza felt free to pick and choose from the enormous stock of examples collected in the scriptures in order to make his own point.[136] But this is entirely consistent with his aim in the *Theological-Political Treatise*: if we are talking about politics then we do not appeal primarily to reason but to experience, which is constituted through the imagination and memory. In other words, we must use exemplary language. Spinoza starts with an imaginative universal, the scriptures, that has an important political and cultural function in Dutch society, and gives his own interpretation of it, one which he believes is consistent with its function: the preservation of a stable state. Of course, Spinoza could hardly have foreseen the traumatic overthrow of the States party by the Orangists at the time of the French invasion, two years after the *Theological-Political Treatise* was published in 1670.[137] But

it about that men do not make their own judgment about everything, according to their own mentality, and that they are not, to that extent, affected with this or that affect' (*TTP*, XX, 6; GIII/240).

[134] 'That is why rule over minds is considered violent, and why the supreme authority seems to wrong its subjects and to usurp their rights whenever it wants to prescribe to each person what he must embrace as true and what reject as false, and further, by what opinions each person's mind ought to be moved in its devotion to God. For these things are subject to each person's control, which no one can surrender even if he wishes to' (*TTP*, XX, 3; GIII/239).

[135] This is still a vexed question in modern scholarship of the Old Testament. The narrative accounts of events were themselves probably reconstructed for a religious purpose. Thus we have an interpretation of an interpretation, ad infinitum. For more on this question, see the anthology edited by Ronald E. Clements, *The World of Ancient Israel: Sociological, Anthropological and Political Perspectives* (Cambridge, 1991).

[136] See *TTP*, V, 41–2; GIII/78. On the necessity for selection in an exemplar, see Lyons, *Exemplum*, introduction. As Lyons points out, the etymology of both *exemplar* and *exemplum* is in the terms *excisio*, which means 'to cut out'.

[137] For a vivid account of the political events leading up to the debacle, in which the De Witt brothers, leaders of the States party, were brutally murdered by a mob, see Rowen, *John De Witt*, Chs. 11 and 12. This occasion also gave rise to the story, probably

at the time he was writing the *Theological-Political Treatise* there is no doubt that he believed it imperative to intervene and to try to persuade people of the dangers involved in the Calvinists' misuse of prophetic language, not only for philosophers but for all citizens of the republican state.

Conclusion: The Hebrews, Spinoza and Us

As prophecy was for the ancient Hebrews, or the scriptures for the Dutch, Spinoza's *Theological-Political Treatise* is to the contemporary political theorist: an exemplar. We have seen that, in his analysis of the Hebrew state, Spinoza argues that prophecy (i.e. revelation) was the means by which Moses was able to persuade individuals led by their passions and self-interest that they should join together into a state governed by a single rule of law. We have also seen that Spinoza chose the example of the ancient Hebrews for the *Theological-Political Treatise* precisely because the seventeenth-century Dutch saw themselves as the 'new Israel'. In both these cases, the notion of 'chosenness' played an important role in the construction of an exemplar. The prophets and preachers used this idea because it effectively appealed to individuals to see themselves within a divine plan, whose realization depended upon their actions. Both prophecy and scripture, then, are exemplars in the sense Spinoza defined it in the *Ethics*: imaginative universals that function to create a standard on the basis of which value can be defined and right conduct urged. What I hope to have shown in the preceeding discussion is that exemplars play as important a role in politics as they do in ethics. But there is one further point I would like to make in conclusion.

From the very first sentence of the *Theological-Political Treatise*, Spinoza emphasizes that the power of reason over human affairs is limited. Over and over he argues that most people are ruled by their passions and best appealed to through the imagination. Thus, if a person wants to convince another of something, then he or she 'should not connect his arguments or give definitions . . . Otherwise he will write only for the learned, i.e. he will be intelligible only to a few men compared with the others'.[138] Rather, he must *accommodate* his arguments to the understanding and experience of the common people through the use of narratives that appeal to the imagination.[139] This does not mean, of course, that reason has no role in politics. Reason teaches that society is the best means to preserve individual existence, and it shows us ways to improve society

apocryphal, that Spinoza, so incensed at the death of the two politicians, attempted to confront the mob with a sign reading 'Ultimi barbarorum' (you are the worst of barbarians) and had to be restrained by his landlord who prudently feared for the philosopher's life (see *ibid.*, p. 220).

[138] *TTP*, V, 37; GIII/77.

[139] *Ibid.* (quoted above) and *TTP*, V, 38–9. For further discussion of *accommodation* as a hermeneutical principle, see Brown, 'Philosophy and Prophecy', pp. 202–5.

for that end.[140] Moreover, it is an insight of reason, and amply demonstrated by experience, that people cannot escape being influenced by the passions and the imagination.[141] But although reason teaches important principles about politics, it cannot itself solve the pre-eminent political problem: how to convince men led by their passions that their interests are best served in society.

In order to do that, the theorist, as well as the politician, must appeal to the imagination to address the very passions and interests that dominate most people's lives. The same principles that apply in the creation of political society through prophecy thus apply to its interpretation. It is necessary to use examples, like that of the Hebrews, not just to prove theoretical points but to exhort people to act in a certain way. In contemporary language, we could say that Spinoza was well aware of the 'performative' dimension of political discourse. He did not choose the Hebrews merely because they were an appropriate example for his theory, but because they already functioned as an exemplar, as a model that ought to be imitated. In the interpretation of this particular exemplar he is himself trying to govern its meaning and use in political life.

Spinoza chose to analyse the Hebrews because of the cultural resonance of scripture in seventeenth-century Dutch society. We choose to analyse Spinoza's work because it has some resonance in some small part of our own society, among academics and other 'learned men'. But if we wish to seriously address the problem Spinoza did, if we intend to use Spinoza's *Theological-Political Treatise* in our time as he intended it to be used in his, not just as an example to be used in intellectual discussion, but as an exemplar to imitate in the service of establishing a free and tolerant political order, then we must look to the Hebrews, and ask ourselves what imaginative notions lie at the foundations of our own society and how we ought to interpret them.

Michael A. Rosenthal GRINNELL COLLEGE

[140] *TTP*, III, 14; GIII/47.

[141] In the *Ethics*, Spinoza argues that 'man is necessarily always subject to the passions, that he follows and obeys the common order of Nature, and accommodates himself to it as much as the nature of things requires' (*E*4p4c).

Name Index